NONPARAMETRIC INFERENCE

NONPARAMETRIC INFERENCE

Z GOVINDARAJULU

University of Kentucky, USA

World Scientific

NEW JERSEY • LONDON • SINGAPORE • BEIJING • SHANGHAI • HONG KONG • TAIPEI • CHENNAI

Published by

World Scientific Publishing Co. Pte. Ltd.

5 Toh Tuck Link, Singapore 596224

USA office: 27 Warren Street, Suite 401-402, Hackensack, NJ 07601

UK office: 57 Shelton Street, Covent Garden, London WC2H 9HE

British Library Cataloguing-in-Publication Data
A catalogue record for this book is available from the British Library.

NONPARAMETRIC INFERENCE

ISBN-13 978-981-270-034-6
ISBN-10 981-270-034-X

Printed in Singapore by B & JO Enterprise

Life is an experience meant to train the individual for a higher, deeper and more expanded state of existence.

— Sathya Sai Speaks, Vol. 28, Page 369

This book is dedicated to my wife, Gayatri for her patience and encouragement throughout the completion of this project.

Preface

Nonparametric statistical procedures are widely used due to their simplicity, applicability under fairly general assumptions and robustness to outliers in the data. Hence they are popular statistical tools in industry, government and various other disciplines. Since there is an extensive literature on nonparametric statistics ranging from theory to applications, this book focuses on a selective choice of topics that constitutes a foundation for nonparametric inference. Thus, topics such as multivariate methods, robustness of nonparametric procedures and survival analysis are not included in this book. The objectives of this book are:

(i) To gather extensive theoretical results including some recent ones, pertaining to nonparametric statistics, and thus provide a solid foundation of nonparametric inference for students taking a graduate course on nonparametric statistics and to serve as an easily accessible source for research workers in this area.

(ii) To cover adequately certain aspects of nonparametric estimation.

The manuscript can be divided into five parts:

I. Statistical terminology and basic tools of nonparametric statistical inference (Chapters 1, 2).

II. Nonparametric statistical estimation (Chapters 3, 4, 5, 6, 21, and 22).

III. Nonparametric testing statistical hypotheses (Chapters 7–20, and 22).

IV. Asymptotic theory for nonparametric statistics (Chapters 23–27).

V. Appendices (I–VI) in which certain numerical tables (which typically existing books do not have) are included.

A list of all references that are cited in the book is given at the end. Although the chapters are arranged in a natural sequence, they can be read essentially independent of other chapters. Thus the reader has the flexibility to select certain chapters that are of interest to him/her. Problem sets with some dealing with numerical data are included at the end of each chapter. Sections requiring familiarity with basic measure theory or group theory, such as Sections 1.3, 5.5, 5.6, 7.3, 8.10 can easily be omitted by a reader who had only advanced calculus prerequisite.

This book contains several results published by the author and/or his former students in research journals. For instance, see Chapter 20 dealing with nonparametric testing for random effects in ANOVA models. Some results on asymptotic theory pertaining to nonparametric statistics are appearing here for the first time in a book form. Also some new insights, more details, and elaborations of the asymptotic results are provided so that they can more easily be read and understood than from the published papers. Chapters 24–27, dealing with asymptotic theory germane to nonparametric statistics, can be omitted by a casual reader without losing the continuity of thought.

The competitors of this book might be Gibbons & Chakraborti(2003), Hájek, Sidák & Sen(1999) and Hettmansperger & McKean(1998). However, this book's approach is systematic and comprehensive and, hence, more likely to appeal to a broader audience. Also it should be noted that the present book is the result of the author's teaching a graduate level course on nonparametrics at Case-Western Reserve University and the University of Kentucky over the years.

This book will be very useful to research workers in the field of nonparametric statistics. Furthermore, it can be adopted as a prescribed text in a course suitable for advanced undergraduate and graduate students. For instance, Chapters 2–5, 8, 9, 11–15 and 18 can be covered in a semester-long course.

Z. Govindarajulu
Professor of Statistics
University of Kentucky
Lexington, KY 40506-0027
October 2005

Acknowledgements

I am grateful to Dr. James A. Boling, Interim Vice-President for research at the University of Kentucky for providing me with funds for technical assistance. I thank Dr. Steven Koch, Dean of the college of Arts & Science for approving my sabbatical leave during 2004–2005 and for providing me with a sabbatical leave grant which enabled me to travel to UC-Berkeley and other campuses in the Bay area for consultation with experts in the area of nonparametrics. My thanks are due to Professor Charles Stone of UC-Berkeley for a very useful discussion I had regarding nonparametric regression. I am also thankful to Professor C. Srinivasan of the University of Kentucky for help in proof reading. My sincere thanks go also to the anonymous reviewers who have critically read certain chapters of the book, brought to my attention some typographical errors, and made helpful comments and suggestions. I am grateful to Ms. Yubing Zhai, Commissioning editor, and to Ms. Tan Rok Ting and Ms. E H Chionh, production editors of WSPC for their generous help and moral support throughout the completion of this project. I also thank Ms. Yolande Koh and Ms. Irene Ong of WSPC for their excellent typesetting and correcting the last batch of typographical errors. My thanks are also due to a variety of people, namely Dr. Rytis Juras of Lithuania and Dr. Yuhua Su of Yale University for their help in preparing the subject index and Shankar Bhamidi of U.C. Berkeley for other help. This work would not have been possible without the support of the Departments of Statistics at the University of Kentucky and the University of California at Berkeley for which I am very thankful.

I thank Brian Moses for typing diligently the first 22 chapters of the book, Mrs. Sarah Nielson for typing Chapter 23, Miss Anna Fuller of UC-Berkeley for typing the rest of the chapters, some problem sets and for reformatting the first 22 chapters, and Qiangfeng Jiang of the Computer Science Department at the University of Kentucky for typing the rest of the problem sets, reformatting the references and the appendices and preparing the author index. He has also corrected several typographical errors. I also thank several batches of graduate students in statistics at the University of Kentucky who served as guinea pigs in a course on nonparametrics in which old versions of the present book were used as lecture notes.

I would like to thank several publishers, especially, The Institute of Mathematical Statistics and the American Statistical Association, for their kind and generous permissions for the use of their publications as source for some of the material in this book.

Contents

List of Tables

Chapter 1

Statistical Terminology

1.1 Sufficient Statistics

In this chapter certain important statistical terms that will be used in subsequent chapters will be defined. Usually, we do not know the distribution characterizing the phenomena of the experiment. However, we can often choose a sufficiently large class of distributions $\{F_\theta(\alpha)\}$ invariably indexed by an unknown parameter θ. The range of θ is Ω which is called the *parameter space.* The statistician has to decide upon the particular probability distribution which explains most the phenomena of the experiment. That is, the statistician has to make a decision about the value of the parameter, by means of the observable random variable X. However, in many situations the outcome X is a complicated set of numbers. If at all feasible, he would like to condense his data and come out with a magic number which contains all the relevant information about the parameter θ.

Definition 1.1.1. *A statistic T is said to be a sufficient statistic for the family $\mathcal{F} = \{F_\theta, \theta \in \Omega\}$ or sufficient for θ if the conditional distribution of X given $T = t$ is independent of θ.*

If one is permitted to observe T instead of X, this does not restrict the class of available decision procedures. Since the conditional distribution of X given T can be constructed (theoretically) by means of a random mechanism, there is no loss of generality in restricting ourselves to a sufficient statistic.

Factorization criterion. In order to find whether a certain statistic T is sufficient for θ or not, a simple check is provided by the following factorization criterion: If X is discrete, then a necessary and sufficient condition for T to be sufficient for θ is

$$p_\theta(x) = P_\theta(X = x) = g_\theta\,(T(x))\,h(x)$$

where the first factor depends on θ but depends on x only through T, while the second factor is independent of θ.

For a proof of this and some examples see Lehmann (1959, pp. 18–19). If X is absolutely continuous, then a necessary and sufficient condition for T to be sufficient for θ is a factorization of the density of the form

$$f_\theta(x) = g_\theta\,[T(x)]\,h(x)\,.$$

Incidentally, θ, x and T could be vectors.

Definition 1.1.2. *A sufficient statistic T is said to be minimal sufficient if the data cannot be reduced beyond T without losing sufficiency.*

For the binomial case $T = \sum_1^n X_i$ can be shown to be minimal. (Suppose that $U = l(T)$ is sufficient and that $l(k_1) = \ldots = l(k_r) = u$. Then $P\{T = t|U = u\}$ depends on p.) This shows that a sufficient statistic determined by inspection usually turns out to be minimal sufficient.

Example 1.1.1. Let $X_1, X_2, \ldots$ be independent and identically distributed uniform variables on $[0, \theta]$. Then max X_i is sufficient for θ, because

$$f_\theta(x) = \theta^{-n} u(\max\ x_i, \theta) \text{ where } u(a, b) \text{ is 1 or 0}$$

according as $a \leq b$ or $a > b$.

1.2 Properties of Estimators

Consistency. An estimator (statistic T) is to be a consistent estimator of the parameter θ if T converges in probability to θ as the size of the sample tends to infinity.

Unbiasedness. An estimator T is said to be an unbiased estimate of θ if $E(T) = \theta$.

Asymptotic unbiasedness. An estimator T is said to be asymptotically unbiased for θ if lim $E(T) = \theta$.

Efficiency. Let θ be a parameter and $f(x;\theta)$ be the underlying density function. For any estimate $\hat{\theta}$ of θ we have

$$E(\hat{\theta}-\theta)^{\geq}\left[1+b'(\theta)\right]^2/E\left[\left(\frac{\partial lnf}{\partial\theta}\right)^2\right]$$

where $b'(\theta) = \frac{d}{d\theta}b(\theta)$ and $b(\theta) = E(\hat{\theta}-\theta)$. The right-hand expression is called *Cramér-Rao lower bound for the variance.* If we confine ourselves to the class of unbiased estimates, then the ratio of Cramér-Rao lower bound to the variance of $\hat{\theta}$, denoted by $e(\hat{\theta})$, is called the *efficiency* of $\hat{\theta}$. Also, $\hat{\theta}$ is said to be efficient if $e(\hat{\theta}) = 1$.

Asymptotic efficiency. $\hat{\theta}$ is said to be asymptotically efficient if $\lim_{n\to\infty} e(\hat{\theta}) = 1$. An efficient estimate exists only under restrictive conditions, whereas an asymptotically efficient estimate exists under certain general regularity conditions.

Precision. The precision of an estimate is defined as the reciprocal of its variance.

Relative efficiency. The efficiency of $\hat{\theta}$ relative to another estimate $\hat{\hat{\theta}}$ is defined as the ratio of the precision of $\hat{\theta}$ to the precision of $\hat{\hat{\theta}}$.

Asymptotic relative efficiency is defined as the limit of the relative efficiency as the sample size (usually) tends to infinity.

The above concepts which are due to R.A. Fisher are very useful, especially in parametric statistical inference. The above procedure of asymptotic relative efficiency is not adequate for nonparametric statistical problems.

1.3 [1]Principle of Invariance

The principle of invariance is an important concept in statistical inference. If a test procedure or an estimator is optimum for a certain problem, then we expect the test procedure or the estimator to be invariant under a change of scale or measurement. Let us state the principle of invariance formally.

[1]This section can be omitted from a first reading of the book.

Let the class of distributions be denoted by $\{F_\theta | \theta \in \Omega\}$ defined over a measurable space $\mathcal{X}$. Let $\mathcal{G}$ be a class of transformations sx which map $\mathcal{X}$ into itself. $\mathcal{G}$ is said to be an invariant class of transformations if

1. $\mathcal{G}$ is a group; that is, it satisfies:

   (a) If s_1 and s_2 are in $\mathcal{G}$, then the product transformation $s_1 s_2$ is also in $\mathcal{G}$,

   (b) If s belongs to $\mathcal{G}$, then the inverse transformation s^{-1} is in $\mathcal{G}$.

2. The class of distributions $\{F_\theta | \theta \in \Omega\}$ is closed under $\mathcal{G}$; that is, if X has d.f. $F_\theta(x)$, then sX for s in $\mathcal{G}$ has the d.f. $F_{\bar{s}\theta}(x)$ where $\bar{s}\theta$ is in Ω.

The second restriction can be interpreted as follows: If a transformation s is applied to the outcome of an experiment, then the d.f.'s that describe the transformed outcome should be the ones in the original class of d.f.'s. Thus, in this sense, the class of transformations $\mathcal{G}$ does not alter the problem, but leaves it 'invariant'. The first restriction ensures that the inverse of each transformation is in $\mathcal{G}$ and that the composite transformation is also in $\mathcal{G}$.

The transformation s in $\mathcal{G}$ also induces a transformation $\bar{s}$ on the parameter space which maps Ω into Ω. One can easily prove that $\bar{s}$ maps Ω onto Ω in the form of a one-to-one correspondence and that the class $\bar{\mathcal{G}}$ of transformations $\bar{s}$ is a group. Then the class $\bar{\mathcal{G}}$ is said to form a group homomorphic to $\mathcal{G}$.

We know that $\mathcal{G}$ leaves the probability model in a problem unchanged. Suppose $g(\theta)$ is a real parameter to be estimated. If we wish the structure of the parameter to be unchanged, we impose the additional restrictions on $\mathcal{G}$:

3. For each s in $\mathcal{G}$, $g(\theta) = g(\theta')$ implies that $g(\bar{s}\theta) = g(\bar{s}\theta')$ for all θ, θ' in Ω.

The third condition means that if a transformation is applied to an outcome, $g(\theta)$ is also transformed and the new value does not depend on which θ corresponds to the original value of $g(\theta)$. Thus, the transformation s on $\mathcal{X}$ or $\bar{s}$ on Ω induces a transformation $\bar{s}_g$ on the values of the parameter $g(\theta)$. Then, we have the following equation:

$$\bar{s}_g g(\theta) = g(\bar{s}\theta)$$

which implies that if $g(\theta)$ is the parameter for X, then $\bar{s}_g g(\theta)$ is the parameter for sX. If (3) is fulfilled, then $\mathcal{G}$ is said to be *invariant for the parameter* $g(\theta)$.

Definition 1.3.1. *An estimator T is said to be an invariant estimator for $g(\theta)$ if $\bar{s}_g T(x) = T(sx)$, for all s in $\mathcal{G}$ and all x in $\mathcal{X}$. In other words, if a transformation s in $\mathcal{G}$ changes the parameter values, then the values of the invariant estimators are changed in exactly the same manner.*

In estimation of $g(\theta)$, if the statistician confines himself to the class of invariant estimators and in this class, looks for one that has some optimum property like uniformly minimum variance (risk) then we impose one further restriction:

4. For each s in $\mathcal{G}$,

$$\{T(x) - g(\theta)\}^2 = \{\bar{s}_g T(x) - \bar{s}_g g(\theta)\}^2 = \{T(sx) - g(\bar{s}\theta)\}^2 .$$

Example 1.3.1 (Fraser, 1957). Consider the following estimation problem. Let Y_i, $i = 1, 2, \ldots, N$ be random variables defined by the equations:

$$Y_i = \alpha + \beta x_i + U_i, \quad i = 1, 2, \ldots, N$$

where U_i are independent random variables each being uniform on $\left(-\frac{1}{2}, \frac{1}{2}\right)$, and x_i's are known constants. The class of d.f.'s corresponds to all values (α, β) in the two dimensional space. We wish to estimate the parameters α and β. Without loss of generality, we can assume that $\sum_1^N x_i = 0$, since we can write

$$\begin{aligned} \alpha + \beta x_i &= \alpha + \beta \bar{x} + \beta(x_i - \bar{x}) \\ &= \alpha' + \beta'(x_i - \bar{x}) . \end{aligned}$$

Consider the group of transformations $\mathcal{G}$ where

$$\mathcal{G} = \{s y_i = y_i + a_s + b_s x_i (i = 1, 2, \ldots, N) |\ a_s, b_s \text{ are real } \} .$$

$\mathcal{G}$ satisfies our requirements. It is a group (a symmetric group). Each d.f. is transformed by an element of $\mathcal{G}$ into another of the d.f.'s for the problem. The induced class of transformations on the parameter space is given by

$$\bar{\mathcal{G}} = \left\{ \begin{array}{l} \bar{s}\alpha = \alpha + a_s \\ \bar{s}\beta = \beta + b_s \end{array} \middle| \ a_s, b_s \text{ are real} \right\} .$$

Let $\{T_1(y_1, \ldots, y_N), T_2(y_1, \ldots, y_N)\}$ be a statistic for (α, β). If (T_1, T_2) is an invariant estimate, then

$$T_1(s\ \underline{y}) = \bar{s} T_1(\underline{y})$$

and

$$T_2(s\,\underline{y}) = \bar{s}T_2(\underline{y}) \text{ when } \underline{y} = (y_1, \ldots, y_N)\,.$$

That is, for a typical transformation s, the regression equations become

$$\begin{aligned} T_1(y_1 + a_s + b_s x_1, \ldots, y_N + a_s + b_s x_N) &= T_1(y_1, \ldots, y_N) + a_s \\ T_2(y_1 + a_s + b_s x_1, \ldots, y_N + a_s + b_s x_N) &= T_2(y_1, \ldots, y_N) + b_s\,. \end{aligned}$$

If the loss function is of the form

$$w_1(T_1 - \alpha)^2 + w_2(T_2 - \beta)^2$$

where the weights w_1 and w_2 are positive; then it is easy to see that the loss function is also invariant because

$$w_1(\bar{s}T_1 - \bar{s}\alpha)^2 + w_2(\bar{s}T_2 - \bar{s}\beta)^2 = w_1(T_1 - \alpha)^2 + w_2(T_2 - \beta)^2\,.$$

The estimator which minimizes the risk corresponding to the loss function is found to be the center of gravity of the set of all values of (α, β) for which the probability density at a given outcome $(y_1, \ldots, y_N)$ is positive and not equal to zero.

Chapter 2

Order Statistics

2.1 Domain of Nonparametric Statistics

Let us briefly review the origin and the domain of nonparametric statistics. In order to extend the field of application of statistics, statistical workers pursued two approaches:

1. Instead of normal distribution, the functional form of the distribution has been altered in some simple form. (This will still be parametric statistics);

2. To restate the standard problems in quite general terms and then look for adequate statistical procedures. In this case the class of probability distributions is quite large. This field of investigation has been given the title 'nonparametric statistics'. Fraser (1957) defines it as "that portion of the statistical inference for which the parametric space cannot be simply represented as a subset of a real space of finite number of dimensions". Even this is not a clear-cut definition. Let us be happy with the following definition:

> Nonparametric statistics is concerned with the treatment of standard statistical problems, when the familiar assumption of normality is replaced by general assumptions concerning the distribution function.

Distribution-free statistics. A statistic is distribution-free if its distribution under the null hypothesis does not depend upon the distribution of the underlying population of observations.

2.2 Order Statistics

Let a random sample (independent and identically distributed random variables) of observations consisting of $X_1, X_2, \ldots, X_N$ be drawn. If these are arranged in the order of magnitude, but not in the order in which they come, like

$$X_{1,N} \leq X_{2,N} \leq, \ldots, \leq X_{N,N} ,$$

$(X_{1,N}, X_{2,N}, \ldots, X_{N,N})$ is called the *order statistics*. Note that the $X_{i,N}$ are neither independent nor identically distributed.

Definition of ranks. If the random sample of N observations is arranged in order of increasing magnitude and if the smallest observation is assigned the value 1, second smallest the value 2, and the largest the value N, then

$$(X_1, X_2, \ldots, X_N) = (r_1, r_2, \ldots, r_N)$$

where r_i is the rank assigned to X_{iN} $(i = 1, 2, \ldots, N)$, the r_i's are called the *ranks*.

Example 2.2.1. Consider $(X_1, X_2, X_3, X_4) = (+2.1, -1.3, 0.5, 4.5)$. The ordered observations will be:

$$-1.3 < 0.5 < 2.1 < 4.5 .$$

Hence,

$$r_1 = 3,\ r_2 = 1,\ r_3 = 2,\ r_4 = 4 .$$

Note that $(r_1, r_2, r_3, \ldots, r_N)$ is a permutation of the integers $(1, 2, \ldots, N)$. If $(X_1, X_2, \ldots, X_N)$ is replaced by $(R_1, R_2, \ldots, R_N)$ where R_i is the rank of $X_{i,N}$, then $(R_1, R_2, \ldots, R_N)$ is called the *rank order* and it is a random variable. *Order statistics* (O.S.) play a dominant role in nonparametric statistics. We will study these in some detail. They are of two kinds: (i) order statistics in samples drawn from continuous populations; in this case, ties occur with zero probability; (ii) order statistics in samples drawn from discrete populations. In the latter case, ties occur with positive probabilities. We discuss first the O.S. in samples from continuous populations. Let F be the d.f. of the underlying population and $f(x)$ its p.d.f. if it exists. X denotes the random variable having $F(x)$ for its d.f.

2.3 Distribution Theory of Order Statistics

Result 2.3.1. *In a random sample of size N $(N > 1)$ drawn from a continuous population, the probability that two or more observations are equal is zero or, more strongly,*

$$\sum_{i=1}^{N} \sum_{\substack{j=1 \\ i \neq j}}^{N} P(X_{i,N} = X_{j,N}) = 0\,.$$

PROOF: For any $\epsilon > 0$, we have

$$\begin{aligned} P(X_1 = X_2) &= \lim_{\epsilon \to 0} P(-\epsilon \le X_1 - X_2 \le \epsilon) \\ &= \lim_{\epsilon \to 0} \int_{-\infty}^{\infty} [F(y+\epsilon) - F(y-\epsilon)]\, dF(y) \\ &= \int_{-\infty}^{\infty} \left[\lim_{\epsilon \to 0} \{F(y+\epsilon) - F(y-\epsilon)\}\right] dF(y) \end{aligned}$$

Since $F(x)$ is continuous, the integrand of the integral on the right side tends to zero as ϵ tends to zero. Hence,

$$P(X_1 = X_2) = 0$$

which implies that $P(X_{i,N} = X_{j,N}) = 0$. Consequently,

$$\sum_{i=1}^{N} \sum_{\substack{j=1 \\ i \neq j}}^{N} P(X_{i,N} = X_{j,N}) = 0\,.$$

Result 2.3.2. *The d.f. of $X_{i,N}$ is given by*

$$H_{i,N}(x) = P(X_{i,N} \le x) = \sum_{k=i}^{N} \binom{N}{k} F^k(x)\,[1 - F(x)]^{N-k}\,.$$

PROOF: $H_{i,N} = P(X_{i,N} \le x) = P$ (at least i X's $\le x$)

$$= \sum_{k=i}^{N} \binom{N}{k} F^k(x)\,[1 - F(x)]^{N-k}\,.$$

Corollary 2.3.2.1. *The d.f. of* $X_{N,N}$ *is* $H_{N,N}(x) = F^N(x)$.

Corollary 2.3.2.2. *The d.f. of* $X_{1,N}$ *is* $H_{1,N}(x) = 1 - [1 - F(x)]^N$.

Result 2.3.3. *The density function of* $X_{i,N}$ *is given by*

$$h_{i,N}(x)dx = i\binom{N}{i} F^{i-1}(x)\,[1-F(x)]^{N-i}\,dF(x)\,.$$

PROOF: Consider

$$\begin{aligned}
\frac{d}{dx}H_{i,N}(x) &= \frac{d}{dx}\sum_{k=i}^{N}\binom{N}{k}F^k(x)\,[1-F(x)]^{N-k} \\
&= \frac{d}{dx}i\binom{N}{i}\int_0^{F(x)} u^{i-1}(1-u)^{N-i}du \\
&= i\binom{N}{i}F^{i-1}(x)\,[1-F(x)]^{N-i}\,\frac{dF(x)}{dx}\,.
\end{aligned}$$

One can obtain the result independently of the result for $H_{i,N}(x)$. Divide the real line into three mutually exclusive intervals I_1, I_2 and I_3 with associated probabilities p_1, p_2, and p_3, respectively, such that $p_1 + p_2 + p_3 = 1$:

$$\begin{aligned}
I_1 &: \quad (-\infty < X \le x) \\
I_2 &: \quad (x < X \le x + dx) \\
I_3 &: \quad (x + dx < X < \infty)\,.
\end{aligned}$$

Then, the density function of $X_{i,N}$ is obtained by computing the joint probability that in a sample of size N, $i-1$ observations are in I_1, one in I_2 and the remaining $N-i$ observations are in I_3. This probability is obtained from the multinomial law as

$$\frac{N!}{(i-1)!(N-i)!}p_1^{i-1}p_2p_3^{N-i}$$

where

$$\begin{aligned}
p_1 &= \int_{-\infty}^{x} dF(x) = F(x), \\
p_2 &= P(x < X \le x + dx) = dF(x) + O(dx),
\end{aligned}$$

and

$$p_3 = P(x + dx < X < \infty) = 1 - F(x) + O(dx)\,.$$

Now, neglecting terms of order higher than $dF(x)$, the desired expression for the density function of $X_{i,N}$ follows.

Result 2.3.4. *The joint d.f. of $X_{i,N}$ and $X_{j,N}$ $(i < j)$ is given by*

$$H_{i,j,N}(x,y) = \begin{cases} H_{j,N}(y) & \text{if } x > y\,, \\ \displaystyle\sum_{\substack{N \geq r \geq 1 \\ N \geq r+s \geq j}}\sum \frac{N!}{r!s!(N-r-s)!} F^r(x)\,[F(y)-F(x)]^s\,[1-F(y)]^{N-r-s} & \text{if } x < y\,. \end{cases}$$

PROOF: If $x > y$, then

$$\begin{aligned} H_{i,j,N}(x,y) &= P(X_{i,N} \leq x,\ X_{j,N} \leq y) \\ &= P(X_{j,N} \leq y) = H_{j,N}(y)\,. \end{aligned}$$

If $x < y$, then assume that r sample observations are less than x and s observations are lying between x and y. Then $N - r - s$ observations are greater than y. The probability of obtaining such a sample is

$$\frac{N!}{r!s!(N-r-s)!} F^r(x)\,[F(y)-F(x)]^s\,[1-F(y)]^{N-r-s}$$

$$\begin{aligned} &H_{i,j,N}(x,y) \\ &= P(X_{i,N} \leq x,\ X_{j,N} \leq y) \\ &= P(\text{at least } i \text{ observations are } \leq x \text{ and at least } j \text{ observations are less than or equal to } y) \\ &= \sum_{\substack{N \geq r+s \geq j \\ N \geq r \geq i}}\sum \frac{N!}{r!s!(N-r-s)!} F^r(x)\,[F(y)-F(x)]^s\,[1-F(y)]^{N-r-s}\,. \end{aligned}$$

This completes the proof of the result.

Result 2.3.5. *The joint density function of $X_{i,N}$ and $X_{j,N}$ is given by*

$$h_{i,j,N}(x,y)dx\,dy = \begin{cases} \frac{N!}{(i-1)!(j-i-1)!(N-j)!} F^{i-1}(x)\,[F(y)-F(x)]^{j-i-1}\,[1-F(y)]^{N-j}\,dF(x)dF(y) & \text{if } x < y \\ 0 & \text{otherwise.} \end{cases}$$

PROOF: When $x < y$, since $1 \le j-r \le j-i$, $H_{i,j,N}(x,y)$ can be rewritten as

$$\begin{aligned} H_{i,j,N}&(x,y) \\ &= \sum_{r=i}^{j-1} \binom{N}{r} F^r(x)\,[1-F(x)]^{N-r} \\ &\qquad \cdot \sum_{s=j-r}^{N-r} \left[\frac{1-F(y)}{1-F(x)}\right]^{N-r-s} \left[1-\frac{1-F(y)}{1-F(x)}\right]^s \binom{N-r}{s} \\ &= \sum_{r=i}^{j-1} \binom{N}{r} F^r(x)\,[1-F(x)]^{N-r}\,(j-r)\binom{N-r}{j-r} \\ &\qquad \cdot \int_0^{\frac{F(y)-F(x)}{1-F(x)}} u^{j-r-1}(1-u)^{N-j}\,du\,. \end{aligned}$$

Now,

$$\begin{aligned} \frac{\partial}{\partial y}&H_{i,j,N}(x,y)dy \\ &= \sum_{r-i}^{j-1} \binom{N}{r}(j-r)\binom{N-r}{j-r} F^r(x)\,[F(y)-F(x)]^{j-r-1}\,[1-F(y)]^{N-j}\,dF(y) \\ &= \binom{N}{j}[1-F(y)]^{N-j}\,F^{j-1}(y)dF(y) j \sum_{r=i}^{j-1}\binom{j-1}{r}\left[\frac{F(x)}{F(y)}\right]^r\left[1-\frac{F(x)}{F(y)}\right]^{j-r-1} \\ &= \binom{N}{j}[1-F(y)]^{N-j}\,F^{j-1}(y)dF(y) ji\binom{j-1}{i}\int_0^{F(x)/F(y)} v^{i-1}(1-v)^{j-i-1}dv\,. \end{aligned}$$

Therefore,

$$\begin{aligned} h_{i,j,N}&(x,y)dy\,dx = \\ &ij\binom{N}{j}\binom{j-1}{i}F^{i-1}(x)\,[F(y)-F(x)]^{j-i-1}\,[1-F(y)]^{N-j}\,dF(y)dF(x)\,. \end{aligned}$$

when $x > y$,

$$\frac{\partial}{\partial y}H_{i,j,N}(x,y)dy = h_{j,N}(y)dy$$

and

$$\frac{\partial}{\partial x}\frac{\partial}{dy}H_{i,j,N}(x,y)dy\,dx = 0\,.$$

This completes the proof of the result.

Remark 2.3.1. *Alternately, one can divide the real line into the following mutually exclusive intervals:*

$$\begin{aligned} I_1 &: \quad (-\infty, x) \\ I_2 &: \quad (x, x+dx) \\ I_3 &: \quad (x+dx, y) \\ I_4 &: \quad (y, y+dy) \\ I_5 &: \quad (y+dy, \infty)\,. \end{aligned}$$

Then, proceed as in Result 2.3.3 and obtain the desired expression for the joint density function of $X_{i,N}$ and $X_{j,N}$.

Remark 2.3.2 (Savage, 1963). *The joint d.f. of $X_{i,N}$, $i = 1, 2, \ldots, N$ is given by*

$$H_{1,2,\ldots,N}(x_1, x_2, \ldots, x_N) = \sum \frac{N!}{\prod_{k=1}^N i_k!} \prod_{k=1}^N \left[F(x_k) - F(x_{k-1})\right]^{i_k}$$

$$\begin{aligned} i_j &\ge 0 \text{ for } j = 1, 2, \ldots, N \\ 0 &\le i_N \le 1 \\ 0 &\le i_N + i_{N-1} \le 2 \\ 0 &\le i_N + i_{N-1} + \cdots + i_2 \le N-1,\ i_1 + i_2 + \cdots + i_N = N\,, \end{aligned}$$

where i_k is the number of X's lying in the interval (x_{k-1}, x_k), $= 1, \ldots, N$ with $x_0 = -\infty$.

Proof: $H_{1,2,\ldots,N}(x_1, x_2, \ldots, x_N) = P$ [No X's $> x_N$, at most one $X > x_{N-1}$, at most two X's are $> x_{N-2}, \ldots,$ at most i X's are $> x_{N-i}, \ldots,$ at most $N-1$ X's are $> x_1$].

Let i_k be the number of X's lying in the interval (x_{k-1}, x_k) for $k = 1, 2, \ldots, N$, with $x_0 = -\infty$. These points divide the real line into $(N+1)$ mutually exclusive intervals. The probability that there are i_k observations in (x_{k-1}, x_k), $k = 1, 2, \ldots, N$ is

$$\frac{N!}{\prod_{k=1}^N i_k!} \prod \left[F(x_k) - F(x_{k-1})\right]^{i_k}\,. \tag{2.3.1}$$

Now, the desired expression for the joint d.f. of all the order statistics follows by summing (2.3.1) on all possible i_k's.

Result 2.3.6 (Savage, 1963). *The joint density function of* $X_{i,N}$, $i = 1, 2, \ldots, N$ *is given by*

$$N! \prod_{i=1}^{N} dF(x_i), \text{ for } -\infty < x_1 < x_2 < \cdots < x_N < \infty$$

and zero otherwise.

PROOF: Consider

$$\begin{aligned} D &= \prod_{k=1}^{N} [F(x_k) - F(x_{k-1})]^{i_k} \\ &= \prod_{k=1}^{N-1} [F(x_k) - F(x_{k-1})]^{i_k} [F(x_N) - F(x_{N-1})]^{i_N} \\ \sum_{0 \le i_N \le 1} D &= \prod_{k=1}^{N-1} [F(x_k) - F(x_{k-1})]^{i_k} \\ &\quad + \prod_{k=1}^{N-1} [F(x_k) - F(x_{k-1})]^{i_k} [F(x_N) - F(x_{N-1})] \,. \end{aligned}$$

When we take the partial differential of $\sum_{i_N=0}^{1} D$ with respect to x_N, we get

$$D_1 = \prod_{k=1}^{N-1} [F(x_k) - F(x_{k-1})]^{i_k} \, dF(x_N) \,.$$

Now, sum on i_{N-1} such that $0 \le i_{N-1} \le 1$ since $0 \le i_N + i_{N-1} \le 2$ and $i_N = 1$. Thus, giving the two values for i_{N-1}, we get two terms for D_1. Considering the partial differential with respect to x_{N-1} we get

$$D_2 = \prod_{k=1}^{N-2} [F(x_k) - F(x_{k-1})]^{i_k} \, dF(x_{N-1}) dF(x_N) \,.$$

Continuing this process of considering differentials up to x_1, we finally get $\prod_{k=1}^{N} dF(x_k)$ for the joint partial differential of

$$D = \sum_{k=1}^{N} [F(x_k) - F(x_{k-1})]^{i_k} \text{ w.r.t. } x_N, x_{N-1}, \ldots, x_1 \,.$$

Also, since $i_1, i_2, \ldots, i_N$ have taken only two values, namely, 0 and 1, the multiplier of $\prod_{k=1}^{N} dF(x_k)$ is $N!$. Thus, the joint density function of $X_{1,N}$, $X_{2,N}$, …, and $X_{N,N}$ is

$$N! \prod_{k=1}^{n} dF(x_k), \quad -\infty < x_1 < x_2 < \cdots < x_N < \infty .$$

Remark 2.3.3. *One can also obtain the joint density function of all the order statistics by considering one observation in each of the intervals* $(x_1, x_1 + dx_1), (x_2, x_2 + dx_2), \ldots, (x_N, x_N + dx_N)$.

2.3.1 Distribution of Sample Range and Mid Range

Result 2.3.7. *Let* $R = X_{NN} - X_{1N}$. *Then the probability distribution function of* R *is given by*

$$f_R(r) = N(N-1) \int_{-\infty}^{\infty} [F(t) - F(t-r)]^{N-2} f(t-r) f(t) dt .$$

PROOF: The joint density of X_{1N} and X_{NN} is given by

$$N(N-1)\,[F(y) - F(x)]^{N-2} f(x) f(y) \text{ for } -\infty < x < y < \infty$$

and zero elsewhere. Now let $r = y - x$ and $t = y$. Then the joint density of r and t is

$$N(N-1)\,[F(t) - F(t-r)]^{N-2} f(t-r) f(t) .$$

Hence the marginal density of R is

$$f_R(r) = N(N-1) \int_{-\infty}^{\infty} [F(t) - F(t-r)]^{N-2} f(t-r) f(t) dt .$$

Similarly, put $m = (v+u)/2$, that is, $v = 2m - u$, integrate on u and obtain (since $u \le v$ implies that $2u < 2m$)

$$f_M(m) = 2N(N-1) \int_{-\infty}^{m} [F(2m-u) - F(u)]^{N-2} f(2m-u) f(u) du .$$

Note: It may be easier to obtain the d.f.'s of R and M and from them the p.d.f.'s.

In the uniform case,

$$f_R(r) = N(N-1) r^{N-2} (1-r), \quad 0 < r < 1 ,$$

and

$$f_M(m) = 2N(N-1) \cdot \int_{\max(0,2m-1)}^{m} (2m-2u)^{N-2} du = N2^{N-1} \{\min(m, 1-m)\}^{N-1} .$$

Alternatively, the joint density of X_{1N}, X_{NN} in the uniform case is

$$f(x,y) = N(N-1)(y-x)^{N-2}, \quad 0 \le x \le y \le 1 .$$

Now let $u = \frac{x+y}{2}$, $v = y$. Then $x = 2u - v$ and the joint density of u and v is

$$g(u,v) = N(N-1)(-2u+2v)^{n-2}|J|$$

where

$$J = \begin{vmatrix} \frac{\partial x}{\partial u} & \frac{\partial x}{\partial u} \\ \frac{\partial y}{\partial u} & \frac{\partial y}{\partial u} \end{vmatrix} = \begin{vmatrix} 2 & -1 \\ 0 & 1 \end{vmatrix} = 2 .$$

So,

$$g(u,v) = N(N-1)2^{N-1}(v-u)^{N-2}$$

where

$$0 \le \frac{v}{2} \le u \le v \le 1 .$$

So,

$$\begin{aligned} g_U(u) &= \int_u^{2u} N(N-1)2^{N-1}(v-u)^{N-2} dv = N2^{N-1}u^{N-1}, \ 0 \le u \le \frac{1}{2} \\ &= \int_u^1 N(N-1)2^{N-1}(v-u)^{N-2} dv = N2^{N-1}(1-u)^{N-1}, \ \frac{1}{2} \le u \le 1 . \end{aligned}$$

2.3.2 The Distribution of the Median

If $X_{1N} \le \cdots \le X_{NN}$ denote order statistics in a random sample of size N drawn from a continuous distribution $F(x)$ having density $f(x)$, the sample median T is defined as

$$T = \begin{cases} X_{(N+1)/2,N} & \text{if } N \text{ is odd,} \\ \frac{1}{2}\{X_{N/2,N} + X_{(N+2)/2,N}\} & \text{if } N \text{ is even.} \end{cases}$$

Then the probability density of T is given by Result 2.3.8.

Result 2.3.8. *The probability density function of T, the median, is*

$$
\begin{aligned}
f_T(u) &= \frac{N!}{\{(\frac{N-1}{2})!\}^2}\left[F(u)\{1-F(u)\}\right]^{\frac{N-1}{2}} f(u) && \textit{if } N \textit{ is odd,} \\
&= 2\frac{N!}{\{(\frac{N-1}{2})!\}^2}\int_u^{\infty} F^{\frac{N-2}{2}}(2u-v)\left[1-F(v)\right]^{\frac{N-2}{2}} f(2u-v)f(v)dv \\
& && \textit{if } N \textit{ is even.}
\end{aligned}
$$

PROOF: If N is odd, set $i = (N+1)/2$ in Result 2.2.2. If N is even, the joint density of $X_{N/2,N}$ and $X_{(N+2)/2,N}$ is given by

$$
\begin{aligned}
h_{N/2,(N+2)/2,N}(x,y) &= 2\frac{N!}{((N-2)/2!)^2}F^{\frac{N-2}{2}}(x)\left[1-F(y)\right]^{\frac{N-2}{2}} f(x)f(y), \\
& \qquad \text{for } x < y, \\
&= 0 \qquad \text{elsewhere.}
\end{aligned}
$$

Making the change of variables, $u = (x+y)/2$ and $v = y$ and integrating on v, we obtain the other expression.

Special Case. Let $F(x)$ be the standard uniform distribution. Then

$$
\begin{aligned}
f_T(u) &= \frac{N!}{\left(\frac{(N-1)}{2}!\right)^2}\left[u(1-u)\right]^{(N-1)/2}, \ 0 < u < 1, && \text{if } N \text{ is odd,} \\
&= 2\frac{N!}{\left(\frac{(N-2)}{2}!\right)^2}\sum_{j=0}^{(N-2)/2}\binom{(N-2)/2}{j}(1-2u)^j\frac{u^{N-1-j}}{N-1-j}, \\
& && \text{if } u < \tfrac{1}{2} \text{ and } N \text{ is even,} \\
&= 2\frac{N!}{\left(\frac{(N-2)}{2}!\right)^2}\sum_{j=0}^{(N-2)/2)}\binom{(N-2)/2}{j}(2u-1)^j\frac{(1-u)^{N-1-j}}{N-1-j}, \\
& && \text{if } u > \tfrac{1}{2} \text{ and } N \text{ is even.}
\end{aligned}
$$

PROOF: It suffices to prove the case when N is even. Consider

$$\int_{\substack{u \\ 0<2u-v<1}}^{1} (2u-v)^{(N-2)/2}(1-v)^{(N-2)/2}dv = \int_{\max(u,2u-1)}^{\min(1,2u)} [(2u-v)(1-v)]^{(N-2)/2}\,dv\,.$$

Case 1. $0 < u < 1/2$. *Then the integral on the right side becomes*

$$\begin{aligned}\int_{u}^{2u} (2u-v)^{(N-2)/2}(1-v)^{(N-2)/2}dv &= \int_{0}^{u} t^{\frac{N-2}{2}}(1-2u+t)^{(N-2)/2}dt \\ &= \sum_{j=0}^{(N-2)/2} \binom{\frac{N-2}{2}}{j}(1-2u)^{j}\frac{u^{N-1-j}}{N-1-j}\,.\end{aligned}$$

Case 2. $1/2 \le u \le 1$.

$$\begin{aligned}\int_{u}^{1} \{(2u-v)(1-v)\}^{(N-2)/2}\,dv &= \int_{0}^{1-u} \{(2u-1+s)s\}^{\frac{N-2}{2}}\,ds \\ &= \sum_{j=0}^{(N-2)/2} \binom{\frac{N-2}{2}}{j}(2u-1)^{j}\frac{(1-u)^{N-2-j+1}}{N-1-j}\,.\end{aligned}$$

Coverages: The intervals $(-\infty, X_{1,N}), (X_{1,N}, X_{2,N}), \ldots, (X_{N,N}, \infty)$ are respectively called *sample blocks* $B^{(1)}, \ldots, B^{(N+1)}$. The function $U_1 = F(X_{1,N}), U_2 = F(X_{2,N}) - F(X_{1,N}), \ldots, U_{N+1} = 1 - F(X_{N,N})$ of these blocks are called the *coverages*. Usually U_{N+1} is omitted since the sum of the U_i's is unity. The coverage for a given sample block is the amount of probability in the population distribution contained in that sample block. It is easy to see that the U_i's are random variables.

2.3.3 Sampling Distribution of the Coverages

Lemma 2.3.1. *If X is a continuous random variable having $F(x)$ for its distribution function, then the probability integral transform $F(X)$ is a uniform random variable on [0,1].*

PROOF: For any x, define x^- and x^+ as follows:

$$x^- = \inf_{F(x')=F(x)} x' \quad \text{and } x^+ = \sup_{F(x')=F(x)} x'\,.$$

Since $F(x)$ is continuous $F(x^-) = F(x^+) = F(x)$. Also, from the definition of x^- and x^+ we have the following set inclusion relations:

$$\{x'|x' \leq x^-\} \subseteq \{x'|F(x') \leq F(x)\} \subseteq \{x'|x' \leq x^+\}.$$

Consequently, we obtain

$$P(X \leq x^-) \leq P[F(X) \leq F(x)] \leq P(X \leq x^+).$$

Since the extreme members of the inequality are each equal to $F(x)$, it follows that $P[F(x) \leq F(x)] = F(x)$; also, since $F(x)$ takes all values in [0,1], we obtain

$$P[F(X) \leq y] = y, \quad 0 \leq y \leq 1.$$

This completes the proof of the result.

Result 2.3.9. *$F(X_{1,N}), F(X_{2,N}), \ldots, F(X_{N,N})$ form an O.S. of a sample size N drawn from the standard uniform distribution and the joint density of $F(X_{i,N})$, $i = 1, 2, \ldots, N$ is given by $N! dy_1\, dy_2 \cdots dy_N$, $0 < y_1 < y_2 \cdots y_N < 1$ and zero otherwise.*

PROOF: It is known that $F(X)$ is uniform on (0,1) if X is continuous. Hence, $F(X_1), F(X_2), \ldots, F(X_N)$ constitutes a random sample of size N drawn from the standard uniform population. Then, $F(X_{i,N})$, $i = 1, 2, \ldots, N$ will be the O.S. in the preceding sample because $F(X_{1,N}) = \min[F(X_1), \ldots, F(X_N)]$, and $F(X_{N,N}) = \max[F(X_1), \ldots, F(X_N)]$. The joint density of $F(X_{1,N})$, …, $F(X_{N,N})$ is

$$N! \prod_1^N dy_i, \quad 0 < y_1 < \cdots < y_N < 1.$$

Result 2.3.10. *The joint density of the coverages is given by*

$$N! \prod_{i=1}^N du_i \text{ for } u_i > 0, \quad i = 1, \ldots, N \text{ and } \sum_1^N u_i < 1,$$

and the density is zero elsewhere.

PROOF: Make the following substitution:

$$\begin{aligned} F(X_{1,N}) &= U_1 \\ F(X_{2,N}) &= U_1 + U_2 \\ &\vdots \\ F(X_{N,N}) &= U_1 + U_2 + \cdots + U_N. \end{aligned}$$

Then the joint density function of $U_1, U_2, \ldots, U_N$ is given by

$$N! du_1\, du_2 \cdots du_N,$$

in the simplex $S_N = \left\{(u_1, \ldots, u_N) : u_k > 0, k = 1, 2, \ldots, N, \sum_1^N u_k < 1\right\}$ and the density is 0 outside S_N. Notice that the joint p.d.f. of the U_i is symmetric in the N-variables.

Result 2.3.11. *The sum of any k of the coverages has a beta distribution.*

PROOF: Because of the symmetry of the joint p.d.f. of the U_i, it suffices to consider consecutive sums of the U_i's. $F(X_{k,N})$ is the sum of the first k coverages $U_1, U_2, \ldots, U_k$ and $F(X_{k,N})$ has a beta distribution.

Result 2.3.12. *If $V_1, V_2, \ldots, V_s$ are the sums of $k_1, \ldots, k_s$ respectively of the coverages where no coverage belongs to more than one V_i, then the distribution of $(V_1, V_2, \ldots, V_s)$ is an s-variate Dirichlet distribution given by*

$$[N!/\Gamma(k_1)\cdots\Gamma(N - k_1 \cdots - k_s)]\, v_1^{k_1-1} v_2^{k_2-1} \cdots v_s^{k_s-1} \\ \times (1 - v_1 - v_2 \cdots - v_s)^{N - \sum k_i},$$

for $0 < v_i$ $(i = 1, \ldots, s)$ and $v_1 + v_2 + \cdots + v_s < 1$.

Remark 2.3.4. *One can define multi-dimensional coverages in an analogous manner.*

2.4 Moments of Order Statistics

Let $X_{1N} \le \cdots \le X_{NN}$ be the O.S. in a random sample of size N drawn from a continuous population having $F(x)$ and $f(x)$ for its d.f. and p.d.f., respectively. Then let $\mu_{iN}^{(k)} = E(X_{iN}^k)$, $1 \le i \le N$, and $\mu_{i,j,N} = E(X_{iN}, X_{jN})$, $1 \le i \le j \le N$ and $\sigma_{ijN} = \text{cov}(X_{iN}, X_{,N})$. Only for certain distributions, one can obtain explicit expressions for the moments of order statistics.

Special Case 1. Let $F(x) = x$, $0 \le x \le 1$. Then

$$\begin{aligned} \mu_{iN}^{(k)} &= \frac{N!}{(i-1)!(N-i)!} \int_0^1 x^{k+i-1}(1-x)^{N-i} dx \\ &= \frac{N!}{(i-1)!(N-i)!} \frac{(k+i-1)!(N-i)!}{(N+k-1)!} \\ &= \frac{(k+i-1)\cdots i}{(N+k-1)\cdots(N+1)}, \quad k = 1, 2, \ldots . \end{aligned}$$

Thus,

$$E(\mu_{iN}) = \frac{i}{N+1},$$

and

$$E(\mu_{iN}^{(2)}) = \frac{i(i+1)}{(N+2)(N+1)}.$$

Hence,

$$\text{var } X_{iN} = \sigma_{iiN} = \frac{i(N+1-i)}{(N+1)^2(N+2)}, \quad 1 \le i \le N.$$

Next consider, for $i \le j$,

$$\begin{aligned} \mu_{ijN} &= E(X_{iN}X_{jN}) \\ &= \frac{N!}{(i-1)!(j-i-1)!(N-j)!} \iint_{x<y} x^i y(y-x)^{j-i-1}(1-y)^{N-j} dx\, dy. \end{aligned}$$

First, let us evaluate

$$\begin{aligned} \int_0^y x^i (y-x)^{j-i-1} dx &= y^j \int_0^1 u^i (1-u)^{j-i-1} du \\ &= y^j \cdot i!(j-i-1)!/j!. \end{aligned}$$

Hence,

$$\begin{aligned} \mu_{ijN} &= \frac{N!i}{j!(N-j)!} \int_0^1 y^{j+1}(1-y)^{N-j} dy \\ &= i(j+1)/(N+1)(N+2). \end{aligned}$$

Thus,

$$\sigma_{i,j,N} = \text{cov}(X_{iN}, X_{jN}) = i(N-j+1)/(N+1)^2(N+2) \text{ for } i \le j.$$

Special Case 2. Let $F(x) = 1 - \exp(-x)$, $0 < x < \infty$. Then, from Renyi's representation (see Corollary 2.7.3.1), we obtain

$$X_{iN} = \frac{\delta}{N} + \cdots + \frac{\delta_i}{N-i+1},$$

where the δ_i are i.i.d. standard exponential random variables. Consequently,

$$\begin{aligned} \mu_{iN} &= \frac{1}{N} + \cdots + \frac{1}{N-i+1} \approx \int_{N-i+1}^{N} \frac{dx}{x} = \log\left(\frac{N}{N-i+1}\right) \\ \text{var } X_{iN} &= \frac{1}{N^2} + \cdots + \frac{1}{(N-i+1)^2} \approx \int_{N-i+1}^{N} \frac{dx}{x^2} = \frac{i-1}{N(N-i+1)}. \end{aligned}$$

For $i \le j$,

$$\operatorname{cov}(X_{iN}X_{jN}) = \sigma_{ijN} = \operatorname{var} X_{iN} = \frac{i-1}{N(N-i+1)}\,.$$

Result 2.4.1 (Barnett, 1966). *Let $X_{1N} \le \cdots \le X_{NN}$ denote O.S. in a random sample of size N drawn from the Cauchy density given by*

$$f(x) = \frac{1}{\pi(1+x^2)}, \quad -\infty < x < \infty\,.$$

Then, $E\left\{|X_{kN}|^i\right\} < \infty$ for all $i < k < N-i+1$.

PROOF: One can easily show that

$$f(x) \le \frac{1}{\pi x^2} \text{ and } 1 - F(x) \le \frac{1}{\pi}\int_x^\infty \frac{dy}{(1+y^2)} \le \frac{1}{\pi}\int_x^\infty \frac{dy}{y^2} = \frac{1}{\pi x} \text{ for } x > 0\,.$$

Consider

$$\begin{aligned} E\left\{|X_{kN}|^i\right\} &= C_{k,N}\int_{-\infty}^{\infty} |x|^i F^{k-1}(1-F)^{N-k} f(x)dx \\ &= C_{k,N}\left[\int_{-\infty}^{-1} + \int_{-1}^{+1} + \int_1^\infty\right], \end{aligned}$$

where

$$C_{k,N} = \frac{N!}{(k-1)!(N-k)!}\,.$$

Clearly, the integral on $(-1,1)$ is finite since x is bounded. Consider

$$\begin{aligned} \int_{-\infty}^{-1} &= \int_1^\infty x^i\,[1-F(x)]^{k-1}\,[F(x)]^{n-k} f(x)dx \\ &\le A\int_1^\infty x^{i-k+1-2}dx < \infty \text{ provided } k > i\,. \end{aligned}$$

Similarly, we can show that the integral on $(1,\infty)$ is finite provided $k < N-i+1$.

Corollary 2.4.1.1. *Since $|\mu_{ijN}|^2 \le \mu_{i,N}^{(2)}, \mu_{j,N}^{(2)}, \mu_{i,j,N} < \infty$ for $2 < i,j < N-1$.*

For example: First moments exist for all $X_{k,N}$ such that $1 < k < N$. Second moments exist for all $X_{k,N}$ such that $2 < k < N-1$. Thus, one can obtain BLUE of location and scale parameters of the Cauchy distribution provided the first and second moments of the relevant order statistics are available.

Recall that $\mu_{i,N}^{(k)} = E(X_{i,N}^k)$ and $\mu_{i,j,N} = E(X_{i,N}X_{j,N})$, $1 \le i \le j \le N$. Then we write $\mu_{i,N}^{(2)} = \mu_{i,i,N}$. We have the following general recurrence formulae and certain identities among the moments of O.S.

Result 2.4.2. *For any continuous distribution*

$$i\mu_{i+1,N}^{(k)} + (N-i)\mu_{i,N}^{(k)} = N\mu_{i,N-1}^{(k)}, \quad 1 \le i \le N-1, \quad k = 1,2,\ldots .$$

PROOF: Follows from the identity $F + 1 - F = 1$.

Result 2.4.3. *For an arbitrary continuous distribution and for* $1 < i \le j \le N$,

$$(i-1)\mu_{i,j,N} + (j-i)\mu_{i-1,j,N} + (N-j+1)\mu_{i-1,j-1,N} = N\mu_{i-1,j-1,N-1} .$$

PROOF: Follows from the identity $F(x) + [F(y) - F(x)] + 1 - F(y) = 1$.

Result 2.4.4. *For an arbitrary distribution*

$$\sum_{i=1}^{N}\sum_{j=1}^{N} cov(X_{i,N}, X_{j,N}) = N var\, X .$$

PROOF: Consider $\text{var}(X_{1,N} + X_{2,N} + \cdots + X_{N,N}) = \text{var}(X_1 + \cdots + X_N) = N$ var X where X_i's denote the unordered $X_{i,N}$'s.

Result 2.4.5. *For an arbitrary continuous distribution and for* $r, s \ge 0$,

$$\sum_{i=1}^{N-1}\sum_{j=i+1}^{N} E(X_{i,N}^r\, X_{j,N}^s) = (1/2)N(N-1)E[X_{1,2}^r\, X_{2,2}^s] .$$

PROOF: L.H.S.

$$= \sum_{i=1}^{N-1} \sum_{j=i+1}^{N} \frac{N!}{(i-1)!(j-i-1)!(N-j)!}$$

$$\iint_{-\infty<x<y<\infty} \left\{x^r y^s F^{i-1}(x) \left[F(y)-F(x)\right]^{j-i-1} \left[1-F(y)\right]^{N-j} dF(x)dF(y)\right\}$$

$$= \sum_{i=1}^{N-1} \frac{N!}{(i-1)!(N-i-1)!} \iint_{x<y} x^r y^s F^{i-1}(x) \left[1-F(x)\right]\right]^{N-i-1} dF(x)dF(y)$$

$$= N(N-1) \iint_{x<y} x^r y^s dF(x)dF(y) = \binom{N}{2} E(X_{1,2}^r \, X_{2,2}^s) \,.$$

Corollary 2.4.5.1. *For* $r \geq 0$,

$$\sum_{i=1}^{N-1} \sum_{j=i+1}^{N} E(X_{i,N}^r \, X_{j,N}^r) = \binom{N}{2} \left[E(X^r)\right]^2 .$$

PROOF: Follows from the above result and the relation $E(X_{1,2}^r \, X_{2,2}^r) = [E(X^r)]^2$.

Result 2.4.6. *If g is any differentiable function, such that differentiation of g with respect to its argument and its expectation with respect to an arbitrary continuous distribution, are interchangeable, then*

$$E\left\{g'(X_{i,N})\right\} = -\sum_{j=1}^{N} E\left[g(X_{i,N}) f'(X_{j,N}) / f(X_{j,N})\right], \quad i = 1(1)N .$$

PROOF: Consider

$$\begin{aligned} E\left[g(X_{i,N}+t)\right] &= N! \iint_{-\infty<x_1 \cdots <x_N<\infty} g(x_i+t) \prod_{j=1}^{N} f(x_j) dx_j \\ &= N! \iint_{-\infty<x_1 \cdots <x_N<\infty} g(x_i) \prod_{j=1}^{N} f(x_j - t) dx_j . \end{aligned}$$

The result follows after differentiating both sides with respect to t and setting $t = 0$.

Corollary 2.4.6.1.

1. *If* $g(x) = x$,

$$\sum_{j=1}^{N} E\left[X_{i,N} f'(X_{j,N})/f(X_{j,N})\right] = -1, \quad i = 1(1)N\,.$$

2. *If* $g(x) = x$ *and* $f(x)$ *is the standard normal density, then*

$$\sum_{j=1}^{N} \mu_{i,j,N} = 1, \quad i = 1(1)N\,.$$

3. *If* $g(x) = 1$ *and* $f(x)$ *denotes the normal density having mean 0, then*

$$\sum_{j=1}^{N} \mu_{j,N} = 0\,.$$

Result 2.4.7. *For an arbitrary continuous distribution and even* N,

$$\begin{aligned} \mu_{1,N,N} &= \sum_{i=1}^{(N-2)/2} (-1)^{i-1} \binom{N}{i} \mu_{i,i}\mu_{N-i,N-i} \\ &\quad + (1/2)(-1)^{(N-2)/2} \binom{N}{N/2} \mu^2_{N/2,N/2}\,. \end{aligned}$$

PROOF: Consider

$$\mu_{1,N,N} = N(N-1) \iint_{-\infty<x<y<\infty} xy\,[F(y) - F(x)]^{N-2}\, dF(x) dF(y)\,.$$

The integrand in the above integral is symmetric in x and y. Hence,

$$\mu_{1,N,N} = (1/2)N(N-1) \int_{-\infty}^{\infty} \int_{-\infty}^{\infty} xy\,[F(y) - F(x)]^{N-2}\, dF(x) dF(y)\,.$$

Now, expand $[F(y) - F(x)]^{N-2}$ in powers of $F(y)$ and $F(x)$ and obtain

$$\begin{aligned}
\mu_{1,N,N} &= (1/2)N(N-1)\left[\sum_{i=0}^{N-2}(-1)^i\binom{N-2}{i}\left\{\int_{-\infty}^{\infty} xF^i(x)dF(x)\right\}\right. \\
&\qquad \left.\cdot\left\{\int_{-\infty}^{\infty} yF^{N-i-2}(y)dF(y)\right\}\right] \\
&= (1/2)\sum_{i=0}^{N-2}(-1)^i\binom{N}{i+1}\mu_{i+1,i+1}\mu_{N-i-1,N-i-1} \\
&= \sum_{i=0}^{(N-4)/2}(-1)^i\binom{N}{i+1}\mu_{i+1,i+1}\mu_{N-i-1,N-i-1} \\
&\quad + (1/2)(-1)^{(N-2)/2}\binom{N}{N/2}\mu^2_{N/2,N/2}
\end{aligned}$$

after combining like terms. That is,

$$\begin{aligned}
\mu_{1,N,N} &= \sum_{i=1}^{(N-2)/2}(-1)^{i-1}\binom{N}{i}\mu_{i,i}\mu_{N-i,N-i} \\
&\quad + (1/2)(-1)^{(N-2)/2}\binom{N}{N/2}\mu^2_{N/2,N/2}\,.
\end{aligned}$$

This completes the proof of the result.

For example, the result for $N = 4$ and 6, respectively, are

$$\mu_{1,4,4} = 4\mu_{1,1}\mu_{3,3} - 3\mu^2_{2,2}$$

and

$$\mu_{1,6,6} = 6\mu_{1,1}\mu_{5,5} - 15\mu_{2,2}\mu_{4,4} + 10\mu^2_{3,3}\,.$$

If the population mean is zero, that is, $\mu_{1,1} = 0$, then the first term in each expression will disappear.

Result 2.4.8. *For an arbitrary continuous distribution symmetric about zero, the distributions of $-X_{i,N}$ and $-X_{N-i+1,N}$ are identical; and the distributions of $(-X_{i,N}, -X_{j,N})$ and $(X_{N-j+1,N}, X_{N-i+1,n})$ are identical, $1 \le i \le j \le N$.*

Proof: Follows by writing down their density functions and using the symmetry of the distribution.

Corollary 2.4.8.1.

1. $\mu_{i,N} = -\mu_{N-i+1,N}$ *and* $\mu_{i,j,N} = \mu_{N-j+1,N-i+1,N}$, $1 \le i \le j \le N$.
2. *If the mean of the continuous and symmetric population is zero and* N *is odd, then* $\mu_{(N+1)/2,N} = 0$.

Result 2.4.9. *In order to find the first, second and mixed (linear) moments of O.S. in a sample of size* N *drawn from an arbitrary continuous population symmetric about zero, given these moments for all sample sizes less than or equal to* $N-1$, *one has to evaluate at most one single integral and* $(N-4)/2$ *double integrals when* N *is even; and at most one single integral and* $(N-3)/2$ *double integrals when* N *is odd.*

PROOF: See Govindarajulu (1963), Theorem 4.12.

Result 2.4.10 (Savage, 1963). *For a population having no moments, at least one of* $\mu_{1,N}$ *and* $\mu_{N,N}$ *does not exist.*

PROOF: Suppose both $\mu_{1,N}$ and $\mu_{N,N}$ exist. Then, consider

$$\begin{aligned}\mu_{1,N} + \mu_{N,N} &= N\int_{-\infty}^{\infty} x\left\{[1-F(x)]^{N-1} + [F(x)]^{N-1}\right\}dF(x)\\ &\ge N\int_{-\infty}^{\infty} x(2^{1-N} + 2^{1-N})dF(x) = N2^{2-N}E(X)\,.\end{aligned}$$

Hence, $\mu_{1,N} + \mu_{N,N} < \infty$ implies that $E(X) < \infty$, which is a contradiction.

Result 2.4.11 (Ludwig, 1960). *If* $X_{1,N} < X_{2,N} < \cdots < X_{N,N}$ *are O.S. in a sample of size* N *drawn from an absolutely continuous population having* $F(x)$, $f(x)$, μ, σ^2 *for its d.f., p.d.f., mean and variance, respectively, then*

$$\begin{aligned}[(\mu_{j,N} - \mu_{i,N})/\sigma]^2 \le \frac{2N}{\binom{2N}{N}} &\left[\binom{2j-2}{j-1}\binom{2N-2j}{N-j}\right.\\ &+\binom{2i-2}{i-1}\binom{2N-2i}{N-i}\\ &\left.-2\binom{i+j-2}{i-1}\binom{2N-i-j}{N-j}\right].\end{aligned}$$

PROOF: Consider

$$(\mu_{j,N} - \mu) - (\mu_{i,N} - \mu) = \int_{-\infty}^{\infty} (x-\mu)\left\{h_{j,N}(x) - h_{i,N}(x)\right\} f(x)dx\,.$$

Also, from Schwarz inequality, we have

$$\left[\int ab\,dx\right]^2 \le \int a^2\,dx \cdot \int b^2\,dx\,.$$

Apply Schwarz inequality with $a = (x-\mu)f^{1/2}(x)$, $b = [h_{j,N}(x) - h_{i,N}(x)]$ $f^{1/2}(x)$ and obtain

$$\begin{aligned}[(\mu_{j,N} - \mu_{i,N})]^2 \;\le\; & \int_{-\infty}^{\infty} (x-\mu)^2 f(x)dx \\ & \cdot \int_{-\infty}^{\infty} \left\{N\binom{N-1}{j-1}F^{j-1}(x)\,[1-F(x)]^{N-j}\right. \\ & \left. -N\binom{N-1}{i-1}F^{i-1}(x)\,[1-F(x)]^{N-i}\right\}^2 f(x)dx\end{aligned}$$

and the result follows after integration and a little simplification.

Corollary 2.4.11.1.

$$\mu_{N,N} - \mu_{1,N} \le \sigma N\left[2(2N-1)^{-1}\left\{1-\binom{2N-2}{N-1}^{-1}\right\}\right]^{1/2}$$

which was earlier obtained by Plackett, Morigutti, Hartley and David (see Ludwig, 1960, for appropriate references).

Corollary 2.4.11.2.

$$\mu_{i+1,N} - \mu_{i,N} \le \sigma N\left[\binom{N-1}{i-1}\binom{N-1}{i}\Big/\binom{2N-4}{2i-2}(2N-1)(2N-3)\right]^{1/2}.$$

Result 2.4.12 (Ludwig, 1960).

$$(\mu_{j,N} - \mu)^2 \le \sigma^2\left[\frac{N\binom{2N-2j}{N-j}\binom{2j-2}{j-1}}{\binom{2N-1}{N-1}} - 1\right].$$

PROOF: Apply Schwarz inequality with $a = (x-1)f^{1/2}(x)$ and

$$b = \left\{1 - N\binom{N-1}{j-1}F^{j-1}(1-F)^{N-j}\right\}f^{1/2}(x)\,.$$

Corollary 2.4.12.1.

$$\mu - \mu_{1,N} \le (N-1)(2N-1)^{-1/2}\sigma ,$$

which was also obtained by Gumbel (1958) and Hartley and David (1954).

Result 2.4.13 (Ludwig, 1960).

$$\mu - \mu_{1,2} = \mu_{2,2} - \mu = \left[\sigma^2 - \{var(X_{1,2}) + var(X_{2,2})\}/2\right]^{1/2} .$$

PROOF: Since $\sum_{i=1}^{N} E[X_{i,N} - \mu]^2 = N\sigma^2$, it follows that

$$\sum_{i=1}^{N} \left[\{E(X_{i,N} - \mu)\}^2 + \text{var } X_{i,N}\right] = N\sigma^2 .$$

Put $N = 2$ and the result follows.

Result 2.4.14 (Ludwig, 1960). *If X is a non-negative and continuous random variable, then*

$$\mu_{1,N} \ge (\mu - \delta\sigma)(1-\delta^{-2})^N \text{ for every } \delta > 1 .$$

PROOF:

$$\begin{aligned}
\mu_{1,N} &\ge \int_{\mu-\delta\sigma}^{\infty} x d\left\{[1-F(x)]^N\right\} \\
&= -x\,[1-F(x)]^N \,|_{\mu-\delta\sigma}^{\infty} + \int_{\mu-\delta\sigma}^{\infty} [1-F(x)]^N\, dx \\
&\ge (\mu-\delta\sigma)\,[1-F(\mu-\delta\sigma)]^N \ge (\mu-\delta\sigma)(1-\delta^{-2})^N ,
\end{aligned}$$

since

$$\begin{aligned}
1 - F(\mu-\delta\sigma) = P(X > \mu - \delta\sigma) &= 1 - P(X - \mu < -\delta\sigma) \\
&\ge 1 - P(|X-\mu| \ge \delta\sigma) \\
&\ge 1 - \delta^{-2}, \quad \delta > 1 .
\end{aligned}$$

Result 2.4.15. *Let $X_{1,N} < \cdots < X_{N,N}$ be the order statistics in a random sample of size N drawn from the continuous distribution function $F(x)$. Then*

1. *$X_{1,N} < \cdots < X_{i-1,N}$ for given $X_{i,N} = t$ constitute the order statistics in a random sample of size $i-1$ drawn from the continuous distribution function*

$$F_1(x) = \begin{cases} F(x)/F(t), & -\infty < x \le t \\ \\ 1 & x > t\,. \end{cases}$$

For given $X_{i,N} = t_1$ and $X_{j,N} = t_2$, $X_{i+1,N} < \cdots < X_{j-1,N}$ constitute the O.S. in a random sample of size $j - i - 1$ from the continuous distribution function

$$F_2(x) = \begin{cases} 0, & x < t_1 \\ F(x)/\left[F(t_2) - F(t_1)\right], & t_1 \le x \le t_2 \\ 1, & x > t_2\,. \end{cases}$$

For given $X_{j,N} = t, X_{j+1,N} < \cdots < X_{N,N}$ constitute the O.S. in a random sample of size $N - j$ drawn from the continuous distribution function

$$F_3(x) = \begin{cases} \left[F(x) - F(t)\right] / \left[1 - F(t)\right], & x \ge t \\ 0, & x < t\,. \end{cases}$$

PROOF: The joint density of $X_{1,N}, \ldots, X_{i,N}$ is

$$\frac{N!}{(N-i)!}\left\{\prod_{j=1}^{i} f(x_j)\right\}\left[1 - F(x_i)\right]^{N-i}, \quad -\infty < x_1 < \cdots < x_i < \infty\,.$$

The conditional density of $X_{1,N}, \ldots, X_{i-1,N}$ for given $X_{i,N} = t$ is

$$(i-1)!\prod_{j=1}^{i}\left\{f(x_i)/F(t)\right\}, \quad -\infty < x_1 < \cdots < x_{i-1} < t\,.$$

Proofs for the other two results are analogous.

For an expansion of moments of O.S. and joint moments of O.S., see Sugiura (1962, 1964).

2.5 Order Statistics: Discrete Populations

In certain experiments like taste-testing, there arises the necessity of testing the homogeneity of k mutually independent frequencies of occurrences of a certain event and also testing the significance of the largest or the smallest of a set of observed frequencies. If the number of trials is sufficiently large, one can carry out the test by employing the normal approximation. For example, in wine-tasting experiments, the number of trials is not large enough to apply the normal approximation. One can base the test criterion on the largest, smallest or the range of a set of observations taken from a discrete population. Thus, order statistics in random samples taken from discrete populations are of interest. Siotani (1957, 1959) obtained the distributions of the smallest, largest and the range of a random sample drawn from a discrete parent population, and applied these results to the binomial case. Khatri (1962) derived an elegant expression for the joint probability function of any two order statistics. In the following we shall give these results. Let $F(x) = P(X \le x)$ and $f(x) = P(X = x)$. Let $1 \le j \le N$,

$$f_{i,N}(x) = P(X_{i,N} = x), \quad f_{i,j,N}(x,y) = P(X_{i,N} = x, \ X_{j,N} = y).$$

Then

$$\begin{aligned} f_{i,N}(x) &= P(X_{i,N} \le x) - P(X_{i,N} \le x-1) \\ &= \sum_{r=i}^{N} \binom{N}{r} F^r(x)\,[1-F(x)]^{N-r} - \sum_{r=i}^{N} \binom{N}{r} F^r(x-1)\,[1-F(x-1)]^{N-r} \\ &= \frac{N!}{(i-1)!(N-i)!} \int_{F(x-1)}^{F(x)} u^{i-1}(1-u)^{N-i} du\,. \end{aligned}$$

Notice that $H_{i,N}(x) = P(X_{i,N} \le x)$ will be the same as that in continuous samples. If the original random variable can assume only a finite number of values, $0, 1, \ldots, M$, then we have the following results of Siotani (1957).

Result 2.5.1 (Siotani, 1957). *We have*

1. $\mu_{NN} = \sum_{x=0}^{M-1} \left[1 - \{F(x)\}^N\right]$,

2. $\mu_{N,N}^{(2)} = 2 \sum_{x=0}^{M-1} x \left[1 - F^N(x)\right] + \mu_{N,N}$,

3. $\mu_{1,N} = \sum_{x=0}^{M-1} [1 - F(x)]^N$,

and

4. $\mu_{1,N}^{(2)} = 2 \sum_{x=0}^{M-1} x\,[1 - F(x)]^N + \mu_{1,N}$.

PROOF:

$$\begin{aligned}\mu_{NN} = E(X_{NN}) &= \sum_{x=0}^{M} x f_{NN}(x) \\ &= \sum_{x=0}^{M} x\left[F^N(x) - F^N(x-1)\right] \\ &= M - \sum_{x=0}^{M-1} F^N(x);\end{aligned}$$

$$\begin{aligned}\mu_{NN}^{(2)} &= \sum_{x=0}^{M} x^2 f_{NN}(x) = \sum_{0}^{M} x^2\left[F^N(x) - F^N(x-1)\right] \\ &= M^2 - \sum_{x=0}^{M-1}(2x+1)F^N(x) \\ &= \sum_{x=0}^{M-1}(2x+1)\left[1 - F^N(x)\right] = 2\sum_{x=0}^{M-1} x\left[1 - F^N(x)\right] + \mu_{NN}\,.\end{aligned}$$

Now the proofs for (iii) and (iv) are similar to those of (i) and (ii) respectively.

From the preceding results one can obtain the distribution function, mean and variance of the sample range. Next, for the joint probability function of two order statistics, we have for $x \le y$,

$$f_{ijN}(x,y) = H_{ijN}(x,y) - H_{ijN}(x-1,y) - H_{ijN}(x,y-1) + H_{ijN}(x-1,y-1)\,.$$

A form more useful for theoretical purposes has been obtained by Khatri (1962).

Result 2.5.2 (Khatri, 1962). *The joint probability function of X_{iN} and X_{jN} is given by*

$$f_{ijN}(x,y) = C_{ijN}\iint u^{i-1}(v-u)^{j-i-1}(1-v)^{N-j}\,du\,dv$$

where

$$C_{ijN} = N!/(i-1)!(j-i-1)!(N-j)!$$

and the range of integration is: $u \le v$, $F(x-1) \le u \le F(x)$, $F(y-1) \le v \le F(y)$.

Remark 2.5.1. *The range of integration takes simple forms when $x = y$ and $x < y$.*

PROOF: Consider the configuration shown for $x < y$ and obtain

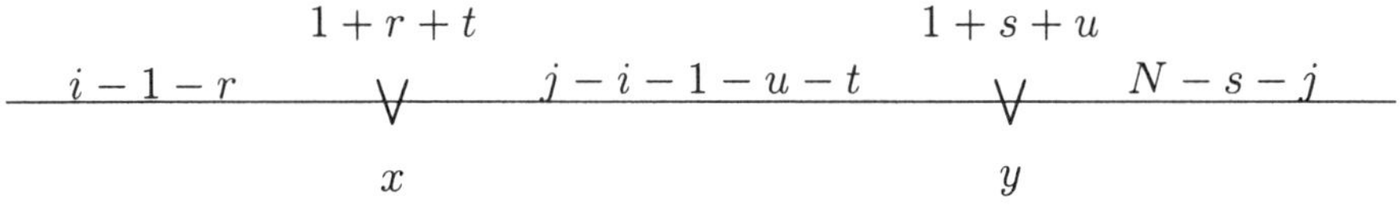

$$f_{ijN}(x,y) = \sum_{r=0}^{i-1}\sum_{s=0}^{N-j}\sum_{u,t}$$

$$\cdot\ \{N!/(i-1-r)!(1+r+t)!(j-i-1-u-t)!(1+s+u)!(N-s-j)!\}$$
$$\cdot\ F^{i-1-r}(x-1)\,[F(y-1)-F(x)]^{j-i-1-u-t}\,[1-F(y)]^{N-s-j}$$
$$\cdot\ [f(x)]^{1+r+t}\,[f(y)]^{1+s+u}$$

where $\sum_{u,t}$ denotes summation over non-negative integral values of u, t subject to $u + t \leq j - i - 1$.

We can also rewrite $f_{i,j}(x,y)$ as

$$f_{ij}(x,y) = C_{ijN}\sum_{r=0}^{i-1}\sum_{s=0}^{N-j}\sum_{u,t}\binom{i-1}{r}\binom{N-j}{s}\frac{j-i-1)!}{(j-i-1-u-t)!u!t!}$$

$$\cdot\ [F(x-1)]^{i-1-r}\,[1-F(y)]^{N-j-s}\,[F(y-1)-F(x)]^{j-i-1-u-t}$$
$$\cdot\ [f(x)]^{1+r+t}\,[f(y)]^{1+s+u}\int_0^1\int_0^1 v^r(1-v)^t w^s(1-w)^u dv\,dw\,.$$

Interchanging the summation and integral signs and setting $z = F(y) - vf(y)$, $z' = F(x-1) + wf(x)$ we obtain

$$f_{ijN}(x,y) = C_{ijN}\int_{F(x-1)}^{F(x)}\int_{F(y-1)}^{F(y)} z'^{i-1}(z-z')^{j-i-1}(1-z)^{N-j}dz\,dz'\,.$$

When $x = y$ analogously, we obtain

$$f_{ijN}(x,x) = C_{ijN}\iint_{F(x-1)\leq z'\leq z\leq F(x)} z'^{i-1}(z-z')^{j-i-1}(1-z)^{(N-j)}dz\,dz'\,.$$

Combining the above two results, the general desired result follows.

2.6 Representation of Exponential Order Statistics as a Sum of Independent Random Variables

Lemma 2.6.1. *A necessary and sufficient condition for a continuous random variable X to be exponentially distributed is*

$$P(X \leq x+y|X \geq y) = P(X \leq x), \quad x, y > 0\,.$$

Proof of Necessity: Assume that the d.f. of X, namely, $F(x) = 1 - e^{-\lambda x}$ for some $\lambda > 0$. Then

$$\begin{aligned} P(X \leq x+y|X > y) &= [F(x+y) - F(y)]\,/\,[1 - F(y)] \\ &= \left[e^{-\lambda x} - e^{-\lambda(x+y)}\right]/e^{-\lambda y} = 1 - e^{-\lambda x}\,. \end{aligned}$$

Proof for Sufficiency: We assume that $P(X \leq x + y)|X > y) = P(X \leq x)$ and we will show that $F(x) = 1 - e^{-\lambda x}$ for some $\lambda > 0$. From $P(X \leq x+y|X > y) = P(X \leq x)$, we get

$$F(x+y) - F(y) = [1 - F(y)]\,F(x)\,.$$

That is,

$$D(x+y) = D(x)D(y)\,,$$

where

$$D(x) = 1 - F(x)\,.$$

This is Cauchy's equation.

Solution to the Cauchy Equation

We have for all $x, y, D(x+y) = D(x)D(y)$, where $D(x) = 1 - F(x)$. Setting $y = 0$ we obtain $D(0) = 1$. That is, $F(0) = 0$, or X is a positive random variable. Setting $y = x$ and by induction one can obtain that

$$D(nx) = D^n(x)\,.$$

That is,

$$D(n) = D^n(1)\,.$$

Also, letting $x = m/n$ where m and n are integers we have $D\left(n\frac{m}{n}\right) = \left(D\left(\frac{m}{n}\right)\right)^n$, or $D(m) = D^m(1) = \left(D\left(\frac{m}{n}\right)\right)^n$. That is, $D\left(\frac{m}{n}\right) = [D(1)]^{m/n}$ or

$D(x) = D^x(1)$ whenever x is a rational number. Since irrational numbers are limits of sequences of rational numbers (because of the continuity of F), we can establish that $D(x) = D^x(1)$ for all real x. Now set $D(1) = e^{-\lambda}$ for some $\lambda > 0$ and this completes the proof of the assertion.

Remark 2.6.1. *If the random variable X is interpreted as the failure of a certain component, then Lemma 2.7.1. says: If time to failure of the component has an exponential distribution, then given it has survived y units of time, the probability that it will fail before $x+y$ units is free of y. In other words if a component has survived y units of time, then it is as good as a new component.*

Strong Markov Property of the Exponential Distribution

Exponential distribution also has the following strong Markov property.

Result 2.6.1. *Let X have an exponential distribution and Y be any continuous nonnegative random variable having d.f. $F(x)$ and is independent of X. Then, for any $t \geq 0$,*

$$P(X > Y + t | X > Y) = P(X > t) .$$

PROOF:

$$\begin{aligned} P(X > Y + t | X > Y) &= P(X > Y + t)/P(X > Y) \\ &= \frac{\int_0^\infty P(X > y + t | Y = y) dF(y)}{\int_0^\infty P(X > y | Y = y) dF(y)} \\ &= \int_0^\infty e^{-\lambda(y+t)} dF(y) / \int_0^\infty e^{-\lambda y} dF(y) \\ &= e^{-\lambda t} = P(X > t) . \end{aligned}$$

Result 2.6.2 (Sukhatme, 1937; Renyi, 1953). *Let $X_{1N} \leq \cdots \leq X_{NN}$ denote the order statistics in a random sample of size N drawn from the negative exponential distribution, $f(x) = \exp(-x)$, $0 < x < \infty$. Then, the random variables*

$$Y_i = (N - i + 1)(X_{iN} - X_{i-1,N}), \quad i = 1, \ldots, N$$

are independent and identically distributed as $f(x)$.

PROOF: Since Renyi's (1953) proof is somewhat long, we shall present a simpler proof based on a result of Sukhatme (1937).

The joint p.d.f. of the X_{iN} is

$$N!\ \exp\left(-\sum_{i=1}^{N} x_i\right),\quad 0 < x_1 < \cdots < x_N < \infty\,,$$

which may be written as

$$N!\ \exp\left[-\sum_{i=1}^{N}(N-i+1)(x_i - x_{i-1})\right],$$

where $x_0 = 0$ since

$$\begin{aligned}
\sum(N-i+1)(x_i - x_{i-1}) &= (N+1)\sum_{i=1}^{N}(x_i - x_{i-1}) - \sum_{i=1}^{N} i(x_i - x_{i-1}) \\
&= (N+1)x_N - \sum_{i=1}^{N} ix_i + \sum_{i=1}^{N} ix_{i-1} \\
&= (N+1)x_N - \sum_{i=1}^{N} ix_i + \sum_{j=0}^{N-1}(j+1)x_j \\
&= (N+1)x_N - \sum_{i=1}^{N} ix_i + \sum_{j=0}^{N-1} jx_j + \sum_{j=0}^{N-1} x_j \\
&= (N+1)x_N - Nx_N + \sum_{j=0}^{N-1} x_j \\
&= \sum_{j=1}^{N} x_j\,.
\end{aligned}$$

Making the transformation

$$Y_i = (N-i+1)(X_{iN} - X_{i-1,N}),\quad i = 1,\ldots,N\,,$$

the joint p.d.f. of the Y_i is (since the Jacobian is $1/N!$)

$$\exp\left(-\sum_{1}^{N} y_i\right),\quad y_i \geq 0$$

from which the desired result follows.

This result has important applications in-life-testing. If the X_{iN}, $i = 1, \ldots, N$ denote the successive life times of N items whose failure times follow an exponential law with mean 1, then the intervals between successive failures, namely, $(X_{i,N} - X_{i-1,N})$, $i = 1, \ldots, N$ are independent and exponentially distributed with means $1/(N - i + 1)$, $i = 1, \ldots, N$, respectively.

Corollary 2.6.2.1. *Let $X_{1N} \leq \cdots \leq X_{NN}$ denote standard exponential order statistics in a random sample of size N. Then*

$$X_{1N} = N^{-1}\delta_1, \quad X_{2N} = N^{-1}\delta_1 + (N-1)^{-1}\delta_2$$
$$\vdots \qquad\qquad\qquad\qquad \vdots$$
$$X_{NN} = N^{-1}\delta_1 + (N-1)^{-1}\delta_2 + \cdots + 2^{-1}\delta_{N-1} + \delta_N \,,$$

where $\delta_1, \ldots, \delta_N$ are i.i.d. standard exponential random variables.

PROOF: Since from Result 2.7.2, $(N-i+1)(X_{iN} - X_{i-1,N})$ for $i = 1, \ldots, N$ are i.i.d. standard exponential, the result follows after setting

$$\delta_i = (N - i + 1)(X_{i,N} - X_{i-1,N}), \quad i = 1, \ldots, N$$

and solving for the X_{iN}'s. Note that $X_{0,N} = 0$.

Remark 2.6.2. *The exponential O.S. can be expressed as linear combinations of i.i.d. exponential r.v.'s.*

Exponential Order Statistics as a Markov Chain

We have

$$\delta_i = (N - i + 1)(X_{iN} - X_{i-1,N})$$

or

$$X_{iN} = \sum_{r=1}^{i} \delta_r/(N - r + 1)\,.$$

Consequently,

$$P(X_{iN} \leq x | X_{1N}, \ldots, X_{i-1,N}) = P(X_{iN} \leq x | X_{i-1,N})\,.$$

That is, $X_{1N}, \ldots, X_{NN}$ form an (additive) Markov Chain (see also Renyi, 1953).

2.7 Representation of General Order Statistics

In this section we will provide a representation for the order statistics in samples drawn from an arbitrary continuous population as functions of independent and identically distributed exponential random variables with mean unity. Consider the following lemma.

Lemma 2.7.1. *If X is a r.v. having a continuous d.f. $F(x)$, then $-\ln F(X)$ has an exponential distribution with mean 1.*

PROOF:

$$\begin{aligned} P[-\ln F(X) \le y] &= P\left[F(X) \ge e^{-y}\right] \\ &= 1 - e^{-y} \end{aligned}$$

since $F(X)$ is uniformly distributed on $[0,1]$. This completes the proof of the lemma.

Result 2.7.1. *If $X_{1,N} \le X_{2,N} \le \cdots \le X_{N,N}$ are the order statistics in a random sample of size N drawn from a continuous population having $F(x)$ for its d.f., then*

$$X_{N-k+1,N} = F^{-1}\left[\exp\left\{-\left(\frac{\delta_1}{N} + \frac{\delta_2}{N-1} + \cdots + \frac{\delta_k}{N-k+1}\right)\right\}\right],$$

for $k = 1, 2, \ldots, N$ where the δ_i are independent and identically distributed exponential random variables having mean unity.

PROOF: If $X_1, X_2, \ldots, X_N$ denotes the random sample drawn from $F(x)$, let $\eta_k = F(X_k)$ and $\xi_k = -\ln \eta_k$, $k = 1, 2, \ldots, N$. Then

$$X_{1,N} \le X_{2,N} \le \cdots \le X_{N,N}$$

induces the orderings:

$$F(X_{1,N}) \le F(X_{2,N}) \le \cdots \le F(X_{N,N})$$

and

$$-\ln F(X_{N,N}) \le -\ln F(X_{N-1,N}) \le \cdots \le -\ln F(X_{1,N}).$$

Consider the distribution of ξ_k,

$$\begin{aligned} P(\xi_k \le x) &= P[-\ln \eta_k \le x] \\ &= P[-\ln F(X_k) \le x] \\ &= 1 - e^{-x} \end{aligned}$$

after applying Lemma 2.8.1. Thus, the ξ_k, $(k = 1, \ldots, N)$ constitute a set of independent and identically distributed exponential random variables with mean unity. Therefore, using Renyi's representation, the kth ordered ξ, can be written as

$$-\ln F(X_{N-k+1,N}) = N^{-1}\delta_1 + (N-1)^{-1}\delta_2 + \cdots + (N-k+1)^{-1}\delta_k \,,$$

where the δ_i's are independent and identically distributed exponential r.v.'s with mean 1. It now follows that

$$X_{N-k+1,N} = F^{-1}\left[\exp\left\{-\left(\frac{\delta_1}{N} + \frac{\delta_2}{N-1} + \cdots + \frac{\delta_k}{N-k+1}\right)\right\}\right], \quad k = 1, 2, \ldots, N \,.$$

This completes the proof of the asserted result.

Corollary 2.7.1.1. *If F is uniform, then*

$$X_{N-k+1,N} = \exp\left[-\left(\frac{\delta_1}{N} + \frac{\delta_2}{N-1} + \cdots + \frac{\delta_k}{N-k+1}\right)\right], \quad k = 1, 2, \ldots, N \,.$$

2.8 Angel and Demons' Problems

Some problems, concerning order statistics and labeled an Angel's problem and Demon's problem, were proposed and empirically solved by Youden (1953). These problems were analytically solved by Kendall (1954).

Angel's Problem

Youden's version: "N members are drawn at random from a normal population with unit variance. A benevolent angel tells us which is nearest to the true mean, and the others are rejected. What is the variance of the retained member of the sample?" However, the practical situation which suggested it, is not angelic.

Kendall's version of the problem: "A number of laboratory assistants are given a standard experiment to perform. They replicate it and, knowing what the true result ought to be, each submits only his best result. What effect does this have on estimates of experimental error"?

Solution: For $N = 1$, the variance is trivially unity. For $N = 2$, the solution was obtained by E.P. King, one of Youden's colleagues at the National Bureau of Standards; for $N = 3$, the solution was obtained by H. Fairfield; for $N = 4$ and 5 exact solutions and approximations for higher values of N were provided by Kendall (1954).

From considerations of symmetry, it is clear that the mean of the variable under consideration is the parent mean which we may take to be zero. The problem reduces to finding $E(X_{1,N}^2)$ where X is the smallest of the N members chosen from the half normal population, the density function and d.f. of which is given by

$$f(x) = (2/\pi)^{\frac{1}{2}} e^{-x^2/2}, \quad x > 0 \text{ and } F(x) = \int_0^x f(t)dt\,.$$

The required "variance", $\mu_{1,N}^{(2)}$ is given by

$$\begin{aligned} \mu_{1,N}^{(2)} &= N\int_0^\infty x^2(1-F)^{N-1}dF(x) \\ &= N(2/\pi)^{N/2}\int_0^\infty x^2 e^{-x^2/2}\left(\int_x^\infty e^{-t^2/2}dt\right)^{N-1} dx\,. \end{aligned}$$

The problem is to evaluate this integral. The exact values of $\mu_{1,N}^{(2)}$ are given by Govindarajulu (1962) and Kendall (1954) for sample sizes up to and including 5. They are

$$\begin{aligned} \mu_{1,2}^{(2)} &= 1-(2/\pi) = 0.36338 \\ \mu_{1,3}^{(2)} &= 1-(3-3^{\frac{1}{2}})(2/\pi) = 0.19280 \\ \mu_{1,4}^{(2)} &= 1-(12/\pi)+(16/\pi 3^{\frac{1}{2}}) = 0.12070 \\ \mu_{1,5}^{(2)} &= 1-(20/\pi)+\left[\text{arc tan } (5/3)^{\frac{1}{2}}-(\pi/6)\right](240/3^{\frac{1}{2}}\pi^2) \\ &= 0.08308\,. \end{aligned}$$

Kendall (1954), by expanding x as power series in F and integrating termwise, obtained an approximation for $\mu_{1,N}^{(2)}$ which is given by

$$\mu_{1,N}^{(2)} \approx \pi/(N-1)(N+4)\,.$$

The Demon's Problem

Given a small sample of N values from a normal population, what is the probability that their mean lies between the Nth and the $(N-1)^{\text{th}}$ observations in order of magnitude?

Solution: The desired probability is one if $N = 2$. When N = 3, the required probability is $P(X_{2,3} - X_{1,3} \leq X_{3,3} - X_{2,3}) = \frac{1}{2}$, since the distributions of $X_{2,3}$ and $-X_{1,3}$ are identical and the distribution of $X_{2,3}$ is symmetric about zero.[1] Also, the problem is not of much interest for $N > 10$, since the desired probability tends to zero as N exceeds 10. Denote the sample mean by $\bar{X}$. Since $\bar{X}$ is always less than the Nth largest observation, the problem reduces to finding $P(X_{N-1,N} - \bar{X} < 0)$. Let us find the general

$$P(X_{j,N} - \bar{X} < 0) = 1 - P(X_{j,N} - \bar{X} \geq 0)\,.$$

It has been established by McKay (1935) that the deviations of order statistics from their mean in normal samples are independent of the mean itself. If $X = X_{j,N} - \bar{X}$, the characteristic function of $X_{j,N}$ can be obtained by considering the product of the characteristic functions of X and $\bar{X}$. Hence, for the cumulants

$$\begin{aligned} k_r(X) &= k_r(X_{j,N}) \;\; (r = 1, 3, \ldots) \\ k_2(X) &= k_2(X_{j,N}) - N^{-1}\,. \end{aligned}$$

An explicit form for the density function of X can be obtained by using McKay's method. However, Kendall (1954) found it rather troublesome to handle and has adopted a different approach. Kendall's (1954) result for the frequency function of X is given by

$$(2\pi)^{-\frac{1}{2}} \{N/(N-1)\}^{\frac{1}{2}} \exp\{-Nx^2/2(N-1)\} \exp\{-(N-1)D^2/N^2\} \cdot A\{Nx/(N-1)\}$$

where

$$D = \frac{d}{dx}, \quad A(x) = N\binom{N-1}{j-1}\Phi^{j-1}(x)\,[1 - \Phi(x)]^{N-j}$$

and $\Phi(x)$ denotes the standard normal d.f.

It is also of interest to note that in normal samples, since the distribution of $\bar{X}$ is independent of all the deviations $X_{j,N} - \bar{X}$, the distribution of $\bar{X}$ is independent of the sample variance. Kendall (1954) found the moments of $X_{N-1,N}$, hence those of $X_{N-1,N} - \bar{X}$, hence the cumulants of X and hence $P(X < 0)$ from the Edgeworth form of the Gram-Charlier expansion for the d.f. of X, using cumulants up to order six. David (1963) besides

[1]If X and Y have a bivariate continuous distribution $H(x, y)$ where $H(x, y) = H(y, x)$ then $P(X < Y) = \frac{1}{2}$.

other results, gives an asymptotic solution to the Demon's Problem which yields probabilities of order substantially lower than what had previously been hypothesized.

Result 2.8.1 (David, 1963). *Let*

$$P(N,k) = P(X_{k,N} \le \bar{X}_N \le X_{k+1,N}), \quad k = 1, \dots, N-1.$$

Then

$$\begin{aligned} P(4,1) &= [(3 \textit{ arc sec } 3)\pi] - 1, \\ P(5,1) &= [(5 \textit{ arc sec } 4)/2\pi] - 1, \end{aligned}$$

and

$$\lim_{N - \infty} \{P(N,k)/f(N,k)\} = 1$$

where

$$f(N,k) = \left[k^{N-k-1} e^{N-2k} / 2(k!)^2 (N)^{-N-3k-1} (2\pi)^{N-k}\right]^{\frac{1}{2}}.$$

By symmetry $P(N,k) = P(N, N-k)$, David (1963) has numerically evaluated $P(N,1)$ for some small values of N. They are given below.

Table 2.8.1: Values of $P(N,1)$ for selected values of N

N	4	5	6	7	8
Exact	.175	.049	.011*		
f(N,k)	.122	.036	.0088	.0019	.0004
Monte Carlo	.174	.049	.011	.0025	.0015
Kendall's method	.174	.078	.040	.024	.015

*Obtained by numerical integration of the expression verified in Lemma 2.4.2 of David (1963).

From Table 2.9.1 we gather that the probability decreases more rapidly than suggested by the values obtained by Kendall (1954).

2.9 Large Sample Properties of Order Statistics

In this section we shall present some large sample properties concerning order statistics in samples drawn from continuous populations.

Result 2.9.1. *If $X_{k,N}$, is the kth smallest order statistic in a sample of size N drawn from a continuous population having $F(x)$ for its d.f. then, for a fixed k, $NF(X_{k,N})$, $N = k, k+1, \ldots$, is a sequence of random variables converging in distribution to the gamma distribution having k for its parameter.*

PROOF: Consider

$$\begin{aligned} P\{NF(X_{k,N}) \le w\} &= \frac{N!}{(k-1)!(N-k)!} \int_0^{w/N} u^{k-1}(1-u)^{N-k} du \\ &= \frac{N!}{N^k(k-1)!(N-k)!} \int_0^w y^{k-1}\left(1-\frac{y}{N}\right)^{N-K} dy \end{aligned}$$

since the sequence of integrands on the right side converges to the function $y^{k-1}e^{-y}/(k-1)!$ uniformly over the interval $(0, w)$, it follows that

$$\lim_{N\to\infty} P\{NF(X_{k,N}) \le w\} = [(k-1)!]^{-1} \int_0^w e^{-y} y^{k-1} dy \,.$$

This completes the proof of the assertion.

Remark 2.9.1. *The above result holds if $F(X_{k,N})$ is replaced by the sum of any k consecutive coverages $F(X_{1,N})$, $F(X_{2,N}) - F(X_{1,N}), \ldots, 1-F(X_{N,N})$, for example, $1 - F(X_{N-k+1,N})$ or $F(X_{s+k,N}) - F(X_{s,N})$.*

Result 2.9.2. *Let $X_{k,N}$ and $X_{l,N}$ respectively denote the kth smallest and l^{th} smallest order statistics in a sample of size N drawn from a continuous population having $F(x)$ for its d.f. Then, if k and l are fixed integers, the sequence of pairs of random variables $\{NF(X_{k,N}), NF(X_{k+l,N})\}$, $N = M$, $M+1, \ldots$, where $M \ge k+l$ converges in distribution to a distribution having p.d.f.*

$$\begin{aligned} f(u,v) &= [(k-1)!(l-1)!]^{-1} u^{k-1}(v-u)^{l-1}e^{-v}, && 0 < u < v < \infty \\ &= 0, && \textit{elsewhere}\,. \end{aligned}$$

PROOF: The proof is similar to the one used to obtain the preceding result.

Result 2.9.3. *For fixed k and l such that $l+k<N$, the random variables $N\{1-F(X_{N-l+1,N})\}$ and $NF(X_{k,N})$ are asymptotically independent and have a joint gamma distribution with parameters l and k respectively.*

PROOF: The joint density function of $N\{1-F(X_{N-l+1,N})\}$ and $NF(X_{k,N})$ is given by

$$\frac{N!}{N^2(k-1)!(l-1)!(N-k-l)!}(u/N)^{k-1}(v/N)^{l-1}\left(1-\frac{u}{N}-\frac{v}{N}\right)^{N-k-l}$$

where $u,v>0$, $u+v<N$ and $l+k<N$. As $N\to\infty$, this joint density tends to

$$\frac{e^{-u}u^{k-1}}{(k-1)!}\cdot\frac{e^{-v}v^{l-1}}{(l-1)!},\quad u,v>0\,.$$

This completes the proof of the assertion.

Corollary 2.9.3.1. *The random variables $N\{1-F(X_{N,N})\}$ and $NF(X_{1,N})$ have a joint limiting density function given by*

$$e^{-(u+v)},\quad u,v>0\,.$$

Corollary 2.9.3.2. *The limiting density function of $N\{1-F(X_{1,N})-F(X_{N,N})\}$ is given by*

$$\left(\frac{1}{2}\right)e^{-|x|},\quad -\infty<x<\infty\,.$$

PROOF: Let $U=N\{1-F(X_{NN})\}$, $V=NF(X_{1N})$. Then for $x\geq 0$, consider

$$\begin{aligned} P(U-V\geq x)\approx\iint_{u-v\geq x}e^{-(u+v)}dudv &= \int_0^\infty e^{-v}\left[\int_{v+x}^\infty e^{-u}du\right]dv \\ &= \int_0^\infty e^{-(v+x)}e^{-v}dv \\ &= \left(\frac{1}{2}\right)e^{-x}\,. \end{aligned}$$

For $x<0$,

$$P(U-V\leq x)\approx\int_0^\infty e^{-u}\left[\int_{u-x}^\infty e^{-v}dv\right]du=e^x\int_0^\infty e^{-2u}du=\frac{e^x}{2}\,.$$

Hence,

$$\begin{aligned} F_{U-V}(x) &\approx \frac{e^x}{2} \text{ for } x < 0\,, \\ &\approx 1 - \frac{e^{-x}}{2} \text{ for } x > 0\,. \end{aligned}$$

Thus,

$$f_{U-V}(x) \approx \frac{1}{2}e^{-|x|} \text{ for } -\infty < x < \infty\,.$$

Remark 2.9.2. *The joint limiting distribution of $X_{N-k+1,N}$ and $X_{k,N}$ can be obtained from the joint limiting distribution of $N\{1 - F(X_{N-k+1,N})\}$ and $NF(X_{k,N})$.*

Result 2.9.4. *If $X_{k,N}$ is the kth smallest order statistic in a sample of N drawn from a continuous population having $F(x)$ for its d.f., then for a fixed k,*

$$P(X_{k,N} \le w) \sim [(k-1)!]^{-1} \int_0^{NF(w)} e^{-u}u^{k-1}du, \ \textit{as } N \to \infty\,.$$

Proof: From Result 2.9.1 we have

$$\lim_{N\to\infty} P\left(NF(X_{k,N}) \le v\right) = [(k-1)!]^{-1} \int_0^v e^{-u}u^{k-1}du\,.$$

However, since $F(x)$ is continuous

$$P\left(NF(X_{k,N}) \le v\right) = P\left[X_{k,N} \le F^{-1}(v/N)\right\}\,.$$

Hence, letting $w = F^{-1}(v/N)$, we obtain the desired result.

Remark 2.9.3. *Dodd (1923), Fisher and Tippett (1928), Fréchet (1927), Gumbel (1958) and Smirnov (1935) are some of the research workers who have considered the asymptotic results of the preceding type, especially for the smallest and the largest order statistics. For these references the reader is referred to Wilks (1962).*

2.10 Large Sample Properties of Sample Quantiles

Definition 2.10.1. *The sample p^{th} quantile denoted by $X_{k(N),N}$ is defined as follows:*

$$\begin{aligned} X_{k(N),N} &= X_{Np,N} && \textit{if } Np \textit{ is an integer,} \\ &= X_{[Np]+1,N} && \textit{if } Np \textit{ is not an integer,} \end{aligned}$$

where $[Np]$ denotes the largest integer not exceeding Np.

Before we consider the large sample properties of the sample quantiles, we shall present a useful convergence theorem.

Result 2.10.1 (Slutsky). *Let $\{X_n\}$ be a sequence of random variables the distribution function of which converges to $F(x)$ as $n \to \infty$. Let $\{Y_n\}$ be another sequence of random variables converging to a positive constant c. Then at every point of continuity of $F(x)$,*

1. $\lim_{n\to\infty} P(X_n + Y_n \le z) = F(z - c)$,

2. $\lim_{n\to\infty} P(X_n/Y_n \le z) = F(cz)$,

and

3. $\lim_{n\to\infty} P(aX_n + bY_n \le z) = F\left((z - bc)/a\right)$.

PROOF: See Cramér (1946, p. 254) or Wilks (1962, §4.3.6a).

As an application of Result 2.10.1, consider the t-statistic. We will show that if the fourth moment of the underlying population is finite, then $t = n^{(1/2)}(\bar{X}-\mu)/s$ where $s^2 = (n-1)^{-1}\sum(X_i-\bar{X})^2$ tends to a standard normal variable in distribution. Since $\text{var}(s^2) = n^{-1}\left[\mu_4 - (n-3)(n-1)^{-1}\mu_2^2\right]$ which tends to zero as $n \to \infty$, $s \to \sigma$ in probability and hence by Result 2.10.1, t is asymptotically standard normal.

Remark 2.10.1. *In fact, we can show that $s^2 \to \sigma^2$ almost surely without the restriction on the finiteness of the variance of s^2 where we can write*

$$(n-1)s^2 = \sum(X_i - \mu)^2 - n(\bar{X} - \mu)^2$$

since $\bar{X} \to \mu$ almost surely and $\frac{1}{n}\sum(X_i - \mu)^2 \to \sigma^2$ almost surely, s^2 tends to σ^2 almost surely.

Result 2.10.2 (Renyi). *If the density function $f(x) = F'(x)$ exists, is continuous and positive in the neighborhood of ξ_p where $F(\xi_p) = p$, $< 0 < p < 1$ and further, if $k(N) = Np + 0(1)$, then $F_{k(N),N}$ is asymptotically normally distributed with mean p and variance $p(1-p)/N$.*

PROOF: Without any confusion, let us write k in the place of $k(N)$. Let $\xi_{N+1-k,N} = -\ln F(X_{k,N})$. Now, due to Renyi's representation of order statistics as functions of linear combinations of independent and identically distributed exponential random variables with the d.f. $1 - e^{-x}(x > 0)$, we have

$$\xi_{N+1-k,N} = \sum_{j=1}^{N+1-k} \delta_j/(N+1-j) = \sum_{j=k}^{N} \delta_j/j, \quad k = 1, 2, \dots, N,$$

where the δ_j are independent and identically distributed exponential r.v.'s with the d.f. $1 - e^{-x}(x > 0)$. That is,

$$E\delta_j = \text{var}\ \delta_j = 1, \quad j = k, \dots, N.$$

Also,

$$\begin{aligned} E|\delta_j - 1|^3 = \int_0^\infty |x-1|^3 e^{-x} dx \ &\le \int_0^1 e^{-x} dx + \int_1^\infty (x-1)^3 e^{-x} dx \\ &= 1 - e^{-1} + 6e^{-1} = 1 + 5e^{-1} < 3, \end{aligned}$$

$$E\xi_{N+1-k} = \sum_{j=k}^{N} j^{-1},$$

$$\sigma^2_{\xi_{N+1-k,N}} = \text{var}\ \xi_{N+1-k,N} = \sum_{j=k}^{N} j^{-2}$$

and

$$\rho^3 = \sum_{j=1}^{N+1-k} E|\delta_j - 1|^3/(N+1-j)^3 \le 3\sum_{j=k}^{N} j^{-3}.$$

Now consider

$$\begin{aligned} \rho/\sigma_{\xi_{N+1-k,N}} &\le 3^{1/3}\left(\sum_{j=k}^{N} j^{-3}\right)^{1/3} \Big/ \left(\sum_{j=k}^{N} j^{-2}\right)^{1/2} \\ &\le 2\left[k^{-3}(N-k+1)\right]^{1/3} / \left[N^{-2}(N-k+1)\right]^{1/2} \\ &= 2N/k(N-k+1)^{1/6} = 2(k/N)^{-1}(N-k+1)^{-1/6}. \end{aligned}$$

Consequently, $\rho/\sigma_{\xi_{N+1-k,N}}$ tends to zero as N tends to infinity. Hence, we can apply Liapounov's form of the central limit theorem to assert the asymptotic normality of $\xi_{N+1-k,N}$. That is,

$$\lim_{N\to\infty} P\{(\xi_{N+1-k,N} - E\xi_{N+1-k,N})/\sigma\xi_{N+1-k,N} \le w\} = \Phi(w).$$

Let us simplify the normalizing constants. Since

$$\sum_{j=1}^{l} j^{-1} = \ln\, l + \nu + \Delta_l$$

where ν is the Euler's constant and a constant $K > 0$ can be found such that $|\Delta_l| < K/l$. Hereafter K will denote a generic constant. Hence,

$$\begin{aligned} E\xi_{N+1-k,N} &= \sum_{j=1}^{N} j^{-1} - \sum_{j=1}^{k} j^{-1} + k^{-1} \\ &= \ln(N/k) + \epsilon'_N \end{aligned}$$

where $|\epsilon'_N| < K/k$. Also, using an integral approximation, one obtains

$$\int_k^N \frac{dx}{x^2} \le \sigma^2_{\xi_{N+1-k}} = \sum_{j=k}^{N} j^{-2} \le \int_{k-1}^N \frac{dx}{x^2}$$

or

$$k^{-1} - N^{-1} \le \sigma^2_{\xi_{N+1-k,N}} \le (k-1)^{-1} - N^{-1} .$$

That is,

$$\begin{aligned} \sigma^2_{\xi_{N+1-k,N}} &= k^{-1} - N^{-1} + \theta\,[k(k-1)]^{-1}, \quad 0 < \theta < 1, \\ &= k^{-1} - N^{-1} + \epsilon''_N \end{aligned}$$

where $|\epsilon''_N| \le K/k(k-1)$. Also, since $k = Np + 0(1)$,

$$E\xi_{N+1-k,N} = \ln(N/k) + \epsilon'_N = -\ln\, p + \epsilon'_N ,$$

and

$$\sigma^2_{\xi_{N+1-k,N}} = (N-k)/kN + \epsilon''_N = (1-p)/pN + \epsilon''_N .$$

Thus, we can write

$$\frac{\xi_{N+1-k,N} - E\xi_{N+1-k,N}}{\sigma_{\xi_{N+1-k,N}}} = \frac{\xi_{N+1-k,N} + \ln\, p}{[(1-p)/pN]^{1/2}} + \epsilon_N$$

where $\epsilon_N = 0(N^{1/2}/k)$. Consequently, it follows from the continuity of the normal d.f. that

$$\lim_{N\to\infty} P\left\{ \frac{\xi_{N+1-k,N} + \ln\, p}{[(1-p)/pN]^{1/2}} \le w \right\} = \Phi(w)$$

or

$$\lim_{N\to\infty} P\left\{F(X_{k,N}) > pe^{-w\left(\frac{1-p}{Np}\right)^{1/2}}\right\} = \Phi(w)\,.$$

That is,

$$\lim_{N\to\infty} P\left\{\frac{F(X_{k,N}) - p}{[p(1-p)/N]^{1/2}} \le w\right\} = \Phi(w)\,,$$

since

$$e^{-w\left(\frac{1-p}{Np}\right)^{1/2}} = 1 - w\{(1-p)/Np\}^{1/2} + o(N^{1/2})\,.$$

This completes the proof of the desired result.

Result 2.10.3. *If $k(N) = pN + 0(1)$ and if $X_{k(N),N}$ is the sample p-quantile, then*

$$P(X_{k(N),N} \le v) \approx \Phi(w)$$

where

$$w = [F(v) - p]\, N^{1/2} / \,[p(1-p)]^{1/2}\ .$$

PROOF: From Result 2.10.2, we have

$$\lim_{N\to\infty} P\left\{[F(X_{k(N),N}) - p]\, N^{1/2}/\,[p(1-p)]^{1/2} \le w\right\} \approx \Phi(w)\,.$$

Since $F(x)$ is continuous, it follows that

$$P\left\{X_{k(N),N} \le F^{-1}\left(p + w\,[p(1-p)/N]^{1/2}\right)\right\} \approx \Phi(w)\,.$$

Corollary 2.10.3.1. *It is clear from Result 2.10.3 that $F(X_{k(N),N})$ converges in probability to p. Now, if a unique p^{th} quantile, namely, the solution ξ_p of the equation $F(x) = p$ exists, then due to the continuity of $F(x)$, it follows that $X_{k(N),N}$ converges in probability to ξ_p.*

Result 2.10.4. *Let X be an absolutely continuous random variable having $F(x)$ and $f(x)$ for its distribution function and density function respectively. Let $k(N) = Np + 0(1)$ with $0 < p < 1$ and let $f(\xi_p) > 0$ and $f(x)$ be continuous in the neighborhood of ξ_p, the pth quantile of F. Then, $X_{k(N),N}$ is asymptotically normal with mean ξ_p and variance $p(1-p)/Nf^2(\xi_p)$.*

PROOF: Start with the result proved earlier:

$$P\left(X_{k,N} \le F^{-1}\left(p + z\,\{p(1-p)/N\}^{1/2}\right)\right) \to \Phi(z) \text{ as } N \to \infty\,.$$

That is, after expanding F^{-1} around p, we have

$$P(X_{k,N} - \xi_p) f(\xi_p)\sqrt{N}/\left\{p(1-p)\right\}^{1/2} \leq z f(\xi_p)/f\left(F^{-1}(p^*)\right) \to \Phi(z) \text{ as } N \to \infty$$

where p^* lies between p and $p + z\left\{p(1-p)/N\right\}^{1/2}$. That is, $f\left(F^{-1}(p^*)\right)$ lies between $f\left(F^{-1}(p)\right)$ and $f\left(F^{-1}\left(p + z\left\{p(1-p)/N\right\}^{1/2}\right)\right)$.

Now for given $\epsilon > 0$, there exists an $N(\epsilon)$ such that for $N > N(\epsilon)$,

$$\left|\left\{f(\xi_p)/f\left(F^{-1}\left(p + z\left[p(1-p)/N\right]^{1/2}\right)\right)\right\} - 1\right| \leq \epsilon\,.$$

The same $N(\epsilon)$ will ensure that

$$\left|\left\{f(\xi_p)/f\left(F^{-1}(p^*)\right)\right\} - 1\right| \leq \epsilon\,.$$

Since ϵ is arbitrary this completes the proof of the result.

Van der Vaart (1961) provided an alternative proof of the asymptotic normality of the sample quantiles. His proof is based on the following lemma which is an extension of the Laplace-DeMoivre's theorem.

Lemma 2.10.1. *If $\theta(N)$, $0 < \theta(N) < 1$ is bounded away from 0 and 1, if $0 < k(N) < N$ and if*

$$\lim_{N\to\infty} \frac{N^{1/2}\left[\frac{k(N)}{N} - \theta(N)\right]}{\left\{\theta(N)\left[1-\theta(N)\right]\right\}^{1/2}} = -c, \quad |c| < \infty\,,$$

then

$$\lim_{N\to\infty} \sum_{j=k(N)}^{N} \binom{N}{j} \left[\theta(N)\right]^j \left[1-\theta(N)\right]^{N-j} = 1 - \Phi(-c) = \Phi(c)\,.$$

Proof: The proof readily follows from the proof of the case where $\theta(N) = \theta$ does not depend on N. As N tends to infinity, the characteristic function of $(Y - N\theta)/\left[N\theta(1-\theta)\right]^{1/2}$ where Y is a binomial variable, tends to $\exp(-t^2/2)$ whether θ depends on N or not, as long as $\theta(N)$ is bounded away from 0 and 1.

Van der Vaart's Proof of Result 2.10.4: We have shown that

$$P\left[X_{k,N} \leq x\right] = \sum_{j=k}^{N} \binom{N}{j} \left[F(x)\right]^j \left[1 - F(x)\right]^{N-j}\,.$$

Also since $k(N) = [Np] + 1$,

$$0 < \frac{k(N)}{N} - p = \frac{[Np] - Np + 1}{N} = \frac{\eta(N)}{N} \leq \frac{1}{N} .$$

Now, the above Lemma can be applied provided we can choose a function $\theta(N)$ of N such that if $F(x)$ is replaced by $\theta(N)$ in

$$-c_N = \frac{N^{1/2}\left[\frac{k(N)}{N} - F(x)\right]}{\{F(x)\left[1 - F(x)\right]\}^{1/2}} = \frac{N^{1/2}\left[\frac{\eta(N)}{N} + p - F(x)\right]}{\{F(x)\left[1 - F(x)\right]\}^{1/2}} ,$$

then c_N tends to a finite limit c as $N \Rightarrow \infty$. First, put $F(x) = q \neq p$; then $c_N \rightarrow \pm\infty$ according as $q \gtrless p$. Then put

$$x = \xi_p + \frac{\gamma}{\sqrt{N}} \Rightarrow F(x) = p + \frac{\gamma}{\sqrt{N}} f(\xi_p) + o(N^{-1/2}) ;$$

then $F(x)\left[1 - F(x)\right]$ tends to $p(1-p)$, and hence

$$\lim_{N\rightarrow\infty} -c_N = -\gamma f(\xi_p) / \left[p(1-p)\right]^{1/2} = -c, \text{ say.}$$

Then, the lemma may be applied to the distribution of $X_{k,N}$, provided $x = \xi_p + \gamma/\sqrt{N}$ and $k = [Np] + 1$. That is,

$$\lim_{N\rightarrow\infty} \left[X_{k,N} \leq \xi_p + \gamma/\sqrt{N}\right] = \Phi(c) .$$

Writing $\gamma = c\left\{p(1-p)\right\}^{1/2} / f(\xi_p)$, we conclude the proof.

Remark 2.10.2. *If* $k = Np + \delta N^{1/2} + o(N^{1/2})$, *then it is easy to prove along the same lines that*

$$\lim_{N\rightarrow\infty} P\left[(X_{k,N} - \xi_p) f(\xi_p)\sqrt{N} \leq \delta + c\left\{p(1-p)\right\}^{1/2}\right] = \Phi(c) .$$

Normal Approximations for Small Samples

Even though we have a normal approximation to the distribution of the sample quantile, the approximation may not be good unless N is large. Since the distribution of $X_{k,N}$ is related to the binomial distribution, the usual methods of approximating the binomial distribution functions will be preferable. The following approximations should work better.

$$P(X_{k,N} \leq x) = \Phi(c_N)$$

and

$$P(X_{k,N} \le x) = \Phi(c'_N)$$

where

$$c'_N = -N^{1/2} \left[\frac{k(N) - \frac{1}{2}}{N} - F(x) \right] / \{F(x)\,[1 - F(x)]\}^{1/2} .$$

As an illustration let $k = 0.4N + 2N^{1/2}$, $\theta = 0.4 + 3N^{-1/2}$. Comparing the approximations we have

Table 2.10.1

N	100	400	900
True value	.9875	.9801	.9790
Normal approximation	.9794	.9794	.9794
Approximation based on c_N	.9854	.9778	.9772
Approximation based on c'_N	.9890	.9803	.9790

Corollary 2.10.4.1. *$X_{[(N+1)/2],N}$ is asymptotically normally distributed with mean $\xi_{0.5}$ and variance $\left[4Nf^2(\xi_{0.5})\right]^{-1}$.*

A General Definition of Sample Median

Consider

$$M = \alpha X_{[(N+1)/2],N} + (1-\alpha) X_{[N/2]+1,N}$$

for some $0 < \alpha < 1$. When $N = 2m+1$, $M = X_{m+1,N}$. When $N = 2m$, we usually take $\alpha = \frac{1}{2}$. Recall that we defined the pth sample quantile as $X_{k,N}$ where $k = Np$ or $[Np] + 1$ accordingly as Np is an integer or not. However, when $p = \frac{1}{2}$, this does coincide with the above definition of M when $N = 2m$ only for $\alpha = 1$ and when $N = 2m+1$ for all α.

From Corollary 2.10.3.1, we have that the sample quantile converges to the population quantile in probability. Sarkadi (1974) has shown that the expectation of the sample quantile converges to the population quantile under general conditions on the underlying distribution function.

Let $X_{1N} \le \cdots \le X_{NN}$ denote the order statistics in a random sample of size N drawn from an arbitrary and nontrivial distribution $F(x)$ which is right continuous. Let us define its quantile function $H(x)$ by

$$H(x) = \inf \{y : F(y) \ge x\} .$$

If U is uniformly distributed on (0,1), then $H(U)$ has $F(x)$ for its distribution. If $U_{i,N}$, $i = 1, \ldots, N$ denote the uniform order statistics in a random sample of size N, then $H(U_{i,N})$ has the same distribution as $X_{i,N}$, $i = 1, \ldots, N$. Hence

$$E(X_{i,N}) = EH(U_{i,N}) = \frac{N!}{(i-1)!(N-i)!} \int_0^1 H(u)u^{i-1}(1-u)^{N-i}du\,.$$

Let $F_n(x)$ denote the empirical distribution based on the sample of size N drawn from $F(x)$. Then the empirical quantile $H_N(x)$ is given by

$$H_N(x) = X_{r,N}$$

where r is an integer such the $Nx \le r < Nx+1$. Then we have the following result of Sarkadi (1974).

Result 2.10.5 (Sarkadi, 1974). *If $E(X_{i,N})$ exists for some i and N, then*

$$\lim_{N\to\infty} EH_N(y) = H(y), \quad (0 < y < 1)\,.$$

PROOF: See Sarkadi (1974, pp. 342–345).

Result 2.10.6 (Sarkadi, 1974). *Let $\{r_N\}$ be a sequence such that $\lim_{N\to\infty}\{r_N/N\} = y$. If $E(X_{i,N})$ exists for some i and N, and if $r_N = Ny + 0(N^{1/2})$, then*

$$\lim_{N\to\infty} E(X_{r_N,N}) = H(y)\,.$$

PROOF: See Sarkadi (1974, p. 345).

Remark 2.10.3. *Notice that $E(X_{i,N})$ exists for $1 < i < N-1$ for given N.*

Result 2.10.4 can be extended to two or more quantiles. We will state the result about the limiting joint distribution of s sample quantiles, which is due to Mosteller (1946). The bivariate version is attributed to Smirnov (1935).

Result 2.10.7 (Mosteller, 1946). *Let s be a given number such that $0 < p_1 < p_2 < \cdots < p_s < 1$ and let ξ_{p_i} denote the p_i-quantile of the population ($i = 1, 2, \ldots, s$), having $F(x)$ and $f(x)$ for its d.f. and p.d.f. respectively. Assume that $f(\xi_{p_i}) > 0$, and $f(x)$ is continuous in the neighborhood* *of ξ_{p_i}, $i = 1, 2, \ldots, s$. Then the joint distribution of the s sample quantiles $X_{k_1,N}, X_{k_2,N}, \ldots, X_{k_s,N}$ where*

$$k_i = [Np_i] + 1, \quad i = 1, 2, \ldots, s$$

tends to an s-dimensional normal distribution with means $\xi_{p_1}, \xi_{p_2}, \ldots, \xi_{p_s}$ and with variance-covariance matrix.

$$N^{-1} \begin{pmatrix} \frac{p_1(1-p_1)}{f_1^2} & \frac{p_1(1-p_2)}{f_1 f_2} & \cdots & \frac{p_1(1-p_s)}{f_1 f_s} \\ \frac{p_1(1-p_2)}{f_1 f_2} & \frac{p_2(1-p_2)}{f_2^2} & \cdots & \frac{p_2(1-p_s)}{f_2 f_s} \\ \cdots & \cdots & \cdots & \cdots \\ \frac{p_1(1-p_s)}{f_1 f_s} & \frac{p_2(1-p_s)}{f_2 f_s} & \cdots & \frac{p_s(1-p_s)}{f_s^2} \end{pmatrix}$$

where $f_i = f(\xi_{p_i})$, $i = 1, 2, \ldots, s$.

2.11 Quasi-ranges

The ith quasi-range in samples of size N from an arbitrary continuous population is defined as the range of $N - 2i$ sample observations omitting the i largest and the i smallest observations. Sample quasi-ranges have been used in estimating population standard deviation and in setting up confidence intervals for the population standard deviation. In the following, we will give some results on expected values and product moments of sample quasi-ranges. We need the following notation. Let

$$\begin{aligned} W_{i,N} &= X_{N-i,N} - X_{i+1,N}, & i &= 0, 1, \ldots, [N-2]/2, \\ \omega_{i,N} &= E(W_{i,N}) = \mu_{N-i,N} - \mu_{i+1,N}, & i &= 0, 1, \ldots, [N-2]/2. \end{aligned}$$

$W_{0,N}$ will be the sample range and $\omega_{0,N}$ will be the expected range. Also, let

$$\begin{aligned} a_{i,j,N} &= E(W_{i,N} W_{j,N}), & 0 \le i \le j &\le [N-2]/2 \\ a_{i,N}^{(2)} &= a_{i,i,N} & i &= 0, 1, \ldots, [N-2]/2. \end{aligned}$$

Result 2.11.1. *For an arbitrary continuous distribution symmetric about zero,*

$$\omega_{i,N} = 2\mu_{N-i,N}, \quad i = 0, 1, \ldots, [N-2]/2.$$

PROOF: Follows from the fact that $\mu_{i+1,N} = -\mu_{N-i,N}$.

Result 2.11.2. *For an arbitrary continuous distribution, one has*

$$(N-i)\omega_{i-1,N} + i\omega_{i,N} = N\omega_{i-1,N-1}, \quad i = 0, 1, \ldots, [N-2]/2\,.$$

PROOF: We have the recurrence formula,

$$i\mu_{i+1,N} + (N-i)\mu_{i,N} = N\mu_{i,N-1}, \quad i = 1, 2, \ldots, N-1\,.$$

Changing $N - i$ to i in the preceding relation, one obtains

$$(N-i)\mu_{N-i+1,N} + i\mu_{N-i,N} = N\mu_{N-i,N-1}\,, \quad (i = 1, 2, \ldots, N-1)\,.$$

From the preceding two equations one gets

$$\begin{aligned}(N-i)(\mu_{N-i+1,N} - \mu_{i,N}) &+ i(\mu_{N-i,N} - \mu_{i+1,N}) \\ &= N(\mu_{N-i,N-1} - \mu_{i,N-1})\,.\end{aligned}$$

Using the definition of $\omega_{i,N}$, the desired result follows.

Result 2.11.3. *For any distribution symmetric about zero,*

1. $a_{i,j,N} = 2(\mu_{i+1,j+1,N} - \mu_{i+1,N-j,N})$,
2. $cov(W_{i,N}, W_{j,N}) = 2\,[cov(X_{i+1,N}, X_{j+1,N}) - cov(X_{i+1,N}, X_{N-j,N})]$ *for* $0 \le i \le j \le [N-2]/2$.

PROOF: Follows from writing $W_{i,N}$ and $W_{j,N}$ in terms of $X_{i,N}$'s and using the fact that the distributions of $X_{i+1,N}$ and $-X_{N-i,N}$, and the distributions of $X_{i+1,N} \cdot X_{j+1,N}$ and $X_{N-i,N} \cdot X_{N-j,N}$ are identical.

2.12 Problems

2.3.1 Find the density function of the smallest standard exponential order statistic in a random sample of size N.

2.3.2 Find the density function of the smallest order statistic in a random sample of size N drawn from the Weibull density given by

$$\begin{aligned} f(x;m) &= me^{-x^m}x^{m-1} && x > 0,\ m > 0, \\ &= 0 && \text{elsewhere.} \end{aligned}$$

2.3.3 Let $X_{1N} \leq \cdots \leq X_{NN}$ be the order statistics in a random sample of size N drawn from the standard exponential distribution.

(a) Show that the joint density of $(X_{1N}, \ldots, X_{n,N})$ $(n \leq N)$ is

$$\frac{N!}{(N-n)!} \exp\left[-\sum_{1}^{n} y_i + (N-n)y_n\right], \quad 0 < y_1 < \cdots < y_n.$$

(b) Show that $2\left[\sum_1^n X_{i,N} + (N-n)X_{n,N}\right]$ is distributed as χ^2_{2n}.

2.3.4 Obtain the density of sample range and mid-range if the parent population is standard negative exponential.

2.4.1 Let $U_{1N} \leq \cdots \leq U_{NN}$ be the standard uniform order statistics in a random sample of size N. Let

$$V_i = \frac{U_{iN}}{U_{i+1,N}} \quad \text{for } 1 \leq i \leq N-1 \text{ and } V_N = U_{NN}.$$

(a) Find the marginal distribution of V_n, $1 \leq n \leq N$.

(b) Find the joint distribution of $V_1, \ldots, V_N$.

(c) Show that $V_1, \ldots, V_N$ are independent.

(d) Show that V_j^j is distributed as uniform $(0,1)$ for $j = 1, \ldots, N$.

2.5.1 Let $X_{1N} \leq \cdots \leq X_{NN}$ denote the order statistic in sample of size N drawn from the Weibull density

$$f(x,m) = m\exp(-x^m)x^{m-1} \quad \text{for} \quad m > 0,\ x > 0.$$

Show that

$$\mu_{1,N} = E(X_{1N}) = \frac{1}{m}N^{\frac{1}{m}-2}\Gamma\left(2-\frac{1}{m}\right).$$

2.5.2 For an arbitrary distribution if the density $f(x)$ is such that $f'(x) = -xf(x)$, then show that $\mu_{N,N}^{(2)} = 1 + \mu_{N-1,N,N}$.

2.5.3 $E(X_{i,N}^r X_{j,N}^s)$ $(i < j)$ exists if the moment of $\max(r,s)^{\text{th}}$ order exists in the population.

2.5.4 Show that for the population having $F(x) = 1 - (\ln\ x)^{-1}$, $x \geq e$ for its d.f., no order statistics will have moments.

HINT:

$$\begin{aligned} \mu_{r,N} &= C \int_e^\infty x F^{r-1}(x)\,[1 - F(x)]^{N-r}\, f(x)dx \\ &\geq C 2^{1-r} \int_{e^2}^\infty (\ln\ x)^{r-N-2} dx. \end{aligned}$$

2.5.5 Using Result 2.4.11, show that

$$[(\mu_{N,N} - \mu_{1,N})/\sigma]^2 \leq 2N^2/(2N-1).$$

2.6.1 Suppose X is a Bernoulli variable taking values 1 and 0 with probabilities p and 1-p respectively. Then show that

(a) $\mu_{N,N} = 1 - (1-p)^N$

(b) $\mu^{(2)}_{N,N} = \mu_{NN}$

(b) $\mu_{1,N} = p^N$.

2.6.2 Using Result 2.6.2, for the Bernoulli variable defined in Problem 2.6.1, obtain the joint probability function $f_{1,N,N}(x, y)$.

2.7.1 Using Corollary 2.7.2.1, obtain the mean and variance of (a) X_{IN} and (b) X_{NN}.

2.8.1 Using Result 2.8.1, obtain Renyi's representation for the smallest and largest Weibull order statistics in a sample of size N (see Problem 2.3.2 for the Weibull density function).

2.10.1 Using Result 2.10.4, obtain the asymptotic normal distribution of pth sample quantile in a sample of size N from the uniform $(0, 1)$ distribution.

2.10.2 If X is the median of a sample of size N from a continuous d.f. $F(x)$, show that $F(X)$ is asymptotically normally distributed with mean $\frac{1}{2}$ and variance $\frac{1}{4}N$.

2.10.3 If $\tilde{X}$ is the median in a sample of size $2M + 1$from an exponential distribution with parameter λ, show that $\tilde{X}$ has an asymptotic normal distribution with mean $(\ln\ 2)/\lambda$ and variance $(2\lambda^2 M)^{-1}$.

Chapter 3

Ordered Least Squares Estimators

3.1 Introduction

Many contributions have been made to the problem of linear unbiased estimation of the parameters of a distribution using the ordered observations in the sample. Lloyd [1952] gave explicit formulae for estimating the location and scale parameters of a distribution using order statistics in a random sample. Gupta [1952] considered the problem of best linear estimation of the mean and standard deviation of a normal population using a censored sample. Sarhan [1954-56], using Lloyd's formula computed the best linear coefficients of the estimates of the mean and standard deviation of rectangular, triangular, normal and double exponential populations for small sample sizes. Sarhan and Greenberg [1956, 1958a, 1958b] considered estimating the location and scale parameters using order statistics in singly and doubly censored samples from normal and exponential populations. Tate [1959] considered unbiased estimation of functions of location and scale parameters. Plackett [1949] gave some approximate expressions for the variance-covariance matrix of the censored sample and thus obtained some 'almost' best linear estimates of the parameters of a population. Blom [1958] gave some approximations to the means, variances and covariances of order statistics in samples from a continuous population, the inverse of the d.f. of which could be expanded in Taylor series. Govindarajulu and Joshi [1968] considered the best linear estimation of location and scale parameters of the Weibull distribution using ordered observations in censored and uncensored samples.

3.2 Explicit Formulae for Estimators

In the following, the general formulae for the least squares estimates of location and scale parameters will be presented and these be applied to some well known distributions. Let X be a random variable having the density function $f\left((x-\theta)/\sigma\right)/\sigma$ so that the transformed variable $Y=(X-\theta)/\sigma$ has density $f(y)$. That is, Y has 0 and 1 for its location and scale parameters respectively. Also, let

$$X_{r,N} \le X_{r+1,N} \le \cdots \le X_{s,N}, \quad 1 \le r < s \le N$$

be the available portion of the ordered sample of size N drawn from the population with density function $f\left((x-\theta)/\sigma\right)/\sigma$. That is, the first $r-1$ and the last $N-s$ observations are either missing or ignored. In particular, r could be 1 and s could be N. This kind of censoring generally arises in most of the practical problems. We wish to obtain the best linear unbiased estimates of θ and σ on the basis of the censored sample. Let

$$Y_{i,N} = (X_{i,N}-\theta)/\sigma, \quad i = r, 1+r, \ldots, s$$

and

$$\mu_{i,N} = E(Y_{i,N}); \quad \sigma_{i,j,N} = \operatorname{cov}(Y_{i,N}, Y_{j,N})$$

for $r \le i \le j \le s$ and $s \ge 1+r$. We assume that $\mu_{i,N}$ and $\sigma_{i,j,N}$, $r \le i \le j \le s$ $(s \ge 1+r)$ are known. Then,

$$E(X_{i,N}) = \theta + \sigma\mu_{i,N}$$

and

$$\operatorname{cov}(X_{i,N}, X_{j,N}) = \sigma^2\sigma_{i,j,N}, \quad r \le i \le j \le s \quad (s \ge 1+r)\,.$$

Now, since $E(X_{i,N})$ is linear in θ and σ, using Gauss-Markov Theorem, one can unbiasedly estimate θ and σ by linear combinations of $X_{i,N}$. Let

$$\boldsymbol{X}_N = (X_{r,N}, \cdots, X_{s,N})', \mathbf{1} = (1, \ldots, 1)',$$

$$\boldsymbol{\mu} = (\mu_{r,N}, \ldots, \mu_{s,N})' \text{ and } \Omega^{-1} = ((\sigma_{i,j,N}))\,.$$

Notice that Ω^{-1} is an $(s-r+1)\times(s-r+1)$ positive definite symmetric matrix. Then in matrix form, we have

$$E(\boldsymbol{X}_N) = \mathbf{1}\theta + \sigma\boldsymbol{\mu} = ((\mathbf{1}, \boldsymbol{\mu}))\begin{pmatrix}\theta\\ \sigma\end{pmatrix}$$

and

$$\text{var}(\boldsymbol{X}_N) = \sigma^2\Omega^{-1}\,.$$

Denoting $((\mathbf{1}, \boldsymbol{\mu}))$ by Q and (θ, σ) by α', we have by Gauss-Markov theorem (see Lloyd [1952] and also Aitken [1945]) the estimate of $\boldsymbol{\alpha}$. Dwyer [1958] gave an elegant proof of the results of Aitken [1935]. Following Dwyer's [1958] method of proof, we wish to minimize the weighted error sum of squares given by

$$\begin{aligned}
\chi &= (\boldsymbol{X}_N - Q\boldsymbol{\alpha})'\Omega(\boldsymbol{X}_N - Q\alpha) \\
&= \boldsymbol{\alpha}'Q'\Omega Q\boldsymbol{\alpha} - \boldsymbol{\alpha}'Q'\Omega\boldsymbol{X}_N - X_N'\Omega Q\boldsymbol{\alpha} + \boldsymbol{X}_N'\Omega\boldsymbol{X}_N \\
\frac{\partial\chi}{\partial\left\langle\boldsymbol{\alpha}\right\rangle} &= \text{the symbolic derivative of } \chi \text{ with respect to a particular element of the vector } \alpha \\
&= S'Q'\Omega Q\boldsymbol{\alpha} + \boldsymbol{\alpha}'Q'\Omega QS - S'Q'\Omega\boldsymbol{X}_N - \boldsymbol{X}_N'\Omega QS
\end{aligned}$$

where S is a matrix having the same dimensions as α' and having all its elements equal to zero except for a unit element in its ith row and j^{th} column where $\alpha_{ij} = \langle\alpha\rangle$. [See Dwyer and MacPhail (1948, Section 9)]. Similarly, define a matrix W having the same dimensions as χ and having all its elements equal to zero except for a unit element in the ith row and j^{th} column where $\chi_{ij} = \langle\chi\rangle$. Here, since χ is a scalar, $W = W' = 1$. $\partial\left\langle\chi\right\rangle/\partial_\alpha$ is obtained from $\partial\chi/\partial\left\langle\alpha\right\rangle$ using the following rules:

1. Each S becomes W and each S' becomes W'.
2. The Pre (Post) multiplier of S becomes its transpose.
3. The Pre (Post) multiplier of S' becomes a Post (Pre) multiplier of W'.

Thus

$$\frac{\partial\left\langle\chi\right\rangle}{\partial\alpha} = Q'\Omega Q\alpha W' + Q'\Omega Q\alpha W - Q'\Omega\boldsymbol{X}_N W' - Q'\Omega\boldsymbol{X}_N W\,.$$

Setting $\frac{\partial\langle\chi\rangle}{\partial\alpha} = 0$ and $W = W' = 1$ and solving for $\hat{\alpha}$, we obtain

$$\hat{\alpha} = (Q'\Omega Q)^{-1}Q'\Omega\boldsymbol{X}_N = b\boldsymbol{X}_N \tag{3.2.1}$$

where b is a $2 \times (s - r + 1)$ matrix which will be called the coefficient matrix.

$$\begin{aligned}
\text{var}(\hat{\alpha}) &= b\,\text{var}(\boldsymbol{X}_N)b' \\
&= (Q'\Omega Q)^{-1}Q'\Omega\Omega^{-1}\Omega Q(Q'\Omega Q)^{-1}\sigma^2 \\
&= (Q'\Omega Q)^{-1}\sigma^2\,, \qquad (3.2.2)
\end{aligned}$$

where

$$Q'\Omega Q = \begin{pmatrix} \mathbf{1}'\Omega \boldsymbol{I} & \mathbf{1}'\Omega\mu \\ \mathbf{1}'\Omega\mu & \mu\Omega\mu \end{pmatrix}$$

the elements of the matrix being scalars. Its inverse is given by

$$(Q'\Omega Q)^{-1} = \Delta^{-1} \begin{pmatrix} \mu\Omega\mu & -I'\Omega\mu \\ \mathbf{1}'\Omega\mu & \mathbf{1}'\Omega I \end{pmatrix}$$

where Δ is the determinant of the matrix $Q'\Omega Q$. Using this result in (3.2.1) and (3.2.2) one obtains

$$\hat{\theta} = -\mu' D\boldsymbol{X}_N, \quad \hat{\sigma} = \mathbf{1}' D\boldsymbol{X}_N$$

where D is the skew symmetric matrix defined by

$$D = \Omega(\mathbf{1}\boldsymbol{\mu}' - \boldsymbol{\mu}\mathbf{1}')\Omega/\Delta\,.$$

Also,

$$\text{var}\,\hat{\theta} = \mu'\Omega\mu\sigma^2/\Delta, \quad \text{var}\,\hat{\sigma} = \mathbf{1}'\Omega\mathbf{1}\sigma^2/\Delta$$

and

$$\text{cov}(\hat{\theta}, \hat{\sigma}) = -\mathbf{1}'\Omega\mu\sigma^2/\Delta\,.$$

Suppose we are interested in least squares estimates of linear functions of the location and scale parameters. For example, the mean μ_X and the standard deviation σ_X of the population could conceivably be linearly related to θ and σ. Then the question is how to find least squares estimates of μ_X and σ_X, having found the least squares estimates of θ and σ. If

$$\begin{pmatrix} \mu_X \\ \sigma_X \end{pmatrix} = A \begin{pmatrix} \theta \\ \sigma \end{pmatrix},$$

then, the least squares estimates of μ_X and σ_X are given by

$$\begin{pmatrix} \hat{\mu}_X \\ \hat{\sigma}_X \end{pmatrix} = A \begin{pmatrix} \hat{\theta} \\ \hat{\sigma} \end{pmatrix}.$$

PROOF: The model can be rewritten as

$$\begin{aligned} E(\boldsymbol{X}_N) = ((\mathbf{1}, \boldsymbol{\mu})) \begin{pmatrix} \theta \\ \sigma \end{pmatrix} &= ((\mathbf{1}, \boldsymbol{\mu}))\, A^{-1} \begin{pmatrix} \mu_X \\ \sigma_X \end{pmatrix} \\ &= Q^* \begin{pmatrix} \mu_X \\ \sigma_X \end{pmatrix} \end{aligned}$$

where

$$Q^* = QA^{-1}, \quad Q = ((\mathbf{1}, \boldsymbol{\mu}))\,.$$

Then the least squares estimates of $\begin{pmatrix} \mu_X \\ \sigma_X \end{pmatrix}$ are given by

$$\begin{aligned}
\begin{pmatrix} \hat{\mu}_X \\ \hat{\sigma}_X \end{pmatrix} &= (Q^{*\prime}\Omega Q^*)^{-1} Q^{*\prime}\Omega \mathbf{X}_N \\
&= \left((A^{-1})' Q'\Omega Q A^{-1}\right)^{-1} (A^{-1})' Q'\Omega \mathbf{X}_N \\
&= A(Q'\Omega Q)^{-1} Q'\Omega \mathbf{X}_N\,.
\end{aligned}$$

Variance-covariance matrix of $\hat{\mu}_X$ and $\hat{\sigma}_X$ is given by

$$(Q^*\Omega Q^*)^{-1} = A(Q'\Omega Q)^{-1} A'\,.$$

3.3 Estimation for Symmetric Populations

If the population is symmetric about θ and the sample is symmetrically censored (that is, $s = N - r + 1$) we will show that $Q\Omega Q'$ is diagonal. Its off-diagonal elements are each equal to zero. To show this, let us introduce the permutation matrix given by

$$R = \begin{pmatrix} 0 & & 1 \\ & \ddots & \\ 1 & & 0 \end{pmatrix}.$$

R is symmetric and orthogonal and its row totals are all unity. Also

$$R = R' = R^{-1}, \quad R'\mathbf{1} = \mathbf{1}\,.$$

When R pre-multiplies a certain matrix, it has the effect of reversing the order of the rows of the matrix. That is

$$\begin{pmatrix} -Y_{s,N} \\ \vdots \\ -Y_{r,N} \end{pmatrix} = -R \begin{pmatrix} Y_{r,N} \\ \vdots \\ Y_{s,N} \end{pmatrix}.$$

Also, the set $-Y_{s,N}, \ldots, -Y_{r,N}$ may be regarded as ordered observations on the variate $-Y$. Since Y is symmetrically distributed about zero, the distributions of Y and $-Y$ are identical. Thus, since $s = N - r + 1$, the distribution of $(Y_{r,N}, \ldots, Y_{s,N})$ and $(-Y_{s,N}, \ldots, -Y_{r,N})$ are identical. Hence, the vector random variables $\boldsymbol{Y}_N$ and $-R\boldsymbol{Y}_N$ have the same distribution. It follows by considering the mean and variance of $\boldsymbol{Y}_N$ and $-R\boldsymbol{Y}_N$, that is,

$$\boldsymbol{\mu} = -R\boldsymbol{\mu} \text{ and } \Omega^{-1} = R\Omega^{-1}R\,.$$

Inverting the last equation, we obtain

$$\Omega = R\Omega R\,.$$

Using the above relationships, one can obtain

$$\mathbf{1}'\Omega\boldsymbol{\mu} = \mathbf{1}'(R\Omega R)(-R\mu) = -\mathbf{1}'R\Omega R^2\boldsymbol{\mu} = -\mathbf{1}'\Omega\boldsymbol{\mu}$$

since $\mathbf{1}'R = \mathbf{1}'$ and R^2 is equal to the identity matrix. That is, $\mathbf{1}'\Omega\boldsymbol{\mu}$ is zero and thus $Q\Omega Q'$ is diagonal. Then, the estimates become

$$\hat{\theta} = \mathbf{1}'\Omega\boldsymbol{X}_N/\mathbf{1}'\Omega\mathbf{1}, \quad \hat{\sigma} = \boldsymbol{\mu}'\Omega\boldsymbol{X}_N/\boldsymbol{\mu}'\Omega\boldsymbol{\mu}$$

$$\text{var}(\hat{\theta}) = \sigma^2/\mathbf{1}'\Omega\mathbf{1}, \quad \text{var}(\hat{\sigma}) = \sigma^2/\boldsymbol{\mu}'\Omega\boldsymbol{\mu}$$

and their covariance is zero.

3.4 Estimation in a Single Parameter Family

Let $f(x/\lambda)/\lambda$ be the functional form of the density function of the r.v. X. Then $Y = X/\lambda$ has the density function $f(y)$. Then the ordered least squares estimate of λ and its variance are given by

$$\hat{\lambda} = \boldsymbol{\mu}'\Omega\boldsymbol{X}_N/\boldsymbol{\mu}'\Omega\boldsymbol{\mu}$$

and

$$\text{var}(\hat{\lambda}) = \lambda^2/\boldsymbol{\mu}'\Omega\boldsymbol{\mu}$$

where μ and Ω have the usual meanings. If the density function is of the type $f(x-\theta)$, then

$$\hat{\theta} = \mathbf{1}'\Omega(\boldsymbol{X}_N - \boldsymbol{\mu})/\mathbf{1}'\Omega\mathbf{1}$$

and

$$\text{var}(\hat{\theta}) = 1/\mathbf{1}'\Omega\mathbf{1}\,.$$

3.5 Optimum Properties of Ordered Least Squares Estimates

Case 1. Location Parameter

Suppose θ and σ are respectively the mean and the standard deviation in the population. Since the sum of the ordered observations is the same as the sum of the unordered observations, the sample mean can be defined as $\sum X_{i,N}/N$. Hence, it is an unbiased linear combination of the ordered observations. However, our estimate $\hat{\theta}$ has, minimal variance in the class of such estimates, so that its variance is at most equal to that of the sample mean. Thus, $\text{var}(\hat{\theta}) \leq \sigma^2/N$. Now one might ask under what conditions the above relation is a strict inequality. For the following discussion, we assume that complete samples are available.

Let the density function be of the form $f(x-\theta)$ where θ is also the mean of the population. The following result is due to Downton [1953] which is a generalization of a result due to Lloyd [1952] proved for symmetric populations.

Result 3.5.1. *The least squares estimate of θ which is the mean and location parameter of a continuous distribution, has a variance strictly less than the variance of the sample mean if and only if there does not exist a scalar c such that*

$$\Omega^{-1}\mathbf{1} = \mathbf{1} - c\boldsymbol{\mu}\,.$$

PROOF: Since the sum of the ordered observations is equal to the sum of the unordered observations, it follows that

$$\mathbf{1}'\boldsymbol{\mu} = 0, \text{ that } \boldsymbol{\mu} \neq 0 \text{ and that } \mathbf{1}'\Omega^{-1}\mathbf{1} = N\,.$$

Since Ω^{-1} is symmetric and positive definite, Ω^{-1} may be expressed as

$$\Omega^{-1} = VV' \text{ and } \Omega = (V^{-1})'V^{-1}$$

where V is a lower triangular matrix. Also, for any two $(N \times 1)$ vectors B and C we have

$$B'\Omega^{-1}B = B'VV'B = H'H = \sum h_i^2$$

where $H = V'B$. Also,

$$C'\Omega C = C'(V^{-1})'V^{-1}C = K'K = \sum k_i^2$$

where $K = V^{-1}C$. Now by applying Schwarz inequality, we get

$$\left(\sum h_i^2\right)\left(\sum k_i^2\right) \geq \left(\sum h_i k_i\right)^2,$$

the necessary and sufficient condition for equality being

$$h_i = ck_i, \text{ for all } i \text{ and for some scalar constant } c.$$

In matrix form the above inequality can be written as

$$(B'\Omega^{-1}B)(C'\Omega C) \geq (B'C)^2$$

with $B = c\Omega C$ as the condition for equality. Put $B = \Omega\mathbf{1} - \mathbf{1}$ and $C = \boldsymbol{\mu}$. Then we obtain

$$(\mathbf{1}'\Omega\mathbf{1} - 2\mathbf{1}'\mathbf{1} + \mathbf{1}'\Omega^{-1}\mathbf{1})(\boldsymbol{\mu}'\Omega\boldsymbol{\mu}) \geq (\mathbf{1}'\Omega\boldsymbol{\mu} - \mathbf{1}'\boldsymbol{\mu})^2.$$

However,

$$\mathbf{1}'\mathbf{1} = \mathbf{1}'\Omega^{-1}\mathbf{1} = N \text{ and } \mathbf{1}'\boldsymbol{\mu} = 0.$$

Hence

$$(\mathbf{1}'\Omega\mathbf{1} - N)(\boldsymbol{\mu}'\Omega\boldsymbol{\mu}) \geq (\mathbf{1}'\Omega\boldsymbol{\mu})^2$$

or

$$\left\{(\mathbf{1}'\Omega\mathbf{1})(\boldsymbol{\mu}'\Omega\boldsymbol{\mu}) - (\mathbf{1}'\Omega\boldsymbol{\mu})^2\right\}/(\boldsymbol{\mu}'\Omega\boldsymbol{\mu}) = \sigma^2/\text{var}(\hat{\theta}) \geq N,$$

since $\boldsymbol{\mu}'\Omega\boldsymbol{\mu}$ is essentially positive. Thus

$$\text{var}(\hat{\theta}) \leq \sigma^2/N$$

and the necessary and sufficient condition for equality is

$$\Omega^{-1}\mathbf{1} = \mathbf{1} - c\boldsymbol{\mu}$$

for some scalar constant c. However, when this condition is satisfied $\hat{\theta}$ is necessarily the sample mean. For, by pre-multiplying the above condition by $\boldsymbol{\mu}'\Omega$, $\mathbf{1}'\Omega$ and Ω we obtain

$$\boldsymbol{\mu}'\Omega\mathbf{1} - c\boldsymbol{\mu}'\Omega\boldsymbol{\mu} = \boldsymbol{\mu}'\mathbf{1} = 0 \quad c = \boldsymbol{\mu}'\Omega\mathbf{1}/\boldsymbol{\mu}'\Omega\boldsymbol{\mu},$$

$$\mathbf{1}'\Omega\mathbf{1} - c\mathbf{1}'\Omega\boldsymbol{\mu} = \mathbf{1}'\mathbf{1} = N$$

and

$$\Omega\mathbf{1} - c\Omega\boldsymbol{\mu} = \mathbf{1},$$

respectively. Now, consider Δ

$$\begin{aligned}\Delta &= (\boldsymbol{\mu}'\Omega\boldsymbol{\mu})(\mathbf{1}'\Omega\mathbf{1}) - (\mathbf{1}'\Omega\boldsymbol{\mu})^2 \\ &= (\boldsymbol{\mu}'\Omega\boldsymbol{\mu})(\mathbf{1}'\Omega\mathbf{1} - c\mathbf{1}'\Omega\boldsymbol{\mu}) \\ &= N\boldsymbol{\mu}'\Omega\boldsymbol{\mu}\,.\end{aligned}$$

Also,

$$\begin{aligned}\hat{\theta} &= (\boldsymbol{\mu}'\Omega\boldsymbol{\mu}\mathbf{1}' - \boldsymbol{\mu}'\Omega\mathbf{1}\boldsymbol{\mu}')\Omega\boldsymbol{X}_N/\Delta \\ &= \boldsymbol{\mu}'\Omega\boldsymbol{\mu}(\mathbf{1}'\Omega - c\boldsymbol{\mu}'\Omega)\boldsymbol{X}_N/\Delta \\ &= \boldsymbol{\mu}'\Omega\boldsymbol{\mu}\mathbf{1}'\boldsymbol{X}_N/\Delta \\ &= \mathbf{1}'\boldsymbol{X}_N/N\,,\end{aligned}$$

which is the sample mean.

Corollary 3.5.1.1. *If the population is symmetric about θ, then the necessary and sufficient condition for the equality of the ordered least squares estimate of θ and the sample mean is*

$$\Omega^{-1}\mathbf{1} = \mathbf{1}\,.$$

PROOF: We have

$$\mathbf{1} = \Omega\mathbf{1} - c\Omega\boldsymbol{\mu}\,.$$

Multiplying by $\boldsymbol{\mu}'$ on both sides and using the fact that $\boldsymbol{\mu}'\mathbf{1} = 0$ and $\boldsymbol{\mu}'\Omega\mathbf{1} = 0$ and $\boldsymbol{\mu}'\Omega\boldsymbol{\mu} \neq 0$, we infer that $c = 0$. Hence, $\Omega^{-1}\mathbf{1} = \mathbf{1}$.

Result 3.5.2 (Govindarajulu, 1966). *If θ denotes the population mean, then for $i = 1, 2, \ldots, N$,*

$$\sum_{j=1}^{N} \sigma_{i,j,N} = 1 \textit{ for } N = 2, 3, \ldots,$$

if and only if $F(x)$ equals the standard normal distribution function.

PROOF: See Govindarajulu (1966, Theorem 3.3 p. 1013). Now combine Result 3.5.2 and Corollary 3.5.1.1. We have the following result:

Result 3.5.3. *The least squares estimate of θ which is the mean and location parameter of a continuous and symmetric distribution function has a smaller variance than the arithmetic mean, with equality holding only if the underlying distribution is normal.*

Example 3.5.1 (Downton). Consider the density function

$$f(x) = \begin{cases} \frac{p}{\sigma b^p}\left(\frac{x-\theta}{\sigma} + a\right)^{p-1} & \theta - a\sigma \le x < -(a\sigma - b) + \theta\,, \\ 0 & \text{otherwise,} \end{cases}$$

where $p \ge 1$, $a = [p(p+2)]^{\frac{1}{2}}$ and $b = (p+1)\,[(p+2)/p]^{\frac{1}{2}}$,

$$E(X) = \theta \text{ and } \operatorname{var}(X) = \sigma^2\,.$$

Consider

$$T = (X - \theta + a\sigma)/b\sigma\,.$$

Its density function is

$$\begin{aligned} f(t) &= \begin{cases} pt^{p-1} & 0 \le t < 1\,, \\ 0 & \text{otherwise} \end{cases} \\ E(T_N) &= E\begin{pmatrix} T_{1,N} \\ \ddots \\ T_{N,N} \end{pmatrix} = \boldsymbol{\mu}\,, \\ \operatorname{var}(T_N) &= \Omega^{-1}\,. \end{aligned}$$

Downton [1953] has obtained explicit expressions for $\boldsymbol{\mu}$ and Ω^{-1} and the other required quantities. Thus, one can find ordered least squares estimates of θ and σ.

Case 2. Scale Parameter

In the case of the one parameter family of density functions of the form $f(x/\lambda)/\lambda$, where λ is also the mean of the population, Downton [1954] has similarly shown that the ordered least squares estimate of λ has variance strictly less than the variance of the sample mean if and only if

$$\Omega^{-1}\mathbf{1} \ne (\operatorname{var} Y)\boldsymbol{\mu}/E(Y)\,,$$

$$Y = X/\lambda\,.$$

In the following, we shall give a couple of results characterizing the exponential distribution.

Result 3.5.4 (Govindarajulu, 1975). *Let $EX = \theta$, where X is absolutely continuous having $F(x)$ for its distribution function. For $i = 1, 2, \ldots,$*

$$\sum_{j=1}^{N} \sigma_{i,j,n} = \theta \mu_{i,N}, \quad N = i, i+1, \ldots,$$

$$\text{iff } F(x) = 1 - \exp(-x/\theta), \quad x, \theta > 0\,.$$

PROOF: See Govindarajulu (1975, pp. 126–128).

Result 3.5.5 (Govindarajulu, 1975). *Let the distribution of X be of the form $F(x/\theta)$ where θ denotes the mean of the population. Then, the ordered least squares estimate of θ based on the complete sample coincides with the arithmetic mean iff $F(x) = 1 - \exp(-x)$, $x > 0$.*

PROOF: The ordered least squares estimates coincides with the sample mean if $\sum \mathbf{1} = \theta \boldsymbol{\mu}$ with $\theta = \mathbf{1}' \sum \mathbf{1}/\mathbf{1}'\boldsymbol{\mu})$ (see Govindarajulu [1968, Result 4.2.(i)]). Combining this with Result 3.5.2, the proof is completed.

3.6 Examples

Example 3.6.1 (Lloyd, 1952). Let

$$\begin{aligned} f(x, \theta_1, \theta_2) &= (2.3^{\frac{1}{2}}\theta_2)^{-1}, \ \ \theta_1 - 3^{\frac{1}{2}}\theta_2 < x < \theta_1 + 3^{\frac{1}{2}}\theta_2 \\ &= 0, \ \ \text{otherwise}\,. \end{aligned}$$

We wish to find least squares estimates of θ_1 and θ_2 based on a sample of size N drawn from this population. Define $Y_{i,N} = (X_{i,N} - \theta_1)/\theta_2$ and $\mu_{i,N} = E(Y_{i,N})$ and $\text{cov}(Y_{i,N}, Y_{j,N}) = \sigma_{i,j,N}$. Also, $Y_{i,N} = \sqrt{3}(2U_{i,N} - 1)$ where $U_{i,N}$ are standard uniform order statistics. Then we find that

$$\mu_{i,N} = 3^{\frac{1}{2}}(2i - N - 1)/(N+1)\,,$$

$$\sigma_{i,j,N} = 12i(N - j + 1)/(N+1)^2(N+2)\,, \quad (i \le j)\,,$$

$$\frac{12\Omega}{(N+1)(N+2)} = \begin{pmatrix} 2 & -1 & 0 & 0 & 0 & \cdots & 0 & 0 \\ -1 & 2 & -1 & 0 & 0 & \cdots & 0 & 0 \\ 0 & -1 & 2 & -1 & 0 & \cdots & 0 & 0 \\ \cdot & \cdot & \cdot & \cdot & \cdot & \cdot & \cdot & \cdot \\ 0 & 0 & 0 & 0 & 0 & \cdots & -1 & 2 \end{pmatrix},$$

$$I'\Omega = \frac{(N+1)(N+2)}{12}(1,0,0,\ldots,0,1)\,,$$

and

$$\mu'\Omega = \frac{(N+1)(N+2)}{3^{\frac{1}{2}}}\;(-1,0,0,\ldots,0,1)\,.$$

Hence,

$$\hat{\theta}_1 = (X_{1,N} + X_{N,N})/2 \text{ and } \hat{\theta}_2 = (N+1)(X_{NN} - X_{1,N})/3^{\frac{1}{2}} \cdot 2(N-1)\,.$$

The sampling variances are given by

$$\text{var}(\hat{\theta}_1) = 6\theta_2^2/(N+1)(N+2), \quad \text{var}(\hat{\theta}_2) = 2\theta_2^2/(N-1)(N+2),$$

and the covariance is zero.

Example 3.6.2 (Sarhan, 1955, Part III). Consider the exponential density function given by

$$f\left((x-\theta)/\sigma\right)/\sigma = (\sigma)^{-1}e^{-(x-\theta)/\sigma}, \quad \theta \le x < \infty\,.$$

Here,

$$\mu_{i,N} = \sum_{k=N-i+1}^{N} k^{-1}, \quad \sigma_{i,i,N} = \sum_{k=N-i+1}^{N} k^{-2},$$

$$\sigma_{i,m,N} = \sum_{k=N-i+1}^{N} k^{-2}, \quad i \le m\,,$$

$$\Omega = \begin{pmatrix} N^2+(N-1)^2 & -(N-1)^2 & 0 & 0 & 0 \\ & (N-1)^2+(N-2)^2 & -(N-2)^2 & & \\ & & (N-2)^2+(N-3)^2 & -(N-3)^2 \;\cdots & 0 \\ & & & & \cdots \\ & & & & 1 \end{pmatrix}$$

$$(Q'\Omega Q)^{-1} = [N(N-1)]^{-1}\begin{pmatrix} 1 & -1 \\ -1 & N \end{pmatrix},$$

$$(Q'\Omega Q)^{-1}Q'\Omega = [N(N-1)]^{-1}\begin{pmatrix} (N^2-1) & -1 & \cdots & -1 \\ -N(N-1) & N & \cdots & N \end{pmatrix}.$$

Therefore

$$\hat{\theta} - (NX_{1,N} - \bar{X})/(N-1)$$

and

$$\hat{\sigma} = N(\bar{X} - X_{1,N})/(N-1)\,,$$

which are in agreement with the maximum likelihood estimates, where $\bar{X}$ denotes the arithmetic mean. Since, the mean of the population is $\theta + \sigma$, the ordered least squares estimate is

$$(N-1)^{-1}(1,1)\begin{pmatrix} NX_{1,N} & -\bar{X} \\ N\bar{X} & -NX_{1,N} \end{pmatrix} = \bar{X}\,.$$

3.7 Approximations to the Best Linear Estimates

If X is a random variable with d.f. $F\left((x-\theta)/\sigma\right)$ and p.d.f. $f\left((x-\theta)/\sigma\right)/\sigma$ where θ and σ are parameters, expressions for best linear unbiased estimates of θ and σ using ordered observations have been obtained earlier. However, it is not easy to compute the best coefficients and we would like to approximate these. If the $\mu_{i,N}$ and the $\sigma_{i,j,N}$ are difficult to compute, A.K. Gupta (1952), for censored samples of sizes greater than 10 from normal populations, suggested that the variance- covariance matrix of the order statistics be an identity matrix. Sarhan and Greenberg (1956, 1958) found the alternative estimates obtained when Ω^{-1} is the identity matrix to be highly efficient. The alternative estimates of θ and σ for any populations are given by (when Ω^{-1} is replaced by the identity matrix)

$$\begin{aligned} \hat{\theta} &= \left\{(\boldsymbol{\mu}'\boldsymbol{\mu})\mathbf{1}' - (\mathbf{1}'\boldsymbol{\mu})\boldsymbol{\mu}'\right\}/\Delta\,, \\ \hat{\sigma} &= \left\{(\mathbf{1}'\mathbf{1}\boldsymbol{\mu}' - (\mathbf{1}'\boldsymbol{\mu})\mathbf{1}'\right\}/\Delta\,, \end{aligned}$$

where

$$\Delta = (\mathbf{1}'\mathbf{1})(\boldsymbol{\mu}'\boldsymbol{\mu}) - (\mathbf{1}'\boldsymbol{\mu})^2\,.$$

Plackett (1958) proposed that $\sigma_{i,j,N}$ could be asymptotically replaced by

$$p_i(1-p_j)/Nf(\mu_{i,N})f(\mu_{j,N})$$

where $p_i = F(\mu_{i,N})$, $i < j = 1 + r, \ldots, s$. Moreover, the inverse of such a matrix is easy to obtain and has been explicitly given by Hammersley and Morton (1954). (See also Plackett (1958), p. 134). Plackett also showed that the maximum likelihood estimates of θ and σ are asymptotically linear in the ordered observations in the censored sample. Thus, it follows that the asymptotic estimates of θ and σ obtained by substituting $p_i(1 - p_j)/Nf(\mu_{i,N})f(\mu_{j,N})$ in place of $\sigma_{i,j,N}$ for $i, j = 1 + r, \ldots, s$ are asymptotically normal and efficient. If it is difficult to compute the $\mu_{i,N}$ also, then one can use the following asymptotic expansion of $X_{i,N}$ in powers of $\{F(X_{i,N}) - i/(N + 1)\}$ provided i is not near the extremes and compute the asymptotic values for the moments of order statistics.

Result 3.7.1 (Clark and Williams, 1958, and Johnson and David, 1954). *If the inverse function of the population d.f. $F(x)$ can be expanded in Taylor series, then*

$$\begin{aligned} X_{i,N} &= Z_{i,N} + Z'_{i,N} \{F(X_{i,N}) - i/(N + 1)\} \\ &\quad + \frac{1}{2} Z''_{i,N} \{F(X_{i,N}) - i/(N + 1)\}^2 + \cdots \end{aligned}$$

in probability as $N \to \infty$, where

$$F(Z_{i,N}) = i/(N + 1), \quad i = 1, 2, \ldots, N\,,$$

and

$$Z'_{i,N} = \left.\frac{dZ}{dF}\right|_{Z=Z_{i,N}}, \quad Z''_{i,N} = \left.\frac{d^2 Z}{dF^2}\right|_{Z=Z_{i,N}}, \quad \textit{etc.}$$

Asymptotic values for the coefficients of the ordered observations have also been obtained by Jung (1956). Suppose we are interested in estimating $\theta^* = b\theta + c\sigma$ using ordered observations in the complete samples of size N. Let us write

$$\theta^* = \sum_{i=1}^{N} a_{i,N} X_{i,N}\,.$$

We would like to approximate the best coefficients $a_{i,N}$ with the functions of i and N depending on $F(x)$ only, $i = 1, 2, \ldots, N$. Jung approached this problem by considering

$$a_{i,N} \sim N^{-1} a\left(i/(N + 1)\right)\,,$$

where $a(u)$is a continuous differentiable function defined in the interval (0,1). Then we have the following result.

Result 3.7.2 (Jung, 1956). *If $f(y)$ and $yf(y)$ tend to zero as y approaches the finite or infinite end points of the distribution, and if $a(u)$ and its first four derivatives exist and are bounded in the interval (0,1), then*

$$E(\hat{\theta}^*) = M(a) + O(N^{-1}),$$

$$var\,\hat{\theta}^* = \sigma^2 V(a) N^{-1} + 0(N^{-2})$$

where

$$\begin{aligned} M(a) &= \int_{-\infty}^{\infty} (\theta + \sigma x) a\left(F(x)\right) f(x) dx \\ V(a) &= \int_{-\infty}^{\infty}\int_{-\infty}^{\infty} K(x,y) a\left(F(x)\right) a\left(F(y)\right) dy\, dx\,, \\ K(x,y) &= \begin{cases} F(x)\left[1 - F(y)\right], & x < y, \\ \left[1 - F(x)\right] F(y), & x > y, \end{cases} \end{aligned}$$

and the remainder terms $0(N^{-1})$ and $0(N^{-2})$ depend on the upper bounds of the derivatives of $a(u)$.

PROOF: Jung employed Taylor expansions.

Result 3.7.3 (Jung, 1956). *Let $f(y)$ and $yf(y)$ tend to zero as y approaches the finite or infinite end points of the distribution. The asymptotic best linear estimate of θ^* is given by*

$$\hat{\theta}^* = N^{-1} \sum_{i=1}^{N} a_o\left(i/(N+1)\right) X_{i,N}$$

with

$$a_o\left(F(x)\right) \equiv A_o(x) = e_1 \gamma_1'(x) + e_2 \gamma_2'(x)$$

where

$$\begin{pmatrix} e_1 \\ e_2 \end{pmatrix} = \begin{pmatrix} d_{11} & d_{12} \\ d_{21} & d_{22} \end{pmatrix}^{-1} \begin{pmatrix} b \\ c \end{pmatrix}$$

$$\gamma_1(x) = -f'(x)/f(x), \;\; \gamma_2(x) = -1 - x\frac{f'(x)}{f(x)}\,,$$

and

$$d_{l,k} = \int_{-\infty}^{\infty} \gamma_l(x)\gamma_k(x) f(x) dx, \;\; l,k = 1,2\,.$$

The resulting variance is

$$var(\hat{\theta}^*) = N^{-1}\sigma^2 \cdot (b,c)\begin{pmatrix} d_{11} & d_{12} \\ d_{21} & d_{22} \end{pmatrix}^{-1}\begin{pmatrix} b \\ c \end{pmatrix} + 0(N^{-2}).$$

PROOF: We wish to choose the function $a(u)$ such that among all linear estimates $\hat{\theta}$ satisfying $E(\hat{\theta}^*) = M(a)$, we consider that the estimate $\hat{\theta}^* = a_o(u)$ is asymptotically best for which $V(a)$ is minimum. That is, we wish to minimize

$$V(a) = \int_{-\infty}^{\infty}\int_{-\infty}^{\infty} K(x,y)A(x)A(y)dy\, dx$$

with respect to $A(x) = a\,(F(x))$ subject to the conditions

$$\int_{-\infty}^{\infty} A(x)f(x)dx = b\,,$$

and

$$\int_{-\infty}^{\infty} A(x)xf(x)dx = c\,.$$

By calculus of variations, the solution obtained is

$$A_o(x) = e_1A_1(x) + e_2A_2(x)$$

where $A_1(x)$ and $A_2(x)$ are the unique solutions of the integral equations

$$\int_{-\infty}^{\infty} K(x,y)A_1(y)dy = f(x)\,,$$

and

$$\int_{-\infty}^{\infty} K(x,y)A_2(y)dy = xf(x)\,.$$

After differentiating with respect to x on both sides the integral equations become

$$\int_{x}^{\infty} [1 - F(y)]\, A_1(y)dy - \int_{-\infty}^{x} F(y)A_1(y)dy = -\gamma_1(x)$$

and

$$\int_{x}^{\infty} [1 - F(y)]\, A_2(y)dy - \int_{-\infty}^{x} F(y)A_2(y)dy = -\gamma_2(x)\,.$$

Differentiating again with respect to x on both sides, the above equations yield

$$A_1(x) = \gamma_1'(x) \text{ and } A_2(x) = \gamma_2'(x)\,.$$

Now, substitute $e_1\gamma_1'(x) + e_2\gamma_2'(x)$ for $A(x)$ in the side conditions obtained by setting $E(\hat{\theta}^*) = M(a)$. Also integrating by parts once and using the hypothesis that $f(y)$ and $yf(y)$ tend to zero at the finite and infinite end points of the distribution, one can obtain

$$\int \gamma_1'(x)f(x)dx = \int \gamma_1(x)\gamma_1(x)f(x)dx = d_{11}\,,$$

$$\int \gamma_1'(x)xf(x)dx = \int \gamma_1(x)\gamma_2(x)f(x)dx = d_{12}\,,$$

$$\int \gamma_2'(x)f(x)dx = \int \gamma_2(x)\gamma_1(x)f(x)dx = d_{21}\,,$$

and

$$\int \gamma_2'(x)xf(x)dx = \int \gamma_2(x)\gamma_2(x)f(x)dx = d_{22}\,,$$

where the range of integration is $(-\infty, \infty)$. Solving for e_1 and e_2, one obtains

$$\begin{pmatrix} e_1 \\ e_2 \end{pmatrix} = \begin{pmatrix} d_{11} & d_{12} \\ d_{21} & d_{22} \end{pmatrix}^{-1} \begin{pmatrix} b \\ c \end{pmatrix}.$$

Then,

$$\begin{aligned} V_o(a) &= \int_{-\infty}^{\infty}\int_{-\infty}^{\infty} K(x,y)A_o(x)A_o(y)dx\,dy \\ &= \int_{-\infty}^{\infty} A_o(x)\left[\int_{-\infty}^{\infty} K(x,y)\left\{e_1A_1(y) + e_2A_2(y)\right\}dy\right]dx \\ &= \int_{-\infty}^{\infty} A_o(x)\left\{e_1f(x) + e_2xf(x)\right\}dx \\ &= e_1b + e_2c = (b,c)\begin{pmatrix} d_{11} & d_{12} \\ d_{21} & d_{22} \end{pmatrix}^{-1} \begin{pmatrix} b \\ c \end{pmatrix}. \end{aligned}$$

Thus, the variance of $\hat{\theta}^*$ is given by

$$\text{var}(\hat{\theta}^*) = \sigma^2 V_o(a)N^{-1} + O(N^{-2})\,.$$

The asymptotic efficiency of $\hat{\theta}^*$

If the p.d.f. $f\left((x-\theta)/\sigma\right)/\sigma$ is denoted by $g(x;\theta,\sigma)$, it can easily be shown that

$$\begin{pmatrix} d_{11} & d_{12} \\ d_{21} & d_{22} \end{pmatrix} = E\begin{pmatrix} \dfrac{\partial^2 \ln\, g}{\partial\theta^2} & \dfrac{\partial \ln\, g}{\partial\theta}\dfrac{\partial \ln\, g}{\partial\sigma} \\ \dfrac{\partial \ln\, g}{\partial\sigma}\dfrac{\partial \ln\, g}{\partial\theta} & \dfrac{\partial^2 \ln\, g}{\partial\sigma^2} \end{pmatrix}.$$

Hence, it follows that the estimates $\hat{\theta}$ and $\hat{\sigma}$ are asymptotically jointly efficient, and each of them is efficient when the other is regarded as a nuisance parameter, with efficiency measured according to Cramér (1947). If one of the parameters is known, the arguments and the expressions will take simpler forms.

Example 3.7.1. Consider the Student distribution with ν degrees of freedom.

$$f(x) = (\nu\pi)^{-\frac{1}{2}}\frac{\Gamma\left((\nu+1)/2\right)}{\Gamma(\nu/2)}\left[1+(x^2/\nu)\right]^{-(\nu+1)/2} \quad -\infty < x < \infty\,.$$

Computations yield

$$\begin{aligned} \gamma_1(x) &= (\nu+1)x/\nu\left[1+(x^2/\nu)\right] \\ \gamma_2(x) &= (\nu+1)x^2/\nu\left[1+(x^2/\nu)\right] - 1 \\ A_1(x) &= a_1\left(F(x)\right) = (\nu+1)\left\{1-(x^2/\nu)\right\}/\nu\left\{1+(x^2/\nu)\right\}^2 \\ A_2(x) &= a_2\left(F(x)\right) = 2(\nu+1)x/\nu\left\{1+(x^2/\nu)\right\}^2. \end{aligned}$$

Substituting $1+(x^2/\nu) = \xi^{-1}$, we may easily compute

$$d_{11} = (\nu+1)/(\nu+3), \quad d_{12} = d_{21} = 0 \text{ and } d_{22} = 2\nu/(\nu+3)\,.$$

Thus, for the location parameter θ, we obtain ($b=1$, $c=0$)

$$\begin{pmatrix} e_1 \\ e_2 \end{pmatrix} = \begin{pmatrix} (\nu+1)/(\nu+3) & 0 \\ 0 & 2\nu/(\nu+3) \end{pmatrix}^{-1}\begin{pmatrix} 1 \\ 0 \end{pmatrix}.$$

That is,

$$e_1 = (\nu+3)/(\nu+1), \quad e_2 = 0\,,$$

$$\hat{\theta} = N^{-1}\sum_{i=1}^{N} a_o\left(i/(N+1)\right)X_{i,N}$$

and

$$\text{var}(\hat{\theta}) = \left\{\sigma^2(\nu+3)/N(\nu+1)\right\} + 0(N^{-2}),$$

where

$$a_o\left(F(x)\right) = A_o(x) = (\nu+3)\left[1-(x^2/\nu)\right]/\nu\left[1+(x^2/\nu)\right].$$

In similar manner, one can obtain

$$\hat{\sigma} = N^{-1}\sum_{i=1}^{N} a_o^*\left(i/(N+1)\right)X_{i,N}$$

and

$$\text{var}(\hat{\sigma}) = \left\{\sigma^2(\nu+3)/2N\nu\right\} + 0(N^{-2})$$

where

$$a_0^*\left(F(x)\right) = A_o^*(x) = 2(\nu+3)(\nu+1)x/\nu^2\left[1+(x^2/\nu)\right]^2.$$

3.8 Unbiased Nearly Best Linear Estimates

Blom (1958) considered 'nearly' best estimates, namely those estimates with nearly minimum variance, as approximations to the best estimates. He retained the condition of unbiasedness, although later on he considered nearly unbiased and nearly best estimates. In the following we briefly present Blom's (1958) results, without providing the proofs. As before, let the d.f. and the p.d.f. of the random variable X be given by $F\left((x-\theta)/\sigma\right)$ and $f\left((x-\theta)/\sigma\right)/\sigma$ respectively. Let X denote the ith fractile of the reduced random variable $Y = (X-\theta)/\sigma$ with the d.f. $F(x)$, that is, $F(\lambda_i) = p_i = i/(N+1)$. The covariance of any two order statistics $Y_{i,N}$ and $Y_{j,N}$ where $Y_{i,N} = (X_{i,N}-\theta)/\sigma$, $i = 1, 2, \ldots, N$, may be written as

$$\text{cov}(Y_{i,N}, Y_{j,N}) = \frac{p_i(1-p_j)}{(N+2)f(\lambda_i)f(\lambda_j)} + R_{i,j},\quad i \le j$$

where under general conditions the error term $R_{i,j}$ tends to zero when $N \to \infty$ and $i/(N+1) \to b_1$, $j/(N+1) \to b_2$ $(0 < b_1 < b_2 < 1)$. Then, if N is large, we have approximately

$$\text{cov}\left[f(\lambda_i)Y_{i,N}, f(\lambda_j)Y_{j,N}\right] \sim p_i(1-p_j)/(N+2).$$

Setting $Z_{i,N} = f(\lambda_{i+1})Y_{i+1,N} - f(\lambda_i)Y_{i,N}$, $i = 0, 1, \ldots, N$, with $f(\lambda_o) = f(\lambda_{N+1}) = 0$, we find that

$$\mathrm{var}(Z_{i,N}) \sim N(N+1)^{-2}(N+2)^{-1}$$

after writing

$$Z_{iN} \doteq f(\lambda_{i+1})F^{-1}(p_{i+1}) - f(\lambda_i)F^{-1}(p_i) + (U_{i+1,N} - U_{i,N} - \frac{1}{N+1})$$

and

$$\mathrm{cov}(Z_{i,N}, Z_{j,N}) \sim -(N+1)^{-2}(N+2)^{-1}.$$

Note that the above asymptotic expressions for the variances and the covariances of the $Z_{i,N}$ are independent of $F(x)$. Now, consider the estimation of an unknown parameter $\theta^* = b\theta + c\sigma$, where b and c are known constants. Any linear estimate $\hat{\theta}$ of θ^* can be written as

$$\hat{\theta}^* = \sum_{i=1}^{N} a_{i,N}X_{i,N} = \sum_{i=1}^{N} a_{i,N}(\theta + \sigma Y_{i,N}).$$

Now, let us write the $Y_{i,N}$ in terms of the $Z_{i,N}$, and introduce the new set of coefficients $b_{i,N}$ defined, apart from an additive constant, by

$$a_{i,N} = f(\lambda_i)(b_{i,N} - b_{i-1,N}), \quad i = 1, 2, \ldots, N.$$

Also, write, for $i = 0, 1, \ldots, N$,

$$\begin{aligned} C_{1,i} &= f(\lambda_i) - f(\lambda_{i+1}) \\ C_{2,i} &= f(\lambda_i)\mu_{i,N} - f(\lambda_{i+1})\mu_{i+1,N}, \end{aligned}$$

where $\mu_{i,N} = E(Y_{i,N})$. Notice that $\sum_0^N C_{1,i} = \sum_0^N C_{2,i} = 0$. Then, one can write

$$\hat{\theta}^* = \sum_{i=0}^{N} b_{i,N}\left[\theta\left\{f(\lambda_i) - f(\lambda_{i+1})\right\} - \sigma Z_{i,N}\right]$$

since $\sum d_i(b_i - b_{i-1}) = \sum b_i d_i - \sum b_i d_{i+1} = \sum(d_i - d_{i+1})b_i$. Hence,

$$E(\hat{\theta}^*) = \theta\sum_{i=0}^{N} C_{1,i}b_{i,N} + \sigma\sum_{i=0}^{N} C_{2,i}b_{i,N}.$$

Consequently,

$$\mathrm{var}(\hat{\theta}^*) = \sigma^2\left[\sum_{i=0}^{N} b_{i,N}^2 \mathrm{var}(Z_{i,N}) + \sum_{i\neq j}\sum b_{i,N}b_{j,N}\mathrm{cov}(Z_{i,N}, Z_{j,N})\right].$$

Using the approximate expressions for the variances and covariances of the $Z_{i,N}$, we obtain

$$\operatorname{var}(\hat{\theta}^*) \sim \left[\sigma^2/(N+1)(N+2)\right] \sum_{i=0}^{N} (b_{i,N} - \bar{b})^2$$

where $\bar{b} = \sum_{i=0}^{N} b_{i,N}/(N+1)$. Thus the original least squares problem has been reduced to one that is easier to handle, although the demand for exact solution has to be given up since the error terms are discarded. We will minimize $\sum_{i=0}^{N}(b_{i,N} - \bar{b})^2$ subject to the conditions

$$\sum_{i=0}^{N} C_{1,i} b_{i,N} = b \text{ and } \sum_{i=0}^{N} C_{2,i} b_{i,N} = c$$

which are the consequences of $\hat{\theta}^*$ being an unbiased estimate of θ^*.

The solutions are

$$b_{i,N} = \bar{b} + l_1 C_{1,i} + l_2 C_{2,i}$$

where the two Lagrangian multipliers l_1 and l_2 are given by

$$l_1 = e^{11} b + e^{12} c$$
$$l_2 = e^{21} b + e^{22} c\,,$$

where the matrix $D^{-1} = (e^{rs})$ is the inverse of the matrix $D = (e_{rs})$ and

$$e_{rs} = \sum_{i=0}^{N} C_{r,i} C_{s,i}\,, \quad r, s = 1, 2\,.$$

That is

$$\begin{pmatrix} b_{1,N} - b_{0,N} \\ b_{2,N} - b_{1,N} \\ \vdots \\ b_{N,N} - b_{N-1,N} \end{pmatrix} = \begin{pmatrix} C_{1,1} - C_{1,0} & C_{2,1} - C_{2,0} \\ & \vdots \\ C_{1,2} - C_{1,1} & C_{2,2} - C_{2,1} \\ C_{1,N} - C_{1,N-1} & C_{2,N} - C_{2,N-1} \end{pmatrix} D^{-1} \binom{b}{c}.$$

Thus, we can obtain the $a_{i,N} = f(\lambda_i)(b_{i,N} - b_{i-1,N})$, $i = 1, 2, \ldots, N$, and hence the unbiased nearly best linear estimate. In particular, letting $b = 1$,

$c = 0$ and $b = 0$, $c = 1$, we obtain the following unbiased nearly best estimates of θ and σ respectively

$$\hat{\theta} = \sum_{i=1}^{N} a_{1,i,N} X_{i,N}$$

and

$$\hat{\sigma} = \sum_{i=1}^{N} a_{2,i,N} X_{i,N}$$

where

$$a_{r,i,n} = f(\lambda_i)\left[e^{r1}(C_{1,i} - C_{1,i-1}) + e^{r2}(C_{2,i} - C_{2,i-1})\right], \; r = 1, 2\,.$$

The variances are approximately given by

$$\begin{aligned} \operatorname{var}(\hat{\theta}) &\sim \sigma^2 e^{11}(N+1)^{-1}(N+2)^{-1}, \\ \operatorname{var}(\hat{\sigma}) &\sim \sigma^2 e^{22}(N+1)^{-1}(N+2)^{-1}, \end{aligned}$$

and the covariance is given by

$$\operatorname{cov}(\hat{\theta}, \hat{\sigma}) \sim \sigma^2 e^{12}(N+1)^{-1}(N+2)^{-1}\,.$$

In order to obtain the unbiased nearly best linear estimates, the following steps may be taken:

1. Compute $f(\lambda_i)$;
2. Compute $f(\lambda_i)\mu_{i,N}$;
3. Compute the elements e_{rs} of D and hence the elements of D^{-1};
4. Insert the numerical values in the expressions for $a_{r,i,N}$ $(r = 1, 2)$.

Remark 3.8.1. *The above method can be used both when the p.d.f. $f(x)$ is continuous and when it is discontinuous, for example, the rectangular and exponential distributions which are discontinuous at the end points of the range of variation.*

Remark 3.8.2. *If σ is known, then*

$$\hat{\theta} = \sum_{i=1}^{N} [f(\lambda_i)(C_{1,i} - C_{1,i-1})/e_{11}] - \sigma e_{12}/e_{11}\,.$$

If θ is known, then

$$\hat{\sigma} = \sum_{i=1}^{N} [f(\lambda_i)(C_{2,i} - C_{2,i-1})/e_{22}] - \theta e_{12}/e_{22}\,.$$

Notice that $e_{12} = 0$ if the distribution is symmetric about zero.

Unbiased nearly best estimates when the samples are censored

If the samples are censored, the preceding procedure holds true with the following modification. We have $b_{i,N} = b_{i-1,N}$ corresponding to any censored or missing observation $X_{i,N}$ since $a_{i,N}$ is made to be zero. The $C_{1,i}$ and the $C_{2,i}$ should be replaced by

$$C_{1,i}^{*} = \begin{cases} -f(\lambda_r)/r, & 0 \le i \le r-1, \\ C_{1,i}, & r \le i \le s-1, \\ f(\lambda_s)/(N-s+1), & s+1 \le i \le N, \end{cases}$$

$$C_{2,i}^{*} = \begin{cases} -r^{-1} f(\lambda_r)\mu_{r,N}, & 0 \le i \le r-1, \\ C_{2,i}, & r \le i \le s, \\ (N-s+1)^{-1} f(\lambda_s)\mu_{s,N}, & s+1 \le i \le N, \end{cases}$$

and the same procedure be adopted.

The efficiency of nearly best linear estimates

Blom (see Sarhan and Greenberg (1962)) has shown that the unbiased nearly best linear estimates are asymptotically and jointly efficient.

Definition 3.8.1. *If V is the variance-covariance matrix of any two estimates $\hat{\theta}$ and $\hat{\sigma}$ of θ and σ in the d.f. $F((x-\theta)/\sigma)$, then the generalized variance of $\hat{\theta}$ and $\hat{\sigma}$ is defined to be the determinant of V.*

Definition 3.8.2 (Cramér and Rao). *The estimates $\hat{\theta}$and $\hat{\sigma}$ of θ are said to be asymptotically jointly efficient if the determinant of V is asymptotically equivalent to the determinant of the matrix*

$$N^{-1} \begin{pmatrix} E\left(\dfrac{\partial \ln f_1}{\partial \theta}\right)^2 & E\left(\dfrac{\partial \ln f_1}{\partial \theta}\dfrac{\partial \ln f_1}{\partial \sigma}\right) \\ E\left(\dfrac{\partial \ln f_1}{\partial \sigma}\dfrac{\partial \ln f_1}{\partial \theta}\right) & E\left(\dfrac{\partial \ln f_1}{\partial \sigma}\right)^2 \end{pmatrix},$$

where $f_1(x) = f((x-\theta)/\sigma)/\sigma$.

In the case of unbiased nearly best estimates, since

$$f(\lambda_i) - f(\lambda_{i+1}) = N^{-1} f'(\lambda_i)/f(\lambda_i),$$

one can approximately write

$$e_{11} \sim N^{-1} \int_0^1 \left[f'(x)/f(x)\right]^2 du$$

where $u = F(x)$ or

$$e_{11} \sim N^{-1} \int_{-\infty}^{\infty} \left(\frac{\partial \ln f}{\partial x}\right)^2 f(x)dx = (\sigma^2/N)E\left(\frac{\partial \ln f_1}{\partial \theta}\right)^2 .$$

Similarly, by assuming that $xf(x)$ tends to zero at the end points of the distribution, it can be shown that

$$\begin{aligned} e_{12} &\sim N^{-1} \int_{-\infty}^{\infty} \left(\frac{\partial \ln f}{\partial x}\right)\left(1 + x\frac{\partial \ln f}{\partial x}\right) f(x)dx \\ &= (\sigma^2/N)E\left\{\frac{\partial \ln f_1}{\partial \theta}\frac{\partial \ln f_1}{\partial \sigma}\right\} \end{aligned}$$

and

$$e_{22} \sim N^{-1} \int_{-\infty}^{\infty} \left(1 + x\frac{\partial \ln f}{\partial x}\right)^2 f(x)dx = (\sigma^2/N)E\left\{\frac{\partial \ln f_1}{\partial \sigma}\right\}^2$$

where $f_1(x) = \sigma^{-1} f\left((x-\theta)/\sigma\right)$. Thus, the generalized variance of the unbiased nearly best estimates of θ and σ is given by

$$\operatorname{var}(\hat{\theta}) \operatorname{var}(\hat{\sigma}) - \left[\operatorname{cov} \hat{\theta}, \hat{\sigma}\right]^2 \sim \sigma^4 N^{-2}(e_{11}e_{22} - e_{12}^2)^{-1},$$

which after substituting the asymptotic expressions for the e_{rs} attains the Cramér-Rao lower bound. Hence, the unbiased nearly best estimates are asymptotically jointly efficient.

3.9 Nearly Unbiased and Nearly Best Estimates

When the $\mu_{i,N}$ are not available, Blom (1958) proposed to approximate the $\mu_{i,N}$ by

$$\mu_{i,N} \sim F^{-1}\left\{(i - \beta)/(N - 2\beta + 1)\right\},$$

$$\{\text{or } \beta = [(N+1)F(\mu_{i,N}) - i] \,/\, [2F(\mu_{i,N}) - 1]\}$$

where β is a suitably chosen constant, Blom (1958) suggests that approximations of reasonable accuracy may be obtained by using $\beta = 3/8$ if F is the normal d.f. Harter (1961) tabulates the value of β for each i and N in large samples drawn from the standard normal population. If these approximations for the $\mu_{i,N}$ are employed in the procedure for obtaining the unbiased nearly best estimates, we obtain the nearly unbiased, nearly best estimates.

Ogawa (1951) by considering the asymptotic distribution of the sample quantiles, obtained the least squares estimates of the location and scale parameters in terms of the sample quantiles.

Concluding Remarks

Either best, nearly best or nearly unbiased and nearly best estimation using ordered observations in the complete or censored samples provides an alternative procedure to the method of maximum likelihood. However, the classical method of maximum likelihood is applicable to discrete distributions and estimation of any parameters, whereas the least squares estimation using ordered observations concerns only estimation of location and scale parameters. Also, tables of coefficients of best or nearly best linear estimates must be prepared for the various distributions.

3.10 Inversion of a Useful Matrix

Consider the matrix

$$NV_{k\times k} = ((\lambda_i(1-\lambda_j)/f(u_i)f(u_j))) \; .$$

The inverse of this matrix was obtained by Hammersley and Morton (1954) and also used by Plackett (1958). Dyer (1973) gave a simple proof of this which is given in the following.

Result 3.10.1. $(NV)^{-1} = \left((v^{ij})\right)$ *where, for* $= 1, 2, \ldots, k$,

$$\begin{aligned} v^{ii} &= [f(u_i)]^2 \left[(\lambda_{i+1}-\lambda_i)^{-1} + (\lambda_i - \lambda_{i-1})^{-1}\right] \\ v^{i,i+1} &= v^{i+1,i} = -f(u_i)f(u_{i+1})(\lambda_{i+1}-\lambda_i)^{-1} \\ & \qquad v^{ij} = v^{ji} = 0 \text{ for } j = i+2, i+3, \ldots, k \, . \end{aligned}$$

PROOF: It can easily be verified that

$$nV = D_1 T D_2 T' D_1$$

where D_1 is a diagonal matrix whose i^{th} diagonal element is $(1-\lambda_i)/(1-\lambda_1)f(u_i)$, T is a lower triangular matrix with each element on and below the diagonal equal to one, and D_2 is a diagonal matrix whose i^{th} diagonal element is $(1-\lambda_1)^2(\lambda_i-\lambda_{i-1})/(1-\lambda_i)(1-\lambda_{i-1})$. Then $N^{-1}V^{-1} = D_1^{-1}(T')^{-1}D_2^{-1}T^{-1}D_1^{-1}$ and upon simplification we get the desired result.

Result 3.10.2. *$\lambda_i(1-\lambda_j)/Nf(u_i)f(u_j)$ denotes the asymptotic covariance of $X_{[N\lambda_i]+1}$, and $X_{[N\lambda_j]+1}$. Also,*

$$\begin{aligned} N^{-1}u'V^{-1}u &= \sum_{i=1}^{k} u_i^2 v^{ii} + 2\sum_{i=1}^{k-1} u_i u_{i+1} v^{i,i+1} \\ &= \sum_{i=1}^{k} \frac{[u_i f(u_i)]^2}{(\lambda_i-\lambda_{i-1})} \\ &+ \frac{[u_i f(u_i)]^2 - 2u_i\ u_{i+1} f(u_i) f(u_{i+1})}{(\lambda_{i+1}-\lambda_i)}, \end{aligned}$$

where the upper limit on the Index of summation has been extended to k by noting that $\lambda_{k+1}=1$, so that $u_{k+1}\to\infty$ and $u_{k+1}f(u_{k+1})\to 0$.

So, write

$$\begin{aligned} &[u_i f(u_i)]^2 - 2u_i u_{i+1} f(u_i) f(u_{i+1}) \\ &\quad = [u_{i+1} f(u_{i+1}) - u_i f(u_i)]^2 - [u_{i+1} f(u_{i+1})]^2 . \end{aligned}$$

Then

$$(1/N)u'V^{-1}u = \sum_{i=0}^{k} \frac{[u_{i+1} f(u_{i+1}) - u_i f(u_i)]^2}{(\lambda_{i+1}-\lambda_i)}$$

where the u_i are the population quantiles of the standardized density.

3.11 Problems

3.4.1 If the density function is of the type of $f(x-\theta)$, show that the best ordered least squares estimate of θ and its variance are given by the expressions given at the end of Section 3.4.

3.4.2 Suppose the following random sample of size $n = 10$ was drawn from the uniform density

$$\begin{aligned} f(x) &= \frac{1}{\sigma}, && \textit{for } \theta < x < x+\sigma \\ &= 0 && \textit{elsewhere.} \end{aligned}$$

3.08, 3.98, 3.70, 3.48, 4.88, 4.50, 4.26, 3.76, 3.90, 4.72.
Assume that the standard uniform order statistics in a sample of size N have

$$\mu_{i,N} = E(U_{i,N}) = \tfrac{i}{N+1} \text{ and}$$
$$\text{for } i \leq j, \quad \sigma_{i,j,N} = \text{cov}(U_{i,N}, U_{j,N}) = \tfrac{i(N-j+1)}{(N+1)^2(N+2)}$$

find the best linear ordered estimates of θ and σ and their covariance matrix.

(HINT: Use Example 3.6.1 for the inverse of the covariance matrix $((\sigma_{i,j,N}))$. Also note that the true values of θ and σ are 3 and 2 respectively.)

3.4.3 Let 4.828, 1.070, 2.140, 0.093, 0.432, 1.197, 0.226, 2.079, 3.453, 0.741 constitute a random sample of size 10 from the exponential density

$$f(x|\lambda) = \frac{1}{\lambda} e^{-\frac{x}{\lambda}}.$$

Find the best linear unbiased estimate of λ.

(HINT: Use Example 3.6.2 for the expected values of the exponential order statistics, their variance-covariance matrix and its inverse. Also note that the true value of λ is 1.5.)

3.4.4 Let 0.75, 3.95, 3.23, 3.59, 7.20 constitute a random sample of size 5 from a normal population with mean θ and standard deviation σ. Assuming that the vector of expected values $\mu' = (-1.163, -0.495, 0, 0.495, 1.163)$ and

$$\Sigma = \begin{pmatrix} 0.447 & 0.224 & 0.148 & 0.106 & 0.074 \\ & 0.312 & 0.208 & 0.150 & \\ & & 0.287 & & \\ & & & & \\ & & & & \end{pmatrix}$$

of the standard normal order statistics in a sample of size $N = 5$, obtain the best estimates of θ and σ and their covariance matrix. Also note that the true values of θ and σ are 4 and 2 respectively. The rest of the entries in Σ can be obtained from the relations $\sigma_{j,i,N} = \sigma_{i,j,N}$, $\sigma_{i,j,N} = \sigma_{N-j+1,N-i+1,N}$ for $i \leq j$.

3.7.1 For the data in Problem 3.4.2, obtain the approximate values of θ and σ assuming that the variance-covariance matrix of the order statistics is the identity matrix.

3.7.2 For the data in Problem 3.4.3, Obtain the approximate value of λ assuming that the var-cov. matrix of the order statistics is the identity matrix.

3.7.3 For the data in Problem 3.4.4, obtain the approximate values of θ and σ assuming that the var-cov. matrix of the order statistics is the identity matrix.

3.7.4 Carry out Jung's approximation to the estimate of the normal location parameter.

3.7.5 Carry out Jung's approximation to the estimate of the scale parameter λ of the following density function:

$$\begin{aligned} f(x|\lambda) &= \lambda^{-2} x e^{-\frac{x}{\lambda}}, && x, \lambda > 0 \\ &= 0 && \text{elsewhere.} \end{aligned}$$

3.8.1 Suppose X has the density function of the form $(1/\sigma)f(x/\sigma)$. Obtain Blom's unbiased nearly best estimate of σ, based on a random sample of size N.

3.8.2 In particular, suppose the density of X is of the form

$$\frac{1}{\sigma} f\left(\frac{x}{\sigma}\right) = \frac{1}{\sigma} e^{-\frac{x}{\sigma}}, \qquad x, \sigma > 0.$$

Obtain Blom's unbiased nearly best estimate of σ for the data given in Problem 3.4.3 for $N = 10$.

3.10.1 Using Result 3.10.1, evaluate the inverse of the variance-covariance matrix of the k quantiles of the standard uniform order statistics in a sample of size N.
(HINT: let $p_i = \frac{i}{N+1}$ and $f(x) = 1, 0 \leq x < 1$.)

Chapter 4

Interval Estimation and Tolerance Limits

4.1 Confidence Intervals for Quantiles

We will consider estimates of quantiles and functions of the population d.f. These estimates will not depend on the functional form of the d.f. sampled. Estimates of this kind will be called non-parametric estimators. As one can guess the basic random variables involved in nonparametric estimation will be the order statistics in the sample and the elementary coverages obtained from the order statistics. First let us consider infinite populations and determine confidence intervals for a given quantile ξ_p from the order statistics in a sample from the d.f. $F(x)$. We have the following result.

Result 4.1.1 (Wilks, 1962). *If $(X_1, X_2, \ldots, X_N)$ is a random sample from a continuous d.f. $F(x)$ and if $X_{k,N}$ and $X_{l,N}$ are the kth and l^{th} order statistics in the sample, then $(X_{k,N}, X_{l,N})$ is a confidence interval for the pth quantile ξ_p having confidence coefficient $I_p(k, N-k+1) - I_p(l, N-l+1)$, where*

$$I(r,s) = \frac{(r+s-1)!}{(r-1)!(s-1)!} \int_0^p x^{r-1}(1-x)^{s-1}dx\,.$$

PROOF: Since, for $k < l$,

$$\{X_{k,N} \le \xi_p\} = \{X_{k,N} \le \xi_p, X_{l,N} \ge \xi_p\} \cup \{X_{k,N} \le \xi_p \text{ and } X_{l,N} \le \xi_p\}\,.$$

We have

$$P(X_{k,N} \le \xi_p) = P(X_{k,N} \le \xi_p \le X_{l,N}) + P(X_{l,N} \le \xi_p)\,.$$

That is,

$$\begin{aligned}
P(X_{k,N} \le \xi_p \le X_{l,N}) &= P(X_{k,N} \le \xi_p) - P(X_{l,N} \le \xi_p) \\
&= P\left(F(X_{k,N}) \le p\right) - P\left(F(X_{l,N}) \le p\right) \\
&= \sum_{j=k}^{N} \binom{N}{j} p^j (1-p)^{N-j} - \sum_{j=l}^{N} \binom{N}{j} p^j (1-p)^{N-j} \\
&= \sum_{j=k}^{l-1} \binom{N}{j} p^j (1-p)^{N-j} \\
&= I_p(k, N-k+1) - I_p(l, N-l+1)\,, \\
&\doteq \Phi\left(\frac{l-\frac{1}{2}-Np}{\sqrt{Np(1-p)}}\right) - \Phi\left(\frac{k-\frac{1}{2}-Np}{\sqrt{Np(1-p)}}\right),
\end{aligned}$$

which is independent of $F(x)$. This completes the proof.

Since k and l are positive integers, for a given p one can set up a confidence interval for ξ_p by using two order statistics, the confidence coefficient being at least γ for some range of values of γ. Also, the two order statistics are arbitrary. Usually, for given γ, we would choose k and l such that $l-k$ is as small as possible. In setting confidence intervals for the median, namely $\xi_{0.5}$, one can choose k as large as possible such that

$$P(X_{k,N} \le \xi_{0.5} < X_{N-k+1,N}) \ge \gamma\,. \tag{4.1.1}$$

Nair (1940) has tabulated the values of k for $N = 6, 7, \ldots, 81$ and for $\gamma = 0.95$ and 0.99. Nair's table for $N = 6(1)30(5)80$ is given in Appendix II. The exact value of the probability $P(X_{k,N} < \xi_{0.5} < X_{N-k+1,N})$ is $1-2I_{0.5}(N-k+1, k)$, a result surmised independently by Thompson (1936) and Savur (1937). The confidence coefficient can be approximated by $1-2\Phi\left(\frac{k-\frac{1}{2}-N/2}{\sqrt{N}/2}\right)$. From this approximation, we obtain

$$k \le \frac{N+1}{2} + \frac{\sqrt{N}}{2}\Phi^{-1}\left(\frac{1-\gamma}{2}\right). \tag{4.1.2}$$

Noether (1948) gave a method of obtaining confidence limits for quantiles with the help of a confidence band for the unknown distribution function. The method is more transparent if central confidence intervals for the unknown quantile are desired.

4.2 Large Sample Confidence Intervals

Let $k(N) = [Np] + 1$, if Np is not an integer, otherwise$= Np$. We know that

$$P\left[\left|\frac{(X_{k,N} - \xi_p) f(\xi_p)}{\sqrt{p(1-p)/N}}\right| \le z\right] \to 2\Phi(z) - 1\,.$$

Also, one can replace $f(\xi_p)$ by $f(X_{k,N})$ because of Cramér's (1946, p. 254) lemma. Thus,

$$P\left[X_{k,N} - \frac{\sqrt{p(1-p)/N}}{f(X_{k,N})} \cdot Z \le \xi_p \le X_{k,N} + \frac{\sqrt{p(1-p)/N}}{f(X_{k,N})} \cdot Z\right] \doteq 2\Phi(z) - 1\,.$$

So, either knowledge of f at ξ_p or $f(x)$ is required. If $f(x)$ is unknown, one can estimate $f(X_{k,N})$ as follows:

$$f_N(y) = \left[F_N(y+h) - F_N(y-h)\right]/2h$$

where

$$y = X_{k,N} \text{ and } h = gN^{-1/5}\,,$$

and

$$g = \left[\frac{2}{9}\int_{-\infty}^{\infty} \left[f''(y)\right\}^2 dy\right]^{-1/5}\,.$$

The latter computation is based on nonparametric density estimation which is considered in more detail in Chapter 6.

We also give the alternative method of Wilks (1962, §11.2).

4.2.1 Wilks' (1962) Method

If N_1 denotes the number of observations in the sample of size N, which are less than or equal to ξ_p, then N_1 is a random variable having a binomial distribution with parameters N and p. For large N, N_1 has an asymptotic normal distribution having mean Np and variance $Np(1-p)$. Thus,

$$\lim_{N\to\infty} P\left[-z_{1-\gamma} \le \frac{N_1 - Np}{\{Np(1-p)\}^{\frac{1}{2}}} \le z_{1-\gamma}\right] = 1 - 2\gamma$$

where $z_{1-\gamma}$ is the $(1-\gamma)^{\text{th}}$ percentile on the standard normal distribution. Consequently, for large N, an approximate $(1-2\gamma)^{\text{th}}$ confidence interval

$(p_\gamma; \bar{p}_\gamma)$ for p is obtained by the set of all values of p satisfying the above inequality for fixed N, N_1 and $z_{1-\gamma}$; and p_γ and $\bar{p}_\gamma$ are the two solutions of

$$\frac{(N_1 - Np)^2}{Np(1-p)} = z_{1-\gamma}^2 \, .$$

Thus, for large N,

$$P(p_\gamma < p < \bar{p}_\gamma) \approx 1 - 2\gamma$$

which implies that

$$P(X_{[Np_\gamma],N} < \xi_p < X_{[N\bar{p}_\gamma],N}) = 1 - 2\gamma \, .$$

Hence, the order statistics $X_{[Np_\gamma],N}$ and $X_{[N\bar{p}_\gamma],N}$ constitute an $100(1-2\gamma)\%$ confidence interval for ξ_p where $[\cdot]$ denotes the largest integer contained in $(\cdot)$. However, p_γ and $\bar{p}_\gamma$ are functions of N_1 which we cannot observe. So, we replace N_1 by N_1^* such that $X_{N_1^*,N}$ estimates ξ_p; i.e., $N_1^* = [Np] + 1$.

For confidence intervals for quantile intervals and quantiles in finite populations, the reader is referred to Wilks (1962, §11.3 and 8.11.4).

Walsh's (1958) Approach

Let ξ_p denote the pth population quantile. The sample quantile is given by $X_{k,N}$ where $k = [Np] + 1$. Due to the asymptotic normality of sample quantiles, one can set up large-sample confidence intervals for the population quantile provided a consistent estimate of the standard deviation of $X_{k,N}$ is available:

$$P\left[X_{k,N} - \frac{z}{\sqrt{N}}\hat{\sigma}_{X_{k,N}} \le \xi_p \le X_{kN} + \frac{z}{\sqrt{N}}\hat{\sigma}_{X_{k,N}}\right] \doteq 1 - \alpha$$

where $\Phi(z) = 1 - \frac{\alpha}{2}$.

In the following, we shall present $\hat{\sigma}_{X_{k,N}}$ as obtained by Walsh (1958). We shall assume that the population density is analytic and nonzero at all points of interest. These restrictions on the density are met by the usual densities.

Lemma 4.2.1 (Walsh, 1958). *Let p $(0 < p < 1)$ be bounded away from 0 and 1 and let $X_{k,N}$ denote the samle quantile: Consider estimates for the standard deviation of $X_{k,N}$ of the form:*

$$a(X_{k+i,N} - X_{k-i,N})$$

where $i = \epsilon(N+1)^\alpha$, $0 \le \alpha < 1$ and ϵ is a bounded constant. Then the asymptotically unbiased estimator of $\sigma_{X_{k,N}}$ such that the order of the magnitude of *the bias is the same as the order of the magnitude of the standard deviation of the estimate, is given by*

$$\epsilon = 1, \quad i \doteq (N+1)^{4/5}$$

and

$$a = \frac{1}{2}(N+1)^{-3/10}\left[p(1-p)\right]^{1/2}.$$

PROOF: Expanding in Taylor's expansion around p, we obtain

$$\begin{aligned} X_{k\pm i,N} &= F^{-1}(U_{k\pm i,N}) \\ &= F^{-1}(p) + (U_{k\pm i,N} - p)\frac{1}{f(\xi_p)} - \frac{1}{2}(U_{k\pm i,N} - p)^2 \frac{f'(\xi_p)}{f^3(\xi_p)} + \cdots \end{aligned}$$

and taking expectations and using the results of David and Johnson (1954), namely

$$E(X_{k,N}) = \xi_p - \frac{p(1-p)f'(\xi_p)}{2(N+2)f^2(\xi_p)} + O(N^{-2})$$

and

$$\sigma_{X_{k,N}} = \frac{\{p(1-p)\}^{1/2}}{(N+1)^{1/2}f(\xi_p)} + O(N^{-3/2}).$$

we have

$$\begin{aligned} E(X_{k\pm i,N}) = \xi_p \pm \frac{\epsilon}{(N+1)^{1-\alpha}f(\xi_p)} &- \left[\frac{\epsilon^2}{(N+1)^{2-2\alpha}} + \frac{p(1-p)}{2(N+2)}\right]\frac{f'(\xi_p)}{f^3(\xi_p)} \\ &+ O(N^{-3+3\alpha}) + O(N^{-2+\alpha}) \end{aligned}$$

and consequently

$$\begin{aligned} E\left\{a(X_{k+i,N} - X_{k-i,N})\right\} &= \frac{2a\epsilon}{(N+1)^{1-\alpha}f(\xi_p)} \\ &\quad + O(aN^{-3+3\alpha}) + O(aN^{-2+\alpha}). \end{aligned}$$

Analogously, we obtain the standard deviation of

$$a(X_{k+i,N} - X_{k-i,N}) = a(2\epsilon)^{1/2}/(N+1)^{1-\alpha/2} f(\xi_p) + O(aN^{-3/2+\alpha}).$$

Now the problem is to determine ϵ, α and a suitably. The estimate will be asymptotically unbiased for $\sigma_{X_{k,N}}$ provided

$$2a\epsilon/(N+1)^{1-\alpha} = \{p(1-p)\}^{1/2}/(N+1)^{1/2}$$

or

$$a = \left(\frac{1}{2\epsilon}\right)\{p(1-p)\}^{1/2}(N+1)^{1/2-\alpha}.$$

Using this expression for a, we have

$$E\{a(X_{k+i,N} - X_{k-i,N})\} = \sigma_{X_{k,N}} + O(N^{-5/2+2\alpha}) + O(N^{-3/2}).$$

$$\sigma_{a(X_{k+i,N}-X_{k-i,N})} = O(N^{-(1+\alpha)/2}).$$

Thus, increasing α decreases the order of magnitude of the standard deviation of the estimate; it, however, increases the order of magnitude of the bias of the estimate. Hence, the order of the error is minimized when

$$-(1+\alpha)/2 = -5/2 + 2\alpha.$$

That is, $\alpha = 4/5$ appears to be the most desirable choice for α. In the standard deviation of the estimate, the parameter ϵ appears predominantly as the factor $\epsilon^{-1/2}$. In the bias of the estimator, the predominant factor is ϵ^2. Thus, setting $\epsilon^2 = \epsilon^{-1/2}$. We get $\epsilon = 1$ as a compromise choice for ϵ. Thus, we obtained the estimator with the desired properties.

Remark 4.2.1. *Examination of the expansions employed in the derivations suggest that the standard deviation estimates are satisfactory if* $p(1-p)(N+1)^{9/10} \geq 3$.

4.3 Tolerance Limits

A tolerance interval is a confidence interval about the probability concentration in a population determined from a sample. Paulson (1943) has shown that the connection between tolerance intervals and confidence intervals will arise if confidence limits are determined not for a parameter of the distribution but for a future observation. Also, Noether (1951) has shown the close connection between confidence intervals for the unknown probability p of a binomial distribution and tolerance intervals. The notion of a tolerance limit or interval was introduced by Shewhart (1931). One of the usual methods of specifying the quality of a manufactured product is to set limits within which a certain percentage of the manufactured items may be expected to lie. Contributions to the problem.of determining tolerance regions when the population d.f. is known to be normal or absolutely continuous have been made by Wilks (1941), Wald (1942), Wald and Wolfowitz (1946) and Mitra (1957). Significant contributions to distribution-free tolerance regions have been made by Wilks (1941 and 1942), Wald (1943), Robbins (1944), Tukey (1945, 1947 and 1948) and Fraser (1953 and 1956). Wald (1943) and Tukey (1947) extended the method of obtaining distribution-free tolerance limits to multivariate situations. Murphy (1948) presented graphs of minimum probable coverage by sample blocks determined by order statistics in a sample drawn from a continuous but unknown d.f. Somerville (1958) extended Murphy's (1948) results and tabularized the results in a manner which eliminates or minimizes interpolation. In the following, we will restrict ourselves to distribution-free tolerance regions.

Let $(X_1, X_2, \ldots, X_N)$ be a random sample from a continuous population having $F(x)$ for its d.f. Let $L_1(X_1, \ldots, X_N)$ and $L_2(X_1, \ldots < X_N)$ be any two observable and symmetric functions of $(X_1, X_2, \ldots, X_N)$ such that, for $0 < \beta < 1$,

$$P\left[F(L_2) - F(L_1) \geq \beta\right] \geq \gamma\,.$$

Definition 4.3.1. *The limits L_1 and L_2 are called tolerance limits. If, the distribution of the random variable $F(L_2) - F(L_1)$ which denotes the fraction of the population contained within the random limits L_1 and L_2 is independent of $F(x)$, then L_1 and L_2 are called distribution-free tolerance limits.*

Result 4.3.1 (Wilks, 1941 and 1942). *If L_1 and L_2 are order statistics in samples drawn from a continuous d.f. $F(x)$, then L_1, and L_2 are distribution-free tolerance limits.*

PROOF: If $L_1 = X_{r,N}$ and $L_2 = X_{N-s+1,N}$, then the proportion of the population covered by $(X_{r,N}, X_{N-s+1,N})$ is

$$U_{r,s} = F(X_{N-s+1,N}) - F(X_{r,N})$$

which is the sum of $N-s-r+1$ elementary coverages. Also, it is known that $U_{r,s}$ has a beta distribution with $N-s-r+1$ and $s+r$ for its parameters.

Thus, for given positive values of β and γ and $k = r + s$, one can find the smallest N such that

$$P(U_{r,s} \geq \beta) = 1 - I_\beta(N - k + 1, k) \geq \gamma\,,$$

where $I_x(p, q)$ is Karl Pearson's notation for the incomplete beta function. Hence, L_1 and L_2 constitute $100\beta\%$ distribution-free tolerance limits at probability level at least γ. This completes the proof of the result.

Notice that we get one-sided confidence intervals when either $r = 0$ or $s = 0$ for given k.

Scheffé and Tukey (1944) gave an approximate formula for N in terms of β, γ and k:

$$N \doteq \{(1+\beta)/4(1-\beta)\}\,\chi^2_{2k,1-\gamma} + \frac{1}{2}(k-1) \tag{4.3.1}$$

where $\chi^2_{2k,1-\gamma}$ is the $100(1-\gamma)$ percentile of the chi-square distribution with 2k degrees of freedom. For $0.005 < 1-\gamma < 0.1$ and $\beta \geq 0.9$, it is conjectured by Scheffé and Tukey (1944) that the error in (4.6.1) is less than one tenth of one percent and is always positive, that is, N is slightly overestimated by the approximate formula. Solving for β from the approximate formula, one obtains

$$\beta \doteq \left[4N - \chi^2_{2k,1-\gamma} - 2(k-1)\right]\left[\chi^2_{2k,1-\gamma} + 4N - 2(k-1)\right]^{-1}. \tag{4.3.2}$$

That is,

$$\beta \doteq 1 - (2N)^{-1}\chi^2_{2k,1-\gamma}\left\{1 + \frac{1}{4N}\chi^2_{2k,1-\gamma}\right\}.$$

Murphy (1948) points out that Scheffé and Tukey (1944) have also obtained

$$\beta \doteq (4N)^{-2}\left[\{(\chi^2_{2k,1-\gamma} - 2k)^2 + 16N(N-k)\}^{\frac{1}{2}} - (\chi^2_{2k,1-\gamma} - 2k)\right]^2. \tag{4.3.3}$$

Murphy (1948) has graphed values of β for $N = 1(1)10(10)100(100)500$, $\gamma = .90, .95, .99$ and for $k = 1(1)6(2)10(5)30(10)60(20)100$. When $r = s = 1$, that is, $k = 2$, then the sample size required in order to assert with confidence at least γ that 100β percent of the population is covered by the sample range is given by the equation

$$N\beta^{N-1} - (N-1)\beta^N \leq 1 - \gamma\,. \tag{4.3.4}$$

Birnbaum and Zuckerman (1949) have proposed a graphical and an iterative solution of (4.3.4). Values of N for given β and γ have been tabulated by

Dion (1951) and are given in Freeman (1963). These values of N for $\beta = .50$, .70, .80, .90, .95, .98, .99 and for $\gamma = .70$, .80, .90, .95, .98, .99 are presented in Appendix III.

Somerville (1958) has tabulated values of k for $\beta = .50$, .75, .90, .95., .99, $\gamma = .50$, .75, .90, ,95, .99 and for $N = 50(5)100(10)150$, 170, 200(100)1000. Somerville (1958) also tabulated values of γ for $\beta = .50$, .75, .90, .95, .99 and $N = 3(1)20(5)30(10)100$. Values of k for $\beta = .50$, .75, .90, ,95, $\gamma = .75$, .90, .95, .99 and for $N = 50(5)100(10)150$, taken from Somerville (1958) are presented in Appendix IV.

One might ask whether L_1 and L_2 can be any other functions of the sample observations such that $F(L_2) - F(L_1)$ is distribution-free. Robbins (1944) has shown that if $F(x)$ is absolutely continuous, then the only symmetric distribution-free tolerance limits are the order statistics. Scheffé and Tukey (1945) showed that if $F(x)$ is continuous, then L_1 and L_2 being order statistics is sufficient for them to be distribution-free tolerance limits. In the following we present the result due to Scheffé and Tukey (1945).

Result 4.3.2 (Scheffé and Tukey, 1945). *A sufficient condition for the joint distribution of* $F(L_1), F(L_2), \ldots, F(L_r)$ *(*$L_1, L_2, \ldots, L_r$*) are functions of a random sample from the population having the continuous d.f.* $F(x)$*) to be independent of* $F(x)$ *is that* $\{L_j\}$ *be a subset of the order statistics* $\{X_{i,N}\}$ *in the sample.*

PROOF: It will suffice to show that the joint distribution of $F(X_{1,N})$, $F(X_{2,N})$, ..., $F(X_{N,N})$ is independent of $F(x)$. However, it is well known that $F(X_{1,N})$, $F(X_{2,N})$, ..., $F(X_{N,N})$ constitute order statistics in a sample of size N drawn from the uniform distribution on (0,1) and consequently, their joint distribution is independent of $F(x)$. This completes the proof of the result.

Scheffé and Tukey (1945) also proved that the order statistics in a sample drawn from discontinuous distributions can consequently be used as distribution-free tolerance limits, In other words, if the d.f. is not continuous, the statement on tolerance limits should be interpreted as follows:

> The probability is at least γ that the population proportion included in the interval $[X_{r,N}, X_{N-s+1,N}]$ is at least β.

Hanson and Owen (1963) point this out and use it while constructing tolerance limits on an additional finite sample.

Robbins (1944) has shown that if the underlying distribution$F(x)$ is continuous and differentiable, then the only symmetric distribution-free tolerance limits (D.F.T.L.) are the order statistics in the random sample. In the following, with a slight modification of his proof, we will present the proof of this assertion assuming only the continuity of $F(x)$.

Lemma 4.3.1 (Robbins, 1944). *Let $X_1, \ldots, X_N$ denote a random sample from $F(x)$. A necessary condition that the continuous function $g(x_1, \ldots, x_n)$ be a distribution-free upper tolerance limit (D.F.U.T.L.) is that*

$$\hat{g}(x_1, \ldots, x_N) = \prod_{i=1}^{N} \{g(x_1, \ldots, x_N) - x_i\} \equiv 0 \,. \tag{4.3.5}$$

PROOF: We shall prove the necessity of the condition by deriving a contradiction to the assumption that g is a D.F.U.T.L. for which there exist distinct numbers $a_1, \ldots, a_N$ such that $g(a_1, \ldots, a_N) = A \neq a_i$ for some i $(i = 1, \ldots, N)$.

Since the numbers $a_1, \ldots, a_N$ and A are distinct, there will exist a positive number ϵ such that the $(N+1)$ intervals

$$\begin{aligned} I:&\quad A - \epsilon \leq x \leq A + \epsilon \\ J_i:&\quad a_i - \epsilon \leq x \leq a_i + \epsilon, \ (i = 1, \ldots, N), \end{aligned}$$

have no points in common. Further, since g is continuous, there exists an $\epsilon_1 < \epsilon$ such that

$$A - \epsilon \leq g(x_1, \ldots, x_N) \leq A + \epsilon$$

provided that simultaneously $|x_i - a_i| < \epsilon_1$, $(i = 1, \ldots, n)$. Next, let p be any number between 1/3 and 2/3. Corresponding to p, we define the function $F_p(x)$ as follows: In the interval I we set $F_p(x) = p$. In every interval J_i $(i = 1, \ldots, N)$ where $J_i : a_i - \epsilon_1 \leq x \leq a_i + \epsilon_1$, we let $F_p(x)$ increase an amount $\left(\frac{1}{3N}\right)$. Outside the intervals I, $J_1, \ldots, J_N$, we define $F_p(x)$ arbitrarily so that it is continuous and nondecreasing for every x, and has the properties $F_p(-\infty) = 0$ and $F_p(\infty) = 1$. Clearly we can construct such a distribution function.

Let

$$S = \{(x_1, \ldots, x_N) : |x_i - a_i| \leq \epsilon_1, \ i = 1, \ldots, N\} \,.$$

Then, by the construction of $F_p(x)$, we have

$$P\left((X_1, \ldots, X_N) \in S\right) = \left(\frac{1}{3N}\right)^N \,.$$

However, if $(X_1, \ldots, X_N) \in S$, then by construction,

$$A - \epsilon \le g(X_1, \ldots, X_N) \le A + \epsilon$$

and

$$Y = F_p\left(g(X_1, \ldots, X_n)\right) = p\,.$$

Hence, for $F(x) = F_p(x)$ we have

$$P(Y = p) \ge \left(\frac{1}{3N}\right)^N .$$

However, since g is a D.F.U.T.L., this inequality must hold for every $F(x)$. Now choose a set of numbers

$$1/3 < p_1 < p_2 < \cdots < p_m < 2/3, \text{ with } m = 2(3N)^N\,.$$

Then,

$$P(Y = p_i \text{ for some } i = 1, \ldots, m) \ge 2$$

which is a contradiction. This completes the proof of the necessity.

In the following we shall characterize the continuous functions g in terms of the order statistics in the sample.

Result 4.3.3 (Robbins, 1944). *Let $X_{1N} \le \cdots \le X_{NN}$ denote the ordered observations $(X_1, \ldots, X_N)$. Let $i_1, \ldots, i_N$ be a permutation of the integers $1, \ldots, N$, and let $E(i_1, \ldots, i_N) = \{(x_1, \ldots, x_N) : x_{i_1} < x_{i_2} < \cdots < x_{i_N}\}$. Then any continuous function $g(x_1, \ldots, x_N)$ satisfying (4.3.5) is of the form $g(x_1, \ldots, x_N) = x_{i_r}$ for some $1 \le r \le N$, for each $E(i_1, \ldots, i_N)$ and $r = r(i_1, \ldots, i_N)$ depends on the permutation $(i_1, \ldots, i_N)$.*

PROOF: The $N!$ sets E are open and disjoint. Since g is continuous and $\hat{g} \equiv 0$ in each $E(i_1, \ldots, i_N)$, we must have for some r

$$g(x_1, \ldots, x_N) = x_{i_r}$$

when r depends on the permutation $(i_1, \ldots, i_N)$ in such a way that g may be extended continuously over the whole plane. For if there exist two points $(x_1, \ldots, x_N)$ and $(y_1, \ldots, y_N)$ belonging to $E(i_1, \ldots, i_N)$ for which $g(x_1, \ldots, x_N) = x_r$ and $g(y_1, \ldots, y_N) = y_j$, $r \ne j$, then y cannot be continuous. The condition for this is as follows. Two permutations $(i_1, \ldots, i_N)$ and $(j_i, \ldots, j_N)$ may be called *adjacent* if they differ only by an interchange of two adjacent integers. Then, for any two adjacent permutations, either

$r(i_1, \ldots, i_N) = r(j_1, \ldots, j_N)$ or the two values of r are the two interchanged integers. For example, the function

$$g(x_1, x_2, x_3) = \begin{cases} x_{33} & \text{if } x_{33} = x_1, \\ x_{23} & \text{otherwise,} \end{cases}$$

satisfies the requirement. □

Lemma 4.3.2. *Condition (4.3.5) is a sufficient condition for a continuous function to be a D.F.U.T.L.*

PROOF: Since the variables $X_1, \ldots, X_N$ are independent and identically distributed, the probability that $(X_1, \ldots, X_N)$ belongs to $E(i_1, \ldots, i_N)$ is equal to $(1/N!)$ for every permutation $(i_1, \ldots, i_N)$. Let $W = g(X_1, \ldots, X_N)$. Then the conditional distribution of W given that $(X_1, \ldots, X_N)$ belongs to $E(i_1, \ldots, i_N)$ is

$$\begin{aligned} G_r(F(w)) &= P[W \le w | (X_1, \ldots, X_N) \in E(i_1, \ldots, i_N)] \\ &= P[X_{r,N} \le w] = \sum_{j=r}^{N} \binom{N}{j} F^j(w) [1 - F(w)]^{N-j}, \end{aligned}$$

because of Result 4.3.3. That is, $G_r(F(w))$ is a polynomial in $F(w)$. Thus, if $Y = F(W)$, then

$$\begin{aligned} P[Y \le y | (X_1, \ldots, X_N) \in E(i_1, \ldots, i_N)] &= P(U_{r,N} \le y) \\ &= \sum_{j=r}^{N} \binom{N}{j} y^j (1-y)^{N-j} \\ &= G_r(y) \end{aligned}$$

where $U_{r,N}$ denotes the rth smallest order statistic in a random sample of size N drawn from the uniform (0,1) population. Now, the unconditional distribution of Y is

$$H(y) = \sum_{r=1}^{N!} (N!)^{-1} G_r(y)$$

which is independent of F; hence, f is a D.F.U.T.L. This completes the proof of Lemma 4.3.2.

Definition 4.3.2. *A function $g(x_1, \ldots, x_N)$ is symmetric if its value is unchanged by any permutation of its arguments.*

Result 4.3.4 (Robbins, 1944). *The only symmetric D.F.U.T.L.'s in a random sample drawn from a continuous distribution are the N order statistics.*

PROOF: From Lemma 4.3.2 and Result 4.3.3, we have that the only continuous functions g which is a D.F.U.T.L. is some order statistics in each $E(i_1, \ldots, i_N)$. Now, if g is symmetric in its arguments, then g should be the same order statistic in each $E(i_1, \ldots, i_N)$. In other words, the only continuous and symmetric functions g that are D.F.U.T.L. are the order statistics in the sample.

Paulson (1943) pointed out the relation between tolerance limits and confidence limits (in the sense of Neyman). Noether (1951) showed the close connection between confidence intervals for the binomial parameter p and tolerance limits.

4.4 Distribution-free Tolerance Limits

Case 1: $\gamma \to 1$

For given $0 < \beta$, $\gamma < 1$, the interval $(X_{i,N}, X_{j,N})$ is called distribution-free tolerance interval if i and j are chosen such that

$$P\left[F(X_{j,N}) - F(X_{i,N}) > \beta\right] > \gamma .$$

Let us rewrite this statement differently. Let Y denote an observation independent of $(X_{1,N}, \ldots, X_{N,N})$. Then we can rewrite the probability statement as

$$P\left[P\left\{X_{i,N} \leq Y \leq X_{j,N} | \boldsymbol{X}\right\} \geq \beta\right] \geq \gamma .$$

Let

$$A = \left[w : X_{i,N} \leq Y \leq X_{j,N}\right] ,$$

$$P\left[E_{Y|\boldsymbol{X}} I_A > \beta\right] \geq \gamma : \text{ where } \boldsymbol{X} = (X_{1,N}, \ldots, X_{N,N}) .$$

Now let $\gamma = 1$. Then

$$E_{Y|\boldsymbol{X}} I_A > \beta \text{ a.s.}$$

or

$$E_{\boldsymbol{X}} E_{Y|\boldsymbol{X}} I_A > \beta ,$$

i.e.,

$$E_Y E_{\mathbf{X}|Y} I_A > \beta\,.$$

Now,

$$\begin{aligned} E_{\mathbf{X}|Y} I_A &= P\left[X_{i,N} \le Y \le X_{j,N}|Y\right] \\ &= P\left[U_{i,N} \le U \le U_{j,N}|U\right] \end{aligned}$$

where U is uniform on $(0,1)$ and $U_{i,N}$ are the uniform order statistics

$$\begin{aligned} E_{\mathbf{X}|Y} I_A &= P\left[U_{i,N} \le U|U\right] - P\left[U_{j,N} \le U|U\right] \\ &= \sum_{k=i}^{j-1} \binom{N}{k} U^k (1-U)^{N-k} \\ E_Y E_{\mathbf{X}|Y} I_A &= E_U\left[\sum_{k=i}^{j-1} \binom{N}{k} U^k (1-U)^{N-k}\right] \\ &= \sum_{k=i}^{j-1} \binom{N}{k} \cdot \frac{k!(N-k)!}{(N+1)!}\,. \end{aligned}$$

Thus, RHS $\ge \beta$ means

$$j - i \ge (N+1)\beta\,.$$

If $i = r$, $j = N - s + 1$, then $r + s \le (N+1)(1-\beta)$. Notice that $r + s$ increases as γ decreases. In the following we tabulated the largest value of $k = r + s$ for certain chosen values of N and β.

Case 2: N is Sufficiently Large

Based on a normal approximation to the incomplete beta function, Govindarajulu (1977) presents simple methods for solving any one of the four parameters $k = r + s$, β, γ and N in terms of the other three. Numerical examples indicate that the approximations are very reasonable. Certain 'generalized' tolerance limits are defined which enable one to relate the confidence and tolerance limits. Also considered by Govindarajulu (1977) are tolerance limits with a specified precision. The following formulae give the

Table 4.4.1: Limiting values of k as $\gamma \to 1$

β/N	.5	.75	.90	.95	.99
50	25 (16)*	12 (6)	5 (1)	2	0 (0)
100	50 (38)	25 (15)	10 (4)	5 (1)	1 (0)
150	75 (61)	37 (26)	15 (7)	7 (2)	1 (0)
200	100 (84)	50 (36)	20 (11)	10 (4)	2 (0)
300	150 (130)	75 (58)	30 (19)	15 (7)	3 (0)
400	200 (177)	100 (80)	40 (27)	20 (11)	4 (0)
500	250 (224)	125 (103)	50 (35)	25 (14)	5 (1)
700	350 (319)	175 (149)	70 (52)	35 (22)	7 (2)
800	400 (367)	200 (172)	80 (61)	40 (26)	8 (2)
1000	500 (463)	250 (219)	100 (79)	50 (38)	10 (3)

*The values in parentheses are those given by Sommerville (1958) which correspond to $\gamma = 0.99$.

approximations:

$$1-\gamma \doteq \Phi\left\{\frac{k-1/2-N(1-\beta)}{\{N\beta(1-\beta)\}^{1/2}}\right\} \tag{4.4.1}$$

$$\beta \doteq \frac{(1-a+\frac{z^2}{2N} \pm \frac{z}{\sqrt{N}}\left\{\frac{z^2}{4N}+(1-a)a\right\}^{1/2}}{(1+z^2/N)} \tag{4.4.2}$$

or for large N,

$$\beta \doteq (1-a)+\frac{z}{\sqrt{N}}\{a(1-a)\}^{1/2} \tag{4.4.3}$$

and

$$N \doteq \left[-z\sqrt{\beta}+\{\beta z^2+4(k-1/2)\}^{1/2}\right]^2/4(1-\beta) \tag{4.4.4}$$

where

$$a=(k-1/2)/N \text{ and } z=\Phi^{-1}(1-\gamma),$$

Φ denoting the standard normal d.f.

Generalized Tolerance Limits

The relation between tolerance and confidence is more than casual if we consider the generalized tolerance limits given as follows:

1. $P(U_{r,s,N} - \alpha \geq -\Delta) \geq \gamma$;
2. $P(U_{r,s,N} - \alpha \leq \Delta) \geq \gamma$;
3. $P(|U_{r,s,N} - \alpha| \leq \Delta) \geq \gamma$;

when $U_{r,s,N} = F(X_{N-s+1,N}) - F(X_{r,N})$ and $\alpha = EU_{r,s,N} = 1 - \frac{k}{N+1}$.

The above formulation is somewhat appealing since it is analogous to setting up one-sided or two-sided fixed-width confidence interval for α for specified γ. However, α will be known as soon as N and k are specified. Solving for Δ from (1), we have

$$\Delta \geq \alpha + [(1-a) + (z^2/2N) + zN^{-1/2}\{(z^2/4N) + a(1-a)\}^{1/2}] \div (1 + z^2N^{-1})$$

where $z = \Phi^{-1}(1-\gamma)$ and $a = (k - 1/2)/N$. For instance, $N = 100$, $k = 5$, $\gamma = .95$ yield $\Delta \geq 0.043$. Formulation (2) is analogous to (1) except that we take the larger root and solve for Δ and obtain

$$\Delta \geq -\alpha + [1 - a + (z^2/2N) - zN^{-1/2}\{(z^2/4N) + a(1-a)\}^{1/2}] \div (1 + z^2N^{-1}) .$$

For instance, $N = 100$, $k = 5$ and $\gamma = .95$ yield $\Delta \geq .027$.

One can obtain an explicit solution for (3) by considering symmetrical or central tolerance limits.

4.5 Other Tolerance Limit Problems

Danziger and Davis (1964) have considered the following tolerance limit problems. Consider an ordered sample $X_1 \leq X_2 \leq \cdots \leq X_n$ and a second finite random sample $Y_1, \ldots, Y_N$ from an infinite population having a continuous density function, $f(x)$. The one-tolerance limit problem is: for any integer r, such that $1 \leq r \leq n$, and for any integer N_0 such that $0 \leq N_0 \leq N$, find the probability that at least N_0 of the Y_i's are greater than X_r. The two-tolerance limit problem is: for any pair of integers r_1 and r_2 such that $1 \leq r_1 \leq r_2 \leq n$ and for any integer N_0, such that $0 \leq N_0 \leq N$, find the probability that at least N_0 of the Y_i's are greater than X_{r_1} and less than X_{r_2}.

The probability that N_0 of the Y_i's lie above X_r is given by

$$p(N_0) = \binom{N_0 + n - r}{N_0}\binom{N - N_0 + r - 1}{N - N_0} \bigg/ \binom{N+n}{N} . \tag{4.5.1}$$

From the theory of statistically equivalent blocks as defined by Tukey (1947) it can be shown that the probability of at least N_0 of the Y_i's lying between X_{r_1} and X_{r_2} is equal to the probability of at least N_0 of the Y_i's being greater than X_r, where $r = r_1 + n + 1 - r_2$, that is, all two-tolerance limit cases are equal to and are reducible to the one-tolerance limit problem.

From Eq. (4.5.1), iterative computation was performed using

$$p(N) = n!(N+n-r)!/(n-r)!(N+n)!$$

and

$$p(N_0 - 1) = p(N_0)(N - n_0 + r)N_0/(N_0 + n - r)(N - N_0 + 1)$$

such that

$$\sum_{N_k=N_0}^{N} p(N_k) \geq \gamma\,.$$

Danziger and Davis (1964) tabulate for given γ the least value N_0 which lies above the rth lowest of the X's for $N = 5,\ 10,\ 25,\ 50,\ 75,\ 100,\ \infty$, $n = 5,\ 10,\ 25,\ 50,\ 765,\ 100$, $\gamma = .50,\ .75,\ .95,\ 99$ and $r = 1(1)10$. Values for the proportion of the population covered by the specified interval are also included primarily to illustrate the rapidity with which these limiting values are approached from a finite second sample. These limiting values can be obtained from Murphy (1948). These proportions do not always appear to converge monotonically to the limiting values due to the discreteness of N_0 and N.

4.6 Tolerance Regions

Wald (1943) has extended Wilks' (1942) method of constructing tolerance limits to bivariate and multivariate date. In the following we shall summarize Wald's (1943) procedures for the bivariate data. Assume that a random sample of size N, namely $\{(X_i, Y_i),\ i = 1, \ldots, N\}$ is drawn from a bivariate population having the continuous distribution function $F(x, y)$. Consider the problem of finding a rectangle T in the (x, y)-plane, called the *tolerance region*, such that the proportion of the population included in the rectangle T is not less than a given number β with probability at least γ. The rectangle T is constructed as follows: Assume that the points $p_1, \ldots, p_N$ where $p_i = (X_i, Y_i)$ are arranged in order of increasing magnitude of their abscissa values, that is, $X_1 < X_2 < \cdots < X_N$. Draw a vertical line V_{r_1} through the point p_{r_1} and a vertical line V_{s_1} through p_{s_1} where r_1 and s_1 are positive integers such that $1 \le r_1 \le N - s_1 - 2$ and $s_1 \ge 1$. Then consider the set consisting of the points $p_{r_1+1}, \ldots, p_{N-s_1+1}$ that lie between the vertical lines V_{r_1} and V_{N-s_1+1}. Now draw horizontal lines, H_{r_2} and H_{s_2} with $r_2 < s_2$ through the points of S having the r_2 smallest and the s_2 largest ordinates in S, respectively. The tolerance region T is the rectangle determined by the lines V_{r_1}, V_{N-s_1+1}, H_{r_2} and H_{N-s_2+1}. The probability γ that at least the proportion β $(0 < \beta < 1)$ of the universe is included in T can be shown to be (see Wald, 1943, §3)

$$\gamma \le I - I_\beta(N - k + 1, k), \quad k = r_2 + s_2$$

which is independent of r_1 and s_1.

For the case $r_1 = s_1 = 1$ and $r_2 = s_2 = 1$, Wald (1943) has computed the sample size N required for specified β and γ.

Table 4.6.1: Values of N for specified β and γ

β/γ	.97	.975	.98	.985	.99
.99	332	398	499	668	1001
.95	256	309	385	515	771

Wald's (1943) bivariate procedure has straightforward extension to the p-variate case and the final equation relating β and γ depends only on the sum $k = r_p + s_p$.

Next, if the variables (X, Y) are strongly correlated, a rectangular tolerance region seems to be unfavorable since it will cover an unnecessarily large area in the (x, y) plane. Thus, Wald (1943) proposes tolerance regions which are composed of several rectangles. As an illustration, we consider tolerance regions T^* constructed as follows.

Let $m_1, \ldots, m_k$ be k positive integers such that $1 \le m_1, \ldots, m_k \le N$ and $m_{i+1} - m_i \ge 3$ where N is the size of the bivariate sample. Let V_i be the vertical line in the (x, y) plane given by the equation $x = x_{m_i,N}$ (where $X_{1N} \le \cdots \le X_{NN}$ denote the ordered abscissa values of the N points) $(i = 1, \ldots, k)$. The number of sample points which lie between the vertical lines V_i and V_{i+1} is obviously equal to $m_{i+1} - m_i - 1$. Through each point which lies between the vertical lines V_i and V_{i+1}, draw a horizontal line. In this way we obtain $m_{i+1} - m_i - 1$ horizontal lines $W_{i,1}, \ldots, W_{i,m_{i+1}-m_i}$ where the line $W_{i,j+1}$ is above the line $W_{i,j}$. Denote by R_{ij} $(i = 1, \ldots, k-1;$

$j = 1, \ldots, m_{i+1} - m_i - 2$) the rectangle determined by the lines V_i, V_{i+1}, $W_{i,j}$, $W_{i,j+1}$. Let T^* be a region composed of s different rectangles $R_{i,j}$. Let Q^* be given by $\iint_{T^*} dF(x, y)$. Then Wald (1943) has shown that Q^* has a beta distribution with parameters s and $N - s + 1$. For specified β and γ the choice of s or N is determined by the relation

$$1 - I_\beta(s, N - s + 1) \geq \gamma .$$

Tukey (1947) has generalized Wald's (1943) method of elimination of statistically equivalent blocks.

Tolerance Regions for Multivariate Distribution

Let

$$\{X_{1,j}, X_{2,j}, \ldots, X_{k,j}\} : (j = 1, 2, \ldots, N)$$

denote a random sample from a continuous k-dimensional d.f. $F(x_1, x_2, \ldots, x_k)$. Introduce an ordering function

$$h(x_1, x_2, \ldots, x_k)$$

such that

$$W = h(X_1, X_2, \ldots, X_k)$$

is a random variable having a continuous d.f. $H(w)$. Then the random variables

$$W_j = h(X_{1,j}, X_{2,j}, \ldots, X_{k,j}) ,$$

$j = 1, 2, \ldots, N$ constitute a random sample from a population having the distribution of $h(X_1, X_2, \ldots, X_k)$. Let

$$(W_{1,N}, W_{2,N}, \ldots, W_{N,N})$$

denote the order statistics in the sample $(W_1, W_2, \ldots, W_N)$. The coverages

$$\begin{aligned} U_1' &= H(W_{1,N}) \\ U_2' &= H(W_{2,N}) - H(W_{1,N}) \\ &\vdots \\ U_N' &= H(W_{N,N}) - H(W_{N-1,N}) \end{aligned}$$

will be the random variables associated respectively with the mutually exclusive and exhaustive k-dimensional sample blocks $B_k^{(1)}, B_k^{(2)}, \ldots, B_k^{(N+1)}$

into which the k-dimensional Euclidean space is decomposed by the ordering "curves" $w_j = h(x_{1j}, x_{2j}, \ldots, x_{kj})$, $j = 1, 2, \ldots, N$. The coverages $U'_1, U'_2, \ldots, U'_N$ have the same distributional properties as those possessed by the elementary coverages defined in the one-dimensional case. Consider any rule for choosing some t of these sample blocks and let the union of these sample blocks be T_t. Also, let $V_t = U'_1 + U'_2 + \cdots + U'_t$, that is, the sum of the first coverages for the selected blocks. Then V_t is a random variable such that

$$V_t = \int_{T_t} dF(x_1, x_2, \ldots, x_k)\,.$$

Thus, for a given β, $\gamma > 0$, if

$$P(V_t \geq \beta) \geq \gamma\,,$$

then V_t is said to be a 100β percent tolerance region with confidence γ.

Tukey (1948) calls these blocks 'statistically equivalent' blocks. The distribution-free tolerance regions are obtained by procedures in which specified numbers of statistically equivalent blocks are removed. Tukey"s method suggested the removal of the blocks one at a time. Since Tukey's method is time consuming, especially when a large number of blocks have to be removed, Fraser (1951) suggested the removal of a specified number of blocks at each time, Fraser (1953), Fraser and Guttman (1956) and Guttman (1957) defined time optimum tolerance regions, β-expectation tolerance regions and proposed procedures for obtaining these regions.

Fraser and Guttman (1956) also give necessary and sufficient conditions for a distribution-free tolerance region. Further, they show that in the case of sampling from discrete populations, there do not exist distribution-free tolerance regions that are symmetric in the observations, other than the null set or the entire sample space.

Let $\{P_\theta | \theta \in \Omega\}$ be a class of probability measures defined over the measurable space $(\mathcal{X}, \mathcal{A})$. Let $S(x_1, \ldots, x_N)$ be a tolerance region mapping $\mathcal{X}^N$ into Q, the sigma algebra on $\mathcal{X}^N$. Let $P_\theta\left(S(x_1, \ldots, x_N)\right)$ denote the probability measure of the tolerance region $S(x_1, \ldots, x_N)$, which has an induced probability distribution corresponding to the product measure of P_θ over $\mathcal{X}^N$. $S(x_1, \ldots, x_N)$ is said to be distribution-free if its induced distribution is free of $\theta \in \Omega$. Let

$$G_\theta(u) = P\left\{P_\theta\left(S(X_1, \ldots, X_N)\right) \leq u\right\}\,.$$

Let us define the characteristic function of a tolerance region, $\phi_y(x_1, \ldots, x_N)$:

$$\begin{aligned} \phi_y(x_1, \ldots, x_N) &= 1 \text{ if } y \in S(x_1, \ldots, x_N) \\ &= 0 \text{ if } y \notin S(x_1, \ldots, x_N)\,. \end{aligned}$$

Then, we have

$$P_\theta\left(S(x_1,\ldots,x_N)\right) = E\left\{\phi_Y(x_1,\ldots,x_N)\right\},$$

where the expectation is with respect to the distribution of Y, namely P_θ. In the following we shall give analytical conditions under which a tolerance region is distribution-free.

Result 4.6.1 (Fraser and Guttman, 1956, Theorem 3.1). *$S(x_1,\ldots,x_N)$ is a distribution-free tolerance region if and only if there exist a sequence of real numbers $\alpha_1,\alpha_2,\ldots,$ such that*

$$\phi_{y_1}(x_1,\ldots,x_N)-\alpha_1,\ \phi_{y_1}(x_1,\ldots,x_N)\,\phi_{y_2}(x_1,\ldots,x_N)-\alpha_2,\ldots,$$

are respectively unbiased estimates of zero over $\mathcal{X}^{N+1},\mathcal{X}^{N+2},\ldots,$ for the power product measures of $\{P'|\theta\in\Omega\}$. The sequence $\alpha_1,\alpha_2,\ldots$ is the moment sequence for the distribution of $P_\theta\left(S(X_1,\ldots,X_N)\right)$ where X_i have measure P_θ.

PROOF: $S(x_1,\ldots,x_N)$ is distribution-free if $G_\theta(u)$ is free of θ. Now, since a distribution function on a bounded interval is uniquely determined by the corresponding moment sequence (see, for example, Feller, 1966, pp. 223–224), equivalently, we can state that the moment sequence for $G_\theta(u)$ is free of θ. Let α_r denote the rth moment of $G_\theta(u)$. Then

$$\begin{aligned}
\alpha_r &= \int_0^1 u^r\,dG_\theta(u)\\
&= \int_{\mathcal{X}^N}\left[P_\theta\left(S(x_1,\ldots,x_N)\right)\right]^r\prod_{i=1}^N dP_\theta(x_i)\\
&= \int_{\mathcal{X}^N}\left[E\left(\phi_Y(x_1,\ldots,x_N)\right)\right]^r\prod_{i=1}^N dP_\theta(x_i)\\
&= \int_{\mathcal{X}^N}\left[\int_{\mathcal{X}}\phi_y(x_1,\ldots,x_N)\,dP_\theta(y)\right]^r\prod_{i=1}^N dP_\theta(x_i)\\
&= \int_{\mathcal{X}^{N+r}}\prod_{j=1}^r\phi_{y_j}(x_1,\ldots,x_N)\prod_{j=1}^r dP_\theta(y_j)\prod_{i=1}^N dP_\theta(x_i).
\end{aligned}$$

Hence, $\prod_{j=1}^r\phi_{y_j}(x_1,\ldots,x_N)-\alpha_r$ has zero expectation over $\mathcal{X}^{N+r}$. Thus, $G_\theta(u)$ being free of θ is equivalent to the existence of a sequence $\{\alpha_r\}$ such that the above expression estimates zero unbiasedly for all r.

Example 4.6.1 (Fraser and Guttman, 1956). Assume sampling from a discrete population on the real line. Hence, $\mathcal{X} = R$. Let $S(x_1, \ldots, x_N)$ be symmetric in the arguments. Then there do not exist distribution-free tolerance regions, other than the null set or the whole space, $\mathcal{X}^N$.

Let $S(x_1, \ldots, x_N)$ be a distribution-free tolerance region which is symmetric in the x's. If $\phi_y(x_1, \ldots, x_N)$ denotes its characteristic function, then by Result 4.6.1, there exist $\alpha_1, \alpha_2, \ldots,$ such that

$$\prod_{j=1}^{r} \phi_{y_j}(x_1, \ldots, x_N) - \alpha_r$$

in an unbiased estimator of zero over $\mathcal{X}^{N+r}$ for all r. For random samples from $\mathcal{X}^N$, let us define the order statistic

$$t(x_1, \ldots, x_N) = (x_{1N} \leq \cdots \leq x_{NN}).$$

One can easily show that $t(x_1, \ldots, x_N)$ is sufficient for the class of power product measures over $\mathcal{X}^N$. According to Halmos (1946), $t(x_1, \ldots, x_N)$ is also complete for the same measures. Then we have

$$E\left\{\prod_{j=1}^{r} \phi_{X_{N+j}}(X_1, \ldots, X_N) - \alpha_r\right\} = 0.$$

That is,

$$E\left\{E\left[\prod_{j=1}^{r} \phi_{X_{N+j}}(X_1, \ldots, X_N) - \alpha_r | t(\boldsymbol{X}) = t\right]\right\} = 0,$$

the quantity within the square brackets being free of θ since $t(\boldsymbol{X}) = t(X_1, \ldots, X_{N+r})$ is sufficient. Now using completeness of $t(\boldsymbol{X})$ we obtain

$$E\left\{\prod_{j=1}^{r} \phi_{X_{N+j}}(X_1, \ldots, X_N) - \alpha_r | t(\boldsymbol{X}) = t\right\} = 0$$

almost everywhere with respect to the induced measure of $t(x_1, \ldots, x_{N+r})$. However, since $\{P_\theta | \theta \in \Omega\}$ is the class of all discrete distributions, almost everywhere implies everywhere. Now, let $r = 1$. Then, the conditional distribution for given $t(\boldsymbol{X}) = t$ assigns equal probability to all permutations of $(x_1, \ldots, x_{N+r})$. Hence, we have

$$[(N+1)!]^{-1} \sum_P \phi_{x_{i_{N+1}}}(x_{i_1}, \ldots, x_{i_N}) - \alpha_1 = 0\,,$$

everywhere; P denotes summation over all permutations $(i_1, \ldots, i_{N+1})$ of $(1, \ldots, N+1)$. Now, $\phi_y(x_1, \ldots, x_N)$ is symmetric in $x_1, \ldots, x_N$ because $S(x_1, \ldots, x_N)$ is. Hence,

$$\phi_{x_{N+1}}(x_1, \ldots, x_N) + \cdots + \phi_{x_1}(x_2, \ldots, x_{N+1}) = (N+1)\alpha_1,$$

for all $(x_1, \ldots, x_{N+1})$. Now set $x_1 = \cdots = x_{N+1} = x$ and obtain

$$(N+1)\phi_x(x, \ldots, x) = (N+1)\alpha_1\,.$$

However, $\phi_x(x, \ldots, x)$ takes the value zero or one. Hence, $\alpha_1 = 0$ or 1. Thus, the first moment of a random variable restricted to (0,1) is either 0 or 1. Hence, the random variable (that is, the coverage of the tolerance region) takes the value zero or one with probability one. Because we are dealing with discrete measures, this means that $S(x_1, \ldots, x_N)$ is either the null set or $\mathcal{X}^N$.

Fraser and Guttman (1956) also define the β-expectation and similar β-expectation tolerance regions.

Definition 4.6.1. *$S(x_1, \ldots, x_N)$ is a β-expectation tolerance region if*

$$E\left\{P_\theta\left(S(X_1, \ldots, X_N)\right)\right\} \le \beta \text{ for all } \theta \in \Omega\,.$$

For such a region, the average probability content of the region is at most β.

Definition 4.6.2. *$S(x_1, \ldots, x_N)$ is a similar β-expectation tolerance region if*

$$E\left\{P_\theta\left(S(X_1, \ldots, X_N)\right)\right\} = \beta \text{ for all } \theta \in \Omega\,.$$

A similar β-expectation region can also be viewed as a β-confidence region for a future observation drawn from the population sampled, since $P_\theta\left(S(x_1, \ldots, x_N)\right)$ is the probability that an independent observation falls in S for given $x_1, \ldots, x_N$, and $E\left\{P_\theta\left(S(X_1, \ldots, X_N)\right)\right\}$ is the unconditional probability of such an event. Hence, the confidence that the future observation falls in S is β.

Guttman (1959) also proved that the optimum tolerance regions are highly powerful and the power of this procedure is at least 0.90 in general, Kemperman (1956) proposed a very general method for determining tolerance regions by removal of statistically equivalent blocks. The basis for this method was furnished by Tukey (1947 and 1948), Fraser and Wormleighton (1951) and Fraser (1951 and 1953). This method has the advantage of using supplementary information and sample information so as to obtain smaller tolerance regions. Walsh (1962) discussed the relative advantages and disadvantages of the various methods of obtaining distribution-free tolerance regions. He also described techniques of obtaining specialized tolerance regions. Jirina (1952) and Saunders (1960) described the procedures for obtaining tolerance regions. Saunders (1960) also pointed out that the sequential procedure might not reduce the sample size in all cases and Wilks' procedure would be better in some situations.

4.7 Problems

4.1.1 Using Eq. (4.1.2) find the largest value of k when

$$(a)\ N = 30,\ \gamma = 0.90, \quad (b)\ N = 40,\ \gamma = 0.95$$

and compare them with the exact values of k determined from Appendix II.

4.2.1 Obtain Walsh's interval estimate for the first and third quantiles of an unknown distribution when $N = 50$ and $\alpha = 0.05$.

4.3.1 Using Appendix III determine the sample size required in order that

$$P[F(X_{N,N}) - F(X_{1,N}) \geq 0.90] \geq 0.95.$$

4.4.1 Obtain the value of N from Eq. (4.4.4) and compare it with the true value obtained in Problem 4.3.1.

Chapter 5

Nonparametric Estimation

5.1 Problems in Non-parametric Estimation

In both parametric and non-parametric estimation problems, it is easy to obtain an estimate having the property of unbiasedness. However, one would like the estimate to possess other properties, like having small variance. Quite often one can obtain a class of unbiased estimates. The question is how to pick one belonging to this class and having the smallest variance. Or if you have an estimate which you think is unbiased and has minimum variance, how to verify that it has minimum variance in the class of all unbiased estimates. Halmos (1946) considered this problem and proved a general result for discrete and continuous populations. In the following we will present Halmos' (1946) result.

Result 5.1.1 (Halmos, 1946). *Let $X_1, X_2, \ldots, X_N$ be a random sample from a d.f. $F(x)$, and let $l(X_1, X_2, \ldots, X_N)$ be any statistic having mean $p(F)$ and finite variance. Also, let $(i_1, i_2, \ldots, i_N)$ be the ith in the set (suitably indexed) of all $N!$ permutations of the integers $(1, 2, \ldots, N)$ and let*

$$l_i(X_1, X_2, \ldots, X_N) = l(X_{i_1}, X_{i_2}, \ldots, X_{i_N}) .$$

Define

$$\bar{l}(X_1, \ldots, X_N) = \sum_{i=1}^{N!} l_i(X_1, \ldots, X_N)/N! .$$

Then $E\bar{l} = p(F)$ and $\bar{l}$ has smaller variance than that of $l(X_1, X_2, \ldots, X_N)$, unless $l(X_1, X_2, \ldots, X_N)$ is symmetric in $X_1, X_2, \ldots, X_N$ with probability one in which case $\bar{l}(X_1, \ldots, X_N)$ is identical with $l(X_1, X_2, \ldots, X_N)$.

PROOF: Since $X_1, X_2, \ldots, X_N$ is a random sample from $F(x)$,

$$\begin{aligned} E\bar{l}_i &= \int \cdots \int l(x_{i_1}, \ldots, x_{i_N}) dF(x_1), \ldots, dF(x_N) \\ &= \int \cdots \int l(x_{i_1}, \ldots, x_{i_N}) dF(x_{i_1}), \ldots, dF(x_{i_N}) \\ &= \int \cdots \int l(x_1, \ldots, x_N) dF(x_1), \ldots, dF(x_N)\,. \end{aligned}$$

Consequently, $E\bar{l} = p(F)$. Also,

$$\operatorname{var}(\bar{l}) = E\left[(N!)^{-1} \sum_{i=1}^{N!} l_i(X_1, X_2, \ldots, X_N)\right]^2 - p^2(F)\,.$$

However, from Schwarz's inequality, it is known that

$$\bar{l}^2 = \left[\sum_{i=1}^{N!} l_i(X_1, X_2, \ldots, X_N)\right]^2 \le (N!)^{-2}\left[\sum_i 1\right]\left[\sum_i l_i^2(X_1, X_2, \ldots, X_N)\right].$$

That is,

$$E(\bar{l}^2) \le (N!)^{-1} \sum_i E l_i^2(X_1, X_2, \ldots, X_N)\,.$$

Also, since $X_1, X_2, \ldots, X_N$ is a random sample, $El_i^2 = El^2$, $i = 1, 2, \ldots, N!$. Therefore,

$$E\left(\bar{l}^2\right) \le El^2\,.$$

That is,

$$\operatorname{var} \bar{l} \le \operatorname{var} l$$

and equality holds if and only if

$$l_i(X_1, X_2, \ldots, X_N) = \bar{l}(X_1, \ldots, X_N), \quad i = 1, 2, \ldots, N!$$

for all points $(X_1, X_2, \ldots, X_N)$ in the sample space (except possibly for a set of probability zero). This implies that $l(X_1, X_2, \ldots, X_N)$ is symmetric in its arguments. This completes the proof of the assertion.

Remark 5.1.1. *Thus, in any estimation problem, consider unbiased estimates and make them symmetric in the observations by the above averaging process. If, e.g., we consider the class of all absolutely continuous populations, this procedure yields an unbiased estimate having the smallest variance in the class of all unbiased estimates of the parameter. In the following, some non-parametric estimation problems will be presented.*

Example 5.1.1. Let X be a r.v. having an unknown continuous d.f. $F(x)$. We wish to estimate $p(F) = P(X > 0)$ on the basis of a random sample $X_1, X_2, \ldots, X_N$.

Remark 5.1.2. *Starting with an unbiased estimator and symmetizing it in the variables is equivalent to Rao-Blackwellization of an unbiased estimator because order statistics is a sufficient statistic. Taking the conditional expectation of $l(X_1, \ldots, X_N)$ given the order statistic is the same as symmetizing $l(X_1, \ldots, X_N)$ with respect to all permutations of the indices of the X's and taking their average.*

Example 5.1.2. Let

$$\theta = EX$$

and

$$\hat{\theta} = \sum_{i=1}^{N} a_i X_i \,.$$

Then $\hat{\theta}$ will be symmetric in the X_i provided $a_i \equiv a$. Now $\hat{\theta}$ is unbiased implies that $Na = 1$ or $a = \frac{1}{N}$. Hence,

$$\text{var } \bar{X} < \text{var } X_i \text{ for } 1 \le i \le N \,.$$

Example 5.1.3. Let

$$\theta = EX$$

and

$$\text{var } X = \sigma^2 \,.$$

We wish to estimate σ^2 on the basis of a random sample $X_1, \ldots, X_N$. Consider the unbiased estimate

$$\frac{1}{2}(X_i - X_j)^2 \,.$$

Now, let

$$T = a \sum_{i \neq j} \sum (X_i - X_j)^2 \,.$$

$ET = \sigma^2$ implies that

$$aN(N-1)2\sigma^2 = \sigma^2 .$$

Hence,

$$a = 1/2N(N-1) .$$

Thus,

$$T = \frac{1}{2N(N-1)} \sum_{i \neq j} \sum (X_i - X_j)^2 .$$

Now,

$$\sum_i \sum_j (X_i - X_j)^2 = 2N \sum X_i^2 - 2N^2 \bar{X}^2 = 2N \sum_{i=1}^{N} (X_i - \bar{X})^2 .$$

Hence,

$$T = \frac{1}{N-1} \sum (X_i - \bar{X})^2 = \text{sample variance}$$

and

$$\text{var } T < \text{var} \left\{ \frac{1}{2} (X_i - X_j)^2 \right\} .$$

Remark 5.1.3. *By using the completeness of the order statistic, we can infer that the above two estimators are the best unbiased estimators of θ and σ^2, respectively.*

If μ is known, then the best unbiased estimate of σ^2 would be $N^{-1} \sum (X_i - \mu)^2$. However, there is seemingly a contradiction because we are getting two functions of the order statistics that are unbiased for σ^2. It should be argued that $(N-1)^{-1} \sum (X_i - \bar{X})^2$ is unbiased in the class of estimators that are translation-invariant with respect to μ.

Example 5.1.4. Let X be a random variable having an unknown continuous d.f. $F(x)$. We wish to estimate $p(F) = P(X > 0)$ on the basis of a random sample $X_1, \ldots, X_N$. Define

$$Z_i = \begin{cases} 1 & \text{if } X_i > 0, \\ 0 & \text{otherwise.} \end{cases}$$

Then, consider

$$\hat{p} = \sum_{i=1}^{N} a_i Z_i .$$

Make $\hat{p}$ symmetric in the Z_i's. Then $a_i = a$, $(i = 1, 2, \ldots, N)$. Then,

$$\hat{p} = a \sum Z_i \, .$$

$\hat{p}$ is unbiased, implies that $a = N^{-1}$. Also, variance of $\hat{p}$ is $p(F)\{1 - p(F)\}/N$ which is less than or equal to $1/4N$. Since $N\hat{p}$ is binomially distributed, $N^{\frac{1}{2}}\{\hat{p} - p(F)\}/[p(F)\{1 - p(F)\}]^{\frac{1}{2}}$ is asymptotically normally distributed. Consequently, one can obtain approximate confidence intervals for $p(F)$.

Example 5.1.5. Let X and Y be two random variables with continuous d.f.'s F and G respectively which are, in general, unknown. We wish to estimate $p(F, G) = P(X < Y)$. Mann and Whitney (1947) proposed the following statistic which gives an estimate of $p(F, G)$. Let $X_1, X_2, \ldots, X_m$ and $Y_1, Y_2, \ldots, Y_n$ $(N = m + n)$ represent random samples of sizes m and n from F and G populations respectively. For each pair of observations X_i and Y_j, define

$$Z_{i,j} = \begin{cases} 1 & \text{if } X_i < Y_j, \\ \\ 0 & \text{if } X_i > Y_j \, . \end{cases}$$

Mann-Whitney estimate is given by U_Y/mn where

$$U_Y = \sum_{i=1}^{m} \sum_{j=1}^{n} Z_{i,j} \, ,$$

$$E(U_Y) = mn\, E(Z_{i,j}) = mn\, P(X < Y) = mn \int F dG \, .$$

Thus $\hat{p}(F, G) = U_Y/mn$ is an unbiased estimate of $p(F, G)$ and by Halmos' (1946) theorem, it has minimum variance. Consider

$$EU_Y^2 = \sum_i \sum_j \sum_h \sum_k E(Z_{i,j} Z_{h,k}) \, .$$

Now,

$$Z_{i,j} Z_{h,k} = \begin{cases} 1 & \text{if } X_i < Y_j \text{ and } X_h < Y_k, \\ \\ 0 & \text{otherwise,} \end{cases}$$

so that

$$
\begin{aligned}
E(Z_{i,j}Z_{h,k}) &= P(X_i < Y_j \text{ and } X_h < Y_k) \\
&= \begin{cases} \int FdG, & i = h \text{ and } j = k, \\ \int (1-G)^2 dF & i = h \text{ and } j \neq k, \\ \int F^2 dG & i \neq h \text{ and } j = k, \\ \left(\int FdG\right)^2 & i \neq h \text{ and } j \neq k. \end{cases}
\end{aligned}
$$

Also, there are mn terms in which $i = h$ and $j = k$; $m(m-1)n$ terms in which $i \neq h$ and $j = k$; $mn(n-1)$ terms in which $i = h$ and $j \neq k$; and $m(m-1)n(n-1)$ terms in which $i \neq h$ and $j \neq k$. Hence,

$$
\begin{aligned}
EU_Y^2 = mn \int FdG &+ mn(n-1)\int (1-G)^2 dF + m(m-1)n \int F^2 dG \\
&+ m(m-1)n(n-1)\left[\int FdG\right]^2,
\end{aligned}
$$

from which one can readily compute the variance of U_Y. When $G = F$, the expressions for EU_Y and var U_Y are considerably simpler and are respectively given by $mn/2$ and $mn(m+n+1)/12$. The asymptotic normality of U_Y , when $G = F$ was asserted by Mann and Whitney (1947) and Lehmann (1951) has established the asymptotic normality of U_Y when $G \neq F$ under the assumptions: $n = cm$, $c > 0$, $m \to \infty$ and $0 < p(F, G) < 1$. One can estimate from the samples,

$$
\begin{aligned}
P[\text{any two } X\text{'s} < Y] &= \int F^2 dG, \\
P[X < \text{any two } Y\text{'s}] &= \int (1-G)^2 dF
\end{aligned}
$$

by defining $Z_{i,j,k}$ and $Z^*_{i,j,k}$ as follows:

$$
\begin{aligned}
Z_{i,j,k} &= \begin{cases} 1 & \text{if } X_i \text{ and } X_j < Y_k, \\ 0 & \text{otherwise,} \end{cases} \\
Z^*_{i,j,k} &= \begin{cases} 1 & \text{if } Y_i \text{ and } Y_j > X_k, \\ 0 & \text{otherwise,} \end{cases}
\end{aligned}
$$

and average $Z_{i,j,k}$ and $Z^*_{i,j,k}$ on all permutations of the X's and the Y's to obtain the unique unbiased minimum variance estimates of $\int F^2 dG$ and $\int (1-G)^2 dF$.

Confidence intervals can be made shorter by considering either consistent estimates of var$(\hat{p})$ or unbiased estimates of var$(\hat{p})$. In the following, we will provide an unbiased estimate of the variance of $\hat{p}$.

Since

$$E(\hat{p}^2) - p^2 = \text{var}\,\hat{p} > 0\,,$$

we have

$$p^2 = E(\hat{p}^2) - \text{var}\,\hat{p}\,.$$

Also, let

$$I_1 = \int (1-G)^2 dF \text{ and } I_2 = \int F^2 dG\,,$$

and let $\hat{I}_1$ and $\hat{I}_2$ denote unbiased estimates of I_1 and I_2, respectively. Recall that we have

$$\begin{aligned} mn\,\text{var}\,\hat{p} &= p + (n-1)I_1 + (m-1)I_2 - (m+n-1)p^2 \\ &= p + (n-1)I_1 + (m-1)I_2 - (m+n-1)\left\{E(\hat{p}^2) - \text{var}\,\hat{p}\right\}\,. \end{aligned}$$

Consequently, an unbiased estimator for var $\hat{p}$ is given by

$$(m-1)(n-1)(\widehat{\text{var}}\,\hat{p}) = \hat{p} + (n-1)\hat{I}_1 + (m-1)\hat{I}_2 - (m+n-1)\hat{p}^2\,.$$

A consistent estimator for var $\hat{p}$ will be

$$(m-1)^{-1}\hat{I}_1 + (n-1)^{-1}\hat{I}_2 - \frac{(m+n-1)}{(m-1)(n-1)}\hat{p}^2\,,$$

provided $\hat{I}_1$ and $\hat{I}_2$ are also consistent estimators of I_1 and I_2 respectively. Z.W. Birnbaum (1956) considered the interval estimation of $p(F,G)$ and obtained some methods of doing so. D. van Dantzig (1951) provided an upper bound for the variance of $\hat{p}$. Z.W. Birnbaum and Klose (1957) obtained upper and lower bounds for the variance of $\hat{p}$ for the case of any X and Y as well as for the case of stochastically comparable X, Y. These results will be presented in the following.

van Dantzig (1951) obtained the sharp upper bound for the variance of $\hat{p}(F,G)$ given by

$$\sigma_{\hat{p}}^2 \leq \frac{p(1-p)}{\min(m,n)} \leq \frac{1}{4\min(m,n)} \text{ where } p = p(F,G)\,,$$

and hence, showed that for $m,n \to \infty$, the estimate $\hat{p}$ is a consistent estimate of p. In order to obtain an asymptotic confidence interval for p, one can use Lehmann's (1951) theorem on the asymptotic normality of $(\hat{p}-p)m^{\frac{1}{2}}$ together with Dantzig's upper bound on the variance of $\hat{p}$, in order to obtain the sample size and the confidence interval for a given confidence coefficient. There will be some situations in which one or both of the assumptions under Lehmann's theorem are violated. When either m and n are not of the same order or $p(F,G)$ is close to 0 or 1, it will not be safe to rely on the normal approximation. This suggests to obtain a statistic Ψ and for every $\epsilon, \alpha > 0$, a pair of integers $M_{\epsilon,\alpha}, N_{\epsilon,\alpha}$ such that

$$P(p \leq \Psi + \epsilon) \geq 1 - \alpha, \text{ if } m \geq M_{\epsilon,\alpha}, \quad n \geq N_{\epsilon,\alpha}\,.$$

5.2 One-sided Confidence Interval for p

Let us assume that G is known and F is not known. This situation arises, for example, when it is easy to obtain a practically unlimited number of observations on Y, and hence to estimate G as accurately as desired, but only a finite sample $X_1, X_2, \ldots, X_m$ of X can be obtained (this corresponds to $\lim(m/n) = 0$ in the general case). Let $X_{1,m} < X_{2,m} < \cdots < X_{m,m}$ be the ordered X's and

$$F_m(z) = \begin{cases} 0, & z < X_{1,m} \\ k/m, & X_{k,m} \leq z < X_{k+1,m}\ ;(k = 1,\ldots,m-1) \\ 1, & X_{m,m} \leq z \end{cases}$$

be the sample d.f. [or empirical distribution function (e.d.f.)] of X. Consider the statistic

$$\begin{aligned}
\hat{p}_1 &= \int_{-\infty}^{\infty} F_m dG = \sum_{i=0}^{m} \int_{X_{i,m}}^{X_{i+1,m}} F_m dG \\
&= m^{-1} \sum_{i=0}^{m} i \{G(X_{i+1,m}) - G(X_{i,m})\} \\
&= m^{-1} \sum_{i=0}^{m} \{(i+1)G(X_{i+1,m}) - iG(X_{i,m}) - G(X_{i+1,m})\} \\
&= m^{-1} \left\{ \sum_{j=1}^{m+1} jG(X_{j,m}) - \sum_{i=0}^{m} iG(X_{i,m}) - \sum_{j=1}^{m+1} G(X_{j,m}) \right\} \\
&= m^{-1} \left\{ (m+1)G(X_{m+1,m}) - \sum_{j=1}^{m} G(X_{j,m}) - G(X_{m+1,m}) \right\} \\
&= 1 - m^{-1} \sum_{j=1}^{m} G(X_{j,m}) \\
&= 1 - m^{-1} \sum_{j=1}^{m} G(X_j) ,
\end{aligned}$$

since, by definition, $G(X_0, m) = 0$ and $G(X_{m+1,m}) = 1$. Hence,

$$E\hat{p}_1 = 1 - E\left\{ m^{-1} \sum_{j=1}^{m} G(X_j) \right\} = 1 - \int GdF = \int FdG = p .$$

To obtain a one-sided (upper) confidence interval for p, consider

$$p - \hat{p}_1 = \int_{-\infty}^{\infty} (F - F_m)dG \leq \sup_z \{F(z) - F_m(z)\} = D_m^- .$$

Hence,

$$P(p - \hat{p}_1 < \epsilon) \geq P(D_m^- < \epsilon) = P_m(\epsilon) .$$

Wald and Wolfowitz (1939) have shown that D_m^- is independent of $F(x)$. Also, Smirnov (1939) has shown that

$$\lim_{m \to \infty} P_m(z/m^{\frac{1}{2}}) = 1 - e^{-2z^2} .$$

Z.W. Birnbaum and Tingey (1951) have obtained a closed expression for $P_m(\epsilon)$ and tabulated the solutions $\epsilon_{m,\alpha}$ of the equation

$$P_m(\epsilon) = 1 - \alpha,$$

for $\alpha = .10, .05, .01, .001$, and showed that the values $\epsilon_{m,\alpha}$ differ from the approximations obtained from the limiting expression for $P_m(\epsilon)$ by less than .005 as soon as $m > 50$. Hence for all practical purposes, one can write

$$P(p < \hat{p}_1 + \epsilon) \geq 1 - e^{-2m\epsilon^2}.$$

By specifying ϵ and α, one can solve for m. For example, let $\epsilon = .05$ and $\alpha = .01$. Then $M_{\epsilon,\alpha} = 921$. However, the normal approximation and the Chebyshev's inequality together with Dantzig's bound respectively yield 541 and 10,000 for the sample size.

Suppose that both F and G are unknown. Then let F_m and G_n respectively denote the sample d.f.'s of X and Y. Consider

$$\hat{p}_2 = \int_{-\infty}^{\infty} F_m(z)dG_n(z).$$

It is easy to verify that $E(\hat{p}_2) = p(F, G)$. Also,

$$\begin{aligned} p - \hat{p}_2 &= \int_{-\infty}^{\infty} Fd(G - G_n) + \int_{-\infty}^{\infty} (F - F_m)dG_n \\ &= \int_{-\infty}^{\infty} (G_n - G)dF + \int_{-\infty}^{\infty} (F - F_m)dG_n \end{aligned}$$

so that

$$p - \hat{p}_2 \leq \sup_z \{G_n(z) - G(z)\} + \sup_z \{F(z) - F_m(z)\} = D_n^+ + D_m^-.$$

It is well known from Wald and Wolfowitz (1939) that

$$P(D_k^- < v) = P(D_k^+ < v) = P_k(v), \quad k = m \text{ or } n$$

and $P_k(v)$ is a d.f. which depends on the sample size k, but not on the d.f.'s F and G. Therefore

$$P(p \leq \hat{p}_2 + \epsilon) \geq P(D_m^+ + D_n^+ \leq \epsilon) = P_{m,n}(\epsilon)$$

where $P_{m,n}(\epsilon)$ is the convolution of P_m and P_n and, consequently, does not depend on F and G. Now, for given ϵ and α, one can determine $M_{\epsilon,\alpha}$ and $N_{\epsilon,\alpha}$ such that

$$P_{m,n}(\epsilon) \geq 1 - \alpha, \text{ for } m \geq M_{\epsilon,\alpha}, \ \ n \geq N_{\epsilon,\alpha}.$$

A numerical procedure for computing $M_{\epsilon,\alpha}$, $N_{\epsilon,\alpha}$ has been proposed by Z.W. Birnbaum and McCarty (1958). Also, from Smirnov (1939), we have

$$\lim_{k\to\infty} P(D_k^+ \le z/k^{\frac{1}{2}}) = \lim_{k\to\infty} P_k(z/k^{\frac{1}{2}}) = 1 - e^{2z^2} = L(z)\,.$$

Since, for fixed k, $P_k(z/k^{\frac{1}{2}}) = H_k(z)$ is a d.f. and $L(z)$ is a continuous d.f., it follows from a well-known argument (see Fréchet, 1937, p. 276) that $H_k(z) \to L(z)$ uniformly in z. We therefore conclude that

$$\lim_{k\to\infty}\left[P(D_k^+ \le v) - L\left(vk^{\frac{1}{2}}\right)\right] = \lim_{k\to\infty}\left[H_k\left(vk^{\frac{1}{2}}\right) - L\left(vk^{\frac{1}{2}}\right)\right] = 0 \quad (5.2.1)$$

uniformly for $0 \le v \le 1$. Writing

$$P_{m,n}(\epsilon) = P(D_m^+ + D_n^+ \le \epsilon) = \int_0^\epsilon P_n(\epsilon - u)dP_m(u),$$

and

$$Q_{m,n}(\epsilon) = \int_0^\epsilon L\left[(\epsilon - u)n^{\frac{1}{2}}\right] dL\left(um^{\frac{1}{2}}\right),$$

we obtain

$$\begin{aligned}
|P_{m,n}(\epsilon) - Q_{m,n}(\epsilon)| &\le \left|\int_0^\epsilon \left\{P_n(\epsilon - u) - L\left[(\epsilon - u)n^{\frac{1}{2}}\right]\right\} dP_m(u)\right| \\
&+ \left|\int_0^\epsilon \left\{P_m(\epsilon - v) - L\left[(\epsilon - v)m^{\frac{1}{2}}\right] dL\left(vn^{\frac{1}{2}}\right)\right\}\right| \\
&\le \max_{0\le u\le\epsilon}\left|P_n(\epsilon - u) - L\left[(e - u)n^{\frac{1}{2}}\right]\right| \\
&+ \max_{0\le v\le\epsilon}\left|P_m(\epsilon - v) - L\left[(e - v)m^{\frac{1}{2}}\right]\right|.
\end{aligned}$$

Thus, it follows from (5.2.1) that

$$\lim_{m,n\to\infty} |P_{m,n}(\epsilon) - Q_{m,n}(\epsilon)| = 0$$

uniformly for $0 \le \epsilon \le 1$. This justifies the use of $Q_{m,n}(\epsilon)$ in place of $P_{m,n}(\epsilon)$ for sufficiently large m and n. Performing integration, one obtains

$$\begin{aligned}
Q_{m,n}(\epsilon) &= 1 - (n/N)e^{-2m\epsilon^2} - (m/N)e^{-2n\epsilon^2} \\
&\quad - 2(2\pi)^{\frac{1}{2}}N^{-\frac{3}{2}}mn\epsilon^2 e^{-2mn\epsilon^2/N}\left\{\Phi(2mN^{-\frac{1}{2}}\epsilon) - \Phi(2nN^{-\frac{1}{2}}\epsilon)\right\}
\end{aligned}$$

where $N = m+n$ and Φ denotes the standard normal d.f. Setting $m/N = \lambda$, $n/N = 1 - \lambda$ and $\epsilon N^{\frac{1}{2}} = \delta$, we obtain

$$\begin{aligned} Q(\delta;\gamma) &= 1 - \lambda e^{-2(1-\lambda)\delta^2} - (1-\lambda)e^{-2\lambda\delta^2} \\ &\quad - 2(2\pi)^{\frac{1}{2}}\lambda(1-\lambda)\delta e^{-2\lambda(1-\lambda)\delta^2}\left\{\Phi(2\lambda\delta) - \Phi\left(2(1-\lambda)\delta\right)\right\}. \end{aligned}$$

Birnbaum and McCarty (1958) have tabulated the solutions $\delta_{\lambda,\alpha}$ of the equation

$$Q(\delta;\lambda) = 1 - \alpha$$

for $\alpha = .001, .005, .01, .05, .10$ $\lambda = .1(.1).5$. These solutions are presented in Appendix V. The use of the quantities N, λ, δ instead of the original m, n, ϵ has not only the advantage of reducing the computations to a table with double entry, but also makes it possible to design an experiment with a given λ which is often dictated by considerations of cost or time.

Example 5.2.1. Let $\lambda = .2$, $\epsilon = .10$ and $\alpha = .05$. From Appendix V, we have $\delta_{.2,.05} = 3.5667$; thus, solving for N from

$$(.10)N^{\frac{1}{2}} = 3.5667\,,$$

one obtains $N = 1272.13$, from which we get $m \doteq 255$ and $n \doteq 1018$.

Remark 5.2.1. *The sample sizes computed for given $\lambda, \epsilon, \alpha$ by the equation*

$$Q(\delta;\lambda) = 1 - \alpha$$

are conservative for the following reasons:

1. *Instead of finding sample sizes m, n such that $P(p \le \hat{p}_2 + \epsilon) = 1 - \alpha$, we used an inequality and looked for m, n satisfying $P_{m,n}(\epsilon) = 1 - \alpha$. This step certainly yields larger values for m and n.*

2. *$P_{m,n}(\epsilon)$ was replaced by $Q_{m,n}(\epsilon)$ on the contention that the solutions m, n of $P_{m,n}(\epsilon) = 1 - \alpha$ differ little from those of $Q_{m,n}(\epsilon) = 1 - \alpha$, especially if m and n are known to be greater than 50.*

Birnbaum and McCarty (1958) conjecture by considerations of numerical computations and some analytical considerations, although no proof is available, that for $n \ge 1$, and for $0 \le v \le 1$,

$$L\left(vn^{\frac{1}{2}}\right) = 1 - e^{-2nv^2} \le P(D_n^+ \le v)\,,$$

from which it follows since

$$\begin{aligned} P_{m,n}(\epsilon) - Q_{m,n}(\epsilon) &= \int_0^\epsilon \left\{ P_m(\epsilon - u) - L\left[(\epsilon - u) n^{\frac{1}{2}} \right] \right\} dP_m(u) \\ &+ \int_0^\epsilon \left\{ P_m(\epsilon - v) - L\left[(\epsilon - v) m^{\frac{1}{2}} \right] \right\} dL\left(v n^{\frac{1}{2}} \right), \end{aligned}$$

that

$$P_{m,n}(\epsilon) \geq Q_{m,n}(\epsilon), \quad \text{for } 0 \leq \epsilon \leq 1 .$$

Consequently, $Q_{m,n}(\epsilon) = 1 - \alpha$ *would yield sample sizes larger than those of* $P_{m,n}(\epsilon) = 1 - \alpha$.

Remark 5.2.2. *Saunders (1959) also considered the problem of confidence bounds for* $p(F, G)$ *and has obtained some results.*

5.3 Two-sided Confidence Interval for p

A two-sided confidence interval for p can be obtained by proceding in a manner similar to Birnbaum and McCarty (1958). Starting from the expression for $p - \hat{p}_2$, we obtain

$$|p - \hat{p}_2| \leq D_m + D_n$$

where

$$D_m = \sup_{-\infty < x < \infty} |F_m(x) - F(x)|$$

and

$$D_n = \sup_{-\infty < x < \infty} |G_n(x) - G(x)| .$$

The statistics D_m and D_n are due to Kolmogorov (1933). It is known that the distributions of D_m and D_n are independent of F and G, but are dependent on m and n and that

$$\begin{aligned} \lim_{k \to \infty} P\left(D_k < z/k^{\frac{1}{2}} \right) &= \lim_{k \to \infty} P_k^*\left(z/k^{\frac{1}{2}} \right) \\ &= 1 - 2 \sum_{j=1}^{\infty} (-1)^{j-1} e^{-2j^2 z^2} \\ &= (2\pi)^{\frac{1}{2}} z^{-1} \sum_{j=1}^{\infty} e^{-(2j-1)^2 \pi^2 / 8z^2} = L_1(z) . \end{aligned}$$

Since, for fixed k, $P_k^*\left(z/k^{\frac{1}{2}}\right) = H_k^*(z)$ is a d.f. and $L_1(z)$ is a continuous d.f., it again follows using Fréchet's (1937) argument that $H_k^*(z) \to L_1(z)$ uniformly in z. Hence, we infer that

$$\lim_{k\to\infty}\left[P(D_k \le v) - L_1\left(vk^{\frac{1}{2}}\right)\right] = \lim_{k\to\infty}\left[H_k^*\left(vk^{\frac{1}{2}}\right) - L_1^*\left(vk^{\frac{1}{2}}\right)\right] = 0$$

uniformly for $0 \le v \le 1$. Hence, writing

$$\begin{aligned} P_{m,n}^*(\epsilon) &= P(D_m + D_n \le \epsilon) = \int_0^\epsilon P_n^*(\epsilon - u)\,dP_m^*(u)\,, \\ Q_{m,n}^*(\epsilon) &= \int_0^\epsilon L_1\left((\epsilon - u)n^{\frac{1}{2}}\right)dL_1\left(um^{\frac{1}{2}}\right)\,, \end{aligned}$$

we can analogously show that

$$\lim_{m,n\to\infty}\left|P_{m,n}^*(\epsilon) - Q_{m,n}^*(\epsilon)\right| = 0$$

uniformly for $0 \le \epsilon \le 1$. Thus, we could use $Q_{m,n}^*(\epsilon)$ in place of $P_{m,n}^*(\epsilon)$ for sufficiently large m and n. The problem is reduced to solving the equation

$$Q_{m,n}^*(\epsilon) \ge 1 - \alpha\,.$$

It is possible to solve the equation by considering only the first few terms of $L_1(z)$. If we take the first two terms in $L_1(z)$, the equation becomes the one obtained by Birnbaum and McCarty (1958) in one-sided confidence limit estimation for p.

One can also obtain approximate confidence intervals for p by using the normal approximation together with the upper and lower bounds for the variance of the Mann-Whitney statistic. Letting

$$V(t) = G\left\{F^{-1}(t)\right\}\,,$$

the variance of the Mann-Whitney statistic, namely U_Y can be written as

$$\begin{aligned} \text{var}\,U_Y = mn\Bigg[&(n-1)\int_0^1\left\{V(t) - \frac{m-1}{n-1}t\right\}^2 dt \\ &- \frac{(m-1)^2}{3(n-1)} - (n-1)(1-p)^2 + (m-1)(1-p^2) - p(1-p)\Bigg]\,. \end{aligned}$$

Hence, finding bounds for the variance of $U_Y$is equivalent to minimizing or maximizing the integral

$$\int_0^1 \left[V(t) - \frac{(m-1)}{n-1}t\right]^2 dt$$

subject to the condition $\int_0^1 t\, dV(t) = p$.

Rustagi (1957) considered this problem and has obtained some bounds in terms of p. However, Birnbaum and Klose (1957) write

$$\text{var } U_Y = mn\left[(n-1)\text{var } G(X) + (m-1)\text{var } F(Y) - p(1-p)\right],$$

obtain bounds for variance of $G(X)$ and $F(Y)$ and consequently, the bounds for the variance of U_Y. The upper bound due to Birnbaum and Klose agree with the sharp bound previously obtained by van Dantzig. The bounds are:

$$\sigma^2_{U_Y} \le mn\, p(1-p) \max(m,n),$$

$$\sigma^2_{U_Y} \ge \begin{cases} \Delta(m+n-\Delta)\left[\Delta(1-\tilde{p}) - \dfrac{(\Delta-1)^2}{12(m+n-\Delta-1)}\right] \\ \qquad\qquad \text{if } \frac{\Delta-1}{m+n-\Delta-1} \le 2\tilde{p}, \\ \Delta(m+n-\Delta)\cdot \\ \left[\frac{4}{3}\tilde{p}\left\{2(m-1)(n-1)\tilde{p}\right\}^{\frac{1}{2}} - (m+n-2)\tilde{p}^2 + \tilde{p}(1-\tilde{p})\right] \\ \qquad\qquad \text{if } \frac{\Delta-1}{m+n-\Delta-1} \ge 2\tilde{p}, \end{cases}$$

where $\Delta = \min(m,n)$, $\tilde{p} = \min(p, 1-p)$.

Sharp upper and lower bounds under the assumption that X is stochastically larger than Y, that is, $F(x) \le G(x)$, $-\infty < x < \infty$, are also given by Birnbaum and Klose.

5.4 Estimation of Distribution Function

Let X be a random variable having a continuous but unknown d.f. which will be denoted by $F(x)$. On the basis of a random sample $X_1, X_2, \ldots, X_N$, we wish to obtain point and interval estimates of $F(x)$. This problem was first considered by Kolmogorov (1933) and an explicit solution was obtained. Subsequently, contributions by Wald and Wolfowitz (1939), Smirnov (1939), Kolmogorov (1941), Massey (1950), Birnbaum and Tingey (1951) and (1952) were made to the problem of confidence interval estimation of $F(x)$. Darling

(1957) has given an expository paper summarizing the results available up to that time.

Define

$$\epsilon(x) = \begin{cases} 1, & x \geq 0, \\ 0, & x < 0, \end{cases}$$

and

$$F_N(x) = N^{-1} \sum_{j=1}^{N} \epsilon(x - X_j)\,.$$

$F_N(x)$ is called the sample d.f. or the empirical distribution function (e.d.f.) of the data. Obviously $F_N(x)$ denotes the proportion of the X_i's which are less than or equal to x. one can easily compute the first two moments:

$$\begin{aligned} E\{F_N(x)\} &= F(x), \\ \operatorname{cov}\{F_N(x), F_N(y)\} &= E\{F_N(x)F_N(y)\} - F(x)F(y), \end{aligned}$$

$$\begin{aligned} &E\{F_N(x)F_N(y)\} \\ &= N^{-2}\left\{\sum_{i=1}^{N} E\epsilon(x - X_i)\epsilon(y - X_i) + \sum\sum_{i \neq j} E\epsilon(x - X_i)E\epsilon(y - X_j)\right\} \\ &= N^{-1}\left[E\epsilon(x - X)\epsilon(y - X) + (N-1)F(x)F(y)\right]. \end{aligned}$$

Consequently,

$$\operatorname{cov}\{F_N(x), F_N(y)\} = N^{-1}C\left(F(x), F(y)\right)$$

where

$$C(s,t) = \min(s,t) - st = \begin{cases} s(1-t), & s \leq t, \\ t(1-s), & s \geq t, \end{cases} \qquad 0 \leq s,\ t \leq 1\,.$$

Since $NF_N(x)$ is the sum of N independent and identically distributed Bernoulli random variables, it has a binomial distribution with parameters N and $F(x)$. Hence, the following are the consequences:

- Due to the **strong law of large numbers**, $F_N(x) \to F(x)$ with probability one for each x.

- Due to the **multidimensional central limit theorem**, for fixed k and $N \to \infty$,

$$\left\{N^{\frac{1}{2}}\left(F_N(x_i) - F(x_i)\right)\right\}, \quad i = 1, 2, \ldots, k$$

has an asymptotic k-dimensional normal distribution, means 0 and variance-covariance matrix $(C(F(x_i), F(x_j)))$.

Lemma 5.4.1 (Cantelli-Glivenko).

$$P\left[\sup_{-\infty<x<\infty} |F_N(x) - F(x)| \to 0\right] = 1\,.$$

In other words, with probability one, $F_N(x) \to F(x)$ uniformly in x.

PROOF: The following proof is given by Loève (1960, pp. 20–21). Let x_{jk} be the smallest value of x such that

$$F(x-0) \le j/k \le F(x), \text{ where } F(x) = P(X \le x) \text{ and } F(x-0) = P(X < x)\,.$$

Since the frequency of the event $(X \le x_{jk})$ is $F_N(x_{jk})$ and its probability (expected value) is $F(x_{jk})$, it follows from the strong law that $P(A'_{jk}) = 1$ where $A'_{jk} = \{F_N(x_{jk} - 0) \to F(x_{jk} - 0)\}$ and $P(A''_{jk}) = 1$ where

$$A''_{jk} = \{F_N(x_{jk}) \to F(x_{jk})\}\,.$$

Let

$$A_{jk} = A'_{jk} A''_{jk}$$

and

$$A_k = \bigcap_{j=1}^{k} A_{jk} = \left\{\sup_{1\le j\le k} |F_N(x_{jk}) - F(x_{jk})| \to 0\right\},$$

$$P(\bar{A}_k) = P\left(\bigcup_{j=1}^{k} \bar{A}_{jk}\right) \le \sum_{j=1}^{k} P(\bar{A}_{jk}) = 0\,.$$

Consequently, $P(A_k) = 1$. Upon setting $A = \bigcap_{k=1}^{\infty} A_k$, in a similar manner, it follows that $P(A) = 1$. Also, for $x_{jk} < x < x_{j+1,k}$,

$$F(x_{jk}) \le F(x) \le F(x_{j+1,k}), \quad F_N(x_{jk}) \le F_N(x) \le F_N(x_{j+1,k} - 0)\,,$$

while for every x_{jk},

$$0 \le F(x_{j+1,k} - 0) - F(x_{j,k}) \le 1/k\,.$$

Therefore,

$$F_N(x) - F(x) \le F_N(x_{j+1,k}-0) - F(x_{jk}) \le F_N(x_{j+1,k}-0) - F(x_{j+1,k}-0) + k^{-1}$$

and

$$F_N(x) - F(x) \ge F_N(x_{jk}) - F(x_{j+1,k} - 0) \ge F_N(x_{jk}) - F(x_{jk}) - k^{-1}\,.$$

Now, it follows that for every x and k,

$$|F_N(x) - F(x)| \le \sup_{0 \le j \le k-1} |F_N(x_{j+1,k} \pm 0) - F(x_{j+1,k} \pm 0)| + k^{-1}$$

or

$$D_N = \sup_{-\infty < x < \infty} |F_N(x) - F(x)| \le \sup_{1 \le j \le k} |F_N(x_{jk} \pm 0) - F(x_{jk} \pm 0)| + k^{-1}\,.$$

Consequently, $P(D_N \to 0) \ge P(A) = 1$.

Result 5.4.1 (Wald and Wolfowitz). *If $F(x)$ is continuous, the statistics D_N^+ and D_N are distribution-free, where*

$$D_N^+ = \sup_{-\infty < x < \infty} \{F_N(x) - F(x)\}\,, \quad D_N = \sup_{-\infty < x < \infty} |F_N(x) - F(x)|\,.$$

That is, the distributions of D_N^+ and D_N are independent of $F(x)$.

PROOF: Let us prove the result for D_N and in a similar manner it follows for D_N^+.

Let x_y denote the maximum of the inverses of y so that $F(x_y) = y$. As y ranges from 0 to 1, x_y ranges over the possible values of x except sets of zero probability.

$$\begin{aligned} D_N = \sup_{\text{all } x} |F_N(x) - F(x)| &= \sup_{\text{all } x_y} |F_N(x_y) - F(x_y)| \\ &= \sup_{0 \le y \le 1} |F_N(x_y) - F(x_y)| \\ &= \sup_{0 \le y \le 1} |F_N^*(y) - y| \end{aligned}$$

where $F_N^*(y)$ is the sample d.f. based on a random sample of size N drawn from the standard uniform population and

$$F_N^*(y) = \sum_{y_i \le y} (1/N) = \sum_{x_i \le x_y} (1/N) = F_N(x_y)\,.$$

One-sided confidence contours for population d.f.

Let $F(x)$ denote the continuous d.f. of a r.v. X and $F_N(x)$ denotes the sample d.f. determined by a sample $X_1, X_2, \ldots, X_N$. That is, if $X_{1,N} \le X_{2,N} \le \cdots \le X_{N,N}$ denote the order statistics in the sample

$$F_N(x) = \begin{cases} 0, & \text{if } x < X_{1,N} \\ k/N, & \text{if } X_{k,N} \le x < X_{k+1,N}\,, \quad k = 1, 2, \ldots, N-1, \\ 1, & \text{if } X_{N,N} \le x\,. \end{cases}$$

The function

$$F^+_{N,\epsilon}(x) = \min\left\{F_N(x) + \epsilon, 1\right\}\,,$$

also determined by the sample, will be called the *upper confidence contour* for $F(x)$. It is well-known that the probability

$$P_N(\epsilon) = P\left\{F(x) \le F^+_{N,\epsilon}(x), \text{ for all } x\right\}$$

of $F(x)$ being everywhere majorized by $F^+_{N,\epsilon}(x)$ is independent of the d.f. $F(x)$. Wald and Wolfowitz (1939) expressed $P_N(\epsilon)$ in the form of a determinant which is cumbersome to evaluate except for small values of N. Smirnov (1939) obtained the asymptotic expression

$$\lim_{N\to\infty} P_N\left(z/N^{\frac{1}{2}}\right) = 1 - e^{-2z^2}\,.$$

Birnbaum and Tingey (1951) gave an explicit expression for $P_N(\epsilon)$ and a tabulation of values $\epsilon_{N,\alpha}$ accurate to four decimal places such that

$$P_N(\epsilon_{N,\alpha}) = 1 - \alpha$$

for α = .10, .05, .01, .001 and N = 5, 8, 10, 20, 40, 50 which are presented in Appendix VI. Birnbaum and Tingey also observe that for $N = 50$, these values agree very closely with the asymptotic values given by $\bar{\epsilon}_{N,\alpha} = \left[-(2N)^{-1} \ln\, \alpha\right]^{\frac{1}{2}}$. In the following, we will present the result of Birnbaum and Tingey. We need the following lemmas.

Lemma 5.4.2. *For any integer i, $1 \le k \le N$, we have*

$$\begin{aligned} f_{k-1}(y_{k-1}) &= \int_{y_{k-1}}^1 \int_{y_k}^1 \cdots \int_{y_{N-1}}^1 dy_N dy_{N-1} \cdots dy_{k+1} dy_k \\ &= \frac{(1 - y_{k-1})^{N-k+1}}{(N-k+1)!}\,. \end{aligned}$$

PROOF: This formula is well-known and follows either by repeated integration or by induction.

Lemma 5.4.3. *For $0 \leq m \leq k-1$ and any real δ, we have*

$$\sum_{j=0}^{k}(-1)^k \binom{k}{j}(\delta - j)^m = 0\,.$$

PROOF: Consider

$$f_0(x) = \sum_{j=0}^{k}(-1)^j \binom{k}{j} e^{(\delta-j)x} = e^{\delta x}(1-e^{-x})^k\,.$$

Then,

$$f_m(x) = \frac{d^m}{dx^m} f_0(x) = e^{x\delta}\sum_{l=0}^{m}\binom{m}{l} A_{l,m}(e^{-x})(1-e^{-x})^{k-l}$$

where $A_{l,m}(y)$ are polynomials in y. Consequently,

$$f_m(0) = 0, \quad m = 0, 1, \ldots, k-1\,.$$

We can also compute $f_m(x)$ in a different manner and obtain

$$f_m(x) = \sum_{j=0}^{k}(-1)^j \binom{k}{j}(\delta - j)^m e^{(\delta-j)x}\,,$$

from which we have

$$f_m(0) = \sum_{j=0}^{k}(-1)^j \binom{k}{j}(\delta - j)^m, \quad m = 0, 1, 2, \ldots\,.$$

Thus,

$$\sum_{j=0}^{k}(-1)^j \binom{k}{j}(\delta - j)^m = 0, \qquad \text{for } 0 \leq m \leq k-1\,.$$

This completes the proof of the desired lemma.

Lemma 5.4.4. *For any integral $k \geq 0$, we have*

$$\begin{aligned} I(\epsilon, k, N) &= \int_0^{\epsilon}\int_{y_1}^{(1/N)+\epsilon}\cdots\int_{y_k}^{(k/N)+\epsilon} dy_{k+1}\cdots dy_2 dy_1 \\ &= \frac{\epsilon}{(k+1)!}\left(\epsilon + \frac{k+1}{N}\right)^k\,. \end{aligned}$$

PROOF: Let us use the method of induction. This is certainly true for $k = 0$ and 1. Assume it to be true for $k \leq m$. That is,

$$I(\epsilon, k, N) = \frac{\epsilon}{(k+1)!}\left(\epsilon + \frac{k+1}{N}\right)^k, \quad k = 0, 1, \ldots, m.$$

Consider

$$\begin{aligned}
I(\epsilon, m+1, N) &= \int_0^{\epsilon}\int_{y_1}^{(1/N)+\epsilon}\cdots\int_{y_{m+1}}^{\{(m+1)/N\}+\epsilon} dy_{m+2}\cdots dy_2 dy_1 \\
&= \left\{\frac{m+1}{N} + \epsilon\right\} I(\epsilon, m, N) \\
&\quad - \int_0^{\epsilon}\int_{y_1}^{(1/N)+\epsilon}\cdots\int_{y_m}^{(m/N)+\epsilon} y_{m+1} dy_{m+1}\cdots dy_2 dy_1 \\
&= \{(m+1)/N + \epsilon\} I(\epsilon, m, N) - \frac{\{(m/N)+\epsilon\}^2}{2} I(\epsilon, m-1, N) \\
&\quad + \frac{1}{2}\int_0^{\epsilon}\int_{y_1}^{(1/N)+\epsilon}\cdots\int_{y_{m-1}}^{\{(m-1)/N\}+\epsilon} y_m^2 dy_m \cdots dy_1 .
\end{aligned}$$

By repeated integration, one can obtain that

$$\begin{aligned}
I(\epsilon, m+1, N) &= \sum_{j=1}^{m+2}(-1)^{j-1}\left\{\epsilon + \frac{m+2-j}{N}\right\}^j I(\epsilon, m+1-j, N)/j! \\
&= \frac{\epsilon}{(m+2)!}\sum_{j=1}^{m+2}(-1)^{j-1}\binom{m+2}{j}\left(\epsilon + \frac{m+2-j}{N}\right)^{m+1}
\end{aligned}$$

after substituting the result for I_{m+1-j}. Also, from Lemma 5.4.3, we have

$$\sum_{j=0}^{m+2}(-1)^{j-1}\binom{m+2}{j}\left(\epsilon + \frac{m+2-j}{N}\right)^{m+1} = 0.$$

Hence, it follows that

$$I(\epsilon, m+1, N) = \frac{\epsilon}{(m+2)!}\left(\epsilon + \frac{m+2}{N}\right)^{m+1}.$$

This completes the proof of the lemma.

Result 5.4.2 (Birnbaum and Tingey). *For $0 \le \epsilon \le 1$ we have*

$$P_N(\epsilon) = 1 - (1-\epsilon)^N - \epsilon \sum_{j=1}^{[N(1-\epsilon)]} \binom{N}{j} \left(1 - \epsilon - \frac{j}{N}\right)^{N-j} \left(\epsilon + \frac{j}{N}\right)^{j-1}$$

where $[N(1-\epsilon)]$ denotes the largest Integer contained in $N(l-\epsilon)$.

PROOF: Since $P_N(\epsilon)$ does not depend on $F(x)$ we assume that X is a standard uniform random variable. For this r.v. $P_N(\epsilon)$ is the probability that the ordered sample falls into the region

$$X_{j-1,N} \le X_{j,N} \le \frac{j-1}{N} + \epsilon, \quad j = 1, 2, \ldots, 1 + [N(1-\epsilon)]$$

$$X_{j-1,N} \le X_{j,N} \le 1, \qquad j = 2 + [N(1-\epsilon)], \ldots, N,$$

where $X_{0,N} = 0$. Thus,

$$P_N(\epsilon) = N!\, J(\epsilon, [N(1-\epsilon)], N)$$

where

$$J(\epsilon, k, N) = \int_0^{\epsilon} \int_{y_1}^{(1/N)+\epsilon} \cdots \int_{y_k}^{(k/N)+\epsilon} \int_{y_{k+1}}^{1} \int_{y_{k+2}}^{1} \int_{y_{N-1}}^{1} dy_N \cdots dy_{k+2} dy_{k+1} \cdots dy_1 \,.$$

Using Lemma 5.4.2, one obtains

$$J(\epsilon, k, N) = \int_0^{\epsilon} \int_{y_1}^{(1/N)+\epsilon} \cdots \int_{y_k}^{(k/N)+\epsilon} \frac{(1-y_{k+1})^{N-k+1}}{(N-k+1)!} dy_{k+1} dy_k \cdots dy_2 dy_1 \,.$$

We will show by induction that

$$J(\epsilon, k+1, N) = J(\epsilon, k, N) - \frac{\epsilon}{N!} \binom{N}{k+1} \left(1 - \epsilon - \frac{k+1}{N}\right)^{N-k-1} \left(\epsilon + \frac{k+1}{N}\right)^{k},$$

for any integer $0 \le k \le N-1$. The above is certainly true when $k = 0$. Assume the asserted result true for $k \le m$ and consider

$$\begin{aligned} & J(\epsilon, m+1, N) \\ &= \int_0^{\epsilon} \int_{y_1}^{(1/N)+\epsilon} \cdots \int_{y_m}^{(m/N)+\epsilon} \int_{y_{m+1}}^{\{(m+1)/N)+\epsilon\}} \frac{(1-y_{m+2})^{N-m-2}}{N-m-2} dy_{m+2} dy_{m+1} \cdots dy_1 \\ &= J(\epsilon, m, N) - \left(1 - \epsilon - \frac{m+1}{N}\right)^{N-m-1} I(\epsilon, m, N), \end{aligned}$$

after integrating with respect to y_{m+2}. Now, using Lemma 5.4.4, one obtains

$$\begin{aligned} J(\epsilon, m+1, N) &= J(\epsilon, m, N) - \frac{\epsilon}{N!}\binom{N}{m+1}\left(1-\epsilon-\frac{m+1}{N}\right)^{N-m-1} \\ &\quad \cdot \left(\epsilon+\frac{m+1}{N}\right)^{m}. \end{aligned}$$

Noting that $J(\epsilon, 0, N) = \left[1-(1-\epsilon)^N\right]/N!$, one readily obtains that

$$J(\epsilon, m, N) =$$

$$(N!)^{-1}\left[1-(1-\epsilon)^N - \epsilon\sum_{j=1}^{m}\binom{N}{j}\left(\epsilon+\frac{j}{N}\right)^{j-1}\left(1-\epsilon-\frac{j}{N}\right)^{N-j}\right].$$

This completes the proof of the assertion.

Remark 5.4.1. *Setting $F_{N,\epsilon}^{-}(x) = \max\left[F_N(x)-\epsilon, 0\right]$, one can easily verify that*

$$P\left\{F(x) \geq F_{N,\epsilon}^{-}(x) \text{ for all } x\right\} = P_N(\epsilon),$$

and hence also is given by Result 5.4.2.

Remark 5.4.2. *Kolmogorov (1941) pointed out that*

$$P_N(\epsilon) \geq 1 - e^{-2N\epsilon^2}.$$

Consequently, the asymptotic values of $\epsilon_{N,\alpha}$ will be larger than the exact values. This fact was also numerically verified by Birnbaum and Tingey (1951) for $.001 \leq \alpha \leq .1$ and they also point out that the error in using the asymptotic values becomes small for $N \geq 50$.

Miller (1956) tabulated the values of $\epsilon_{N,\alpha}$ accurate to five decimal places for $N = 1(1)100$ and for $\alpha = .10, .05, .025, .01$ and $.005$. For $N > 20$, he uses the approximate formula

$$\epsilon_{N,\alpha} = \bar{\epsilon}_{N,\alpha} - 0.16693N^{-1} - A(\alpha)N^{-3/2}$$

where

$$A(\alpha) = .09037(-\ln_{10}\alpha)^{3/2} + .01515(\ln_{10}\alpha)^2 - .08467\alpha - .11143$$

for $.005 \leq \alpha \leq .10$ and where $\bar{\epsilon}_{N,\alpha} = \left[-(2N)^{-1}\ln\,\alpha\right]^{\frac{1}{2}}$. The approximate formula gives at least four decimal place accuracy. Some of Miller's (1956) values for $\epsilon_{N,\alpha}$ are presented in Appendix VI.

Two-sided confidence contours for the population cumulative distribution

In some problems one might he interested in a two-sided confidence contour for the unknown population d.f. That is, we wish to find an $\epsilon > 0$ such that

$$P\left[F_N(x) - \epsilon \leq F(x) \leq F_N(x) + \epsilon, \text{ for all } x\right] = 1 - \alpha$$

for given α. This probability is equal to $P(D_N < \epsilon)$ where D_N is the Kolmogorov's (1933) statistic given by

$$D_N = \sup_{\text{all } x} |F_N(x) - F(x)| \ .$$

Kolmogorov (1933) derived a system of recurrence formulae which make it possible to compute, for finite N the probabilities

$$P_N\left(k/N^{\frac{1}{2}}\right) = P\left(D_N < k/N^{\frac{1}{2}}\right), \quad \text{for } k = 1, 2, \ldots \ .$$

Using this system of recurrence formulae, Birnbaum (1952) tabulated to five decimal places the value of $P_N\left(k/N^{\frac{1}{2}}\right)$ for $k = 1(1)15$ and $N = 1(1)100$ such that $k/N \leq 1$ and gave some other -useful tables. Massey (1950), using the theory of random walks, reduced the problem of finding $P_N\left(k/N^{\frac{1}{2}}\right)$ to solving a single difference equation. Massey (1950) tabulated $P_N\left(k/N^{\frac{1}{2}}\right)$ to four decimal places for $k = 1(1)9$, $N = 5(5)45$ and $k = 5(1)9$, $N = 50(5)80$. Massey (1951) also tabulated ϵ to 2–3 significant digits such that

$$P\left[F_N(x) - \epsilon \leq F(x) \leq F_N(x) + \epsilon, \text{ for all } x\right] = 1 - \alpha$$

for $N = 1(1)20(5)35$, and $\alpha = .2, .15, .10, .05$ and $.01$. These values together with the values of ϵ for $N = 40(10)100$, and $\alpha = .05$ and $.01$ taken from Miller (1956) are given in Appendix VI.

Obtaining two-sided confidence bands from one-sided bands

Let E_1 be the event that $F(x) \geq h_1(x)$ for $A_1 \leq F(x) \leq A_2$ where $h_1(x) = \max\left[0, F_N(x) - \epsilon_1\right]$ and let E_2 be the event that $F(x) \leq h_2(x)$, $A_1 \leq F(x) \leq A_2$ where $h_2(x) = \min\left[1, F_N(x) + \epsilon_2\right]$, where ϵ_1 and ϵ_2 are non-negative numbers. Also, $\bar{E}_1$ and $\bar{E}_2$ denote the events complementary to E_1 and E_2, respectively. Then, in general, for $A_1 \leq F(x) \leq A_2$ [using the fact that $P(\overline{E_1 \cap E_2}) = P(\bar{E}_1 \cup \bar{E}_2)$]

$$P\left[h_1(x) \leq F(x) \leq h_2(x)\right] = 1 - P(\bar{E}_1 \text{ or } \bar{E}_2) = 1 - P(\bar{E}_1) - P(\bar{E}_2) + P(\bar{E}_1\bar{E}_2) \ .$$

Also, asymptotically (see Malmquist, 1954),

$$P(\bar{E}_1\bar{E}_2) \le P(\bar{E}_1)P(\bar{E}_2) = \{1 - P[F(x) \le h_1(x)]\}\{1 - P[F(x) \ge h_2(x)]\} ,$$

and for rather small sample sizes, we have the much less stringent inequality

$$P(\bar{E}_1\bar{E}_2) \le 2P(\bar{E}_1)P(\bar{E}_2) ,$$

which holds if the confidence coefficient for the two-sided band is at least 0.84. Using the inequality due to Malmquist, for $A_1 \le F(x) \le A_2$, we have

$$P[h_1(x) \le F(x)] \le h_2(x)] \le P[F(x) \ge h_1(x)]\, P[F(x) \le h_2(x)] .$$

Thus, one-sided confidence bands can be obtained from symmetric two-sided bands. Use of the above relation for finding the two-sided confidence band will result in an error of at most 5 percent in the value of [1 – the actual confidence coefficient for the two-sided band], provided the actual confidence is at least 0.84.

Miller (1956) suggests that the confidence coefficient of a two-sided symmetric band be written as $1 - 2\alpha + \gamma$ where $1 - \alpha$ denotes the confidence coefficient of a one-sided confidence band for a given ϵ. When N is even, Miller (1956) finds that

$$\gamma = 0, \quad \frac{1}{2} \le \epsilon \le 1$$

and

$$\gamma = (1 - 2\epsilon)^N \left\{1 + 2^{-N}\binom{N}{N/2}\right\} , \quad (N-1)(2N)^{-1} \le \epsilon \le \frac{1}{2} .$$

Using the table of values for the two-sided confidence band given by Birnbaum (1952), Miller verifies that γ, which is a decreasing function of ϵ, is of the order 10^{-6} for $.005 \le \alpha \le .05$. This observation enables one to use the table in Appendix VI for a two-sided confidence band, provided the confidence coefficient is taken as $1 - 2\alpha$.

Asymptotic distribution of the Kolmogorov's statistic

The history of the proofs of the limiting distributions of Kolmogorov-Smirnov statistics is interesting. The original proofs given by Kolmogorov and Smirnov are long and complicated. The proof by Feller (1948) is simpler but unwieldy. A short and elegant proof based on a heuristic approach was given by Doob (1949) and the validity of the heuristic approach was established by Donsker (1952). In the following, the idea of the proof due to Doob (1949) will be presented. Let

$$U_N(x) = N^{\frac{1}{2}}\left[F_N(x) - F(x)\right] .$$

Since $F(x)$ is continuous and $U_N(x)$ is distribution-free, we may assume that $F(x) = x$, $0 \le x \le 1$. Thus,

$$U_N(x) = N^{\frac{1}{2}}\left[F_N(x) - x\right], \quad D_N N^{\frac{1}{2}} = \sup_{0\le x\le 1} |U_N(x)| .$$

We know that

$$\begin{aligned} E\left[U_N(x)\right] &= 0, & (0 \le x \le 1), \\ E\left[U_N(x_1)U_N(x_2)\right] &= x_1(1-x_2), & (0 \le x_1 \le x_2 \le 1) . \end{aligned} \quad (5.4.1)$$

From the multi-dimensional de Moivre-Laplace theorem it follows that for $k = 1, 2, \ldots$ and arbitrary points $x_1, x_2, \ldots, x_k$, where $0 \le x_1 \le x_2 \le \cdots \le x_k$, the random vector

$$[U_N(x_1), U_N(x_2), \ldots, U_N(x_k)]$$

is asymptotically normally distributed with the expected values, variances and covariances given by Eq. (5.4.1).

Let us consider a normal stochastic process $\{U(x), 0 \le x \le 1\}$ such that for $k = 1, 2, \ldots$ and for arbitrary $x_1, x_2, \ldots, x_k$ $(0 \le x_1 \le \cdots \le x_k \le 1)$, the random vector

$$[U(x_1), U(x_2), \ldots, U(x_k)]$$

has the normal distribution with expected values, variances and covariances given by Eq. (5.4.1). The realizations of such a stochastic process are continuous functions, with probability one.

Doob's heuristic approach asserted that

$$\lim_{N\to\infty} P\left[\sup_{0\le x\le 1} |U_N(x)| < \lambda\right] = P\left[\max_{0\le x\le 1} |U(x)| < \lambda\right] .$$

Doob (1948) established that the right-hand side of the last equality is equal to the asymptotic distribution obtained by Kolmogorov (1933).

Since $E\left[U(x)\right]^2 = x(1-x)$, it follows that the variance of $U_N(x)$ decreases towards the tails of the distribution. Thus, in constructing the contours $C_1(x) \le U_N(x) \le C_2(x)$, it seems reasonable to let the width $C_2(x) - C_1(x)$ decrease towards the ends of the distribution. The form $-C_1(x) = C_2(x) =$

$a\{x(1-x)\}^{\frac{1}{2}}$ with exclusion of points $x = 0$ and $x = 1$ has been suggested by Anderson and Darling (1952). For example,

$$C_2(x) = \begin{cases} (a-b)x + b & 0 \le x \le \frac{1}{x} \\ (b-a)x + b & \frac{1}{2} \le x \le 1 \quad a > b. \end{cases}$$

Then deviations $U_N(x)$ at the extremes of the distribution will have a greater chance of being detected, or the width of the confidence contours can be made smaller at the tails than at the middle of the distribution. There are no general principles for the choice of a and b. We may also be interested in the deviations $U_N(x)$ for a certain part of the distribution $F(x)$. That is, $C_2(x)$ is of the form

$$C_2(x) = \begin{cases} ax + b & A \le x \le B \\ (1-x)N^{\frac{1}{2}} & \text{elsewhere.} \end{cases}$$

After finding the limiting probabilities,

$$\lim_{N\to\infty} P\{U_N(x) \le C_2(x),\ 0 \le x \le 1\},$$

$$\lim_{N\to\infty} P\{C_1(x) \le U_N(x) \le C_2(x),\ 0 \le x \le 1\},$$

with special forms for $C_1(x)$ and $C_2(x)$, one can obtain asymptotic confidence contours by inverting the probability statements. Anderson and Darling (1952) and Malmquist (1954) considered these problems. In the following we will present without proof some of the results of Malmquist (1954).

$$\lim_{N\to\infty} P\left\{ \begin{aligned} [F_N(x) - F(x)]\, N^{\frac{1}{2}} &\le (a-b)F(x) + b; \quad -\infty < x \le x^* \\ &\le (b-a)F(x) + a; \quad x^* \le x \le \infty \end{aligned} \right\}$$
$$= \Phi(a+b) - 2e^{-2ab}\Phi(a-b) + e^{-4b(a-b)}\Phi(a-3b),$$

where Φ is the standard normal d.f. and x^* is such that $F(x^*) = \frac{1}{2}$.

An expression of the above type for $a > b$ gives greater weight to deviations at the extremes of $F(x)$ than does the ordinary expression with

$a = b$:

$$\lim_{N\to\infty} P\left\{\begin{array}{rll} [F_N(x) - F(x)]\, N^{\frac{1}{2}} & \leq (a-b)F(x) + b; & -\infty < x \leq x^* \\ & \geq -(a-b)F(x) - b; & x^* \leq x < \infty \end{array}\right\}$$

$$\begin{aligned} &= 1 - 2\Phi(-a-b) - e^{-2ab}[1 - 2\Phi(-a-3b)] \\ &\quad + e^{-8ab}\left[\Phi(3a-b) - \Phi(a-3b)\right] . \end{aligned}$$

Probabilities of this type can be used when the dispersion in $F(x)$ is of particular interest. Finally, let $F(x') = u/(1+u)$, $F(x'') = v/(1+v)$ and $R = (u/v)^{\frac{1}{2}}$. Then,

$$\lim_{N\to\infty} P\left\{|F_N(x) - F(x)|\, N^{\frac{1}{2}} \leq (a-b)F(x) + b,\ \ x' \leq x \leq x''\right\}$$

$$\begin{aligned} &= \frac{\int_{-p_1u^{-\frac{1}{2}}}^{p_1u^{-\frac{1}{2}}} ds_1}{2\pi(1-R^2)^{\frac{1}{2}}} \int_{-p_2v^{-\frac{1}{2}}}^{p_2v^{-\frac{1}{2}}} ds_2 \exp\left\{\frac{-1}{2(1-R^2)}\left[s_1^2 - 2Rs_1s_2 + s_2^2\right]\right\} \\ &\quad - \frac{\sum_{j=1}^{\infty}(-1)^{j+1}e^{-2j^2ab}}{2\pi(1-R^2)^{\frac{1}{2}}} \int_{A_1}^{A_2} ds_1 \int_{B_1}^{B_2} ds_2 \exp\left\{\frac{-\left[s_1^2 - 2Rs_1s_2 + s_2^2\right]}{2(1-R^2)}\right\} \end{aligned}$$

where

$$\begin{aligned} A_1 &= -(p_1 + 2jax)u^{-\frac{1}{2}}, & B_1 &= -(p_2 + 2jb)v^{-\frac{1}{2}}, \\ A_2 &= (p_1 + 2jax)u^{-\frac{1}{2}}, & B_2 &= (p_2 + 2jb)v^{-\frac{1}{2}}, \\ p_1 &= au + b, & p_2 &= av + b . \end{aligned}$$

This distribution for $a = b$ has been evaluated by Anderson and Darling (1952) and Maniya (1949).

Blackman (1958) has obtained an exact expression for $P(-a < F_N(x) - F(x) < b$ for all $x)$. The expression is too complicated to be presented here.

5.5 Characterization of Distribution-free Statistics

Suppose $X_1, X_2, \ldots, X_N$ is a random sample of a one-dimensional random variable X having the continuous d.f. $F(x)$. It has been observed by Z.W. Birnbaum (1953) that all distribution-free statistics considered in the literature can be written in the form

$$\Psi\{F(X_1), F(X_2), \ldots, F(X_N)\}$$

where Ψ is a measurable[1] symmetric function defined on the unit cube $\{U : 0 \leq U_i \leq 1,\ i = 1, 2, \ldots, N\}$. Birnbaum and Rubin (1954) studied the relationship between the class of statistics which can be of this form and the class of distribution-free statistics. Bell (1960) has extended the results of Birnbaum and Rubin. These will be considered in the following.

We need the following notation consistent with that of Scheffé (1943). Let

- Ω_0 denote the class of all d.f.'s;
- Ω_1 denote the class of all non-degenerate d.f.'s;
- Ω_2 denote the class of all continuous d.f.'s;
- Ω^* denote the class of all strictly monotone continuous d.f.'s
- Ω_3 denote the class of all absolutely continuous d.f.'s (with respect to Lebesgue measure) d.f.'s;
- Ω_4 denote the class of d.f.'s with continuous derivatives;
- Ω_u denote the class of d.f.'s which are uniform within q intervals;
- Ω_e denote the class of all d.f.'s with densities of the form

$$C(\theta_1 \cdots \theta_k) \exp\{-x^{2k} - \theta_1 x_1 - \theta_2 x_2 - \cdots - \theta_k x^k\}.$$

Analogously for the unit interval I, one defines

- $\Omega_0(I)$ =the class of all d.f.'s on I;
- $\Omega_1(I)$ =the class of all non-degenerate d.f.'s on I?
- $\Omega_2(I)$ =the class of all continuous d.f.'s on I; etc.

[1]A real-valued function $\Psi(y)$ is measurable if for any real number a, the set of all y for which $\Psi(y) > a$ is measurable.

Let R, $R^{(N)}$, I, $I^{(N)}$, $\mathcal{B}$, $\mathcal{B}^{(N)}$, $\mathcal{B}_I$ and $\mathcal{B}_I^{(N)}$ respectively denote the real line; Euclidean N-space; the open unit interval; the N-dimensional open-unit cube; and the respective classes of Borel subset of R, $R^{(N)}$, I, $I^{(N)}$.

Definition 5.5.1. *If Ω and Ω' are two arbitrary families of d.f.'s, a real-valued function*

$$S_G = S_G(X_1, X_2, \ldots, X_N)$$

will be called a statistic in Ω with regard to (w.r.t.) Ω', if for every G in Ω, and F in Ω', and $X_1, X_2, \ldots, X_N$ in the N-dimensional sample space for a random variable X which has d.f. F,

1. *$S_G(X^{(N)}) = S_G(X_1, X_2, \ldots, X_N)$ is defined everywhere in the sample space, and*
2. *$S_G = S_G(X^{(N)})$ has a probability distribution; this probability distribution will be denoted by $\mathcal{P}(S_G, F)$.*

For example, consider the Kolmogorov statistic

$$\begin{aligned} D_N &= \sup_{-\infty<x<\infty} |F_N(x) - G(x)| \\ &= \max_{i=1,2,\ldots,N} \{\max[G(X_{i,N}) - (i-1)/N, (i/N) - G(X_{i,N})]\} \end{aligned}$$

where $X_{1,N} < X_{2,N}, \ldots, < X_{N,N}$ are the ordered sample values. D_N satisfies 1. and 2. when $\Omega = \Omega' = \Omega_2$. Hence, D_N is a statistic in Ω_2 w.r.t. Ω_2.

Definition 5.5.2. *If for a statistic $S_G(X^{(N)})$ in Ω w.r.t. Ω', there exists a (measurable) function Ψ defined on the N-dimensional unit cube and symmetric in its arguments, such that for any G in Ω, F in Ω', we have*

$$S_G(X^{(N)}) = \Psi(G(X_1), \ldots, G(X_N))$$

almost everywhere in the sample space $X^{(N)}$ for the r.v. X which has d.f. F, then $S_G(X^{(N)})$ is said to be a statistic of structure (d).

Kolmogorov-Smirnov statistics are examples of statistics of structure (d), since they can be written as

$$D_N^+ = \max_{i=1,2,\ldots,N} \{(i/N) - G(X_{i,N})\}$$

and

$$D_N = \max_{i=1,2,\ldots,N} \{\max[G(X_{i,N}) - (i-1)/N,\ (i/N) - G(X_{i,N})]\}\ .$$

Definition 5.5.3. *If $\Omega = \Omega'$ and the statistics $S_G(X^{(N)})$ has the property that the probability distribution $\mathcal{P}(S_G : H)$ is independent of G for G in Ω, then $S_G(X^{(N)})$ is said to be a distribution-free statistic in Ω.*

If $\Omega = \Omega' = \Omega_2$, the class of all continuous d.f.'s, R denotes a uniform r.v. on (0,1) and if $U_1, U_2, \dots, U_N$ is a random sample of size N from the uniform distribution in (0,1), then we have

$$\mathcal{P}\left(\Psi\left(G(X_1), \dots, G(X_N)\right); G\right) = \mathcal{P}\left(\Psi(U_1, \dots, U_N); R\right) .$$

Result 5.5.1 (Birnbaum and Rubin, 1954). *If a statistic in Ω_2 w.r.t. Ω_2 has structure (d), then it is distribution-free in Ω_2.*

PROOF: Trivial.

All distribution-free statistics considered in the literature happen to possess structure (d), with $\Omega = \Omega' = \Omega_2$. However, every distribution-free statistic, symmetric in $X_1, X_2, \dots, X_N$, with $\Omega = \Omega' = \Omega_2$ does not necessarily have structure (d). This can be seen from the following counter example.

Example 5.5.1 (Birnbaum and Rubin). Let ω_1 and ω_2 be non-empty, mutually exclusive subsets of Ω_2 such that $\omega_1 \cup \omega_2 = \Omega_2$. Define

$$S = \begin{cases} \sup_{\text{all } x} \left[F(x) - F_N(x)\right] = S_1 & \text{for } F \in \omega_1, \\ \sup_{\text{all } x} \left[F_N(x) - F(x)\right] = S_2 & \text{for } F \in \omega_2. \end{cases}$$

Since S_1 and S_2 are distribution-free statistics with the same probability distribution, S is a distribution-free statistic. however, it does not have structure (d).

Let G belong to Ω^*, the class of all strictly monotone continuous d.f.'s. Then, the inverse function G^{-1} is uniquely defined in the open unit interval.

Definition 5.5.4. *If $S_G(X^{(N)})$ is a statistic in $\Omega \subset \Omega^*$ w.r.t. some Ω', then $S_G(X^{(N)})$ is called a strongly distribution-free statistic in Ω w.r.t. Ω' if the probability distribution of $S_G(X^{(N)})$ depends only on the function $\tau = FG^{-1}$ for all G in Ω and F in Ω'.*

In view of the above definition, the following result can easily be established.

Result 5.5.2.

1. *If a statistic in Ω^* w.r.t. Ω^* is strongly distribution-free, then it is distribution-free in Ω^*; and*

2. *If a statistic in Ω^* w.r.t. Ω^* has structure (d), then it is strongly distribution-free.*

PROOF:

1. If $\mathcal{P}\left(S_G(X^{(N)}); F\right)$ depends only on FG^{-1} for all F, G in Ω^*, then, in particular, $\mathcal{P}\left(S_G(X^{(N)}); G\right)$ depends on $GG^{-1} = I$, hence is independent of G. Also, since

$$\mathcal{P}\left\{\Psi\left(G(X_1), \ldots, G(X_N)\right); F\right\} = \mathcal{P}\left\{\Psi(U_1, \ldots, U_N); FG^{-1}\right\}.$$

2. Readily follows.

One might ask whether the properties of strongly distribution-free and of structure (d) are equivalent. This has been answered affirmatively by Birnbaum and Rubin (1954) and it will be stated without proof in the following result.

Result 5.5.3. *If a statistic $S_G(X^{(N)})$ in Ω^* w.r.t. Ω^* is symmetric in $X_1, X_2, \ldots, X_N$ and strongly distribution-free, then it has structure (d).*

The results of Birnbaum and Rubin (1954) have been extended to certain classes of d.f.'s by Bell (1960). His main result, together with some definitions, will be given below.

Definition 5.5.5. *An arbitrary class of d.f.'s Ω' is said to be closed under another arbitrary class of d.f.'s Ω if*

$$FG^{-1}G_1 \text{ is in } \Omega' \text{ whenever } F \text{ is in } \Omega' \text{ and } G,\ G_1 \text{ are in } \Omega.$$

Remark 5.5.1. *It follows that Ω_0, Ω_1, Ω_2 and Ω^* are each closed under Ω^* since for F in $\Omega_0(\Omega_1, \Omega_2, \Omega^*)$ and G, G_1 in Ω^*,*

1. *FG^{-1} is in $\Omega_0(I)\,[\Omega_1(I), \Omega_2(I), \Omega^*(I)]$ and*

2. *$FG^{-1}G_1$ is in $\Omega_0(\Omega_1, \Omega_2, \Omega^*)$.*

Further, under such mappings, numerical values are preserved in the following sense. If F is in Ω_0 and G, G_1 are in Ω^*, then

1. $\mathcal{P}^{(N)}_{FG^{-1}} = \mathcal{P}^{(N)}_{F}\left[G^{-1}(B)\right]$ for all B in $\mathcal{B}^{(N)}_{I}$ and

2. $\mathcal{P}^{(N)}_{FG^{-1}G_1}\left[G_1^{-1}(B)\right] = \mathcal{P}^{(N)}_{FG^{-1}}(B)$ for all B in $\mathcal{B}^{(N)}_{I}$ where

$$\left[G(x^{(N)})\right] = [G(x_1),\ldots,G(x_N)]$$

and

$$G^{-1}(u_1,\ldots,u_N) = \left[G^{-1}(u_1),\ldots,G^{-1}(u_N)\right] .$$

Definition 5.5.6. *A class of d.f.'s is said to be symmetrically complete if every unbiased, symmetric estimator of zero, with respect to the class of power probability distributions of Ω is essentially zero, that is, the conditions*

1. *h is symmetric, and*

2. *$\int_{R^{(N)}} h d\mathcal{P}^{(N)}_{F} = 0$*

for all F in Ω imply that $h = 0[\mathcal{P}^{(N)}_{F}]$ for all F in Ω.

Result 5.5.4 (Bell). *If S_G is a statistic in Ω w.r.t. Ω', then the property of being symmetric and strongly distribution-free is equivalent to having structure (d), whenever the following three conditions are fulfilled.*

1. *$\Omega \subset \Omega^*$;*

2. *Ω' is closed under Ω; and*

3. *Ω' is a symmetrically complete class.*

One would like to ask which classes of statistical interest satisfy the hypotheses of the preceding result due to Bell (1960). Also, it should be remarked that a class of d.f.'s is symmetrically complete if and only if the order statistic is a complete statistic with respect to the class of power probability distributions of the given class of d.f.'s. Also the work of Halmos (1946), Fraser (1953), Lehmann (1951), Bell-Blackwell-Breiman (1960) establish that Ω_0, Ω_1, Ω_2, Ω_3, Ω_4, Ω_u and Ω_e are symmetrically complete (see Bell, 1960, for references on Halmos, Fraser and Lehmann). Therefore, Ω_0, Ω_1, Ω_2, Ω_3 and Ω_4 satisfy both the completeness and closure hypotheses of the preceding result. Consequently, we have the following corollary by Bell (1960) to his result.

Corollary 5.5.4.1. *If S_G is a statistic in Ω w.r.t. Ω', then the property of being symmetric and strongly distribution-free is equivalent to having structure (d) for each of the following cases:*

1. $\Omega \subset \Omega^*$ *and* $\Omega' = \Omega_0$*;*

2. $\Omega \subset \Omega^*$ *and* $\Omega' = \Omega_1$*;*

3. $\Omega \subset \Omega^*$ *and* $\Omega' = \Omega_2$*;*

4. $\Omega \subset \Omega^*$ *and* $\Omega' = \Omega^*$*;*

5. $\Omega = \Omega_3 \cap \Omega^*$ *and* $\Omega' = \Omega_3$*;*

6. $\Omega = \Omega_4 \cap \Omega^*$ *and* $\Omega' = \Omega_4$*;*

5.6 Completeness of the Order Statistic

The concept of completeness was first introduced by Halmos (1946) and extended by Lehmann and Scheffé (1950 and 1955), Fraser (1953) and Bell (I960). Although the term completeness is used for a statistic, it is a property of a class of distributions and when applied to a statistic, it will be in reference to the induced distribution of the statistic.

Definition 5.6.1. *A family of distributions* Ω *is said to be complete if for any real function* $h(x)$*,*

$$Eh(X) = \int h(x)dF(x) = 0$$

implies that $h(x) = 0$ *almost everywhere except on sets of probability zero w.r.t. each* F*.*

That is, a class of distributions is said to be complete if there does not exist an unbiased estimate $h(x)$ of zero other than the trivial unbiased estimate which is zero almost everywhere.

Definition 5.6.2. *A statistic* $T(X)$ *is said to be complete relative to the class of d.f.'s* Ω *if the induced class of d.f.'s* $\{F_T\}$ *is complete.*

Example 5.6.1. Consider the r.v. $X = (X_1, X_2, \ldots, X_N)$ where the X_i are independent and each is normally distributed with mean μ, and variance 1. The class of d.f.'s is obtained from values of μ in $(-\infty, \infty)$. We wish to show that $\bar{X} = N^{-1} \sum_1^N X_i$ is complete. The induced distribution of $\bar{X}$ is normal with mean μ, and variance $1/N$. Thus, we wish to show that the class of densities

$$\left\{ (N/2\pi)^{\frac{1}{2}} \exp\left[-\frac{N}{2}(y-\mu)^2 \right] \mid \mu \in (-\infty, \infty) \right\}$$

is complete. Towards this, assume that

$$\int_{-\infty}^{\infty} h(y)(N/2\pi)^{\frac{1}{2}} \exp\left[-\frac{N}{2}(y-\mu)^2\right] dy = 0\,.$$

That is,

$$\int_{-\infty}^{\infty} h(y)(N/2\pi)^{\frac{1}{2}} \exp(-Ny^2/2)\exp(Ny\mu)dy = 0\,.$$

Letting $N\mu = v$, we obtain

$$\int_{-\infty}^{\infty} h(y)(N/2\pi)^{\frac{1}{2}} \exp(-Ny^2/2)\exp(vy)dy = 0\,.$$

This implies that the Laplace transform of the function $h(y)\exp(-Ny^2/2)$ is zero identically. Since 0 also has the transform which is zero identically and by the uniqueness property of Laplace transform, it follows that $h(y)\exp(-Ny^2/2) = 0$. That is, $h(y) = 0$ almost everywhere.

Definition 5.6.3. *A class of d.f.'s Ω is boundedly complete if for any real function $h(x)$ satisfying $|h(x)| < M$, $E\{h(X)\} = 0$ w.r.t. every member of Ω implies that $h(x) = 0$ except on sets of probability zero.*

We have the following simple result connecting completeness and boundedly completeness.

Result 5.6.1. *If a class of d.f.'s is complete, it is boundedly complete.*

PROOF: Trivial.

However, the converse of this result is not necessarily true. Consider the following example which is due to Girshick, Mosteller and Savage (1946).

Example 5.6.2. Let X be a r.v. taking the values $0, 1, 2, \ldots, i+1, \ldots$ with probabilities $q, p^2, p^2q, \ldots, p^2q^i, \ldots$, respectively, where $q = 1-p$. Consider the class of probability functions obtained by letting p range over (0,1). We will show that this class is boundedly complete, but not complete. A function with zero expectation for all the functions in the class will satisfy

$$h(0)q + h(1)p^2 + h(2)p^2q + \cdots = 0\,.$$

That is,

$$\begin{aligned} h(1) + qh(2) + q^2h(3) + \cdots &= -h(0)qp^{-2} \\ &= -h(0)(q + 2q^2 + 3q^3 + \cdots)\,. \end{aligned}$$

Since the two power series are identical for q in (0,1), it follows that

$$h(1) = 0, \;\; h(2) = -h(0), \ldots, h(i) = -(i-1)h(0), \ldots .$$

This gives the form of any unbiased estimate of zero. If $h(0) = 0$, then $h(x) = 0$ at all non-negative integers. If $h(0) \neq 0$, $h(x)$ will be unbounded. Thus, there are non-degenerate unbiased estimates of zero, none of which are bounded. Hence, the class of probability functions is boundedly complete but not complete.

Unlike the property of sufficiency if we have completeness for a class of d.f.'s, we can sometimes infer completeness for a larger class,

Result 5.6.2. *The completeness of Ω implies the completeness of Ω' if Ω is a subset of Ω' and if none of the added distributions assign positive probability to sets having zero probability w.r.t. each member of Ω. The second condition implies that almost everywhere Ω is equivalent to almost everywhere Ω'.*

PROOF: We first prove the last statement in the theorem. The condition clearly implies that any set having probability zero w.r.t. each member of Ω, also has zero probability w.r.t. each member of Ω'; and the converse is easy. However, this is another way of expressing the equivalence of Ω and Ω'.

Since Ω is complete it follows from the definition of completeness that a function which satisfies certain conditions is zero almost everywhere w.r.t. Ω. Introducing more distributions imposes more conditions. The original conditions (Ω) were sufficient to prove that $h(x) = 0$ almost everywhere. Also, since almost everywhere Ω is equivalent to almost everywhere Ω', completeness of the larger class readily follows. This completes the proof of the asserted result.

Completeness of a class of joint d.f.'s which are products of univariate d.f.'s over a product space can sometimes be obtained from completeness over the component spaces. Towards this, we need a stronger concept of completeness.

Definition 5.6.4 (Fraser). *The class Ω is strongly complete if there exists a probability measure on Ω, such that, for any subset Ω^* of Ω, the members $\Omega - \Omega^*$ form a set having probability zero with respect to the probability measure, the condition*

$$E_F h(X) = 0$$

for all F in Ω^* and any real function $h(x)$ implies that $h(x) = 0$ almost everywhere Ω.

Definition 5.6.5 (Fraser). *If the random variable X is complete w.r.t. Ω_1 and the random variable Y is strongly complete w.r.t. Ω_2, then the r.v. (X, Y) is complete w.r.t. the class of product d.f.'s $\{\Omega_1 \times \Omega_2\}$.*

PROOF: See Fraser (1957, pp. 25–27).

With the above concepts and definitions, we will exhibit the property of completeness of the order statistic.

Definition 5.6.6 (Bell). *The order statistic $T(X_1, X_2, \ldots, X_N)$ is said to be complete w.r.t. the class $\Omega^{(N)}$ N-fold power probability distributions if*

$$E_{F^{(N)}}\{h[T(X_1, X_2, \ldots, X_N)]\} = 0$$

for all F in Ω implies that $h(T[X_1, X_2, \ldots, X_N]) = 0$ except on sets of probability zero w.r.t. $F^{(N)}$. The class Ω is said to be symmetrically complete whenever the latter condition hold.

Completeness of the order statistic in samples drawn from the discrete populations has been established by Halmos (1946). The property was extended by Lehmann and Scheffé (1950 and 1955) to the class of all continuous .d.f.'s and to the class of all exponentials of a certain form. Completeness of the order statistic was extended by Fraser (1954) to the class of all absolutely continuous d.f.'s. Bell (1960) has extended the known results on completeness to probability spaces other than the real line. In the following we will present the proof of the completeness of the order statistic w.r.t. the absolutely continuous distributions on the real line and the discrete distributions on the real line and we will state without proof Bell's (1960) main result. We need the following lemma on homogeneous polynomials.

Lemma 5.6.1 (Halmos). *If $Q(p_1, p_2, \ldots, p_N)$ is a homogeneous polynomial of degree greater than zero, satisfying $Q(p_1, p_2, \ldots, p_N) = 0$ whenever, $0 \le p_i \le 1$ $(i = 1, 2, \ldots, N)$ and $\sum_1^N p_i = 1$, then $Q(p_1, p_2, \ldots, p_N)$ is zero identically.*

PROOF: Proof is by method of induction. For $N = 1$, the lemma is trivially true. If each p_i is replaced by cp_i, due to the homogeneity of Ω, a power of c will factor out leaving the original polynomial. Hence, it is sufficient to have the restriction $p_i \ge 0$ $(i = 1, 2, \ldots, N)$. If Ω is rewritten as a polynomial in p_N, the coefficients will be homogeneous polynomials in $p_1, p_2, \ldots, p_{N-1}$ and by assumption in the method of induction, they will be identically zero. Thus Ω is identically zero.

Definition 5.6.7. *A distribution is said to be uniform within intervals if it assigns probabilities $p_1, p_2, \ldots, p_N$ to the disjoint intervals $I_1, I_2, \ldots, I_N$ on the real line ($\sum_1^N p_i = 1$) and within each interval the distribution is uniform, that is, it has a density function which is constant valued within each interval.*

Result 5.6.3 (Fraser). *The order statistic $T(X_1, X_2, \ldots, X_N)$ is complete for the class of N-fold power d.f.'s $\Omega^{(N)} = \{F^{(N)}\}$ defined in the N-dimensional Euclidean space where $F(x)$ is any distribution uniform within intervals.*

PROOF: It is sufficient to show that any real function $h(T[X_1, \ldots, X_N])$ of the order statistic satisfying

$$E\{h(T)\} = 0$$

w.r.t. all d.f.'s in $\Omega^{(N)}$ is necessarily zero almost everywhere. First let us find a suitable way of expressing a function of the order statistic $T(X_1, \ldots, X_N)$. Clearly, any function of T is a symmetric function of $X_1, \ldots, X_N$. Conversely, any symmetric function is a function of the X_i's which does not depend on the order in which the observations come and consequently is a function of the set: $\{x_1, \ldots, x_N\}$, that is, it is a function of the order statistic. Therefore, considering a symmetric function $h(x_1, \ldots, x_N)$ having zero expectation, we will prove that it is zero almost everywhere. Thus,

$$\begin{aligned} 0 &= E\{h(X_1, \ldots, X_N)\} \\ &= \sum_{i_1=1}^{N} \cdots \sum_{i_N=1}^{N} p_{i_1} \cdots p_{i_N} J(i_1, \ldots, i_N) \end{aligned}$$

where J is the integral performed in a certain N-dimensional rectangle, and is given by

$$J(i_1, \ldots, i_N) = [I(i_1) \cdots I(i_N)]^{-1} \int_{I_{i_1}} \cdots \int_{I_{i_N}} h(x_1, \ldots, x_N) \prod_{i=1}^{N} dx_i$$

and $I(1), \ldots, I(N)$ are the lengths, respectively of the intervals $I_1, \ldots, I_N$. Since $h(x_1, \ldots, x_N)$ is symmetric, $J(i_1, \ldots, i_N)$ is also symmetric. Hence, we have

$$0 = \sum p_1^{a_1} p_2^{a_2} \cdots p_N^{a_N} c(a_1, \ldots, a_N)$$

where the summation is over all non-negative integers $a_1, \ldots, a_N$ such that $\Sigma a_i = N$ and where $c(a_1, \ldots, a_N)$ is an integral multiple of the $J(i_1, \ldots, i_N)$ having a_1 of the i_k's equal to 1, a_2 of the i_k's equal to 2, and so on.

Now, the expression on the right side of the above equation satisfies the conditions of Lemma 5.6.1. Therefore

$$c(a_1, \ldots, a_N) = 0\,.$$

Consequently,

$$J(i_1, \ldots, i_N) = 0\,.$$

That is,

$$\int_{I_{i_1}} \cdots \int_{I_{i_N}} h(x_1, \ldots, x_N) dx_1 \cdots d_{x_N} = 0$$

for all $i_k = 1, \ldots, N$ $(k = 1, 2, \ldots, N)$ and all disjoint intervals $I_1, \ldots, I_N$. After using some measure theoretic details, it follows that $h(x_1, \ldots, x_N) = 0$ almost everywhere.

Corollary 5.6.3.1. *The order statistic $T(X_1, \ldots, X_N)$ is complete for the class of N-fold power absolutely continuous d.f.'s.*

PROOF: From Result 5.6.2, completeness can be extended to the class of N-fold power absolutely continuous d.f.'s, provided this class contains the class of distributions uniform within intervals.

Result 5.6.4 (Halmos). *The order statistic $T(X_1, \ldots, X_N)$ is complete for the class of N-fold power discrete distributions.*

PROOF: The proof follows from Result 5.6.3 if the intervals are replaced by points. If the random variable takes only fewer than N points, the argument in the proof remains valid since the only restriction in Lemma 5.6.1 is that the degree of the polynomial be greater than 0.

Definition 5.6.8. *A probability distribution is said to be non-atomic if there does not exist a set A having $P(X \in A) \neq 0$ and such that for any $B \subset A$, $P(X \in B) = 0$ or $P(X \in B) = P(X \in A)$. If such a set exists, it is called an atom.*

Definition 5.6.9 (Bell). *A class Ω is said to be symmetrically complete for N if $h_N = h(X_1, \ldots, X_N)$ is zero almost everywhere w.r.t. $F^{(N)}$. That is, $h_N = 0$ almost everywhere w.r.t. $F^{(N)}$ for all F in Ω whenever h_N satisfies*

1. h_N *is a symmetric function [measurable on* $(X^{(N)}, S^{(N)})$*] and*

2. $\int h_N dF^{(N)} = 0$ *for all* $F \in \Omega$.

Result 5.6.5 (Bell). *The class of all non-atomic probability distributions on an arbitrary measurable space* (X, S) *is a symmetrically complete class for all* N. *In particular, the class of all continuous d.f.'s on the real line is a symmetrically complete class for all* N. *(*S *denotes the class of Borel sets on the real line.)*

5.7 Problems

5.1.1 Let $X_1, \cdots, X_N$ be a random sample from a population having d.f. $F(x)$. Let

$$\begin{aligned} \chi_i(x) &= 1 \qquad if\ X_i \le x \\ &= 0 \qquad otherwise,\ for\ i = 1, \cdots, N. \end{aligned}$$

Consider

$$\hat{F}(x) = \sum_{i=1}^{N} a_i \chi_i(x).$$

Determine the best values of $a_1, \cdots, a_N$ such that $\hat{F}(x)$ is unbiased for $F(x)$.

5.1.2 Show that $\hat{I}_1$ and $\hat{I}_2$ are consistent estimators of I_1 and I_2 respectively.

5.1.3 Obtain a uniformly minimum variance unbiased estimate of $p(F, G) = \int (F - G)^2 d(\frac{F+G}{2})$, which gives a measure of the discrepancy between F and G.

5.2.1 Suppose $\varepsilon = 0.1$ and $\alpha = 0.05$. Determine the value of m (the sample of size on X) such that

$$P_m(0.1) = 0.95.$$

5.2.2 For $\lambda = 0.5$, $\varepsilon = 0.10$ and $1 - \alpha = 0.95$, from Appendix V determine the value of δ and hence the values of m and n.

5.2.3 Let -1.75, -0.025, -0.385, -0.707, 1.55, 0.665, 0.332, -0.306, -0.105, 0.18 denote a random sample of size 10 from $F(x)$, and let -0.605,

-1.201, 0.281, 1.751, -0.611, 1.485, -0.390, 0.375, -0.440, 0.553 denote an independent random sample of size 10 from $G(y)$. Determine $\hat{p}_2$ and obtain a 95% confidence interval for p of the form $(\hat{p}_2, \hat{p}_2 + \varepsilon)$.

5.4.1 Show that $D_N^+(x) = \sup_x\{F_N(x) - F(x)\}$ is distribution-free (i.e. the distribution of D_N^+, when F is the true d.f., does not depend upon F).

5.4.2 Using Appendix VI determine the value of ε when $N = 10$, and $\alpha = 0.05$.

5.4.3 Calculate $\bar{\varepsilon}_{N,\alpha} = \left[-(2N)^{-1} ln\alpha\right]^{\frac{1}{2}}$ when $N = 10$ and $\alpha = 0.05$. Evaluate Miller's (1956) approximation to $\varepsilon_{N,\alpha}$.

5.4.4 Using Appendix VI determine the value of ε such that

$$P\left(F_N(x) - \varepsilon \le F(x) \le F_N(x) + \varepsilon\right) = 1 - \alpha$$

when $N = 10$ and $\alpha = 0.05$.

Chapter 6

Estimation of Density Functions

6.1 Introduction

Non-parametric estimation of the underlying density function has been considered only in the 1950's. Fix and Hodges (1951) used an estimate of the form of difference quotient of the sample distribution function. Rosenblatt (1956) considered the estimation of the density function and obtained a class of consistent estimates and their asymptotic mean square error. Rosenblatt (1956) has further shown that all estimates of the density function satisfying mild conditions are biased. Whittle (1956) considered the estimation of a probability density function by linear smoothing of the observed density and also derived an equation denoting the optimum weighting function. Whittle (1956) discussed the asymptotic behavior of the mean squared deviation of the estimate. Parzen (1962) reconsidered this problem and has shown how one may construct a family of estimates of the density function, and of its mode, which are consistent and asymptotically normal. However, Parzen (1962) does not examine the question of which estimate to use. A possible application is the problem of estimating the hazard, or conditional rate of failure conditional on $X > x$, which is defined as the ratio of the density function to $1 - F(x)$, where $F(x)$ is the d.f. In the following some of the results due to Rosenblatt (1956) and Parzen (1962) will be presented.

Result 6.1.1 (Rosenblatt). *Let $X_1, X_2, \ldots, X_N$ be independent and identically distributed random variables with continuous density function $f(x)$. Let $S(y : X_1, \ldots, X_N)$ symmetric in $X_1, X_2, \ldots, X_N$ be an estimate of $f(y)$. The function $S(y; x_1, \ldots, x_N)$ is assumed to be jointly Borel measurable in*

$(y, x_1, \ldots, x_N)$. It is further assumed $S(y;\ x_1, x_2, \ldots, x_N) \geq 0$ since $f(y) \geq 0$. Then, $S(y;\ X_1, \ldots, X_N)$ is not an unbiased estimate of $f(y)$.

PROOF: $(X_1, X_2, \ldots, X_N)$ is a sufficient statistic for the problem. Since $S(y;\ X_1, X_2, \ldots, X_N)$ is assumed to be symmetric in $X_1, X_2, \ldots, X_N$,

$$\int_a^b S(y; X_1, X_2, \ldots, X_N) dy$$

is a symmetric estimate of $F(b) - F(a) = \int_a^b f(y) dy$. Moreover,

$$\int_a^b S(y;\ X_1, X_2, \ldots, X_N) dy$$

is an unbiased estimate of $F(b) - F(a)$ since

$$\begin{aligned} E \int_a^b S(y; X_1, X_2, \ldots, X_N) dy &= \int_a^b ES(y;\ X_1, X_2, \ldots, X_N) dy \\ &= \int_a^b f(y) dy = F(b) - F(a) \end{aligned}$$

by Fubini's theorem. However, the only unbiased estimate of $F(b) - F(a)$ symmetric in $X_1, X_2, \ldots, X_N$ is $F_N(b) - F_N(a)$. This follows from the completeness of the order statistic. Thus,

$$F_N(b) - F_N(a) = \int_a^b S(y;\ X_1, X_2, \ldots, X_N) dy$$

for all a and b and almost all $X_1, X_2, \ldots, X_N$. This implies that $F_N(y)$ is absolutely continuous in y for almost all $X_1, X_2, \ldots, X_N$ which is impossible.

6.2 Difference Quotient Estimate

An obvious estimate of $f(y)$ is the difference quotient

$$S(y;\ X_1, X_2, \ldots, X_N) = f_N(y) = \{F_N(y + h) - F_N(y - h)\} / 2h$$

of the sample d.f. $F_N(y)$ where $h = h_N$ is a function of the sample size N and $h_N \to 0$ as $N \to \infty$. Fix and Hodges (1951) have used an estimate of this form in their problem on non-parametric discrimination. Now, since

$$\text{cov}\left(F_N(y), F_N(y')\right) = N^{-1}\left[F\left(\min(y, y')\right) - F(y)F(y')\right],$$

it follows that

$$\mathrm{var}\,\{f_N(y)\} = (4h^2N)^{-1}\left[F(y+h) - F(y-h) - \{F(y+h) - F(y-h)\}^2\right].$$

Now, consider the mean square error of $f_N(y)$ which measures the goodness of the estimate $f_N(y)$ locally at y:

$$\begin{aligned}
&E\,|f_N(y) - f(y)|^2 \\
&\quad = \mathrm{var}\,\{f_N(y)\} + [Ef_N(y) - f(y)]^2 \\
&\quad = (4h^2N)^{-1}\left[F(y+h) - F(y-h) - \{F(y+h) - F(y-h)\}^2\right] \\
&\qquad + \left[(2h)^{-1}\left[F(y+h) - F(y-h)\right] - f(y)\right]^2 .
\end{aligned}$$

Assume that the first three derivatives of f exist at y. Then

$$\begin{aligned}
&F(y+h) - F(y-h) \\
&\quad = \int_{y-h}^{y+h} f(u)du \\
&\quad = \int_{y-h}^{y+h}\left[f(y) + (u-y)f'(y) + \frac{1}{2}(u-y)^2 f''(y) + 0|u-y|^3\right] du \\
&\quad = 2h\; f(y) + \frac{1}{3}f''(y)h^3 + 0\left(|h|^4\right).
\end{aligned}$$

If $f''(y) \neq 0$,,

$$|E\; f_N(y) - f(y)|^2 \sim \left(\frac{h^2}{6}f''(y)\right)^2 = \frac{h^4}{36}\left|f''(y)\right|^2$$

and

$$\mathrm{var}\,(f_N(y)) \sim f(y)/2hN\,.$$

Thus, as $h \to 0$, the asymptotic mean square error of $f_N(y)$ is

$$E\,|f_N(y) - f(y)|^2 \sim \frac{f(y)}{2hN} + \frac{h^4}{36}\left|f''(y)\right|^2 + o\left((hN)^{-1} + h^4\right).$$

Now, one chooses h optimally. If $h = gN^{-\alpha}$, it is clear that the optimal choice of α is $1/5$. Then the optimal value of g is the one that minimizes

$$\frac{f(y)}{2g} + \frac{g^4}{36}\left|f''(y)\right|^2 .$$

This value of g is

$$g = \left[\frac{9f(y)}{2\left|f''(y)\right|^2}\right]^{\frac{1}{5}}.$$

This choice of α and g yields

$$E\left|f_N(y) - f(y)\right|^2 \sim \frac{5}{4} 9^{-1/5} 2^{-4/5} \{f(y)\}^{4/5} \left|f''(y)\right|^{2/5} N^{-4/5}.$$

The choice of g would be based on guesses as to the magnitude of $f(y)$, $f''(y)$.

In order to find an optimum value of h as a function of N but independent of y, we consider a global measure of how good f_N is as an estimate of f. The integrated mean square error

$$\int_{-\infty}^{\infty} E\left|f_N(y) - f(y)\right|^2 dy$$

is a measure of this type. Let us further assume that $f(y)$ and $f''(y)$ are bounded functions that are square integrable. Then, one is led to

$$\int_{-\infty}^{\infty} E\left|f_N(y) - f(y)\right|^2 dy \sim (2hN)^{-1} + \frac{h^4}{36}\int_{-\infty}^{\infty} \left|f''(y)\right|^2 dy + o\left((hN)^{-1} + h^4\right)$$

as $h \to 0$ and $N \to \infty$. The optimal choice of h is $h = gN^{-1/5}$,

$$g = \left[\frac{2}{9}\int_{-\infty}^{\infty} \left|f''(y)\right|^2 dy\right]^{-1/5}$$

and

$$\int_{-\infty}^{\infty} E\left|f_N(y) - f(y)\right|^2 dy \sim \frac{5}{4} 2^{-4/5} 9^{-1/5} \left[\int_{-\infty}^{\infty} \left|f''(y)\right|^2 dy\right]^{1/5} N^{-4/5}$$

as $N \to \infty$.

6.3 Class of Estimates of Density Function

One may study a general class of estimates of the density function of which the difference quotient of the sample d.f. will be a special case.

Consider

$$f_N(x) = h^{-1}\int_{-\infty}^{\infty} W\left(\frac{x-y}{h}\right) dF_N(y) = (Nh)^{-1}\sum_{j=1}^{N} W\left(\frac{x - X_j}{h}\right)$$

where $W(y)$ is a real function. This estimate will be equal to the difference quotient of the sample d.f. if

$$W(y) = \tfrac{1}{2}, \qquad |y| \leq 1,$$
$$= 0, \qquad |y| > 1.$$

Note that all estimates of this form are themselves density functions if $W(y)$ is non-negative; that is, $f_N(x) \geq 0$ and $\int_{-\infty}^{\infty} f_N(x)dx = 1$. An estimate with any desired regularity properties can be obtained by choosing a weight function $W(y)$ with these regularity properties. Thus, $f_N(x)$ will be analytic if $W(y)$ is.

We will examine the conditions under which the new estimates are asymptotically unbiased. Now, letting $h = h_N$, a function of N, we have

$$E\left(f_N(x)\right] = E\left[W\left(\frac{x-X}{h_N}\right)/h_N\right] = \int_{-\infty}^{\infty} h_N^{-1} W\left(\frac{x-y}{h_N}\right) f(y)dy\,.$$

We wish to choose h_N and $W(y)$ such that the right hand expression tends to $f(x)$. Towards this, we have the following lemma and result.

Lemma 6.3.1 (Bochner). *Let $W(y)$ be a Borel (Baire) function satisfying the conditions*

$$\sup_{-\infty<y<\infty} |W(y)| < \infty,$$

$$\int_{-\infty}^{\infty} |W(y)|\, dy < \infty,$$

and

$$\lim_{y\to\infty} |yW(y)| = 0\,.$$

Let $f(y)$ be any function (need not be a density function) satisfying

$$\int_{-\infty}^{\infty} |f(y)|\, dy < \infty\,.$$

Let $\{h_N\}$ be a sequence of positive constants such that

$$\lim_{N\to\infty} h_N = 0\,.$$

Define

$$g_N(x) = h_N^{-1} \int_{-\infty}^{\infty} W(y/h_N) f(x-y)dy \,.$$

Then, at every point x of continuity of $f(\cdot)$,

$$\lim_{N\to\infty} g_N(x) = f(x) \int_{-\infty}^{\infty} W(y)dy \,.$$

PROOF: Consider

$$g_N(x) - f(x) \int_{-\infty}^{\infty} W(y)dy = \int_{-\infty}^{\infty} \{f(x-y) - f(x)\} \frac{1}{h_N} W(y/h_N)dy \,.$$

Let $\delta > 0$ and split the region of integration into two regions:

$$|y| \le \delta \text{ and } |y| > \delta \,.$$

Thus,

$$\begin{aligned}
&\left| g_N(x) - f(x) \int_{-\infty}^{\infty} W(y)dy \right| \\
&\le \max_{|y|\le\delta} |f(x-y) - f(x)| \int_{|z|\le\delta/h_N} |W(z)|\, dz \\
&+ \int_{|y|\ge\delta} \frac{|f(x-y)|}{y} \frac{y}{h_N} W(y/h_N)dy + |f(x)| \int_{|y|\ge\delta} \frac{1}{h_N} W(y/h_N)dy \\
&\le \max_{|y|\le\delta} |f(x-y) - f(x)| \int_{-\infty}^{\infty} |W(z)|\, dz \\
&+\delta^{-1} \sup_{|z|\ge\delta/h_N} |zW(z)| \int_{-\infty}^{\infty} |f(y)|\, dy + |f(x)| \int_{|z|\ge\delta/h_N} |W(z)|\, dz \,,
\end{aligned}$$

which tends to zero as one lets N tend to ∞ and then lets δ tend to zero.

Result 6.3.1 (Parzen). *The estimates of the density function defined by*

$$f_N(x) = (Nh_N)^{-1} \sum_{j=1}^{N} W\left((x - X_j)/h_N\right)$$

are asymptotically unbiased at all points x at which the probability density function is continuous if $\lim_{N\to\infty} h_N = 0$ and $W(y)$ satisfies the conditions stipulated in the preceding lemma and in addition satisfies

$$\int_{-\infty}^{\infty} W(y)dy = 1 \,.$$

Definition 6.3.1. *A function $W(y)$ is said to be a* **weighting function** *if it is symmetric about zero and satisfies the following conditions:*

1. $\sup_{-\infty<y<\infty} |W(y)| < \infty$;

2. $\int_{-\infty}^{\infty} W(y)dy = 1$;

3. $\int_{-\infty}^{\infty} |W(y)|\, dy < \infty$;

and

4. $\lim_{N\to\infty} |yW(y)| = 0$.

Some examples of weighting functions are presented in Table 6.3.1.

Result 6.3.2 (Parzen). *Under the conditions of Result 6.3.1, the estimates satisfy*

$$\lim_{N\to\infty} Nh_N var\{f_N(x)\} = f(x) \int_{-\infty}^{\infty} W^2(y)dy$$

and the estimates $f_N(x)$ are consistent provided $\lim_{N\to\infty} Nh_N = \infty$.

PROOF:

$$\text{var}\,[f_N(x)] = N^{-1}\text{var}\left[h_N^{-1} W((x-X)/h_N)\right] .$$

From Bochner's result, we have

$$\begin{aligned} h_N E\left[\left\{h_N^{-1} W\left(\frac{x-X}{h_N}\right)\right\}^2\right] &= h_N^{-1} \int_{-\infty}^{\infty} W^2\left(\frac{x-y}{h_N}\right) f(y)dy \\ &\to f(x) \int_{-\infty}^{\infty} W^2(y)dy\,, \end{aligned}$$

since

$$\sup_{-\infty<x<\infty} |W(y)| < \infty \text{ and } \int_{-\infty}^{\infty} |W(y)|\, dy < \infty$$

imply that

$$\int_{-\infty}^{\infty} W^2(y)dy < \infty .$$

Also, since

$$E\left[h_N^{-1} W\left((x-X)/h_N\right)\right] \to f(x) ,$$

Table 6.3.1

$W(y)$		$\int_{-\infty}^{\infty} W^2(y)dy$
$\frac{1}{2}$,	$\lvert y\rvert \le 1$	$\frac{1}{2}$
0,	$\lvert y\rvert \ge 1$	
$1-\lvert y\rvert$,	$\lvert y\rvert \le 1$	$\frac{2}{3}$
0,	$\lvert y\rvert \ge 1$	
$(2\pi)^{-\frac{1}{2}}e^{-\frac{1}{2}y^2}$		$\left[2(\pi^{\frac{1}{2}})\right]^{-1}$
$\frac{1}{2}e^{-\lvert y\rvert}$		$\frac{1}{2}$
$\left\{\pi(1+y^2)\right\}^{-1}$		π^{-1}

it follows that

$$Nh_N \text{var}\left\{f_N(x)\right\} \to f(x)\int_{-\infty}^{\infty} W^2(y)dy\,,$$

at all points of continuity of $f(x)$. Now, consider the mean square error of $f_N(x)$,

$$E\left|f_N(x)-f(x)\right|^2 = \text{var}\left\{f_N(x)\right\} + \left|Ef_N(x)-f(x)\right|^2\,.$$

We know from Parzen's (1962) result that the bias tends to zero as N tends to infinity. The estimates $f_N(x)$ will be consistent if var $\{f_N(x)\}$ tends to zero. Hence, if addition to $h_N \to 0$, h_N satisfy the condition

$$\lim_{N\to\infty} N\,h_N = \infty\,,$$

then it follows that $f_N(x)$ are consistent estimates of $f(x)$. This completes the proof of the desired result.

Hereafter, we confine ourselves to the class of consistent estimates. That is, $W(y)$ will be a weighting function and h_N satisfy the two restrictions. Next, we explore the asymptotic normality of the estimates $f_N(x)$.

Result 6.3.3 (Parzen). *If the weighting function $W(y)$ is non-negative and the sequence of constants, h_N are such that*

$$\lim_{N\to\infty} h_N = 0 \text{ and } \lim_{N\to\infty} N\,h_N = \infty\,,$$

then

$$\lim_{N\to\infty} P\left[\frac{f_N(x) - Ef_N(x)}{\sigma\left[f_N(x)\right]} \le s\right] = \Phi(s)\,.$$

PROOF: The estimate $f_N(x)$ can be written as

$$f_N(x) = N^{-1}\sum_{k=1}^{N} V_{N,k},\;\; V_{N,k} = h_N^{-1}W\left\{(x - X_k)/h_N\right\}$$

which is the average of independent random variables identically distributed as

$$V_N = h_N^{-1}W\left\{(x - X)/h_N\right\}\,.$$

Then, from Loève (1960, p. 316), it follows that a necessary and sufficient condition for

$$\lim_{N\to\infty} P\left[\left\{f_N(x) - Ef_N(x)\right\}/\sigma[f_N/x] \le s\right] = \Phi(s)\,,$$

to hold is that, for every $\epsilon > 0$,

$$\lim_{N\to\infty} NP\left[\left|\frac{V_N - EV_N}{\sigma(V_N)}\right| \ge \epsilon N^{\frac{1}{2}}\right] = 0\,.$$

A sufficient condition (due to Liapunov) for the above condition to hold is that, for some $\delta > 0$,

$$\left\{E|V_N - EV_N|^{2+\delta}/N^{\frac{1}{2}\delta}\sigma^{2+\delta}[V_N]\right\} \to 0 \text{ as } N \to \infty\,.$$

Or, since V_N is non-negative,

$$\left\{E|V_N|^{2+\delta}/N^{\delta/2}\sigma^{2+\delta}[V_N]\right\} \to 0 \text{ as } N \to \infty\,.$$

Now,

$$E|V_N|^{2+\delta} = \int_{-\infty}^{\infty} \left| h_N^{-1} W\left(\frac{x-y}{h_N}\right)\right|^{2+\delta} f(y)dy$$
$$\sim h_N^{-(1+\delta)} f(x) \int_{-\infty}^{\infty} |W(y)|^{2+\delta} dy$$

and

$$\sigma^2[V_N] \sim h_N^{-1} f(x) \int_{-\infty}^{\infty} W^2(y)dy\,.$$

Thus,

$$\frac{E|V_N|^{2+\delta}}{N^{\frac{1}{2}\delta}\sigma^{2+\delta}[V_N]} = \frac{h_N^{1+\delta} E|V_N|^{2+\delta}}{(N\,h_N)^{\frac{1}{2}\delta} h_N^{1+\frac{1}{2}\delta} \sigma^{2+\delta}[V_N]}\,,$$

which, in view of the asymptotic values for the third absolute moments and the variance of V_N, tends to zero as $N \to \infty$ since

$$\int_{-\infty}^{\infty} |W(y)|^{2+\delta}\, dy < \infty\,.$$

An idea of the closeness of the normal approximation can be obtained from the Berry-Esseen bound (see Loève, 1960, p. 288), namely, for an appropriate constant C,

$$\sup_{-\infty<s<\infty} |P[\{f_N(x) - Ef_N(x)\}/\sigma[f_N(x)] \le s] - \Phi(s)|$$
$$\le C \frac{E|V_N|^3}{N^{\frac{1}{2}}\sigma^3[V_N]} \sim [N\,h_N f(x)]^{-\frac{1}{2}}$$
$$\times \left[\left\{ \int_{-\infty}^{\infty} |W(y)|^3 dy \right\} / \left\{ \int_{-\infty}^{\infty} W^2(y)dy \right\}^{3/2} \right].$$

Evaluation of the mean square error

We may now obtain an approximation for the mean square error of the estimate $f_N(x)$ imposing some regularity conditions on the weighting function $W(y)$ and the density function $f(x)$.

Result 6.3.4 (Rosenblatt and Parzen). *If $f(x)$ has continuous derivatives of the first three orders and the weighting function $W(y)$ is such that*

$$\int_{-\infty}^{\infty} |y|^3 W(y)dy < \infty\,,$$

then

1. $var\ f_N(x) \sim (N\,h_N)^{-1} f(x) \int_{-\infty}^{\infty} W^2(y)dy;$

2. $Ef_N(x) - f(x) \sim -\frac{1}{2} h_N^2 f''(x) \int_{-\infty}^{\infty} y^2 W(y)dy;$

and

3. $$E\left[f_N(x) - f(x)\right]^2 \sim (N\,h_N)^{-1} f(x) \int_{-\infty}^{\infty} y^2 W(y)dy + \frac{1}{4} h_N^4 \left[f''(x) \int_{-\infty}^{\infty} y^2 W(y)dy\right]^2$$

PROOF:

$$\text{var } f_N(x) = N^{-1} \text{var } V_N, \quad V_N = h_N^{-1} W\left(\frac{x - X}{h_N}\right)$$

and

$$\text{var } V_N \sim h_N^{-1} f(x) \int_{-\infty}^{\infty} W^2(y)dy\,.$$

This completes the proof for (1). Now,

$$\begin{aligned} Ef_N(x) - f(x) &= h_N^{-1} \int_{-\infty}^{\infty} W(y/h_N) f(x - y)dy - f(x) \\ &= \int_{-\infty}^{\infty} W(u) f(x - uh_N)du - f(x)\,. \end{aligned}$$

Expanding $f(x - uh_N)$ in Taylor series about x and performing the integration, we obtain

$$\begin{aligned} &\text{Bias in } f_N(x) \\ &= -h_N f'(x) \int_{-\infty}^{\infty} uW(u)du - \frac{1}{2} h_N^2 f''(x) \int_{-\infty}^{\infty} u^2 W(u)du + 0\left(|h_N|^3\right) \\ &= -\frac{1}{2} h_N^2 f''(x) \int_{-\infty}^{\infty} u^2 W(u)du + 0\left(|h_N|^3\right)\,. \end{aligned}$$

Now, the approximate expression for the mean square error follows from the fact

$$E\left[f_N(x) - f(x)\right]^2 = \text{var } f_N(x) + \left[Ef_N(x) - f(x)\right]^2\,.$$

This completes the proof of the result.

An optimum choice of $h_N = gN^{-1/5}$ where the optimum choice for g is given by

$$g = f(x) \int_{-\infty}^{\infty} W^2(y)dy \Big/ \left\{ f''(x) \int_{-\infty}^{\infty} y^2 W(y)dy \right\}^{2/5}$$

and then, the mean-square error is

$$E[f_N(x) - f(x)]^2 \sim 5 \left\{ [f(x)/4N] \int_{-\infty}^{\infty} W^2(y)dy \right\}^{4/5} \left| \frac{1}{2} f''(x) \int y^2 W(y)dy \right|^{2/5}$$

and thus, the estimates $f_N(x)$ have order of consistency $N^{4/5}$ in the sense that

$$N^{4/5} E\left[f_N(x) - f(x)\right]^2$$

tends to a finite positive limit as N tends to infinity.

6.4 Estimate with Prior on Ordinates

Whittle (1958) considered the estimation of the density function by linear smoothing of the observed density on the hypothesis:

> "that the curve $f(x)$ being estimated is one of a whole population of curves and that the population correlation coefficient of $f(x)$ and $f(x+\epsilon)$ tends to unity as ϵ tends zero."

The above hypothesis is still a formulation of a regularity requirement which has the following advantages

1. It is by no means a stringent demand, and allows the curve $f(x)$ to have many kinds of discontinuity and irregularity.

2. It leads to weighting functions which make automatic allowance for sample size.

3. If one is interested in only a finite interval of x, then the weighting functions will make automatic allowance for the end effects of the finite interval.

4. The weighting functions derived from the above hypothesis are *arbitrary transformations of the data.*

Also, it is not always realistic to assume that the sample size is non-random. This problem of estimating the density function when N is random has been considered by Whittle (1958). He obtained a neat theory assuming that N is distributed as a Poisson variable with mean M. The hypothesis will be justified if observations are independent, and if the 'stopping rule' does not depend upon the sample in any way. Then, it is convenient to work with an unnormalized density

$$\psi(x) = Mf(x),$$

where ψ will be estimated by statistics of the form

$$\hat{\psi}(x) = \int W_x(y) dF_N(y)$$

for which

$$E\left\{\hat{\psi}(x)\right\} = \int W_x(y)\psi(y)dy$$

and

$$\operatorname{var}\left\{\hat{\psi}(x)\right\} = \int W_x^2(y)\psi(y)dy$$

where $W_x(y)$ denotes the *a priori* distribution of the ordinates $\psi(x)$. If M is large, it matters little whether one assumed that N is fixed or Poisson distributed since the coefficient of variation of sample size is $M^{-\frac{1}{2}}$.

Optimization of the Estimator

The estimator $\hat{\psi}(x)$ will be optimized by minimizing

$$\begin{aligned}
\Delta^2 &= E_p E_s \left[\hat{\psi}(x) - \psi(x)\right]^2 \\
&= E_p \left[\int W_x^2 \psi \, dy + \left(\int W_x \psi dy\right)^2 - 2\psi(x) \int W_x \psi \, dy + \psi^2(x)\right]
\end{aligned}$$

where E_s denotes expectation with respect to sampling fluctuations and E_p denotes expectation with respect to the *a priori* distribution of ordinates. If

$$E_p\left[\psi(x)\right] = \mu(x)$$

and

$$E_p\left[\psi(x) \cdot \psi(y)\right] = \mu(x, y).$$

Then, on minimizing Δ^2 w.r.t. $W_x(y)$, we obtain the following integral equation:

$$\mu(y)W_x(y) + \int \mu(y, z)W_x(z)dz = \mu(y, x).$$

It can easily be verified that if W^* is any other weight function, then $\Delta^2(W^*) \geq \Delta^2(W)$. Also, since $\mu(x) = 0(M)$, $\mu(x,y) = 0(M^2)$ for large M, we get

$$\int \mu(y,z)W_x(z)dz \sim \mu(y,x)\,,$$

with a solution

$$W_x(y) \sim \delta(y-x)$$

where

$$\delta(t) = \begin{cases} 1 & \text{if } t = 0, \\ 0 & \text{otherwise.} \end{cases}$$

It is thus at least plausible that $W_x(y)$ tends to a δ function when M becomes large. If we normalize as follows:

$$\xi_x(y) = W_x(y)\,\{\mu(y)/\mu(x)\}^{1/2}$$

and

$$\gamma(x,y) = \mu(x,y)/\,\{\mu(x)\mu(y)\}^{\frac{1}{2}}\,,$$

then, the integral equation obtained on minimizing Δ^2 w.r.t. $W_x(y)$ becomes

$$\xi_x(y) + \int \gamma(y,z)\xi_x(z)dz = \gamma(y,x)\,.$$

From the symmetrical form of the equation, it is obvious that

$$\xi_x(y) = \xi_y(x)\,.$$

The effect of the normalization is to throw the smoothed estimate into a form $\int \xi(y)dF_N^*(y)$, where the variance of the increment $dF_N^*(y)$ is independent of y, although the expectation of the increment may depend on y.

The analysis for the case of non-random sample size N is similar and the corresponding integral equation will be

$$m(y)W_x(y) + (N-1)\int m(y,z)W_x(z)dz = Nm(y,x)$$

where the m's and μ's correspond to f and ψ. Notice here that $\hat{f}(x) = \int W_x(y)dF_N(y)$. Also, the mean square error and variance of $\hat{\psi}(x)$ are given by

$$\begin{aligned} \Delta^2 &= \mu(x,x) - \int \mu(x,z)W_x(z)dz \\ &= \mu(x)W_x(x) = \mu(x)\xi_x(x) \end{aligned}$$

and

$$\sigma^2[\hat{\psi}] = E_p \int W_x^2(y)\psi(y)dy$$
$$= \mu(x) \int \xi_x^2(y)dy\,.$$

The mean square error of $\hat{f}(x)$ is given by D^2 where

$$D^2 = (N-1)^{-1}\left[\mu(x)W_x(x) - \mu(x,x)\right]$$
$$= N^{-1}\left[\mu(x)W_x(x) - \int \mu(x,y)W_x(y)dy\right].$$

Invariance properties of the estimates

The scale on which a variate is measured is usually arbitrary and for the sake of internal consistency it is reasonable to demand that the smoothings applied to different versions of the density function (density functions set up on number of scales) be equivalent.

Let x and x' be two scales such that $x' = h(x)$ (h being a monotone function) and let the corresponding density functions be respectively denoted by $\psi(x)$ and $\psi^*(x')$ so that

$$\psi^*(x') = \psi(x)\frac{dx}{dx'}\,.$$

Then, the smoothed estimates are

$$\hat{\psi}(x) = \int W_x(y)dF_N(y)$$

and

$$\hat{\psi}^*(x') = \int W_{x'}^*(y')dF'_N(y')$$

where

$$F'_N(y') = F_N(y) \text{ and } y' = h(y)\,.$$

Starting from $\hat{\psi}(x)$, we can obtain alternate estimate of $\psi^*(x')$ as follows:

$$\psi_1^*(x') = \hat{\psi}(x)\frac{dx}{dx'} = \int W_x(y)\frac{dx}{dx'}dF'_N(y')\,.$$

So, the criterion of internal consistency is that $\psi^*(x')$ and $\hat{\psi}^*(x')$ be identical. That is,

$$W_{x'}^*(y') = W_x(y)\frac{dx}{dx'}\,.$$

The weighting function should be such that the above relation is fulfilled for any non-singular transformation to a new scale x', that is, a transformation for which $\frac{dx}{dx'}$ is nowhere zero or infinity in the range of interest. The optimal weighting function given by Whittle (1958) satisfies this criterion.

Special Case

If $\gamma(x, y)$ depends only upon $(x - y)$, so that the integral equation (normalized) becomes

$$\xi_x(y) + \int \gamma(y - z)\xi_x(s)ds = \gamma(y - x) .$$

By Fourier transform, we find that

$$\xi_x(y) = (2\pi)^{-1} \int_{-\infty}^{\infty} e^{it(y-x)} w(t) \{1 + w(t)\}^{-1} dt$$

where

$$w(t) = \int_{\infty}^{\infty} e^{-itx} \gamma(x) dx .$$

Consequently,

$$\begin{aligned} \Delta^2 &= \frac{\mu(x)}{2\pi} \int w(t) \{1 + w(t)\}^{-1} dt \\ \sigma^2 \left[\hat{\psi}(x)\right\} &= \frac{\mu(x)}{2\pi} \int \left[\frac{w(t)}{1 + w(t)}\right]^2 dt \end{aligned}$$

and

$$\left[\text{Bias in } \hat{\psi}(x)\right]^2 = \frac{\mu(x)}{2\pi} \int w(t) \{1 + w(t)\}^{-2} dt .$$

Whittle (1958) also considered the asymptotic value of the mean square error of $\hat{\psi}(x)$. Parzen (1962) considered the problem of estimating the mode of the density function and obtained some interesting results. However, they are not presented here.

6.5 Problems

6.2.1 Suppose that

$$
\begin{aligned}
f''(y) &= (1+y)^{-2} && for\ y > 1 \\
&= y^{-2} && for\ y < -1 \\
&= 0 && for\ -1 \le y \le 1.
\end{aligned}
$$

Obtain the integrated mean-square error of $f_N(y)$, the difference quotient estimate of the density function.

6.2.2 Suppose that

$$
\begin{aligned}
f''(y) &= 0 && for\ |y| > 1 \\
&= (2+y)^{-2} && for\ -1 < y < 0 \\
&= (1+y)^{-2} && for\ 0 < y < 1.
\end{aligned}
$$

Obtain the integrated mean-square error of $f_N(y)$, the difference quotient estimate of the density function.

6.3.1 Show that when $W(y) = \frac{1}{2}$ for $|y| \le 1$ and zero when $|y| > 1$, the window estimate of the density function coincides with the difference-quotient estimate of the density.

6.3.2 Evaluate the mean-square error of $f_N(x)$ for the special case of

$$
\begin{aligned}
W(y) &= \tfrac{1}{2} && for\ |y| \le 1 \\
&= 0 && for\ |y| > 1.
\end{aligned}
$$

Chapter 7

Review of Parametric Testing

7.1 Preliminaries of Hypothesis Testing

The other phase of statistical inference is hypothesis testing which some people feel is the more important aspect of statistical inference. Although the modern trend is to view testing statistical hypotheses from the point of decision theory, we will view it from the classical point of view, having choice of accepting or rejecting a given hypothesis. Also, we will confine ourselves to two decision (action) problems. In the parametric case, the parameter of interest, namely θ (which could be a vector) labels the class of distributions and the parameter θ will be in Ω. A subset of Ω, namely ω, represents the situation found in the past, the status quo, while the complement, namely $\Omega - \omega$, represents new situation or change. The 'status quo' hypothesis is called the null *hypothesis* and is denoted by H_o. The hypothesis denoting the change is called the *alternative hypothesis* and is denoted by H_1 or H_A. The problem is to decide which set contains the parameter (or the distribution) that best explains the outcome. In general, let $\boldsymbol{X}_N = (X_1, X_2, \ldots, X_N)$ be a finite set or vector of r.v.'s having $F_N(X_1, \ldots, X_N)$ as their joint d.f. Then

$$H_0 : F_N \text{ is in } \mathcal{F}_0 \text{ and } H_1 : F_N \text{ is in } \mathcal{F}_1$$

where $\mathcal{F}_o$ and $\mathcal{F}_1$ are two classes of d.f.'s with no members in common. In parametric problems the classes of d.f. $\mathcal{F}_0$ and $\mathcal{F}_1$ may be characterized by a finite number of parameters.

Definition 7.1.1. *The hypothesis is said to be simple (composite) if the number of d.f.'s in the class is one (many).*

Definition 7.1.2. *The test of a hypothesis-testing problem is a rule specifying for what values of the outcome* $\boldsymbol{X}_N$, H_o *is to be accepted and for what values (namely, all other values) of* $\boldsymbol{X}_N$, H_1 *is to be accepted. Thus, the rule will be:*

> *Reject* H_o *if* $\boldsymbol{X}_N$ *belongs to* W *where* W *is a subset of the* N*-dimensional Euclidean space denoted by* R^N. *Notice that* W, *called the* ***critical region***, *defines the test uniquely.*

Definition 7.1.3. *Size of the test is defined as*

$$\sup_{F \text{ in } \mathcal{F}_o} P(\boldsymbol{X}_N \text{ is in } W | F_N = F) = \alpha\,.$$

Power of the test against a specific alternative G belonging to $\mathcal{F}_1$ is defined and is denoted by

$$\beta_W(G) = P_G(\boldsymbol{X}_N \text{ in } W)\,.$$

The quantity $\beta_W(G)$ as G varies over all the members of $\mathcal{F}_1$ (and $\mathcal{F}_0$) is called the *power function* of the test.

Definition 7.1.4. *A test* W *of size* α *is said to be a* ***similar*** *test if for all* F *in* $\mathcal{F}_0$,

$$P[\boldsymbol{X}_N \text{ in } W) = \alpha\,.$$

Definition 7.1.5. *A test* W *of size* α *is said to be* ***unbiased*** *if for all* G *in* $\mathcal{F}_1$

$$\beta_W(G) \geq \alpha\,.$$

Definition 7.1.6. *A test* W *of* H_o *against* H_1 *is said to be* ***uniformly most powerful*** *(UMP) if, for any other test, say* W' *having the same (or smaller) size* α *one has*

$$\beta_W(G) \geq \beta_{W'}(G), \quad \text{for all } G \text{ in } \mathcal{F}_1\,.$$

Notice that the term 'uniformly' is redundant if H_1 *is simple.*

Definition 7.1.7. *A test* W *is said to be* ***locally most powerful*** *(LMP) if for any other test* W' *with the same (or smaller) size* α,

$$\beta_W(G_k) \geq \beta_{W'}(G_k)$$

for a sequence $\{G_k\}$ converging to an F belonging to $\mathcal{F}_0$. If the d.f.'s are indexed by a parameter θ and θ_o denotes the hypothesized value of the parameter under H_o, then W is said to be LMP if

$$\beta_W(F_\theta) \geq \beta_{W'}(F_\theta)$$

for all θ in the neighborhood of θ_o.

Consider a sequence of hypothesis-testing problems $\left\{\mathcal{F}_o^{(N)}, \mathcal{F}_1^{(N)}, N > 0\right\}$ where $\mathcal{F}_o^{(N)}$ and $\mathcal{F}_1^{(N)}$ are classes of N-dimensional d.f.'s representing H_o and H_1 respectively. Let $\{W_N, N > 0\}$ denote a sequence of subsets of $\{R^{(N)}\}$.

Definition 7.1.8. *The sequence of tests $\{W_N\}$ is said to be consistent if for all G_N, where G_N belongs to $\mathcal{F}_1^{(N)}$,*

$$\lim_{N\to\infty} \beta_{W_N}(G_N) = 1\,.$$

In other words, one may roughly say that a sequence of tests is said to be consistent if the probability of rejecting the null hypothesis when in fact the alternative hypothesis is true, approaches unity as the sample size increases indefinitely.

In hypothesis-testing one is interested in obtaining a uniformly most powerful test. If H_o and H_1 are simple, Neyman and Pearson (1933) have obtained a fundamental result for obtaining a most powerful test.

Result 7.1.1 (Neyman-Pearson lemma). *Consider $H_o : F_N = F$ and $H_1 : F_N = G$.*

1. *If both F and G are absolutely continuous with density functions f and g respectively, the most powerful test (MPT) of size α is the region*

$$W = \left\{\boldsymbol{x}_N \text{ belonging to } R^N \text{ such that } \frac{g(\boldsymbol{x}_N)}{f(\boldsymbol{x}_N)} > C_{N,\alpha} \text{ or } f(\boldsymbol{x}_N) = 0\right\}$$

where $C_{N,\alpha}$ is a suitable constant chosen so that the test has the desired size α.

2. *If both F and G are discrete d.f.'s having their discontinuity points among the values $x_N^{(1)}, x_N^{(2)}, \ldots, x_N^{(k)}, \ldots$ with*

$$P_F(\boldsymbol{X}_N = \boldsymbol{x}_N^{(k)}) = f_k; \;\; P_G(\boldsymbol{X}_N = \boldsymbol{x}_N^{(k)}) = g_k\,.$$

Then, a MPT of size α is the region

$$W = \left\{ \boldsymbol{x}_N^k \text{ in } R^N (k \geq 1) : \frac{g_k}{f_k} > C_{N,\alpha} \text{ or } f_k = 0 \right\},$$

where $C_{N,\alpha}$ is a suitable constant chosen to give the test the proper size. In the latter (discrete) case, α can take on an, at most, countable number of values.

PROOF FOR (A): In general, let $\phi(x)$ denote the test. That is,

$$\phi_W(x) = \begin{cases} 1 & \text{if } x \text{ is in } W, \\ 0 & \text{if } x \text{ is in } \bar{W}. \end{cases}$$

Let $\phi_{W'}(x)$ be any other test having size α. Then consider the difference in the powers

$$\begin{aligned} \beta_W - \beta_{W'} &= E_g\left[\phi_W(X) - \phi_{W'}(X)\right] \\ &= \int_W \left[\phi_W(x) - \phi_{W'}(x)\right] g(x)dx + \int_{\bar{W}} \left[\phi_W(x) - \phi_{W'}(x)\right] g(x)dx\,. \end{aligned}$$

Now, on W,

$$\phi_W(x) - \phi_{W'}(x) \geq 0 \text{ and } g(x) \geq C_{N,\alpha} f(x)$$

and on $\bar{W}$,

$$\phi_W(x) - \phi_{W'}(x) \leq 0 \text{ and } g(x) \leq C_{N,\alpha} f(x)\,.$$

Hence,

$$\begin{aligned} \beta_W - \beta_{W'} &\geq C_{N,\alpha} \int_W \left[\phi_W(x) - \phi_{W'}(x)\right] f(x)dx \\ &\quad + C_{N,\alpha} \int_{\bar{W}} \left[\phi_W(x) - \phi_{W'}(x)\right] f(x)dx \\ &= C_{N,\alpha} \int \left[\phi_W(x) - \phi_{W'}(x)\right] f(x)dx \geq 0\,. \end{aligned}$$

This completes the proof of (a). Replacing the integrals by summations one can analogously establish (b).

7.2 Use of Sufficient Statistic

In point estimation we had a general theorem due to Rao and Blackwell concerning the use of a sufficient statistic. Here we will present a related result in hypothesis-testing.

Result 7.2.1. *If $\phi(x)$ is a test function for a hypothesis-testing problem involving the class of d.f.'s $\{F_\theta(x)|\theta \text{ in } \Omega\}$ and if $T(x)$ is a sufficient statistic, then $E\{\phi(x)|T\}$ is a test function having the same power as $\phi(x)$.*

PROOF: The power function of $\phi(x)$ is

$$\beta_\phi(\theta) = E_\theta\{\phi(X)\}\ .$$

Also, from the definition of conditional expectation, the power function of $E\{\phi(x)|t\}$ is given by

$$E_T E\{\phi(X)|T\} = E\{\phi|X\} = \beta_\phi(\theta)\,.$$

It remains to show that $E\{\phi(X)|T\}$ is a test function, that is,

$$0 \le E\{\phi(X)|T\} \le 1\,.$$

However, this follows from the representation of conditional expectation as an average with respect to conditional probability.

A similar test has constant power, namely α for all values of the parameter in the null hypothesis. If we confine ourselves to similar tests, we are requiring the test to make incorrect decisions at the full allowable rate for all d.f.'s in the null hypothesis. Except for this criticism, there are two things that are in favor of similar tests: (i) the mathematical form of a similar test can be described quite easily. (ii) If for a problem a sufficient and boundedly complete statistic $T(x)$ exists, then a similar test function has a very simple form. Under the null hypothesis, the conditional expected value of the test function given the sufficient statistic, is a constant value α for almost all values of the statistic. The test can be treated as a conditional test and be constructed in each subspace of values of x having $T(x) = t$. Its size given the sufficient statistic must be α for the null hypothesis. Its power can be maximized for any simple alternative by maximizing the conditional power, given the sufficient statistic. Neyman-Pearson lemma can be applied to the reduced problem for each given value of $T(x)$. We have the following result.

Result 7.2.2 (Lehmann-Scheffé, 1950). *If $T(x)$ is a sufficient and boundedly complete statistic for $\{F_\theta | \theta \text{ in } \omega\}$, then any similar size α test $\phi(x)$ has conditional size α, given $T(x) = t$ for almost all values of t; that is,*

$$E\{\phi(X)|T(x) = t\} = \alpha$$

for almost all values of t.

A test satisfying the above condition is said to have *Neyman structure.*

PROOF: Let $\phi(x)$ be a similar size α test. Then

$$\begin{aligned} E_\theta\{\phi(X)\} &= \alpha \text{ for } \theta \text{ in } \omega, \\ E_\theta\{E(\phi(x)|T)\} &= \alpha \text{ for } \theta \text{ in } \omega, \\ E_\theta\{E(\phi(X)|T) - \alpha\} &= 0 \text{ for } \theta \text{ in } \omega. \end{aligned}$$

The conditional expectation does not depend on θ since T is a sufficient statistic. Also, $E(\phi(X)|T) - \alpha$ is a function of T only and has zero expectation for each θ and is bounded. Therefore, the bounded completeness of $T(x)$ implies that $E\{\phi(X)|T\} - \alpha = 0$ for almost all values of T. This completes the proof of the result.

A procedure for obtaining a most powerful similar test.

Consider the following hypothesis-testing problem:

$$H_o: \quad \theta \text{ in } \omega,$$

$$H_1: \quad \theta = \theta_1 .$$

We wish to find a test function $\phi(x)$ such that

$$E\{\phi(X)|T\} = \alpha$$

for all θ in ω and $E_{\theta_1}(\phi(X)|T)$ is maximum. Since the expression to be maximized can be written as

$$E_{\theta_1}\{E_{\theta_1}(\phi(X)|T)\}$$

it is sufficient to maximize $E_{\theta_1}(\phi(X)|T)$ subject to the restriction that $E(\phi(X)|T) = \alpha$, for all T. Thus, finding a best similar problem over $\mathcal{X}$

is equivalent to finding the best test on the subspace of points for which $T(x) = t$ and this is achieved by applying Neyman-Pearson lemma to the conditional d.f.'s given $T(x)$. For an example of this procedure, the reader is referred to Fraser (1957), Example 3.5, page 89.

There is a result connecting similarity with unbiasedness. The condition of unbiasedness is based on inequalities; it is not as easy to handle mathematically as similarity.

Result 7.2.3 (Lehmann, 1959). *If Γ is the common boundary of ω and $\Omega - \omega$ and if the power function $\beta_\phi(\theta)$ is a continuous function of θ for any test ϕ, then an unbiased size α test of ω against $\Omega - \omega$ is a similar test of size α for the d.f.'s of Γ.*

PROOF: The common boundary is the set of points which constitute limit points of sequences in ω as well as in $\Omega - \omega$. Since $\beta_\phi(\theta) \leq \alpha$ for all θ in ω, $\beta_\phi(\theta) \leq \alpha$ for θ in Γ by the continuity. Also, $\beta_\phi(\theta) \geq \alpha$ for θ in $\Omega - \omega$ and $\beta_\phi(\theta) \geq \alpha$ for θ in Γ. Hence it follows that $\beta_\phi(\theta) = \alpha$ for θ in Γ.

Remark 7.2.1. *It follows that the class of unbiased tests of ω against $\Omega - \omega$ is a subset of the class of tests similar on Γ. Thus, if a test is a most powerful similar test of Γ against $\Omega - \omega$ and if it is unbiased of size α, then necessarily it is the most powerful test in the smaller class of unbiased tests.*

7.3 [1]Principle of Invariance

Some transformations when applied to the outcome of an experiment transform the original problem into a problem statistically the same. The invariance restriction is then to consider a test function which has the same values for the transformed outcomes as for the corresponding original outcomes. Such tests are called *invariant tests.*

Let the class of d.f.'s over the space $\mathcal{X}$ be denoted by $\{F_\theta | \theta \in \Omega\}$. Then recall that a class $\mathcal{G}$ of continuous (measurable) transformations sx from $\mathcal{X}$ into $\mathcal{X}$ is called *invariant* for the probability structure if

1. $\mathcal{G}$ is a group, and

2. The class of d.f.'s $\{F_\theta | \theta \in \Omega\}$ is closed under $\mathcal{G}$.

The class of transformations $\bar{s}$ on Ω forms a group homomorphic to $\mathcal{G}$.

[1]This section can be omitted at a first reading of the book

Definition 7.3.1. *A test function $\phi(x)$ is said to be invariant with respect to $\mathcal{G}$ if*

$$\phi(sx) = \phi(x), \text{ for all } s \text{ in } \mathcal{G} \text{ and all } x \text{ in } \mathcal{X}.$$

A weaker form of invariance is useful in certain problems.

Definition 7.3.2. *$\phi(x)$ is said to be almost invariant for $\mathcal{G}$ if, for each s in $\mathcal{G}$,*

$$\phi(sx) = \phi(x)$$

for almost all x.

A test function always induces a partition of the sample space $\mathcal{X}$. If the sets $\{A\}$ form a partition of $\mathcal{X}$ and $\{B\}$ forms another partition of $\mathcal{X}$, then the totality of the sets $A \cap B$ also forms a partition of $\mathcal{X}$, called the intersection partition.

Definition 7.3.3. *The maximal invariant partition is the intersection partition of all invariant partitions. In other words, the maximal invariant partition is the finest partition in the sense that no set of the partition can have a proper subset belonging to an invariant partition.*

Definition 7.3.4. *$m(x)$ is said to be a maximal invariant function if its induced partition is the maximal invariant partition.*

Example of a maximal invariant partition.

For any point x belonging to $\mathcal{X}$, associate a set containing it defined by

$$T_x = \{x' | x' = sx\}; \quad s \text{ in } \mathcal{G},$$

that is, all points obtained from x by transformations in $\mathcal{G}$. For any point $x^* = s^* x$ in T_x consider T_{x^*}. Then,

$$\begin{aligned} T_{x^*} &= \{x' | x' = sx^*\} \\ &= \{x' | x' = ss^* x\} \\ &= T_x \end{aligned}$$

since $\mathcal{G}$ is a group. Thus, the sets T_x form a partition of $\mathcal{X}$. The sets T_x are closed under the transformations s and certainly no proper subset of any T_x would be closed. Hence, this is a maximal invariant partition.

Result 7.3.1 (Fraser, 1957). *Any invariant statistic $\phi(x)$ can be expressed as a continuous (measurable) function of the maximal invariant function $m(x)$.*

PROOF: See Fraser (1957, Theorem 3.7).

From the above result, it is clear that any invariant function is equivalent to a function of the maximal invariant function. Thus, in any hypothesis-testing problem, if we restrict ourselves to invariant test functions, we examine the test functions based on the maximal invariant function. We then have the following procedure. Find a group $\mathcal{G}$ of transformations which is invariant for the problem; for this group, find a maximal invariant function. Calculate the induced d.f.'s for the maximal invariant function and consider the hypothesis-testing problem for these d.f.'s; then look for the best test for this transformed problem. The resulting test expressed in terms of the original outcome by means of the maximal invariant function will be the best invariant test.

For the transformations $\bar{\mathcal{G}}$ on Ω, one can define a maximal invariant partition. Let $\bar{m}(\theta)$ denote the corresponding maximal invariant function. It is then natural to suspect that, for the transformed problem in terms of $m(x)$, the d.f.'s can be indexed by the parameter $\bar{m}(\theta)$.

Result 7.3.2 (Fraser). *If $\phi(x)$ is invariant with respect to $\mathcal{G}$, then the d.f. of $\phi(X)$ is constant over each set of the maximal invariant ($\mathcal{G}$) partition of Ω; that is, the d.f.'s of $\phi(X)$ depend on θ only through $\bar{m}(\theta)$.*

PROOF: We wish to show that for any (Borel) set B,

$$P_\theta\left\{\phi(X) \in B\right\} = P_{\theta'}\left\{\phi(X) \in B\right\}$$

whenever θ and θ' belong to the same set of the maximal invariant partition on Ω, that is, whenever $\theta' = \bar{s}\theta$. We have

$$\begin{aligned} P_{\theta'}\left\{\phi(X) \in B\right\} &= P_{\bar{s}_\theta}\left\{\phi(X) \in B\right\} \\ &= P_\theta\left\{\phi(sx) \in B\right\} \\ &= P_\theta\left\{\phi(X) \in B\right\} \end{aligned}$$

since $\phi(x)$ is invariant under $\mathcal{G}$.

7.4 Problems

7.2.1 If $X_1, \cdots, X_N$ are iid Bernoulli variables taking values 1 and 0 with probabilities θ and $1-\theta$ respectively. Show that $T = \sum_1^N X_i$ is sufficient and complete for the family of distributions.

7.2.2 Let $X_1, \cdots, X_N$ be iid normal $(0, \theta)$ variables. Show that $T = \sum_1^N X_i^2$ is sufficient and complete.

7.2.3 If $X_1, \cdots, X_N$ is a random sample from normal (θ, θ^2). Show that $T = \left(\sum_1^N X_i, \sum_1^N X_i^2\right)$ is sufficient but not complete.
(HINT: Show that

$$E_\theta \left\{ \left(\sum X_i \right)^2 - \frac{N+1}{2N} \sum_1^N X_i^2 \right\} = 0 \qquad \textit{for all } \theta$$

and the function $(\sum X_i)^2 - \frac{N+1}{2N} \sum_1^N X_i^2$ is not identically equal to zero.)

7.2.4 Let X be a discrete random variable such that $P(X = -1) = 2\theta - \theta^2$; $P(X = x) = (1-\theta)^3 \theta^x$, $x = 0, 1, \cdots$. Then show that the family of distributions, $\{P_\theta, 0 < \theta < 1\}$ is boundedly complete but not complete.

7.3.1 Let $X_1, \cdots, X_N$ be a random sample from a distribution $F(x)$.
(i) let $T_1 = (X_1 - \overline{X}, \cdots, X_N - \overline{X})$, where $\overline{X} = \frac{1}{N} \sum_1^N X_i$. Show that T_1 is invariant under all translations of thc form $X' = X - \theta$ for some real θ.
(ii) let $T_2 = \left(\frac{X_1 - \overline{X}}{X_{(N)} - X_{(1)}}, \cdots, \frac{X_N - \overline{X}}{X_{(N)} - X_{(1)}} \right)$. Show that T_2 is invariant under all translation and scale alternatives of the from $X' = \frac{X - \theta}{\sigma}$.

7.3.2 Let $X = (X_1, \cdots, X_N)$ have the density $f(x_1 - \theta, \cdots, x_n - \theta)$ where $-\infty < \theta < \infty$. We wish to test

$$H_0 : \ f = f_0 \qquad \textit{against}$$
$$H_1 : \ f = f_1$$

using a test that is invariant under the group G of transformations

$$gx = (x_1 + c, \ldots, x_n + c), \qquad -\infty < c < \infty.$$

The induced transformations in the parameter space are of the form $\overline{g}\theta = \theta + c$.

(i) Show that a maximal invariant under G is

$$Y = (X_1 - X_n, \cdots, X_{n-1} - X_n).$$

(ii) The density of Y under H_i is

$$\int_{-\infty}^{\infty} f_i(y_1 + z, \cdots, y_{n-1} + z, z)dz \qquad (i = 0, 1).$$

(iii) The uniformly most powerful among all invariant tests has the rejection region of the form

$$\frac{\int_{-\infty}^{\infty} f_1(x_1 + u, \cdots, x_n + u)du}{\int_{-\infty}^{\infty} f_0(x_1 + u, \cdots, x_n + u)du} > c.$$

(Hint: With respect to Y the problem of testing H_0 versus H_1 becomes a test of a simple hypothesis against a simple alternative. The most powerful test is free of θ and consequently UMP among all invariant tests.)

(iv) Suppose the X_i are i.i.d and
H_0: f_0 is uniform on $(-1, 1)$ and
H_1: f_1 is standard normal.
Evaluate the UMP invariant test criterion.

7.3.3 Let $X = (X_1, \cdots, X_N)$ and let G denote the set of all permutations of the coordinates of X. Then the set of induced order statistics $(X_{(1)}, \cdots, X_{(N)})$ is maximal invariant.

Chapter 8

Goodness of Fit Tests

8.1 Introduction

An important part of statistical inference is to obtain information about the population from which the sample is drawn. Most of the parametric tests are based on the assumption that the underlying distribution is normal. However this assumption needs to be validated before we go further with other aspects of statistical inference. First let us assume that the null hypothesis is completely specified. Thus we wish to test the null hypothesis

$$H_0 : F(x) = F_0(x) \text{ for all } x$$

against the alternative

$$H_1 : F(x) \neq F_0(x) \text{ for some } x\,.$$

Available is a random sample $X_1, \ldots, X_N$ from the population.

8.2 Chi Square Test

The sample is classified into k categories and let f_i, $i = 1, 2, \ldots, k$ denote the cell frequencies and let $p_i(\theta)$ denote the probability of an observation falling into the ith category under H_0 ($i = 1, \ldots, k$) such that

$$\sum^k p_i = 1 \text{ and } \sum_1^k f_i = N\,.$$

Karl Pearson (1900) proposed the statistic

$$T = \sum_{i=1}^{k} (f_i - Np_i)^2 / Np_i$$

(which is equivalent to $T' = \sum_{i=1}^{k} (f_i - Np_i)^2 / f_i$ in probability) and we reject H_0 when T is large. When H_0 is true T is asymptotically distributed as chi-square with $k-1$ degrees of freedom. Let us prove the assertion. The likelihood function $L(p_1, \ldots, p_k)$ is proportional to

$$\prod_{i}^{k} p_i^{f_i}, \; 0 \le f_i \le N, \; \sum f_i = N, \; \sum p_i = 1 .$$

Consider

$$l(p_1, \ldots, p_i) = \log\, N! - \sum \log\, f! + \sum f_i \,\log\, p_i - \lambda \left(\sum p_i - 1 \right) .$$

Then it is easy to see that the maximum likelihood estimates of p_i are $\hat{p}_i = f_i/N$, $i = 1, \ldots, k$. The likelihood ratio is

$$\begin{aligned} \Lambda &= \sup_{H_0} L(p_1, \ldots, p_k) / \sup_{H_0 \cup H_1} L(p_1, \ldots, p_k) \\ &= \prod_{i=1}^{k} (p_i / \hat{p}_i)^{f_i} . \end{aligned}$$

Thus,

$$-2\, \log\, \Lambda = -2 \sum_{i=1}^{k} f_i \left\{ \log\, p_i - \log \left(\frac{f_i}{N} \right) \right\} .$$

Consider the Taylor's expansion

$$\begin{aligned} \log\, p_i = \log\, \hat{p}_i \;\; &+ \;\; (p_i - \hat{p}_i) \frac{1}{\hat{p}_i} + \frac{(\hat{p}_i - p_i)^2}{2!} \left(-\frac{1}{\hat{p}_i^2} \right) \\ &+ \left\{ -\frac{(\hat{p}_i - p_i)^2}{2!} \frac{1}{p_i^{*2}} + \frac{(\hat{p}_i - p_i)^2}{2! \hat{p}_i^2} \right\} \end{aligned}$$

where p_i^* lies between p_i and $\hat{p}_i$.

Hence,

$$\begin{aligned} (\log\, p_i - \log\, \hat{p}_i) &= \left(p_i - \frac{f_i}{N} \right) \frac{N}{f_i} - \left(p_i - \frac{f_i}{N} \right)^2 \frac{N^2}{2 f_i^2} + \epsilon_i \\ &= \frac{(Np_i - f_i)}{f_i} - \frac{(Np_i - f_i)^2}{2 f_i^2} + \epsilon_i . \end{aligned}$$

Now $f_i/N \to p_i$ in probability due to the weak law of large numbers. Thus

$$-2\ \log\ \Lambda = -\sum_{i=1}^{k} f_i\{\log\ p_i - \log\ \hat{p}_i\} = \sum_{i=1}^{k} \frac{(f_i - Np_i)^2}{f_i} - 2\sum_{i=1}^{k} \epsilon_i f_i$$

where

$$\epsilon_i f_i = \frac{N}{2}(\hat{p}_i - p_i)^2 \left(p_i^{*^2} - \hat{p}_i^2\right) / \hat{p}_i p_i^{*^2} .$$

Now since $\sqrt{N}\,(\hat{p}_i - p_i)$ is bounded in probability and $\hat{p}_i$ and p_i^* tend to p_i in probability, it follows easily that $\epsilon_i f_i \to 0$ in probability and consequently $\sum_{i=1}^{k} \epsilon_i f_i \to 0$ in probability as $N \to \infty$. Now the proof is complete since $-2\ \log\ \Lambda$ is distributed as χ^2_{k-1}.

Alternate Method of Proof

The joint probability function of the frequencies $f_1, \ldots, f_k$ is multinomial, that is,

$$P(f_1 = v_1, \ldots, f_k = v_k) = \frac{N!}{\pi_1^k v_i!} \pi p_i^{v_i}, \quad \sum f_i = N .$$

However unconditionally the $f_1, \ldots, f_k$ are independent Poisson random variables with parameters $Np_1, \ldots, Np_k$, respectively. Hence $Z_i = (f_i - Np_i)\sqrt{Np_i}$ are asymptotically normal and $\sum Z_i^2 = \sum_1^k (f_i - Np_i)^2/Np_i$ is asymptotically chi-square with k degrees of freedom.

$\sum Z_i^2$, conditional on $\sum \sqrt{p_i} Z_i = 0$ is chi-square with $k-1$ degrees of freedom after using Cochran's theorem.

Neyman (1949) has proposed a 'modified' chi-square test given by

$$T' = \sum_{i=1}^{k} N \frac{(\hat{p}_i - p_i)^2}{\hat{p}_i} .$$

Now one can easily show that T and T' are asymptotically equivalent. Towards this consider

$$T - T' = \sum_{i=1}^{k} N(\hat{p}_i - p_i)^2 \left(\frac{1}{\hat{p}_i} - \frac{1}{p_i}\right) = \sum_{i=1}^{k} N \frac{(\hat{p}_i - p_i)^3}{p_i \hat{p}_i} .$$

However, since $\sqrt{N}(\hat{p}_i - p_i)$ is bounded in probability and $\hat{p}_i$ tends to p_i in probability, we have

$$|T - T'| \le \sum_{i=1}^{k} \frac{K \cdot N^{-1/2}}{p_i^2} \to 0 \text{ as } N \to \infty.$$

8.3 Kolmogorov-Smirnov (K-S) Test

The test criterion is

$$\begin{aligned} D_N &= \sup_x |F_N(x) - F_0(x)| \\ &= \max_{1 \le i \le N} \left\{ \max \left[F_0(X_{iN}) - \left(\frac{i-1}{N}\right), \frac{i}{N} - F_0(X_{iN}) \right] \right\} \end{aligned}$$

where $F_0(x)$ denotes the hypothesized distribution and $X_{1N} < \cdots X_{NN}$ denote the ordered observations. We reject the null hypothesis for large values of the statistic. We also know that

$$\begin{aligned} \psi(z) = \lim_{N \to \infty} P\left(\sqrt{N}\, D_N \le z\right) &= 1 - 2\sum_{j=1}^{\infty} (-1)^{j-1} \exp(-2j^2 z^2) \\ &\doteq 1 - 2\, \exp(-2z^2)\,. \end{aligned} \tag{8.3.1}$$

The critical values of D_N are given in Appendix VI. Notice that D_N is distribution-free whereas the chi-square test is asymptotically distribution-free. Also, for small values of z (say $< .05$), we have from Kolmogorov (1933)

$$\psi(z) = \left\{ (2\pi)^{1/2}/z \right\} \exp(-\pi^2/8z^2)\,. \tag{8.3.2}$$

It is impossible to evaluate the power of the K-S test since the alternative hypothesis is very composite. However, one can get a lower bound for the power. Let $F_1(x)$ denote the alternative at which we wish to evaluate the power of the K-S test. Then letting

$$\Delta = \sup_x |F_1(x) - F_0(x)|\ .$$

Power at $F_1(x)$

$$\begin{aligned} &= P\left(\sup_x |F_N - F_0| > z/\sqrt{N}; F_1\right) \ge P\left(\sup_x |F_N - F_1| > z/\sqrt{N} + \Delta\right) \\ &\approx 2\sum_{j=1}^{\infty} (-1)^{j-1} \exp\left\{ -2j^2 \left(z + \sqrt{N}\,\Delta\right)^2 \right\}, \end{aligned}$$

after using the inequality

$$|F_N(x) - F_1(x)| \le |F_N - F_0| + |F_1 - F_0|\,.$$

Massey, Jr. (1951) has plotted the lower bound for the power function for various values of Δ and for $\alpha = .05$ and $.01$. It should be pointed out that the lower bounds are quite conservative. Massey (1951) has also shown that the K-S test is consistent and biased. If the underlying distribution is discrete, one can use the K-S test conservatively.

When certain parameters of the distribution are to be estimated from the sample, then the K-S test is no longer distribution-free. Massey (1951) and Birnbaum (1952) point out that the K-S, when the commonly tabulated critical values are used, will be conservative (that is, the probability of type I error will be smaller than the nominal level of significance). Lillifors (1967) has shown that the test will be extremely conservative. He has tabulated the Monte Carlo critical values of the K-S test for normality when the mean and variance are estimated from the sample. He points out that the Monte Carlo critical values are somewhat pulled in and are in most cases approximately two thirds the standard values. The asymptotic Monte Carlo critical points coincide with those obtained by Kac, Kiefer and Wolfowitz (1955). In the following table we shall give the critical points using K-S test for normality.

Table 8.3.1: The critical values of the Kolmogorov-Smirnov one-sample test

	level of of significance for $D = \sup \left\|F_N(x) - \Phi\left(\frac{x-\bar{x}}{s}\right)\right\|$		
N = sample size	.10	.05	.01
5	.31	.34	.40
10	.24	.26	.29
15	.20	.22	.26
20	.17	.19	.23
25	.16	.18	.20
30	.14	.16	.19
over 30	$.80/\sqrt{N}$	$.89/\sqrt{N}$	$1.03/\sqrt{N}$

Lower bound for the Power of one-sided K-S test

Birnbaum (1953b) has given lower and upper bounds for the power of one-sided Kolmogorov test of fit for continuous distribution functions. We shall give only the lower bound.

Let $X_1, \ldots, X_N$ be a random sample from a continuous d.f. $F(x) \in \Omega$. Let F_0 and G belong to Ω. We wish to test the hypothesis, $H_0 : F(x) = F_0(x)$

for all x against the alternative $H_1 : F(x) = G(x)$. Let $X_{1N} \le \cdots \le X_{NN}$ denote the order statistics in the sample and $F_N(x)$ denotes the empirical d.f. Let $J(u) = F_0^{-1}(u) = \inf_x \{x : F_0(x) \ge u\}$. Then J, the inverse function of F_0 is uniquely defined. Also, $P(F_0(x) \le F_N(x) + \epsilon, \text{ for all } x) = P_N(\epsilon)$ is known to be independent of $F_0(x)$ and a closed expression for $P_N(\epsilon)$ has been obtained by Birnbaum and Tingey (1951) and is given in Chapter 5, Section 6. We can use $P_N(\epsilon)$ to test H_0 against H_1. Let $\epsilon_{N,\alpha}$ denote the solution of the equation $P_N(\epsilon) \ge 1 - \alpha$. Then we reject H_0 if and only if the inequality

$$F_0(x) < F_N(x) + \epsilon_{N,\alpha} \tag{8.3.3}$$

fails to hold for all real x. Hence, the power of the test is $1 - p$ where

$$p = P\left(F_0(x) < F_N(x) + \epsilon_{N,\alpha} \text{ for all } x;\ G(x)\right). \tag{8.3.4}$$

One can easily show that (8.3.4) is satisfied for all x if and only if it is satisfied for all sample points X_{iN}, $i = 1, \ldots, N$. That is, if and only if

$$F_0(X_{iN}) < \frac{i-1}{N} + \epsilon_{N,\alpha} \text{ for } i = 1, \ldots, N. \tag{8.3.5}$$

Consequently,

$$\begin{aligned} p &= P\left(F_0(X_{iN}) < \frac{i-1}{N} + \epsilon_{N,\alpha} \text{ for } i = 1, \ldots, N;\ G(x)\right) \\ &= P\left(X_{iN} < J\left(\frac{i-1}{N} + \epsilon_{N,\alpha}\right),\ \text{for } i = 1, \ldots, N;\ G(x)\right) \\ &= P\left(G(X_{iN}) < GJ\left(\frac{i-1}{N} + \epsilon_{N,\alpha}\right),\ \text{for } i = 1, \ldots, N;\ G(x)\right) \\ &= P\left(U_{iN} < L\left(\frac{i-1}{N} + \epsilon_{N,\alpha}\right),\ \text{for } i = 1, \ldots, N\right) \end{aligned} \tag{8.3.6}$$

where $U_{1N} \le \cdots \le U_{NN}$ denote the uniform order statistics on (0,1) and

$$\begin{aligned} L(v) &= G\{J(v)\} = G\left\{F_0^{-1}(v)\right\}, & 0 < v < 1, \\ &= \lim_{v \to 0} L(v), & v \le 0, \\ &= \lim_{v \to 1} L(v), & v \ge 1. \end{aligned}$$

Thus,

$$p = N! \int_0^{L(\epsilon)} \int_{u_1}^{L(\frac{1}{N}+\epsilon)} \cdots \int_{u_{i-1}}^{L(\frac{i-1}{N}+\epsilon)} \cdots \int_{u_{N-1}}^{L(\frac{N-1}{N}+\epsilon)} du_n \cdots du_1 \quad (8.3.7)$$

where, for the sake of simplicity, ϵ represents $\epsilon_{N,\alpha}$. In order to evaluate a lower bound for the power, we shall obtain an upper bound for p. Towards this, let

$$\delta = \sup_x \{F_0(x) - G(x)\} = F_0(x_0) - G(x_0) . \quad (8.3.8)$$

Purely on intuitive grounds, we expect the power to be close to its minimum when G is close to G^* where

$$G^*(x) = \begin{cases} F_0(x_0) - \delta & \text{for } x \le x_0 , \\ 1 & \text{for } x > x_0 . \end{cases} \quad (8.3.9)$$

In order to verify this conjecture, let

$$L^*(v) = G^*\{J(v)\} = \begin{cases} F_0(x_0) - \delta & \text{for } 0 < v \le F_0(x_0) \\ 1 & \text{for } F_0(x_0) < v < 1 . \end{cases}$$

Now, let

$$F_0(x_0) = v_0 \text{ and } F_0(x_0) - \delta = u_0$$

so that by (8.3.8) we have

$$L(v_0) = G\{F_0^{-1}(v_0)\} = G(x_0) = u_0 \quad (8.3.10)$$

and

$$L^*(v) = \begin{cases} u_0 & \text{for } 0 < v \le v_0 \\ 1 & \text{for } v_0 < v < 1 . \end{cases}$$

Let j be the largest integer contained in $N(u_0 + \delta - \epsilon)$,

$$j = [N(u_0 + \delta - \epsilon)] = [N(v_0 - \epsilon)] . \quad (8.3.11)$$

Then we have

$$\frac{i-1}{N} + \epsilon \le \frac{j}{N} + \epsilon \le u_0 + \delta = v_0 \text{ for } i - 1 \le j .$$

Hence, using (8.3.10),

$$L\left(\frac{i-1}{N}+\epsilon\right) \leq L(v_0) = u_0 \text{ for } i-1 \leq j\,. \tag{8.3.12}$$

This implies that replacing in (8.3.7) the functions L by L^* in the upper limits of integration will not decrease these limits. Hence,

$$p \leq N! \int_0^{u_0} \cdots \int_{u_j}^{u_0} \int_{u_{j+1}}^{1} \cdots \int_{u_{n-1}}^{1} du_n \cdots du_{j+2} du_{j+1} \cdots du_1\,. \tag{8.3.13}$$

If $j \leq 0$, all upper limits of integration in (8.3.13) are l and we get the trivial inequality, $p \leq 1$. Performing integration on $u_n, u_{n-1}, \ldots, u_{j+2}$, we have

$$p \leq \frac{N!j}{(N-j-1)!} \int_0^{u_0} \cdots \int_{u_j}^{u_0} (1-u_{j+1})^{N-j-1} du_{j+1} \cdots du_1\,. \tag{8.3.14}$$

By induction on j one can show that the integral in (8.3.14) is equal to

$$1 - \sum_{i=0}^{j} \binom{N}{i} u_0^i (1-u_0)^{n-i} = I_{u_0}(j+1, N-j)$$

where $I_x(a,b)$ denotes the incomplete beta function. Thus

$$p \leq 1 - \sum_{i=0}^{j} \binom{N}{i} u_0^i (1-u_0)^{N-i} = I_{u_0}(j+1, N-j)\,. \tag{8.3.15}$$

Hence, the lower bound for the power when G is true is

$$\sum_{i=0}^{j} \binom{N}{i} u_0^i (1-u_0)^{N-i} = 1 - I_{u_0}(j+1, N-j) \tag{8.3.16}$$

where

$$u_0 = F_0(x_0) - \delta,\ j = [N\{F_0(x_0 - \epsilon\}]\,.$$

Notice that this lower bound cannot be improved since, for any given $F_0(x) \in \Omega$, x_0 and ϵ, one can construct a $G(x)$ arbitrarily close to $G^*(x)$.

Now, if N is large one can use a normal approximation to the binomial and obtain the lower bound as (letting, $j = N(u_0 + \delta - \epsilon) - \eta_N$),

$$\Phi\left[\left\{N^{1/2}(\delta-\epsilon) - N^{-1/2}\eta_N\right\}\{u_0(1-u_0)\}^{-1/2}\right]$$

$$> \Phi\left[\left\{N^{1/2}(\delta-\epsilon) - \frac{1}{2}N^{-1/2}\right\}\{u_0(1-u_0)\}^{-1/2}\right]. \tag{8.3.17}$$

The asymptotic value of $\epsilon_{N,\alpha} = \left[-(2N)^{-1} \log \alpha\right]^{1/2}$ (see, for instance, Section 6 of Chapter 5) and hence the large-sample lower bound to the power is

$$\Phi\left[\left\{N^{1/2}\delta - \sqrt{(1/2)\log(1/\alpha)} - \frac{1}{2}N^{-1/2}\right\}\{u_0(1-u_0)\}^{-1/2}\right]. \qquad (8.3.18)$$

If only δ is known, but not u_0, then (8.3.18) may be replaced by its minimum with regard to u_0,

$$\Phi\left[2N^{1/2}\delta - \{-2 \log \alpha\}^{1/2} - N^{-1/2}\right]. \qquad (8.3.19)$$

8.4 Cramér-von-Mises Test

For a given nonnegative function $h(y)$ $0 \le y \le 1$, consider the test criterion of Cramér and von-Mises:

$$N \int_{-\infty}^{\infty} [F_N(x) - F_0(x)]^2 h(F_0(x))\, dF_0(x)$$

which is also distribution-free. With $h(y) \equiv 1$, the test was proposed and its limiting distribution was given by Smirnov (1939). Let

$$N\omega_N^2 = N \int_0^1 [F_N^*(y) - y]^2\, dy$$

where $F_N^*(y)$ denotes the empirical distribution function based on a random sample of size N drawn from the uniform distribution on (0,1). Also carrying out the integration one can show that

$$N\omega_N^2 = (12N)^{-1} + \sum_{i=1}^{N} [\{(2i-1)/2N\} - F_0(X_{iN})]^2$$

where $X_{1N} \le \cdots \le X_{NN}$ denote the ordered X_i's. Notice that $\frac{2i-1}{2N}$ is the mid point of the interval $\left(\frac{i-1}{N}, \frac{i}{N}\right)$.

Marshall (1958) has obtained the distribution of $N\omega_N^2$ for $N = 1, 2, 3$. The distribution of $N\omega_N^2$ is tabulated by Darling and Anderson (1952). Anderson and Darling (1954) propose a new test by considering $h(y) = \{y(1-y)\}^{-1}$. Marshall (1958) points out that the exact distribution of $N\omega_N^2$ or the allied test of Anderson and Darling (1954) tends to its asymptotic distribution fairly rapidly even for N as small as 40. Asymptotic percentiles of Cramér-von-Mises test taken from Anderson and Darling (1952) are given below.

Table 8.4.1: Asymptotic percentiles: $\alpha = P(N\omega_N^2 > z)$.

α	.20	.10	.05	.02	.01	.001
z	.241	.347	.461	.620	.743	1.168

Tiku (1965) gives a chi-square approximation to the distribution of $N\omega_N^2$. The modified version of Cramér-von-Mises test is given by

$$W_N^2 = N \int [F_N(x) - F_0(x)]^2 \, \psi\left(F_0(x)\right) dF_0(x)$$

where

$$\psi(u) = [u(1-u)]^{-1} \, .$$

Let $U_{iN} = F_0(X_{iN})$ where $X_{1N} \leq \cdots \leq X_{NN}$ denote the ordered observations in the sample. Then writing

$$\begin{aligned} \frac{1}{N} W_N^2 &= \int_0^1 [F_N^*(u) - u]^2 \, \psi(u) du \\ &= \int_0^{U_{1N}} u^2 \psi(u) du + \cdots + \int_{U_{NN}}^1 (1-u)^2 \psi(u) du \end{aligned}$$

and carrying out straightforward integrations and collecting terms, we obtain

$$W_N^2 = -N - \frac{1}{N} \sum_{j=1}^{N} (2j-1) \left[\log \, U_{jN} + \log(1 - U_{N-j+1,N})\right] \, .$$

We reject H_0 when W_N^2 is large. Anderson and Darling (1954) give the asymptotic percentiles of W_N^2.

Anderson and Darling (1954) give the asymptotic distribution of W_N^2. In particular they obtain

$$\lim_{N\to\infty} \, E(W_N^2) = E(W_N^2) = \sum_{j=1}^{\infty} \frac{1}{j(j+1)} = 1 \, ,$$

Table 8.4.2: Asymptotic percentiles of $W_N^2 : \alpha = P(W_N^2 > z)$.

α	.10	.05	.01
z	1.933	2.492	3.857

and

$$\lim_{N\to\infty} \operatorname{var}(W_N^2) = 2\sum_{j=1}^{\infty} \frac{1}{j^2(j+1)^2} = \frac{2}{3}(\pi^2 - 9) - 0.590.$$

The asymptotic distribution of the Cramér-von-Mises test statistic when certain parameters (especially, the mean and variance in the normal case) are estimated from the sample has been considered by Darling (1955), Kac, Kiefer and Wolfowitz (1955), Sukhatme (1972) and Gregory (1977). For latest results and other references the reader is referred to Gregory (1977).

8.5 Shapiro-Wilk (S-W) Test

Shapiro and Wilk (1965) have proposed a criterion that is based on the regression coefficient of the ordered observations on the vector of expected values of the standardized order statistics. In the following we shall present their results. Suppose we wish to test $H_0 : F(x) = F_0(x) = F^*\left(\frac{x-\theta}{\sigma}\right)$ where θ and σ are unknown. Then $X_{1N} \le \cdots \le X_{NN}$ denote the ordered sample from $F(x)$ and let $Y_{iN} = (X_{iN} - \theta)/\sigma$, $i = 1, \ldots, N$. Assume that $F^*(y)$ is symmetric about zero. Also, let $\mu_{iN} = E(Y_{iN})$, $i = 1, \ldots, N$, and $\operatorname{cov}(Y_{iN}, Y_{jN}) = \sigma_{i,j,N}$, $1 \le i, j \le N$.

Let $\Sigma = ((\sigma_{ijN}))$ and $\Omega = \Sigma^{-1}$. Then the best linear ordered estimates of θ and σ are

$$\hat{\theta} = \mathbf{1}'\Omega\boldsymbol{X}/\mathbf{1}'\Omega\mathbf{1}, \; \hat{\sigma} = \boldsymbol{\mu}'\Omega\boldsymbol{X}/\boldsymbol{\mu}'\Omega\boldsymbol{\mu}$$

since, for symmetric populations $\mathbf{1}'\Omega\boldsymbol{\mu} = 0$, where

$$\mathbf{1} = (1, \ldots, 1)' \text{ and } \boldsymbol{\mu} = (\mu_{1,N}, \ldots, \mu_{N,N})'$$

and $\boldsymbol{X} = (X_{1,N}, \ldots, X_{N,N})'$. Also, let $s^2 = \sum_{i=1}^N (X_i - \bar{X})^2$ which is an unbiased estimate of $(N-1)\sigma^2$. Then the Shapiro-Wilk (1965) statistic is given by

$$W = \frac{R^4\hat{\sigma}^2}{C^2S^2} = \frac{b^2}{S^2} = \frac{(a'X)^2}{S^2} = \frac{\left(\sum_1^N a_i X_{iN}\right)^2}{\sum(X_i - \bar{X})^2}$$

where

$$\begin{aligned} R^2 &= \boldsymbol{\mu}'\Omega\boldsymbol{\mu}, \\ C^2 &= \boldsymbol{\mu}'\Omega\Omega\boldsymbol{\mu}, \\ a' &= (a_1, \ldots, a_N) = \boldsymbol{\mu}'\Omega/(\boldsymbol{\mu}'\Omega\Omega\boldsymbol{\mu})^{1/2}, \\ b &= R^2\hat{\sigma}/C\,. \end{aligned}$$

That is, up to a normalizing constant C, b equals the best linear unbiased estimator of the slope of a linear regression of $\boldsymbol{X}$ on $\boldsymbol{\mu}$. The constant C is defined in such a way that the coefficients are normalized. Small values of W are significant. In the following we shall list some of the properties of the statistic W obtained by Shapiro and Wilk (1965).

Property (i). W is invariant under location and scale changes.

PROOF: Let $X_{iN} = cZ_{iN} + d$ for some fixed constants c and d. Then

$$\sum a_i X_{iN} = c\sum a_i Z_{iN} + d\sum a_i\,.$$

However,

$$\begin{aligned} \sum a_i &= \boldsymbol{\mu}'\Omega\mathbf{1} = 0 \\ W &= \left(\sum a_i Z_{iN}\right)^2 \Big/ \sum(Z_{iN} - \bar{Z})^2\,. \end{aligned}$$

Property (ii). $E(W^r) = Eb^{2r}/ES^{2r}$.

PROOF: It follows from the fact that S^2 is independent of any linear combination of the X_{iN}, for normal samples and that W is independent of S^2.

Property (iii). The maximum value of W is 1.

PROOF: Assume that $\bar{X} = 0$, since W is location-invariant. Then

$$W = \left(\sum a_i X_{iN}\right)^2 \Big/ \sum X_{iN}^2 \le \sum a_i^2 = 1 \text{ by definition.}$$

Property (iv). The minimum value of W is $Na_1^2/(N-1)$.

PROOF: (Due to Mallows). Since W is location and scale invariant, it suffices to maximize $\sum X_{iN}^2$ subject to the constraints $\sum X_{iN} = 0$, $\sum a_i X_{iN} = 1$. Since this is a convex region and $\sum X_{iN}^2$ is a convex function, its maximum will occur at one of the $(N-1)$ vertices of the region:

$$\left(\frac{N-1}{Na_1}, \frac{-1}{Na_1}, \dots, \frac{-1}{Na_1}\right)$$

$$\left(\frac{N-2}{N(a_1+a_2)}, \frac{N-2}{N(a_1+a_2)}, \frac{-2}{N(a_1+a_2)}, \dots, \frac{-2}{N(a_1+a_2)}\right)$$

$$\vdots$$

$$\left(\frac{1}{N(a_1+\cdots+a_{N-1})}, \frac{1}{N(a_1+\cdots+a_{N-1})}, \dots, \frac{-(N-1)}{N(a_1+\cdots+a_{N-1})}\right)$$

It can be verified numerically, for the values of the specific coefficients $\{a_i\}$ that the maximum of $\sum X_{iN}^2$ occurs at the first of these vertices and W achieves the minimum at this point.

Property (v).

$$E(W^{1/2}) = R^2\Gamma\left((N-1)/2\right)/C\Gamma(N/2)\sqrt{2}$$

and

$$E(W) = R^2(R^2+1)/C^2(N-1)$$

where $R^2 = \boldsymbol{\mu}'\Omega\boldsymbol{\mu}$, and $C^2 = \boldsymbol{\mu}'\Omega\Omega\boldsymbol{\mu}$.

PROOF: From Property (ii) we have

$$E(W^{1/2}) = Eb/ES \text{ and } EW = Eb^2/ES^2\,.$$

Now the result follows after noting that

$$\begin{aligned} ES &= \sigma\sqrt{2}\,\Gamma(N/2)/\Gamma\left((N-1)/2\right), \\ ES^2 &= (N-1)\sigma^2, \\ Eb &= R^2E\hat{\sigma}/C = R^2\sigma/C \end{aligned}$$

and

$$Eb^2 = R^4E(\hat{\sigma}^2)/C^2 = R^4\left\{\text{var}\,\hat{\sigma} + (E\hat{\sigma})^2\right\}/C^2 = \sigma^2R^2(R^2+1)/C^2\,,$$

since $\text{var}(\hat{\sigma}) = \sigma^2/\boldsymbol{\mu}'\Omega\boldsymbol{\mu} = \sigma^2/R^2$.

Property (vi). For $N = 3$, the probability density of W is

$$f_W(x) = \frac{3}{\pi}(1-x)^{-1/2}x^{-1/2},\ 3/4 \le x \le 1\,.$$

PROOF: See Shapiro and Wilk (1965, p. 595).

For $N \ge 20$, Shapiro and Wilk (1965) recommend the following approximations for R^2 and C^2:

$$C^2 = -2.722 + 4.083N$$

and

$$R^2 = -2.411 + 1.981N\,.$$

Shapiro and Wilk (1965) have computed the empirical percentage points of W for each N by Monte Carlo methods and plotted them. The number of samples used is 5000 for $N = 3(1)20$ and $\left[\frac{100.000}{N}\right]$ for $N = 21(1)50$. They also present selected null distribution percentage points of W which are based on fitted Johnson (1949) S_B approximation. These are given in their Table 6.

Via Monte Carlo trials, the authors study the power of various goodness of fit tests for different alternatives. On the basis of this study the authors claim that the W test is comparatively quite sensitive to a wide range of non-normality even with samples as small as $N = 20$. In particular, it seems to be sensitive to asymmetry, long tailedness and, to some degree to short-tailedness. The weighted Cramér-von-Mises (WCM) test has the highest power of all tests considered for long tailed alternatives such as the Cauchy. For a more elaborate comparative study of the power properties of various goodness of fit tests the reader is referred to Shapiro, Wilk and Chen (1968).

Durbin (1961) proposed a distribution-free test given by:

$$D = \max_{1\le i\le N}\left(\frac{i}{N} - \sum_{j=1}^{i} g_j\right)$$

where $g_j = (N+2-j)(c_j^* - c_{j-1}^*)$, $j = 1,\ldots,N$, $0 \le c_0^* \le c_1^* \le \cdots \le c_N^*$ obtained by ordering $c_1 = u_1$, $c_2 = u_2 - u_1, \ldots, c_{N+1} = 1 - u_N$ and $u_i = F_0(X_{iN})$, $i = 1,\ldots,N$. The general conclusions reached by Shapiro et al (1968) is that (i) the W test provides a generally superior omnibus measure of non-normality and (ii) tests based on distance (KS, CM, WCM, D) are typically very insensitive.

The numeral values of the elements of Ω are known for $N \leq 20$. Typically in goodness of fit testing we have $N \geq 20$. Thus, it is desirable to approximate Ω. Shapiro and Francia (1972) set $\Omega \equiv I$, the identity matrix and propose

$$W' = \left(\sum_1^N b_i X_{iN}\right)^2 \Big/ \sum (X_{iN} - \bar{X})^2$$

where

$$b' = (b_1, \ldots, b_N) = \frac{\boldsymbol{\mu}'}{(\boldsymbol{\mu}'\boldsymbol{\mu})^{1/2}},$$

the values of $\boldsymbol{\mu}$ are given by Harter (1961) for $N = 2(1)100(25)300(50)400$. That is, W' is obtained by considering the slope of the regression line by simple least squares rather than the generalized least squares. The null distribution of W' was approximated by an empirical study. They are given in Table 8.5.1 and are taken from Shapiro and Francia (1972). The sampling

Table 8.5.1: Empirical percentage points of W'

Sample Size	Level of Significance		
N	.01	.05	.10
35	.92	.94	.95
50	.94	.95	.96
59	.94	.96	.97
69	.95	.97	.97
79	.96	.97	.98
89	.96	.97	.98
99	.97	.98	.98

studies seem to indicate that the sensitiveness of the W and W' tests are similar.

Sarkadi (1975) has shown that the Shapiro-Francia test of normality is consistent, against all alternatives having finite second moment.

Shapiro-Wilk's criterion for testing exponentiality takes the form:

$$W = N(\bar{X} - X_{1N})^2/(N-1)S^2, \; S^2 = \sum (X_{iN} - \bar{X})^2 .$$

Shapiro and Wilk (1972) tabulate the percentage points of W-exponential tests, employing an empirical sampling procedure. Notice here that the W-exponential procedure is a two-tailed one in the sense that for an unspecified alternative to the exponential, the statistic may shift to either smaller or larger values. However, if one wishes to restrict himself to a specified class of alternatives, then the one-tailed version of W can be employed.

8.6 General Version of S-W Test

Let $X_1, \ldots, X_N$ be a random sample from $F(x-\theta)/\sigma$ where θ and σ are unknown. We wish to test the null hypothesis $H_0 : F(y) = F_0(y)$, against the alternative $H_1 : F(y) \neq F_0(y)$ for some y.

Case 1

Let us assume that θ and σ are also specified by the null hypothesis. Then let $X_{1N} \leq \cdots \leq X_{NN}$ denote the ordered X's and let $U_{iN} = F_0\left((X_{iN}-\theta)/\sigma\right)$, $i = 1, \ldots, N$, $\mu_{iN} = EU_{iN} = i/(N+1)$, $\Sigma = ((\sigma_{ijN}))$ = the variance-covariance matrix of $U = (U_{1N}, \ldots, U_{NN})'$. It is well known that $\sigma_{ijN} = i(N-j+1)/(N+1)^2(N+2)$ for $i \leq j$ and

$$\Sigma^{-1} = \Omega = (N+1)(N+2)\begin{pmatrix} 2 & -1 & 0 & & \cdots & 0 \\ -1 & 2 & -1 & 0 & \cdots & 0 \\ & & & & & \\ & & & & & -1 \\ 0 & & & -1 & & 2 \end{pmatrix}$$

Shapiro-Wilk's criteria take the form of

$$\sqrt{W} = b'U, \text{ where } b = \frac{\mu'\Omega}{(\mu'\Omega\Omega\mu)^{1/2}}.$$

Now,

$$\mu'\Omega = (N+1)(N+2)(-1, 0, \ldots, 0, 1).$$

Hence,

$$\sqrt{W} = b'U = \frac{1}{\sqrt{2}}(U_{NN} - U_{1N}).$$

Thus one can use $R = (U_{NN} - U_{1N})$ as a test criteria and reject H_0 for small values of R.

$$P(R \leq u) = N(N-1)\int_0^u r^{N-2}(1-r)dr = u^{N-1}\left(N - (N-1)u\right).$$

$$ER = (N-1)/(N+1) \text{ and } \operatorname{var} R = 2(N-1)/(N+1)^2(N+2).$$

One can easily tabulate the critical points of R. Thus in order to test a simple null hypothesis, one set of tables is all that is required. In order to study the asymptotic distribution of R, consider

$$\begin{aligned} & P\left(\left(R - \frac{N-1}{N+1}\right)\frac{N+1}{\sqrt{2}} \le z\right) \\ = \;& P\left(R \le \frac{N-1+2\sqrt{z}}{N+1}\right) \\ = \;& \left(1 + \frac{z\sqrt{2}-2}{N+1}\right)^{N-1} \left\{N - \frac{N-1}{N+1}(z\sqrt{2}+N-1)\right\} \\ \approx \;& (3 - z\sqrt{2})\ \exp(z\sqrt{2}-2), \ -\infty < z < \sqrt{2}\,, \end{aligned}$$

from which one can easily obtain the asymptotic percentage points of R.

Case 2

Assume that θ and σ are unknown and are not specified by the null hypothesis. Then one can obtain some "good" (especially, consistent) estimates of θ and σ, which are denoted by $\hat{\theta}$ and $\hat{\sigma}$, respectively. For instance, in the case of testing for normality, $\hat{\theta} = \bar{X}$, $\hat{\sigma} = s = S/\sqrt{N-1}$. In the case of negative exponential case, $\hat{\theta} = X_{1N}$ and $\hat{\sigma} = \bar{X} - X_{1N}$. Then we will define $U_{iN} = F_0\left((X_{iN} - \hat{\theta})/\hat{\sigma}\right)$, $i = 1, \ldots, N$ and compute $R = U_{NN} - U_{1N}$ and carry out the test procedure as in Case 1.

8.7 Asymptotic Test Based on Spacings

Background Information

Let $X_{1N} \le \cdots \le X_{NN}$ be the OS in a random sample of size N drawn from the density $f(x; \theta, \sigma) = \frac{1}{\sigma}e^{-(x-\theta)/\sigma}$.

Let $l_i = X_{i+1,N} - X_{i,N}$, $i = 1, 2, \ldots, N-1$. Let $Z_{iN} = (X_{iN} - \theta)/\sigma$ and $\mu_{iN} = EZ_{iN}$, $i = 1, \ldots, N$. Consider

$$l_i^* = l_i/(\mu_{i+1,N} - \mu_{iN})\,.$$

Then l_i^*/σ are mutually independent and have the standard exponential distribution. Further $2l_i^*/\sigma$ is distributed as χ_2^2. Now let $X_{1N} \le \cdots \le X_{N,N}$ be the OS in a sample from the population $F((x-\theta)/\sigma)$, $\mu_{iN} = EZ_{iN}$ where $Z_{iN} = (X_{iN} - \theta)/\sigma$. R. Pyke (1965) has shown that the "leaps" l_i

are asymptotically independent and are asymptotically negative exponential with scale parameter σ. Hence, for $r+s+1 \leq m \leq N$,

$$L(r,s,m,N) = \frac{1}{r} \sum_{j=m-r}^{m-1} l_j^* \Big/ \left[\frac{1}{s} \sum_{j=1}^{s} l_j^* \right]$$

is asymptotically distributed as $F_{2r,2s}$ or $T = (r/s)L/\left[1+(r/s)L\right] = \sum_{j=m-r}^{m-1} l_j^* / \sum_{j=1}^{m-1} l_j^*$ has a beta distribution. The optimum test (namely the one that has the largest power is given by choosing $s = m/2$ or $(m-1)/2$, whichever is an integer, and $r = s-1$). Notice that the test is invariant under location and scale changes. So if you wish to test $H_0 : F\left(\frac{x-\theta}{\sigma}\right) = F_0\left(\frac{x-\theta}{\sigma}\right)$, then the μ_{iN} are computed from $F_0(x)$.

Comments

The test is far superior to Pearson's chi-square or Kolmogorov-Smirnov test, especially, when we are testing the validity of the Weibull distribution or the Gumbel's extreme value distribution (Type I). Its only serious competitor is the Cramér-von-Mises test. One would prefer L-test to von-Mises test even if they had identical power.

8.8 Sherman's Test

A somewhat different criteria for measuring the discrepancy between theoretical and empirical distributions has been proposed by Sherman (1950):

$$\omega_N = \frac{1}{2} \sum_{k=1}^{N+1} \left| F_0(X_{iN}) - F_0(X_{i-1,N}) - \frac{1}{N+1} \right|$$

where $X_{0,N} = -\infty$ and $X_{N+1,N} = \infty$. The statistic is suggested by the fact that if N points are selected independently from a uniform distribution on a unit interval there arise $(N+1)$ subintervals, each of expected length $1/(N+1)$. Then the distribution function of ω_N, namely $G_N(x)$, is derived by Sherman (1950) and is given by

$$G_N(x) = \begin{cases} 0, & \text{if } x < 0, \\ \sum_{i=1}^{N} b_i x^i, & \text{if } 0 < x \leq N/(N+1), \\ 1, & \text{if } x > N/(N+1), \end{cases}$$

where

$$b_k = \sum_{i=0}^{r} (-1)^{i+k+1} \binom{N+1}{i+1} \binom{i+k}{i} \binom{N}{k} \left(\frac{N-i}{N+1}\right)^{N-k}$$

and r is determined by

$$\frac{N-r-1}{N+1} \le x \le \frac{N-r}{N+1}.$$

Sherman (1950) also obtains

$$E(\omega_N) = \left(\frac{N}{N+1}\right)^{N+1} \to \frac{1}{e}$$

and

$$\text{var}(\omega_N) = \frac{2N^{N+2} + N(N-1)^{N+2}}{(N+2)(N+1)^{N+2}} - \left(\frac{N}{N+1}\right)^{2N+2} \sim \frac{2e-5}{e^2} \cdot \frac{1}{N},$$

and Sherman (1950) proves that

$$\omega_N^{(1)} = \{\omega_N - E(\omega_N)\} / \{\text{var}(\omega_N)\}^{1/2}$$

and

$$\omega_N^{(2)} = \left(\frac{e^2}{2e-5} \cdot N\right)^{1/2} (\omega_N - e^{-1})$$

are asymptotically standard normal. He also shows the consistency of the test (see Sherman, 1950, Theorem 3). Sherman (1957) gives certain percentiles (exact and asymptotic) of ω_N. These are presented in Table 8.8.1.

The percentiles of $\omega_N^{(1)}$ are reasonably close to the limiting values but not those of $\omega_N^{(2)}$. In either case, the convergence is slow. N has to be at least 100 before the percentiles of $\omega_N^{(1)}$ and $\omega_N^{(2)}$ are within one percent of the limiting values. The exact percentiles of ω_N taken from Sherman (1957) are given in Table 8.8.1.

8.9 Riedwyl Test

Let F be a continuous distribution. Let

$$d_i = \frac{i}{N} - F_N\left[F_0^{-1}(i/N)\right],\ i = 1, 2, \ldots, N-1.$$

Table 8.8.1: Giving certain percentiles of $\omega_N^{(1)}$ and $\omega_N^{(2)}$

Percentile	.99		.95		.90	
N	$\omega_N^{(1)}$	$\omega_N^{(2)}$	$\omega_N^{(1)}$	$\omega_N^{(2)}$	$\omega_N^{(1)}$	$\omega_N^{(2)}$
5	2.47	1.90	1.72	1.23	1.31	0.86
10	2.44	2.08	1.70	1.37	1.31	1.00
15	2.42	2.14	1.69	1.43	1.30	1.06
20	2.41	1.18	1.68	1.47	1.30	1.10
∞	2.33	2.33	1.65	1.65	1.28	1.28

Then Riedwyl (1967) proposes several one-sided and two-sided test criteria. For two-sided criteria he proposes

$$T_N = \sum_{1}^{N-1} d_i^2 \text{ and } S_N = \sum |d_i| \,.$$

For one-sided criteria he proposes

$$T_N^+ = \sum d_i^2 \epsilon(d_i, 0), \;\; S_N^+ = \sum d_i \epsilon(d_i, 0), \;\; \text{and } S_N^* = \sum_{1}^{N-1} d_i$$

where $\epsilon(a, 0) = 1$ if $a \geq 0$ and 0 if $a < 0$. Riedwyl (1967) provides exact probabilities of the test statistics for small sample sizes. Notice that

$$E_N^+ = \max_i d_i = \text{Smirnov test criterion}\,,$$

$$E_N = \max_i |d_i| = \text{Kolmogorov-Smirnov test criterion}\,.$$

The asymptotic distributions of Riedwyl statistics are not yet known. However, Riedwyl (1967) expects that S_N^* is asymptotically normal with mean 0 and variance $N(N-1)(N+1)/12$.

8.10 Characterization of Distribution-free Tests

Birnbaum (1953a) characterizes the class of distribution-free tests of fit for continuous d.f.'s by a structure called "structure (d)". The class includes most of the goodness of fit tests proposed earlier.

Definition 8.10.1 (Birnbaum, 1953a). *A test of fit is said to have structure (d) if it is based on a statistic of the form* $\psi(F_0(X_1)\cdots F_0(X_N))$ *where* $\psi(U_1,\ldots,U_N)$ *is a measurable symmetric function of* $U_1,\ldots,U_N$.

Recall that a statistic $S(X_1,\ldots,X_N;F_0)$ is said to be distribution-free if (i) it is a measurable function of $X_1,\ldots,X_N$, and (ii) if the common d.f. of the X_i's is F, then, $P(S(X_1,\ldots,X_N;F_0)\le s|F=F_0)$ is free of F_0. One can easily verify that if a statistic has structure (d) then it is distribution-free in the family of all continuous d.f.'s. However, one can construct a counter example [see, for instance, Birnbaum and Rubin (1954), p. 594] in order to show that not every distribution-free symmetric statistic has structure (d). For example, let ω_1 and ω_2 be nonempty, mutually exclusive subsets of Ω such that $\omega_1\cup\omega_2=\Omega$, the class of all continuous d.f's. Let $S=\sup_x[F(x)-F_n(x)]=S_1$ if $F\in\omega_1$ and $S=\sup_x[F_N(x)-F(x)]=S_2$ if $F\in\omega_2$ where $F_n(x)$ denotes the empirical d.f. S is distribution-free since S_1 and S_2 are. However, S is not a statistic of structure (d).

For every continuous $G(x)$, one can define its inverse function by

$$G^{-1}(u)=\inf_x\{x:G(x)\ge u\}\,.$$

If F and G are two continuous d.f's, then $\tau(u)=FG^{-1}(u)$ for $0\le u\le 1$ constitutes a monotone mapping of the unit-interval into itself. $F(x)=G(x)$ if and only if $\tau(u)\equiv u$. Hence $\tau(u)-u$ may be taken as a measure of the discrepancy between F and G.

Definition 8.10.2 (Birnbaum, 1953a). *Let* Ω *and* Ω' *be two families of continuous d.f.'s.* $S(X_1,\ldots,X_N;G)$ *is said to be strongly distribution-free in* Ω *with respcct to* Ω' *if*

1. *for every* $G\in\Omega$, *it is a measurable function of* $X_1,\ldots,X_N$ *and*

2. *if* $(X_1,\ldots,X_N)$ *is a random sample from* $F\in\Omega'$ *and* $G\in\Omega$, *then the probability distribution of* $S(X_1,\ldots,X_N;G)$ *depends only on* FG^{-1}, *that is*

$$P(S(X_1,\ldots,X_N;G)\le s|F)=h(s,FG^{-1})\,.$$

Clearly, if $S(X_1,\ldots,X_N;G)$ *is strongly distribution-free in* Ω *with respect to* Ω' *and if* $\Omega\subset\Omega'$, *then* $S(X_1,\ldots,X_N;G)$ *is distribution-free in* Ω. *If* $S(X_1,\ldots,X_N;G)$ *in* Ω *has structure (d) with respect to* Ω, *then it is strongly distribution-free since*

$$\psi(G(X_1),\ldots,G(X_N);F)=\psi(U_1,\ldots,U_N;FG^{-1})\,.$$

Theorem 8.10.1 (Birnbaum, 1953a). *If Ω is the class of all strictly monotone continuous distribution functions, $\Omega' = \Omega$, and $S(X_1, \ldots, X_N; G)$ is symmetric in $X_1, \ldots, X_N$ and strongly distribution-free, then it has structure (d).*

PROOF: See Birnbaum and Rubin (1954, pp. 595–596).

Criteria for making a choice

Besides consideration of availability of tables, ease of computation, and simplicity in use, the statistician should examine the various properties which make some of the tests theoretically more or less advantageous. The choice of a test depends on the kind of discrepancy between the hypothesis and alternative that we wish to detect. For example the chi-square test is sensitive to discrepancies in the histogram, while the K-S test is more likely to detect vertical discrepancies between the distribution functions. Hence, one should define a distance function on the space of d.f.'s which is appropriate for the specific problem. Then, it is possible to study the powers of various tests with regard to this distance and to select the one having the largest power. Examples of distances based on $\tau(u) = FG^{-1}(u)$ are:

1. $$\left[\int_0^1 \{\tau(u) - u\}^2 \, du\right]^{1/2} = \left[\int_{-\infty}^{\infty} (F(x) - G(x))^2 \, dG(x)\right]^{1/2}$$

2. $$\int_0^1 |\tau(u) - u| \, du = \int_{-\infty}^{\infty} |F(x) - G(x)| \, dG(x)$$

3. $$\sup_{0 \leq u < 1} |\tau(u) - u| = \sup_{-\infty x < \infty} |F(x) - G(x)|$$

4. $$\sup_{0 \leq u < 1} \{\tau(u) - u\} = \sup_{-\infty x < \infty} [F(x) - G(x)] \, .$$

Notice that (3) defines a Hausdorff metric, and the rest define directed distances $\overrightarrow{FG}$.

It is of interest to note that the power of a strongly distribution-free test depends only on $\tau(u)$. Testing a simple null hypothesis is somewhat unrealistic because one can always reject the null hypothesis using a consistent test procedure based on a large sample size. This difficulty can be avoided by defining a distance $\delta(F, G)$ on Ω and considering the alternative consisting of all $G \in \Omega$ such that $\delta(F, G) \geq \delta_1$. Then, the null hypothesis, $H_0 : F$ will be rejected only if there is empirical evidence that the true distribution differs from F by a distance of at least δ_1.

8.11 Problems

8.2.1 The following are 50 pairs of digits in two successive columns from a random number table:

52, 37, 82, 69, 98, 76, 33, 50, 88, 90, 50, 27, 45, 81, 66, 74, 30, 59, 67, 60, 60, 80, 53, 69, 37, 27, 49, 87, 94, 26, 94, 73, 31, 41, 21, 71, 52, 18, 73, 73, 11, 31, 69, 79, 06, 46, 52, 54, 59, 50.

We wish to test the hypothesis that the above 50 pairs of digits constitute observations from a uniform $(0, 99)$ distribution. Suppose we classify each pair into one of the classes $(0, 9), (10, 19), \cdots, (90, 99)$ and obtain the frequencies. Carry out a chi-square goodness of fit test of the hypothesis at $\alpha = 0.05$.

8.2.2 Suppose a discrete random variable X has been observed 100 times and the following data has been obtained.

X	0	1	2	3	4	5	6	7	≥ 8
frequency	19	19	22	22	8	4	4	1	1

We wish to test the null hypothesis that the random variable has a Poisson distribution at $\alpha = 0.05$.
(HINT: Note that the Poisson parameter has not been specified and hence it should be estimated from the data by $\overline{X} = \frac{200}{100} = 2.0$. Assume that the theoretical Poisson probabilities when $\lambda = 2$ are 0.135, 0.271, 0.271, 0.180, 0.090, 0.036, 0.012, 0.004, 0.001.)

8.3.1 We have the following random sample of size $N = 10$:

1.5, 4.95, 4.23, 3.59, 8.1, 6.33, 5.66, 4.39, 4.79, 8.60.

Use one-sample Kolmogorov-Smirnov statistic test

$$\begin{aligned} & H_0: && \text{the population is normal } (5, 2^2) \\ \text{against } & H_1: && \text{the population is not normal } (5, 2^2). \end{aligned}$$

Use $\alpha = 0.10$.

8.3.2 We have the following random sample of size $N = 10$:

2.18, 0.40, 4.84, 9.25, 2.17, 8.46, 3.84, 5.12, 3.68, 5.66

Use one-sample Kolmogorov-Smirnov statistic test

$$\begin{aligned} & H_0: && \text{the population is normal } (4, 3^2) \\ \text{against } & H_1: && \text{the population is not normal } (5, 3^2). \end{aligned}$$

Use $\alpha = 0.10$.

8.4.1 Show that $N\omega_N^2 = \sum_{i=1}^{N} \left(U_{iN} - \frac{2i-1}{2N}\right)^2 + \frac{1}{12N}$ where U_{iN}, $i = 1, \ldots, N$ are the standard uniform order statistics in a random sample of size N.

(HINT: Let $G_N(U)$ denote the empirical d.f. based on a random sample of size N. from the uniform [0,1] population. Write

$$N\omega_N^2 = N\sum_{i=1}^{N+1}\left[\left(\frac{i-1}{N}\right)^2(U_{iN}-U_{i-1,N}) - \left(\frac{i-1}{N}\right)(U_{iN}^2-U_{i-1,N}^2)\right] + \frac{N}{3}$$

where $U_{N+1,N}=1$, and $U_{0,N}=0$. Use the identities:

$$\sum_{i=1}^{N+1}\left\{\left(\frac{i}{N}\right)U_{iN}^2 - \left(\frac{i-1}{N}\right)U_{i-1,N}^2\right\} = \left(\frac{N+1}{N}\right)$$

and

$$\sum_{i=1}^{N+1}\left\{\left(\frac{i}{N}\right)^2 U_{iN} - \left(\frac{i-1}{N}\right)^2 U_{i-1,N}\right\} = \left(\frac{N+1}{N}\right)^2 .)$$

8.4.2 Show that $N\omega_N^2$ is equivalent in probability to

$$\begin{aligned} S_N &= N\int_{-\infty}^{\infty}[F_N(x)-F(x)]^2\,dF_N(x) \\ &= N\int_0^1 [G_N(u)-u]^2\,dG_N(u) = \sum_{i=1}^{N}\left[\frac{i}{N}-U_{iN}\right]^2 . \end{aligned}$$

(HINT: Writing $U_{iN}-\frac{2i-1}{2N} = U_{iN}-\frac{i}{N}+\frac{1}{2N}$ and squaring, we obtain

$$N\omega_N^2 - S_N = \frac{1}{N}\sum U_{iN} - \frac{1}{2} - \frac{1}{6N} = \frac{1}{N}\sum_{i=1}^{N}U_i - \frac{1}{2} - \frac{1}{6N} \to 0$$

in probability since $\frac{1}{N}\sum U_i \to \frac{1}{2}$ in probability due to the weak law of large numbers, where the U_i are the unordered U_{iN}.)

8.4.3 Show that $E(N\omega_N^2) = 1/6$ and

$$\mathrm{var}(N\omega_N^2) = (45)^{-1} - (60N)^{-1} .$$

(HINT: Let $N\{G_N(u)-u\} = \sum_{i=1}^{N} \chi_i(u)$ where

$$\chi_i(u) = \chi_i^*(u) - u \text{ and } \chi_i^*(u) = 1 \text{ when } U_i \le u$$

and is equal to zero when $U_i > u$. Then

$$E(N\omega_N^2) = N \int_0^1 E\left[G_N(u) - u\right]^2 du$$

and

$$E\left\{(N\omega_N^2)^2\right\} = 2N^2 \int\int_{u<v} E\left\{\left[G_N(u)-u\right]^2\left[G_N(v)-v\right]^2\right\} du\, dv$$

where

$$\begin{aligned} N^4 E&\left\{\left[G_N(u)-u\right]^2\left[G_N(v)-v\right]^2\right\} \\ &= E[N\chi_1^2(u)\chi_1^2(v) + N(N-1)\chi_1^2(u)\chi_2^2(v) \\ &\quad +2N(N-1)\chi_1(u)\chi_1(v)\chi_2(u)\chi_2(v)]\,. \end{aligned}$$

Also, for $u \le v$,

$$E\chi_1^2(u)\chi_1^2(v) = u - 2uv - 2u^2 + v^2u + 5u^2v - 3u^2v^2\,.)$$

8.4.4 Show that $EW_N^2 = 1$.
(HINT: $-\ln U_{iN}$ is distributed as the $(N-i+1)^{th}$ smallest standard exponential order statistic in a random sample of size N).

8.5.1 Let 0.75, 3.93, 3.23, 3.59, 7.20 constitute a random sample of size 5 from an unknown distribution. Using Shapiro-Wilk test criterion, test at $\alpha = 0.05$ whether the data comes from a normal population. (HINT: Use the values of μ and Σ from Problem 3.4.4 and use 0.962 as the critical value.)

8.5.2 For the data in Problem 8.3.2, use the Shapiro-Francia test criterion (which is obtained by setting $\Omega \equiv$ the identity matrix) for normality and use 0.842 as the 5% critical value.

(HINT: $\mu_{10,10} = 1.539$, $\mu_{9,10} = 1.001$, $\mu_{8,10} = 0.656$, $\mu_{7,10} = 0.376$, $\mu_{6,10} = 0.123$ and the other values can be obtained by symmetry.) The above values are taken from D. Teichroew (1957). Tables of expected values of order statistics and products of order statistics from samples of size 20 and less from the normal distribution. *Ann. Math Statist.* **27** 410–426.

8.6.1 We have the following data (with a sample size $N = 10$)

4.83, 1.07, 2.14, 0.09, 0.43, 1.20, 0.23, 2.08, 3.45, 0.74.

We wish to test whether the data arise from an exponential population. Use Shapiro-Wilk's criterion for testing exponentiality (use $\alpha = 0.10$).
(HINT: Estimate location and scale parameters as suggested in case 2 and use the asymptotic distribution suggested in case 1.)

8.7.1 For the data in Problem 8.6.1, carry out a test for exponentiality based on the criterion L with $m = N = 10$, $s = 5$ and $r = 4$ and $\alpha = 0.05$.

8.8.1 Suppose we have the following data in a random sample of size $N = 20$ from an unknown distribution:

0.04, 0.49, 0.35, 0.24, 0.94, 0.75, 0.63, 0.38, 0.45, 0.86,
0.25, 0.10, 0.61, 0.96, 0.27, 0.93, 0.72, 0.65, 0.33, 0.71.

We wish to test whether this data arose from a standard uniform distribution at $\alpha = 0.05$. Carry out the test using Sherman's $\omega_N^{(1)}$ criterion.

Chapter 9

Randomness Tests Based on Runs

9.1 Introduction

The simplest nonparametric tests are based on the number of runs or length of runs. Although they have low power, they are intuitively appealing. Let us first define a run.

Definition 9.1.1. *Given a sequence of two or more types of symbols, a run is defined to be a succession of one or more identical symbols which are followed and preceded by a different symbol or no symbol at all.*

Lack of randomness is exhibited by a tendency of the symbols to have a definite pattern in the sequence. Both the number of runs and the length of runs (which are interrelated) should indicate the existence of some sort of pattern. So tests for randomness can therefore be based on either criterion or some combination thereof. Too many runs, or runs of excessive lengths can make us reject the hypothesis of randomness. If the alternative is somewhat more specific, then one should be able to provide a better criterion for testing the hypothesis. Later on we shall develop a locally most powerful test of randomness against trend alternatives.

9.2 Total Number of Runs

Let n_1 and n_2 denote the number of elements of type I and type II respectively, with $n = n_1 + n_2$. Also let R_1 $[R_2]$ denote the number of runs of type

I [type II] and R denote the total number of runs. We need the following results in order to obtain the distribution of R.

Lemma 9.2.1. *The number of distinguishable ways of distributing n identical objects into r distinguishable cells with no cell empty is $\binom{n-1}{r-1}$, $n \geq r$.*

PROOF: Suppose the n identical objects are all white balls that are placed in a row. Affect the division into r cells by inserting each of $r-1$ black balls between any two white balls in the line. Since there are $n-1$ positions in which each black ball can be placed, the total number of arrangements is $\binom{n-1}{r-1}$.

Theorem 9.2.1. *Let R_1 and R_2 denote the respective numbers of runs of n_1 objects of type I and n_2 objects of type II in a random sample of size $n = n_1 + n_2$. The joint probability distribution of R_1 and R_2 is*

$$S_{R_1,R_2}(r_1, r_2) = c\binom{n_1-1}{r_1-1}\binom{n_2-1}{r_2-1}\Big/\binom{n_1+n_2}{n_1} \tag{9.2.1}$$

for $r_1 = 1, 2, \ldots, n_1$, $r_2 = 1, 2, \ldots, n_2$, $r_1 = r_2$ or $r_1 = r_2 \pm 1$, and $c = 2$ if $r_1 = r_2$ and $c = 1$ if $r_1 = r_2 \pm 1$.

PROOF: The n_1 objects can be put into r cells in $\binom{n_1-1}{r_1-1}$ ways and for each choice of this, n_2 objects can be partitioned into r_2 parts in $\binom{n_2-1}{r_2-1}$ ways. The blocks of objects of type I and type II must alternate and hence $r_1 = r_2 \pm 1$ or $r_1 = r_2$. If $r_1 = r_2 - 1$, a type II run must come first. If $r_1 = r_2$ the sequence can begin with a run of either type so that the number of distinct arrangements have to be doubled; i.e., $c = 2$.

Corollary 9.2.1.1. *The marginal distributions of R_1 and R_2 are*

$$f_{R_1}(r_1) = \binom{n_1-1}{r_1-1}\binom{n_2+1}{r_1}\Big/\binom{n_1+n_2}{n_1}, \quad r_1 = 1, 2, \ldots, n_1, \tag{9.2.2}$$

$$f_{R_2}(r_2) = \binom{n_2-1}{r_2-1}\binom{n_1+1}{r_2}\Big/\binom{n_1+n_2}{n_2}, \quad r_2 = 1, 2, \ldots, n_2. \tag{9.2.3}$$

PROOF:

$$
\begin{aligned}
f_{R_1}(r_1) &= \sum_{r_2} f_{R_1,R_2}(r_1, r_2) \\
\binom{n_1+n_2}{n_1} f_{R_1}(r_1) &= 2\binom{n_1-1}{r_1-1}\binom{n_2-1}{r_1-1} + \binom{n_1-1}{r_1-1}\binom{n_2-1}{r_1-2} \\
&\quad + \binom{n_1-1}{r_1-1}\binom{n_2-1}{r_1} \\
&= \binom{n_1-1}{r_1-1}\left[\binom{n_2}{r_1-1} + \binom{n_2}{r_1}\right] \\
&= \binom{n_1-1}{r_1-1}\binom{n_2+1}{r_1}
\end{aligned}
$$

after using the Pascal triangular equality.

Theorem 9.2.2. *The probability distribution of R, the total number of runs of $n = n_1 + n_2$ objects of n_1 of type I and n_2 of type II in a random sample is*

$$
\binom{n_1+n_2}{n_1} f_R(r) = \begin{cases} 2\binom{n_1-1}{r/2-1}\binom{n_2-1}{r/2-1) } \\ \text{if } r \text{ is even}, \\ \binom{n_1-1}{(r-1)/2}\binom{n_2-1}{(r-3)/2} + \binom{n_1-1}{(r-3)/2}\binom{n_2-1}{(r-1)/2} \\ \text{if } r \text{ is odd}, \; r = 2, \ldots, n_1 + n_2 \, . \end{cases} \tag{9.2.4}
$$

PROOF: If r is even, then $r_1 = r_2 = r/2$. If r is odd, then either $r_1 = (r-1)/2$ and $r_2 = (r+1)/2$ or $r_1 = (r+1)/2$ and $r_2 = (r-1)/2$. By summing the joint probability function, we get the desired result.

Example 9.2.1. Consider $n_1 = 5$, $n_2 = 6$.

$$\begin{aligned}
f_R(11) &= \binom{4}{4}\binom{5}{5}\Big/\binom{11}{5} = \frac{1}{462} = 0.002 \\
f_R(10) &= 2\binom{4}{4}\binom{5}{4}\Big/\binom{11}{5} = \frac{10}{462} = 0.022 \\
f_R(9) &= \left\{\binom{4}{4}\binom{5}{3} + \binom{4}{3}\binom{5}{4}\right\}\Big/\binom{11}{5} = \frac{30}{462} = 0.065 \\
f_R(2) &= 2\binom{4}{0}\binom{5}{0}\Big/\binom{11}{5} = \frac{2}{462} = 0.004 \\
f_R(3) &= \left\{\binom{4}{1}\binom{5}{0} + \binom{4}{0}\binom{5}{1}\right\}\Big/\binom{11}{5} = \frac{9}{462} = 0.019 \\
f_R(4) &= 2\binom{4}{1}\binom{5}{1}\Big/\binom{11}{5} = \frac{40}{462} = 0.087\,.
\end{aligned}$$

For the critical region, $R \leq 2$ or $R \geq 11$, the level of significance $(\alpha) = 0.006$, for $R \leq 3$ or $R \geq 10$, $\alpha = 0.047$ and for $R \leq 4$ or $R \geq 9$, $\alpha = 0.199$.

Moments of R

It is cumbersome to compute the moments of R from $f_R(r)$. In the following w shall give a simple method of calculating ER and var R.

Let

$$R = \sum_{i=1}^{n} Y_i$$

where

$$Y_1 \equiv 1$$

and

$$Y_i = \begin{cases} 1 & \text{if the } i^{\text{th}} \text{ element is different from } (i-1)^{\text{th}} \text{ element,} \\ 0, & \text{otherwise,} \end{cases}$$

for $i = 2, \ldots, n$. Then Y_i is a Bernoulli variable having expectation $n_1 n_2 / \binom{n}{2}$ for $i = 2, \ldots, n$. Hence,

$$E(R) = 1 + \sum_2^n EY_i = 1 + (n-1)n_1 n_2 \Big/ \binom{n}{2} = 1 + (2n_1 n_2)/n\,. \qquad (9.2.5)$$

Consider

$$\begin{aligned} ER^2 = E\left(\sum_1^n Y_i\right)^2 &= E\left(\sum_1^n Y_i^2 + \sum_{i\neq j}\sum Y_i Y_j\right) \\ &= 1 + (n-1)EY_i + 2\sum_{i=2}^n EY_i + \sum_{2\le i\neq j\le n}\sum EY_iY_j \\ &= 1 + 6\frac{n_1n_2}{n} + \sum_{2\le i\neq j\le n}\sum EY_iY_j\,. \end{aligned} \qquad (9.2.6)$$

For $|i-j| = 1$,

$$\begin{aligned} E(Y_iY_j) &= P(Y_i = 1 \text{ and } Y_j = 1) \\ &= \frac{n_1n_2(n_1-1) + n_2n_1(n_2-1)}{n(n-1)(n-2)} = \frac{n_1n_2}{n(n-1)}\,. \end{aligned}$$

For $|i-j| = 2$,

$$\begin{aligned} E(Y_iY_j) &= P(Y_i = 1)P(Y_j = 1|Y_i = 1) \\ &= \frac{2n_1n_2}{n(n-1)} \cdot \frac{2(n_1-1)(n_2-1)}{(n-2)(n-3)}\,. \end{aligned}$$

One can show that the same formula holds for $|i-j| > 2$. (As an exercise, the reader is encouraged to check the formula for $j = i+3$.)

Substituting these quantities we obtain

$$\begin{aligned} ER^2 &= 1 + \frac{6n_1n_2}{n} + \frac{2(n-2)n_1n_2}{n(n-1)} + \\ &\quad \{(n-1)(n-2) - 2(n-2)\}\frac{4n_1n_2(n_1-1)(n_2-1)}{n(n-1)(n-2)(n-3)} \\ &= 1 + \frac{6n_1n_2}{n} + \frac{2n_1n_2(n-2)}{n(n-1)} + \frac{4n_1n_2(n_1-1)(n_2-1)}{n(n-1)}\,. \end{aligned}$$

Thus,

$$\begin{aligned}\text{var } R &= ER^2 - (ER)^2 \\ &= \frac{6n_1n_2}{n} + \frac{2n_1n_2(n-2)}{n(n-1)} - \frac{4n_1n_2}{n} \qquad (9.2.7)\\ &\quad + \frac{4n_1n_2(n_1-1)(n_2-1)}{n(n-1)} - \frac{4n_1^2n_2^2}{n^2} \\ &= \frac{2n_1n_2}{n^2(n-1)}\left\{ \begin{array}{l} n(2n-3) + 2n(n_1n_2 - n_1 - n_2 + 1) \\ \qquad\qquad - 2nn_1n_2 + 2n_1n_2 \end{array}\right\} \\ &= 2n_1n_2(2n_1n_2 - n_1 - n_2)/n^2(n-1)\,. \qquad (9.2.8)\end{aligned}$$

Now, letting $\lambda = n_1/n$, we obtain

$$E(R/n) \sim 2\lambda(1-\lambda) \text{ and } \text{var}(R/n^{1/2}) \sim 4\lambda^2(1-\lambda)^2\,. \qquad (9.2.9)$$

Theorem 9.2.3 (Wald and Wolfowitz, 1940).

$$P\left[\frac{R - 2n\lambda(1-\lambda)}{2\lambda(1-\lambda)\sqrt{n}} \le z\right] \to \Phi(z) \text{ as } n_1, n_2 \to \infty\,.$$

Proof: Since the asymptotic distribution of the subpopulation of even R is the same as that of odd R, it will suffice to consider the case when R is even. Now, the proof of this theorem is essentially the same as the classical proof that the binomial law converges to the normal distribution. Apply Stirling's approximations to the factorials.

Remark 9.2.1. *As in the binomial case, the normal approximation to the distribution of R can be improved by applying a correction for continuity.*

Example 9.2.2. Suppose that 25 observations on X and 25 observations on Y are made. Let the ranks enjoyed by the X's in the combined sample be

$$1, 5, 6, 7, 12, 13, 14, 15, 16, 17, 19, 20, 21, 25, 26, 27, 28, 31, 32, 38, 42, 43, 44, 45, 50\,.$$

In this case $R = R_1 + R_2$, $R_1 = 9$, $R_2 = 8$,

$$\begin{aligned} E(R) &= 1 + \frac{2(25)}{2} = 26\,, \\ \text{var } R &= \frac{2 \cdot 25 \cdot 25}{50^2 \cdot 49}\{2 \cdot 25 \cdot 25 - 50\} = \frac{0.600}{0.49} = 12.2448\,, \\ \sigma_R &= 3.5\,, \end{aligned}$$

and

$$P(R \leq 17) = P\left(\frac{R-26}{3.5} \leq \frac{17-26}{3.5}\right) \equiv P\left(Z \leq \frac{-8.5}{3.5}\right) = P(Z \leq -2.43) < 0.01\,.$$

We reject the null hypothesis that the X's and Y's have the same distribution provided $\alpha > 0.02$.

Example 9.2.3. Suppose we wish to test the hypothesis that the population median $\xi = \xi_0$. Let $N = 2m+1$ be the size of the sample. If an observation is $< \xi_0$, code it to be zero. If it exceeds ξ_0, code it to be 1.

Consistency of the Test Procedure. Wald and Wolfowitz (1940) have also established the consistency of the test based on R. However, their proof is quite long. For a shorter proof of the consistency property, the reader is referred to Blum and Weiss (1957, p. 245).

Definition 9.2.1. *The distribution of X and Y, namely $F(x)$ and $G(x)$, are said to satisfy condition A if, for any arbitrarily small positive δ, there exist a finite number of closed intervals, such that the probability of the sum I of these intervals is $> 1-\delta$ according to at least one of the distribution functions $F(x)$ and $G(x)$, and such that F and G have positive continuous derivatives $f(x)$ and $g(x)$ in I.*

Theorem 9.2.4 (Wald and Wolfowitz, 1940). *If $F(x)$ and $G(x)$ satisfy condition A, and if $F(x) \neq G(x)$ for all x, then*

$$\lim_{n_1, n_2 \to \infty} P(R < k_\alpha) = 1$$

where k_α is such that $P(R < k_\alpha | H_0) = \alpha$.

PROOF: See Wald and Wolfowitz (1940, pp. 156–161).

Remark 9.2.2. *Independently, Stevens (1939) has proposed a test that is similar to the test based on R. Wald and Wolfowitz (1940) show that the test proposed by Thompson (1938) is not consistent by exhibiting a pair F and $G(F \neq G)$ such that the probability of rejecting the null hypothesis does not tend to one.*

9.3 Length of the Longest Run

Total number of runs in an ordered sequence of two types of objects is one of the criteria for detecting a pattern. One can also use the length of the longest run as another criterion for detecting a similar pattern especially, the tendency for like objects to cluster. Mosteller (1941) has suggested this criterion in order to test for randomness against trend. Exact and asymptotic probability distributions of the runs of specified length have been derived by Mood (1940). In the following we shall present his results. As before we have n_1 objects of type I and n_2 objects of type II. Let R_{ij}, $j = 1, 2, \ldots, n_i$, $i = 1, 2$ denote, respectively, the number of runs of objects of type i which are of length j. For example in the arrangement

$$a\,b\,b\,a\,b\,a\,a\,a\,b\,b\,a\,a\,a$$

we have $R_{11} = 2$, $R_{13} = 2$, $R_{21} = 1$, $R_{22} = 2$ and all other $R_{ij} = 0$. In general we have the following identities between n_i, R_{ij} and R_i ($j = 1, \ldots, n_i$, $i = 1, 2$),

$$\sum_{j=1}^{n_i} jR_{ij} = n_i\,, \quad \sum_{j=1}^{n_i} R_{ij} = R_i\,, \quad i = 1, 2\,.$$

With the above notation we get the following result.

Theorem 9.3.1 (Mood, 1940). *Under the null hypothesis, the point probability function of R_{ij}, $j = 1, \ldots, n_i$, $i = 1, 2$, for given R_1 and R_2 is given by*

$$f(r_{11}, \ldots, r_{1,n_1};\; r_{21}, \ldots, r_{2,n_2}) = \frac{c\, R_1! R_2!}{\binom{n_1+n_2}{n_1} \Pi_{i=1}^{2}\, \Pi_{j=1}^{n_i} r_{ij}!} \tag{9.3.1}$$

where $c = 2$ if $R_1 = R_2$ and $c = 1$ if $|R_1 - R_2| = 1$.

PROOF: The total number of arrangements of the $n_1 + n_2$ objects is still $\binom{n_1+n_2}{n_1}$. For fixed R_i the number of ways of arranging R_{ij} runs of type i and length j is $R_i!/\Pi_{j=1}^{n_i} R_{ij}!$ which is true for $i = 1, 2$. Taking the product of the number of arrangements, the result follows. The value of c is 2 when $R_1 = R_2$ since the sequence can begin with either type and consequently the number of arrangements have to be doubled.

Theorem 9.3.2 (Mood, 1940). *The marginal probability function of* R_{1j}, $j = 1, \dots, n_1$ *for given* R_1 *is*

$$f(r_{11}, \dots, r_{1,n_1}) = R_1! \binom{n_2+1}{R_1} / \Pi_{j=1}^{n_1} r_{1j}! \binom{n_1+n_2}{n_1}.$$

PROOF: One can easily obtain the joint probability function of R_{1j}, ($j = 1, \dots, n_1$) and R_2 as

$$f(r_1, \dots, r_{1,n_1}, R_2) = \frac{c\, R_1! \binom{n_2-1}{R_2-1}}{\binom{n_1+n_2}{n_1} \Pi_{j=1}^{n_1} r_{1j}!}. \tag{9.3.2}$$

Now summing on the possibilities for $R_2 = R_1 \pm 1$, we get the marginal probability function of $R_{11}, \dots, R_{1,n_1}$ for given R_1. Notice that the unconditional probability functions of the R_{ij} ($j = 1, \dots, n_i$, $i = 1, 2$) can be obtained by summing over the possible values of R_1 and R_2. Further, the probability that the longest in any number of runs of type 1 is of length k is

$$\sum_{R_1} \sum_{r_{11}} \sum_{r_{1k}} \frac{R_1! \binom{n_2+1}{R_1}}{\binom{n_1+n_2}{n_1} \Pi_{j=1}^{n_1} r_{1j}!} \tag{9.3.3}$$

where the summation is over all sets of nonnegative integers satisfying

$$\sum_{j=1}^{k} r_{1j} = R_1, \ \sum_{j=1}^{k} j r_{1j} = n_1, \ r_{1k} \geq 1, \ R_1 \leq n_1 - k + 1.$$

If we set $s_{1j} = r_{1j}$, for $j < k$, $s_{1k} = \sum_{j=k}^{n_1} r_{1j}$, $A = \sum_1^{k-1} j r_{1j}$, then Mood (1940) obtains

$$f(s_{11}, \dots, s_{1k}) = \frac{\binom{n_2+1}{s_1} \binom{n_1 - A - (k-1)s_{1k} - 1}{s_{1k-1}}}{\binom{n_1+n_2}{n_1}} \frac{s_1!}{\Pi_{i=1}^{k} s_{1i}} \tag{9.3.4}$$

where $s_1 = s_{11} + \cdots + s_{1k}$.

If we define s_{2j} $(j = 1, 2, \ldots, h)$ and B in terms of s_{2j} just as s_{1i} and A were defined above, we obtain

$$f(s_{11}, \ldots, s_{1k}, s_{21}, \ldots, s_{2h}) = \frac{\binom{n_1 - A - (k-1)s_{1k} - 1}{s_{1k} - 1}\binom{n_2 - B - (h-1)s_{2h-1}}{s_{2h} - 1}}{\binom{n_1 + n_2}{n_1}} \times \frac{s_1! s_2!}{\Pi_{i=1}^k s_{1i}! \Pi_{j=1}^h s_{2j}!} \tag{9.3.5}$$

where $s_1 = \sum_1^k s_{1i}$, $s_2 = \sum_1^k s_{2j}$.

The preceding two distributions should be the most useful for applications. One is free to choose k and h so that the number of variables is appropriate for the data at hand.

Example 9.3.1 (Gibbons, 1971, p. 61). Let $n_1 = 5$ and $n_2 = 6$. Then the longest possible run is of length 5. Let K denote the length of the longest run of type 1. If $K = 5$, then $R_1 = 1$, $R_{11} = R_{12} = R_{13} = R_{14} = 0$. Hence

$$P(K = 5) = \binom{7}{1} \Big/ \binom{11}{5} = 7/462\,.$$

Analogously, we obtain

$$P(K = 4) = \frac{2!}{1!1!} \frac{\binom{7}{2}}{\binom{11}{5}} = \frac{42}{462},$$

$$P(K = 3) = \frac{\frac{3!}{2!1!}\binom{7}{3} + \frac{2!}{1!1!}\binom{7}{2}}{\binom{11}{5}} = \frac{147}{462},$$

$$P(K = 2) = \frac{\frac{4!}{3!1!}\binom{7}{4} + \frac{3!}{1!2!}\binom{7}{3}}{\binom{11}{5}} = \frac{245}{462},$$

and

$$P(K=1)=\frac{5!}{5!}\frac{\binom{7}{5}}{\binom{11}{5}}=\frac{21}{462}\,.$$

If the level of significance is $< .05$, we reject the null hypothesis of randomness when $K = 5$. In general, the critical region consists of those arrangements with at least one run of length t or more. Mood (1940) obtains the first two moments of R_{ij}, R_i and R and establishes the asymptotic normality of the various statistics of interest. We shall give two such results.

Theorem 9.3.3 (Mood, 1940). *If $K > 2$, the random variable*

$$X_i=\{R_i-n\lambda(1-\lambda)\}\,/\sqrt{n}\,,\ i=1,2\,,\lambda=\frac{n_1}{n}\,,n=n_1+n_2 \tag{9.3.6}$$

are asymptotically normally distributed with zero means and variances and covariances

$$\sigma_{12}=\lambda^2(1-\lambda)^2\,,\ \sigma_{ii}=\lambda^2(1-\lambda)^2\,,\ i=1,2\,.$$

Theorem 9.3.4 (Mood, 1940). *Let*

$$X_k=\left\{s_{1k}-n\lambda^k(1-\lambda)\right\}/\sqrt{n}\,. \tag{9.3.7}$$

Then X_k is asymptotically normally distributed with mean 0 and variance

$$\sigma^2_{X_k}=\lambda^{2k-1}(1-\lambda)\left[-k^2(\lambda-1)^2-\lambda\right]+\lambda^k(1-\lambda)\,. \tag{9.3.8}$$

For proofs of these, the reader is referred to Mood (1940, Section 5). Mood (1940) extends all his results to the situation of more than two types of objects.

Mosteller (1941) obtains for the case $n_1 = n_2 = m$,

$$P(R_{1i}\ge 1,\ i\ge k)=\binom{2m}{m}^{-1}\sum_{r_1=1}^{m-k+1}\binom{m+1}{r_1}\cdot\sum_{j=1}^{r_1}(-1)^{j+1}\binom{r_1}{j}\binom{m-1-j(k-1)}{r_1-1} \tag{9.3.9}$$

and

$$P(R_{1i}\ge 1 \text{ or } R_{2i}\ge 1 \text{ or both: } i\ge k)=1-A\binom{2m}{m}^{-1} \tag{9.3.10}$$

where

$$A = \sum_{r_1 > m/k} \left[\sum_{j=0}^{r_1} (-1)^j \binom{r_1}{j} \binom{m-1-j(k-1)}{r_1 - 1} \right]$$

$$\cdot \left[\sum_{r_2 = r_1 - 1}^{r_1+1} c(r_1, r_2) \sum_{i=0}^{r_2} (-1)^i \binom{r_2}{i} \binom{m-1-i(k-1)}{r_2 - 1} \right] \quad (9.3.11)$$

where

$$\begin{aligned} c(r_1, r_2) &= 2 \text{ if } r_1 = r_2 \\ &= 1 \text{ if } |r_1 - r_2| = 1 \\ &= 0 \text{ if } |r_1 - r_2| > 1\,. \end{aligned}$$

A similar result holds for $P(R_{2i} \geq 1,\ i \geq k)$.

Example 9.3.2. Let us consider a random sample of size $2m$, $X_1, \ldots, X_{2m}$ from a continuous distribution $F(x)$. Let $X_{i,2m}$, $i = 1, \ldots, 2m$ denote the ordered observations from smallest to the largest. Now we divide the sample into two sets by considering the $X_{m,2m}$ and $X_{m+1,2m}$. X_i will be called an a if $X_i \leq X_{m,2m}$ and it will be called a b if $X_i \geq X_{m+1,2m}$. If the sample size is odd, say $2m+1$, we treat the data as if there were only $2m$ items by ignoring the sample median. In preparing quality control charts, we set a level of significance α for a given m, and this determines k the length of run of either type necessary for significance at the chosen level. If we are interested in runs above the median when $\alpha = .05$, $m = 20$, then the best value of k is 8. If we are interested in runs either above or below the median, for $\alpha = .05$ and $m = 20$ then the best value of k such that $1 - A\binom{2m}{m}^{-1} \leq .05$ turns out to be 9. Mosteller (1941) has prepared some tables. In the following we shall present one of his tables.

F.N. David (1947) has considered the power of the runs test for randomness against the alternative of Markovian dependence. Bateman (1948) considered the distribution of the longest run under the hypothesis of randomness and the power function when the alternative hypothesis is that of positive dependence (Markovian or more complex), when the number of elements of the two kinds are unequal. Using the conditional power technique as given by David (1947) for the distribution of groups, Bateman (1948) compares the power of the two criteria, length of longest run and total number

Table 9.3.1: Giving the smallest lengths of runs for .05 and .01 significance levels for samples of selected sizes

	Runs on one side of median		Runs on either side of median	
$2m$	$\alpha = .05$	$\alpha = .01$	$\alpha = .05$	$\alpha = .01$
10	5	–	5	–
20	7	8	7	8
30	8	9	8	9
40	8	9	9	10
50	8	10	10	11

of runs, with respect to the alternative hypothesis. In the following we shall present some of his results. Suppose we have two types of elements denoted by a and b of sizes n_1 and n_2. Without loss of generality assume that $n_1 \geq n_2$. Also, let $n = n_1 + n_2$ and let $f_i(t, r)$ denote the number of compositions (compositions of a number are partitions of the number in which the order is taken into consideration) of n_i into t parts such that the greatest part contains exactly r elements ($i = 1, 2$). Then the number of compositions of n_i into t parts, none of which exceeds k in magnitude, is the coefficient of x^{n_i} in $(x + x^2 + \cdots + x^k)^t = x^t(1 - x^k)^t(1 - x)^{-t}$. This coefficient is

$$\sum_{j=0}^{t}(-1)^j(-1)^{n_i-t-jk}\binom{t}{j}\binom{-t}{n_i - t - jk} = \sum_{j=0}^{t}(-1)^j\binom{t}{j}\binom{n_i - jk - 1}{t-1}$$

when $\binom{-k}{r}$ is interpreted as $(-1)^k k(k+1)\cdots(k-r+1)/r!$. Thus, in the f-notation we have

$$\sum_{r\leq k} f_i(t,r) = \sum_{j=0}^{t}(-1)^j\binom{t}{j}\binom{n_i - jk - 1}{t-1}, \tag{9.3.12}$$

from which it readily follows that

$$\begin{aligned} f_i(t,k) &= \sum_{r\leq k} f_i(t,r) - \sum_{r\leq k-1} f_i(t,k-1) \\ &= \sum_{j=0}^{t}(-1)^{j+1}\binom{t}{j}\left[\binom{n_i - j(k-1) - 1}{t-1} - \binom{n_i - jk - 1}{t-1}\right]. \end{aligned} \tag{9.3.13}$$

Now let

$$\psi(t_1, t_2, k) = f_1(t_1, k) \sum_{r \le k} f_2(t_2, r) + f_2(t_2, k) \sum_{r \le k-1} f_1(t_1, r) \qquad (9.3.14)$$

where $|t_1 - t_2| \le 1$ and also let $N(2t,\ k|n_1, n_2)$ denote the number of sequences of $2t$ groups when at least one group contains k elements and no group contains more than k elements. Then

$$\begin{cases} N(2t,\ k|n_1, n_2) &= 2\psi(t, t, k) \\ N(2t+1,\ k|n_1, n_2) &= \psi(t+1, t, k) + \psi(t, t+1, k)\,. \end{cases} \qquad (9.3.15)$$

Then the number of sequences such that at least one group contains k elements and no group contains more than k elements is given by

$$N(k|n_1, n_2) = \sum_{t=1}^{n_1-k+1} \{2\psi(t, t, k) + \psi(t+1, t, k) + \psi(t, t+1, k)\}\ . \quad (9.3.16)$$

Thus, in a sequence of n elements n_1 of which are a's and n_2 of which are b's where $n = n_1 + n_2$, $n_1 \ge n_2$, then the probability that the longest run G consists of k elements is

$$P(G = k|n_1, n_2) = p(k|n_1, n_2) = N(k|n_1, n_2) \Big/ \binom{n}{n_1}\,. \qquad (9.3.17)$$

Bateman (1948) tabulates $p(k|n_1, n_2)$ for $n = 10$ to 15 and $n = 20$. He also gives the joint distribution of the total number of runs and the length of the longest run. Now for statistical applications, the probability that the length of the longest run, G is greater than or equal to k_0 [after using Eq. (9.3.12)] is given by

$$\begin{aligned} &\sum_{k \ge k_0} \psi(t_1, t_2, k) \\ &= \binom{n_1 - 1}{t_1 - 1}\binom{n_2 - 1}{t_2 - 1} \\ &\quad -\Pi_{i=1}^{2} \left\{ \sum_{j=0}^{t_i} (-1)^j \binom{t_i}{j} \binom{n_i - j(k_0 - 1) - 1}{t_i - 1} \right\} \end{aligned} \qquad (9.3.18)$$

for $|t_1-t_2|\le 1$ and equals zero for $|t_1-t_2|>1$. Let us denote this expression by $\psi(t_1,t_2,k\ge k_0)$. Also, from (9.3.17) we have

$$\begin{aligned}P(G\ge k_0|n_1,n_2)&\\&=\sum_{k=k_0}^{n_1} p(k|n_1,n_2)\\&=\binom{n}{n_1}^{-1}\left[\sum_{t=1}^{n_1-k_0-1}\{2\psi(t,t,k\ge k_0)+\psi(t+1,t,k\ge k_0)\right.\\&\qquad\left.+\ \psi(t,t+1,k\ge k_0)\}\right]. \qquad (9.3.19)\end{aligned}$$

Analogously

$$\begin{aligned}P(G\le k_0|n_1,n_2)&\\=\ &\binom{n}{n_1}^{-1}\left[\sum_{t=c}^{n_2}\{2\psi(t,t,k\le k_0)+\psi(t+1,t,k\le k_0)\right.\\&\qquad\left.+\ \psi(t,t+1,k\ge k_0)\}\right], \qquad (9.3.20)\end{aligned}$$

where

$$\psi(t_1,t_2,k\le k_0)=\sum_{k\le k_0}\psi(t_1,t_2,k)=\Pi_{i=1}^2\left\{\sum_{j=0}^{t_i}(-1)^j\binom{t_i}{j}\binom{n_i-jk_0-1}{t_i-1}\right\} \qquad (9.3.21)$$

for $|t_1-t_2|\le 1$ and zero elsewhere and $c=\left[\frac{n_1+k_0-1}{k_0}\right]$.

If G_a denotes the length of the longest run of the a sequence, then (9.3.15) reduces to

$$2f_1(t,k)\binom{n_2-1}{t_1}$$

and

$$f_1(t+1,k)\binom{n_2-1}{t-1}+f_1(t,k)\binom{n_2-1}{t},$$

respectively. Hence,

$$\begin{aligned}N(G_a=k|n_1,n_2)&=\sum_{t=1}^{n_1-k-1} f_1(t,k)\left[\binom{n_2-1}{t-2}+2\binom{n_2-1}{t-1}+\binom{n_2-1}{t}\right]\\&=\sum_{t=1}^{n_1-k-1} f_1(t,k)\binom{n_2+1}{t} \qquad (9.3.22)\end{aligned}$$

and

$$N(G_a \geq k|n_1, n_2) = \sum_{t=1}^{n_1-k+1} \binom{n_2+1}{t} \sum_{j=1}^{t} (-1)^{j+1} \binom{n_1 - j(k-1) - 1}{t-1} \binom{t}{j}. \tag{9.3.23}$$

Now writing

$$\binom{t}{j}\binom{n_2+1}{t} = \binom{n_2+1}{j}\binom{n_2-j+1}{t-j},$$

interchanging the summation and using the relation:

$$\sum_{i=0}^{m} \binom{m}{k+i}\binom{n}{i} = \binom{m+n}{k+n},$$

we obtain

$$N(G_a \geq k|n_1, n_2) = \binom{n}{n_1}^{-1} \left[\sum_{j=1}^{[n_1/k]} (-1)^{j+1} \binom{n_2+1}{j} \binom{n-jk}{n_2} \right]. \tag{9.3.24}$$

Bateman (1948) derives explicit expressions for the distribution of G when the alternatives to randomness hypothesis is that of Markovian dependence and second order dependence. (Notice that Markovian dependence is characterized by the property that the type of element in the ith position is governed by what type of elements were in $(i-1)^{\text{th}}$ and $(i-2)^{\text{nd}}$ positions.) For Markovian dependence Bateman (1948) surmises that the test based on the total number of runs is more powerful than the test based on the length of the longest run. For the complex dependencies, this is not necessarily the case. Bateman (1948) applies the theory of the length of the longest run to solution of DeMoivre's problem called "Runs of luck". This will be dealt with in the following example.

Example 9.3.3. The problem is to find the probability that an event E occurs at best k times in succession in a series of n independent trials, when the probability that the event occurs is equal to p. DeMoivre solved this problem using difference equations. Since $P(n_1) = \binom{n}{n_1} p^{n_1}(1-p)^{n_2}$.

We obtain using (9.3.24) (G_E denoting the longest run of successes)

$$
\begin{aligned}
P(G_E \geq k) &= \sum_{n_1=k}^{n} P(G_E \geq k|n_1)P(n_1) \\
&= \sum_{n_1=k}^{n} \sum_{j\geq 1} (-1)^{j+1} \binom{n_2+1}{j} \binom{n-jk}{n_2} p^{n_1}(1-p)^{n_2} \\
&= \sum_{j\geq 1} (-1)^{j+1} \binom{n-jk}{j-1} p^{jk}(1-p)^{j-1} \\
&\quad \cdot \sum_{n_1 \geq jk} \frac{(n_2+1)}{j} \binom{n-j(k+1)+1}{n_1-jk} \times p^{n_1-jk}(1-p)^{n_2-j+1} .
\end{aligned}
$$

Now write $n_2 + 1 = j + n - j(k+1) + 1 - (n_1 - jk)$, use the identity: $\sum_{k=0}^{n} \binom{n}{k} p^k(1-p)^k = 1$ and obtain $P(G_E \geq k)$

$$
\begin{aligned}
&= \sum_{j\geq 1} (-1)^{j+1} \binom{n-jk}{j-1} p^{jk}(1-p)^{j-1} \left\{1 + \frac{\{n-j(k+1)+1\}}{j}(1-p)\right\} \\
&= \sum_{j=1}^{[n/k]} (-1)^{j+1} \left(p + \frac{\{n-jk+1\}}{j}(1-p)\right) \binom{n-jk}{j-1} p^{jk}(1-p)^{j-1} .
\end{aligned}
$$

This is equivalent to the solution given in Uspensky (1937, p. 77).

9.4 Runs Up and Down

Some information is lost if the number of elements above or below the median or the runs of certain lengths are analyzed for randomness hypothesis. This information might be useful in recognizing a pattern in the time-ordered observations. Instead of using the median as the single focal point, in runs up and down, the magnitude of each element is compared with that of the immediately proceeding element in the sequence. If the next element is larger, a run up is started and it is denoted by a plus sign. If the next element is smaller, a run down is started which will be denoted by a negative sign. Thus, we can observe when the elements are increasing or decreasing and for how long. A test for randomness can be based on the number and length of runs up or down. Since a long sequence of runs up or down indicates a trend in the sequence, the tests based on the analysis of runs up or down should be appropriate for trend alternatives. For example, the sequence of elements 2, 8, 13, 1, 3, 4, 7 yields $+ + - + + +$ which has a run up of length 2, followed by a run down of length 1, followed by a run up of length 3. In general, a sample of size N elements $(X_1, \ldots, X_N)$ yields a derived sequence D_{N-1} of dimension $N - 1$ whose i^{th} element is the sign of $X_{i+1} - X_i$ for $i = 1, \ldots, N - 1$. Let R_i denote the number of runs, either up or down, of length exactly i in the sequence D_{N-1}. Obviously, $\sum_{i=1}^{N-1} R_i = N - 1$. Under the null hypothesis of randomness, all $N!$ permutations of $(X_1, \ldots, X_N)$ are equally likely. The test for randomess will reject the null hypothesis when there are at least r runs of length t or more, where r and t are determined by the specified level of significance. Hence it is desirable to find the joint probability function of $R_1, R_2, \ldots, R_{N-1}$ when the null hypothesis is true. Then

$$P(R_i = r_i,\ i = 1, \ldots, N - 1|H_0) = f_N(r_1, \ldots, r_{N-1}) = U_N(r_1, \ldots, r_{N-1})/N! \tag{9.4.1}$$

where $U_N(r_1, \ldots, r_{N-1})$ denotes the frequency with which $R_i = r_i$ $(i = 1, \ldots, N - 1)$. The probability function f_N can be generated recurrently, that is, f_4 can be generated from f_3 ( ), etc. Gibbons (1971) has generated f_3 ( ). Let $a_1 < a_2 < a_3$ be three given numbers, namely the ordered values of X_1, X_2, X_3. Then

Table 9.4.1:

X sequence	D_2	r_2	r_1	probability function
(a_1, a_2, a_3)	$++$	1	0	$f_3(0,1) = 2/6$
(a_1, a_3, a_2)	$+-$	0	2	
(a_2, a_1, a_3)	$-+$	0	2	$f_3(2,0) = 4/6$
(a_2, a_3, a_1)	$+-$	0	2	
(a_3, a_1, a_2)	$-+$	0	2	$f_3(r_1, r_2) = 0$ for other r_1 and r_2.
(a_3, a_2, a_1)	$--$	1	0	

Notice that only runs of lengths 1 and 2 are possible. If an extra observation X_4 is inserted, it either splits an existing run, lengthens an existing run, or introduces a new run of length 1. U_N can be expressed as a linear function of U_{N-1}. For example, see Gibbons [1971, p. 65, (4.1)]. However, this will become tedious if N is larger than 15.

Levene and Wolfowitz (1944) have derived the first two moments and mixed linear moments of the number of runs of length t and of length t or more where $t \geq 1$. Towards this we need the following additional notation.

Definition 9.4.1 (Levene and Wolfowitz, 1944). *X_i will be called the initial turning point (i.t.p.) if the sign (+ or −) of $(X_{i+1} - X_i)$ is the initial sign of a run.*

With the above notation X_1 is always a i.t.p. and let us adopt the convention that X_N is never a i.t.p. Also, let

$$Z_i = \begin{cases} 1 & \text{if } X_i \text{ is an i.t.p.,} \\ 0 & \text{otherwise;} \end{cases} \tag{9.4.2}$$

$$Z_{t,i} = \begin{cases} 1 & \text{if } X_i \text{ is the i.t.p. of a run of length } t, \\ 0 & \text{otherwise;} \end{cases} \tag{9.4.3}$$

$$W_{t,i} = \begin{cases} 1 & \text{if } X_i \text{ is the i.t.p. of a run of length } t \text{ or more,} \\ 0 & \text{otherwise.} \end{cases} \tag{9.4.4}$$

for $i = 1, 2, \ldots, N$. For the sample (2, 8, 13, 1, 3, 4, 7) we have $Z = (1, 0, 1, 1, 0, 0, 0)$.

Also, let

$$R = \sum_{i=1}^{N} Z_i = \text{ the number of runs in } D_{N-1}, \tag{9.4.5}$$

$$R_t = \sum_{i=1}^{N} Z_{t,i} = \text{ the number of runs of length } t \text{ in } D_{N-1}, \tag{9.4.6}$$

and

$$R' = \sum_{i=1}^{N} W_{t,i} = \text{ the number of runs of length } t \text{ or more in } D_{N-1}. \tag{9.4.7}$$

Assume that $(X_1, \ldots, X_N)$ is a random sample from an unknown continuous distribution $F(x)$. Without loss of generality we can assume $F(x)$ to be the uniform distribution on (0,1). Then

$$P(X_{i-1} < X_i > X_{i+1}) = \int_0^1 \left[\int_{x_{i+1}}^1 \left(\int_0^{x_i} dx_{i-1} \right) dx_i \right] dx_{i+1} = 1/3\,.$$

Hence,

$$EZ_i = P(X_{i-1} < X_i > X_{i+1}) + P(X_{i-1} > X_i < X_{i+1}) = 2/3 \quad (i = 1, 2, \ldots, N-1).$$

Also, $EZ_1 = 1$ and $EZ_N = 0$. Hence,

$$ER = 1 + (N-2)(2/3) = \frac{(2N-1)}{3}. \tag{9.4.8}$$

In order to compute the variance of R, let

$$A_i = \{X_{i-1} < X_i > X_{i+1}\} \text{ and } B_i = \{X_{i-1} > X_i < X_{i+1}\}.$$

Now, since $R = 1 + \sum_2^{N-1} Z_i$,

$$\text{var } R = \sum_2^{N-1} \text{var } Z_i + \sum\sum_{2 \le i \ne j \le N-2} \text{cov}(Z_i, Z_j)$$

$$\text{var } Z_i = (EZ_i)(1 - EZ_i) = 2/9$$

$$\text{cov}(Z_i, Z_j) = EZ_iZ_j - 4/9$$

$$E(Z_iZ_j) = P(A_iA_j) + P(A_iB_j) + P(B_iA_j) + P(B_iB_j). \tag{9.4.9}$$

Case 1: $j = i + 1$:

One can easily verify that $P(A_iA_{i+1}) = P(B_iB_{i+1}) = 0$,

$$\begin{aligned} P(A_iB_{i+1}) &= P(X_{i-1} < X_i > X_{i+1} \text{ and } X_i > X_{i+1} < X_{i+2}) \\ &= P(X_{i-1} < X_i > X_{i+1} < X_{i+2}) \\ &= \int_0^1 \int_0^{x_{i+2}} \left(\int_{x_{i+1}}^1 x_i dx_i \right) dx_{i+1} dx_{i+2} = \frac{5}{24}. \end{aligned}$$

By symmetry we have $P(A_{i+1}B_i) = 5/24$. Hence,

$$\text{cov}(Z_i, Z_{i+1}) = \frac{5}{12} - \frac{4}{9} = \frac{-1}{36}.$$

By symmetry, $\text{cov}(Z_i, Z_{i-1}) = \frac{-1}{36}$.

Case 2: $j = i + 2$:

$$\begin{aligned} P(A_i A_{i+2}) &= P(X_{i-1} < X_i > X_{i+1} \text{ and } X_{i+1} < X_{i+2} > X_{i+3}) \\ &= \int_0^1 \left(\int_{x_{i+1}}^1 x_i dx_i \right) \left(\int_{x_{i+1}}^1 x_{i+2} dx_{i+2} \right) dx_{i+1} \\ &= \frac{1}{4} \int_0^1 \left(1 - x_{i+1}^2\right)^2 dx_{i+1} = \frac{2}{15} \,. \end{aligned}$$

By symmetry, $P(B_i B_{i+2}) = \frac{2}{15}$. Next consider

$$\begin{aligned} P(B_i A_{i+2}) &= P(X_{i-1} > X_i < X_{i+1} \text{ and } X_{i+1} < X_{i+2} > X_{i+3}) \\ &= \int_0^1 \left\{ \int_0^{x_{i+1}} (1 - x_i) dx_i \right\} \left\{ \int_{x_{i+1}}^1 x_{i+2} dx_{i+2} \right\} dx_{i+1} \\ &= \frac{1}{2} \int_0^1 \left(x_{i+1} - \frac{x_{i+1}^2}{2} \right) (1 - x_{i+1}^2) dx_{i+1} = \frac{11}{120} \,. \end{aligned}$$

By symmetry, $P(A_i B_{i+2}) = \frac{11}{120}$; thus,

$$EZ_i Z_{i+2} = \frac{4}{15} + \frac{11}{60} = \frac{9}{20} \,.$$

Hence,

$$\operatorname{cov}(Z_i, Z_{i+2}) = \frac{9}{20} - \frac{4}{9} = \frac{1}{180} \,.$$

Also by symmetry, $\operatorname{cov}(Z_i, Z_j) = \dfrac{1}{180}$. Because of independence,

$$\operatorname{cov}(Z_i, Z_i) = 0 \text{ provided } |j - i| \geq 3 \,.$$

Hence,

$$\text{var } R = \frac{2}{9}(N-2) + 2(N-3)\left(\frac{-1}{36}\right) + 2(N-4)\left(\frac{1}{180}\right) = (16N - 29)/90 \,. \tag{9.4.10}$$

Notice that Levene and Wolfowitz (1944) derive the var R as a special case of the var of R'_t when $t = 1$. They obtain explicit expressions for var R'_t, var R_t, $\operatorname{cov}(R'_t, R_t)$ for various values of t, which are too complicated to be presented here. However, we shall give the ER_t and ER'_t as obtained by Levene and Wolfowitz (1944),

$$EZ_{t,i} = P(-,+^t,-) + P(+,-^t,+) = 2P(-,+^t,-) \tag{9.4.11}$$

$$= 2\int_0^1 \int_{x_{i+t+1}}^1 \int_0^{x_{i+t}} \cdots \int_0^{x_{i+1}} \int_{x_i}^1 dx_{i-1} \cdots dx_{i+t+1} \tag{9.4.12}$$

$$= 2(t^2+3t+1)/(t+3)! \text{ for } i = 2,3,\ldots,N-t-1\,. \tag{9.4.13}$$

$$E(Z_{t,1}) = 2P(+^t,-) \text{ and } EZ_{t,N-t} = 2P(-,+^t)\,.$$

By symmetry,

$$E(Z_{t,1}) = E(Z_{t,N-t}) = 2(t+1)/(t+3)!$$

and

$$E(Z_{t,i}) = 0 \text{ for } i > N-t\,.$$

Hence,

$$\begin{aligned} E(R_t) &= 2E(Z_{t,1}) + (N-t-2)E(Z_{t,i}) \\ &= \left[2N(t^2+3t+1)/(t+3)!\right] \\ &\quad -2(t^3+3t^2-t-4)/(t+3)!\ (t \le N-2)\,. \end{aligned} \tag{9.4.14}$$

Clearly,

$$E(W_{t,i}) = E(Z_{t,N-t})\ (i = 2,\ldots,N-t)$$

and

$$E(W_{t,1}) = 2P(+^t) = 2/(t-1)!\,.$$

Hence

$$E(R'_t) = [2N(t+1)/(t+2)!] - 2(t^2+t-1)/(t+2)!\ (t \le N-1), \tag{9.4.15}$$

setting $t = 1$, we obtain $E(R'_1) = ER = (2N-1)/3$, which agrees with (9.4.8).

Wolfowitz (1944) establishes the asymptotic normality of R, R_t and R'_t, when suitably standardized. He points out that there maybe a loss of information in combining runs of all lengths greater than a certain number. He proposed a test for randomness based on T where

$$T = \sum_{i=1}^{N-1} f(i)R_i\,. \tag{9.4.16}$$

He establishes the asymptotic normality of $(T-ET)/\sigma_T$ as $N \to \infty$, provided $f(\cdot)$ satisfies some regularity conditions.

Theorem 9.4.1 (Wolfowitz, 1942, 1944). *Let $f(i)$ be a function, defined for all positive integral values of i such that*

1. *there exist a pair of positive numbers a and b such that $f(a)/f(b) \neq a/b$, and*

2. *For every $\epsilon > 0$ there exists a positive integer $N(\epsilon)$ such that for all $N \geq N(\epsilon)$,*

$$\sum_{i=N(\epsilon)}^{N-1} |f(i)|\, \sigma_{R_i} \leq \epsilon N .$$

Then,

$$\lim_{N\to\infty} P\left(\frac{T - ET}{\sigma_T} \leq z\right) = \Phi(z), \text{ for every } z .$$

PROOF: See Wolfowitz (1944, Theorem 2) and also Wolfowitz (1942).

Corollary 9.4.1.1. *Set $f(i) \equiv 1$, then $T = R = R_1'$.*

Levene (1952) also establishes the joint normal distribution of the various statistics that are based on the number of runs up and down. He shows that the asymptotic power of these tests depend only on the covariance matrix evaluated under the hypothesis of randomness and the vector of expected values calculated under the alternative hypothesis. Olmstead (1946) gives a recursion formula for computing the exact-distribution of the arrangement of N distinct elements, with runs up or down of length t or more. These are tabulated for $N = 2(1)14$, $t = 1(1)13$.

9.5 Runs of Consecutive Elements

Let $X_1, \ldots, X_N$ be a random sample drawn from a continuous distribution $F(x)$. We wish to test the null hypothesis of randomness. Let $X_{i_1} < X_{i_2} < \cdots < X_{i_N}$ be the ordered X's. Then $(i_1, \ldots, i_N)$ is a permutation of the integers $(1, 2, \ldots, N)$. Let R denote the number of instances in which j is next to $j+1$ in $(i_1, \ldots, i_N)$, that is, in which either of the successions $(j, j+1)$ or $(j+1, j)$ occurs. For example, let $N = 6$ and consider the permutation 2 3 4 6 5 1. Here $R = 3$. Wolfowitz (1942) proposes $N - R$ as a test-criterion for randomness hypothesis and Wolfowitz (1944a) has shown that R is asymptotically Poisson with mean 2. Kaplansky (1945) gives the kth factorial moment and the probability function of R in powers of N^{-1}:

$$\begin{aligned} P(R=r)=p(r,N) &= \frac{2^r e^{-2}}{r!}\left[1-\frac{r^2-3r}{N}+\frac{r^4-8r^3+9r^2+22r-16}{8N(N-1)}\right] \\ &\quad +O(N^{-3}) \end{aligned} \tag{9.5.1}$$

and

$$\begin{aligned} M(k,N) &= E(R^{(k)}) \\ &= E\left(R(R-1)\cdots(R-k+1)\right) \\ &= 2^k\left[1-\binom{k}{1}\frac{(k+1)}{2k}\frac{k}{N}+\binom{k}{2}\frac{(k+2}{k2^2}\frac{k(k-1)}{N(N-1)}\cdots\right]. \end{aligned} \tag{9.5.2}$$

Since 2^k is the kth factorial moment of the Poisson distribution with mean 2, either of these results can be used to establish the asymptotic Poisson character of the distribution of R.

9.6 Problems

9.2.1 Let (6.83, 3.07, 4.14, 2.09, 2.43) and (-3.20, 2.23, 4.08, 5.45, 2.74) be two independent random samples of size 5 from two populations. Test the null hypothesis that the two populations are identical against the alternative hypothesis that they are different, using the sum of the runs of type I and type II as the test criterion.

9.2.2 Suppose that we have 15 observations on X and 15 observations on Y. Let the ranks assigned to the Xs in the combined sample be
1, 3, 5, 8, 9, 11, 12, 13, 15, 18, 20, 21, 23, 25, 27,
Test the hypothesis that the X and Y distributions are the same. (Hint: Use total number of runs, R as the test criterion and use the asymptotic normality given in Theorem 9.2.3.)

9.4.1 Consider the following data in a sample of size $N = 20$:
0.04, 0.10, 0.25, 0.24, 0.33, 0.35, 0.27, 0.34, 0.38, 0.45,
0.49, 0.61, 0.65, 0.63, 0.86, 0.71, 0.75, 0.94, 0.93, 0.96.
Using the test criterion $R =$ number of consecutive runs defined by (9.4.5) test the null hypothesis that the sample is random against the alternative hypothesis that there is a trend in the observations. Use $\alpha = 0.05$.

(HINT: Use the asymptotic normality of R when suitably standardized. Note that a lower tail probability is required since the alternative hypothesis is one-sided and there will be fewer runs if the null hypothesis is not true.)

9.4.2 Susan entered a Weight-Watcher's Program and noted her weight (in pounds) during the past 10 weeks was as follows:
121, 119, 120, 118, 117, 116, 115, 114, 112, 110.
Using the test criterion R, test the null hypothesis that the Weight-Watcher's Program has no affect against the alternative hypothesis that the program is effective. Use $\alpha = 0.05$.

9.4.3 The following data (in feet) gives the highest monthly mean level in Lake Michigan-Huron during 1860-1960[1]

1860	1870	1880	1890	1900	1910	1920	1930	1940	1950
583.4	581.5	581.4	580.8	580.5	580.0	580.8	582.4	579.4	579.6 .

Using R test the null hypothesis of randomness against the alternative that there is a trend in the observations. Use $\alpha = 0.05$.

[1]This data is produced from the graph of Wallis, A.W. and Roberts, H.V. (1956). *Statistics: A New Approach.* New York: The Free Press p. 576.

Chapter 10

Permutation Tests

10.1 Introduction

People began to explore nonparametric statistics in the 30's, although its beginnings go back to Karl Pearson (1900 and 1911). Similar regions for rejecting the null hypothesis were preferred because the probability of rejecting the hypothesis when it is true (probability of type I error) is controlled. Although the existence and construction of similar regions in the parametric case has been achieved under very heavy restrictions, it promises to be relatively simple in the nonparametric case. Then one would choose "the best" test from the class of similar tests. A general method of obtaining similar regions, called randomization method has been put forth by R. A. Fisher (1925). Let us illustrate this by the following example which is due to Pitman (1937b).

10.2 Bivariate Independence

Let (X_i, Y_i), $i = 1, 2, \ldots, m$ be m i.i.d. pairs of observations from some unknown continuous bivariate distribution $H(x, y)$, having marginals $F(x)$ and $G(y)$. We wish to test the hypothesis, $H_0 : H(x, y) = F(x)G(y)$ for all x, y against $H_1 : H(x, y) \neq F(x)G(y)$ for some x, y. The proposed test statistic is

$$T = \sum_{i=1}^{m} X_i X_{i+m} \Big/ \left(\sum_{i=1}^{m} X_i^2 \sum_{j=m+1}^{2m} X_j^2 \right)^{1/2} \tag{10.2.1}$$

where $Y_j = X_{j+m}$ and H_0 is rejected for large values of $|T|$. X_i's can be permuted among themselves in $m!$ ways and the Y_j's can be permuted among themselves in $m!$ ways. Hence, for given $2m$ numbers $X_1, \ldots, X_m$ and $Y_1, \ldots, Y_m$ we can generate $(m!)^2$ points in the sample space, denoted by (X_{i_l}, Y_{j_l}) where $(i_1, \ldots, i_m)$ and $(j_1, \ldots, j_m)$ are permutations of the integers $(1, 2, \ldots, m)$. Thus there will be $(m!)^2$ values for T which are equally probable under H_0. Let the values of T arranged in decreasing order be, $t_1 > \cdots > t_{(m!)^2}$. Now we put the k largest ones, and the k smallest ones in the critical region such that $2k = (m!)^2\alpha$ where α is the level of significance. Notice that T is equivalent to $T^* = \sum_{i=1}^m X_i X_{i+m}$ since the denominator of T is constant under all permutations of the coordinates.

10.3 Two-sample Problems

Let $X_1, \ldots, X_m$ and $Y_1, \ldots, Y_n$ be random samples from univariate populations having distribution functions F and G respectively, which are assumed to be continuous. We wish to test $H_0 : F(x) = G(x)$ for all x against $H_1 : F(x) \neq G(x)$ for some x. Write $Y_i = X_{m+i}$ and let $N = m + n$. Scheffé (1943a) has shown that no similar region exists when F and G are arbitrary distribution functions. If F and G are continuous a similar region necessarily has the randomization structure.

Let the real line be divided into k intervals and m_j and n_j respectively denote the number of X's and Y's that fall into I_j $(j = 1, \ldots, k)$, so that $\sum m_j = m$, $\sum n_j = n$. Karl Pearson (1911) proposed the test statistic

$$S = (mn)^{-1} \sum_{j-1}^{k} (mn_j - nm_j)^2/(m_j + n_j) \tag{10.3.1}$$

and H_0 will be rejected for large values of S. Note that under H_0, $\hat{p} = (m_j + n_j)/(m + n)$. The distribution of S is evaluated over the $\binom{N}{m}$ choices for the X's from the combined sample of size N. Pearson (1911) has shown that S is asymptotically distributed as chi-square with $k - 1$ degrees of freedom. A solution based on the method of randomization for $m = n$ was proposed by Fisher (1936). For arbitrary m and n and independently of Fisher (1936), Pitman (1937a) proposed the statistic

$$T = \left| \sum_{i=1}^{m} X_i/m - \sum_{j=1}^{n} X_{j+m}/n \right| . \tag{10.3.2}$$

Since T is symmetric in X_i's and Y_j's, the distribution of T is evaluated on the $\binom{N}{m}$ choices for the X's from the N pooled observations. Pitman

(1937a) fitted an incomplete beta distribution to T and noted that this approximation coincided with the usual t-test valid when F and G are normal having equal variances. Dixon (1940) proposed the following test criterion. Let $X_{1m} \leq \cdots \leq X_{mm}$ denote the ordered $X_1, \ldots, X_m$ and let r_j denote the number of Y_i's that fall in the interval $(X_{j-1,m}, X_{j,m})$ for $j = 1, \ldots, m+1$ with $X_{0m} = -\infty$, $X_{m+1,m} = \infty$. Then consider the statistic

$$T = \sum_{j=1}^{m+1} \left[(m+1)^{-1} - r_j/n\right]^2 \tag{10.3.3}$$

with large values of T being significant. Dixon (1940) fits a chi-square distribution to T for large values of m and n. He tabulates the critical values for $\alpha = .01$, $.05$, $.1$ and $m, n = 2, 3, \ldots, 10$. The consistency of Dixon's (1940) test has been shown by Blum and Weiss (1957, p. 245).

10.4 Critical Regions Having Structures

Scheffé (1943a) has studied the existence of similar regions for various classes of distributions. Let Ω_1, denote the class of all nontrivial distributions, Ω_2 the class of all continuous distributions, Ω_3 the class of all absolutely continuous distributions (that is, all $F(x)$ for which there exists a probability density function (pdf) $f(x)$ such that $F(x) = \int_{-\infty}^{x} f(t)dt$ and Ω_4 be the class of all $F(x)$ having continuous pdfs. Obviously, $\Omega_1 \supset \Omega_2 \supset \Omega_3 \supset \Omega_4$. Let $X_1, \ldots, X_m$ $(Y_1, \ldots, Y_n)$ denote a random sample from univariate distribution $F(x)\,(G(x))$. Write $X_{m+i} = Y_i$, $i = 1, \ldots, n$ and let $N = m + n$. Let $F^{(N)}(x_1, \ldots, x_N)$ denote the joint distribution function of $(X_1, \ldots, X_N) = E$. We wish to test the null hypothesis, $H_0 : F(x) = G(x)$, all x. We wish to seek a region W (called the "similar critical region") such that $P(E \in W | F \in \Omega_i)$ is the same constant α ("significance level", $\alpha \neq 0$ or 1). Let $P(W|F) = \int_W dF^{(N)}(x_1, \ldots, x_N)$ where $F^{(N)}(x_1, \ldots, x_N) = \prod_1^N F(x_i)$.

Definition 10.4.1 (Scheffé, 1943a). *A region W is said to have property $\prod_i$ if for all $F \in \Omega_i$; $\alpha = P(W|F)$ is free of F and $0 < \alpha < 1$. In the following we shall characterize regions W having the property $\prod_i$.*

Theorem 10.4.1 (Scheffé, 1943a). *There is no similar region when $F \in \Omega_1$.*

PROOF: Assume the contrary. Let L denote the line $x_1 = \cdots = x_N$ and assume that a point E_0 of L belongs to W. Let $E_0 = (a, \ldots, a)$. Let F_h be any $F \in \Omega_1$ such that $P(X = a|F_h) = h$, $(0 < h < 1)$. Then

$$\alpha = P(W|F_h) \geq P(E_0|F_h) = P(X_i = a, i = 1, \ldots, N|F_h) = h^N .$$

According to the hypothesis a is free of h, and h can be chosen to be arbitrarily close to unity. Thus, $\alpha = 1$ which is a contradiction. If no points of L belong to W then the above argument is valid to $\bar{W}$. Since $P(\bar{W}|F) = 1 - \alpha$ is free of $F \in \Omega$, and $\bar{W}$ contains a point E_0 of L. hence $1 - \alpha = 1$, i.e., $\alpha = 0$.

Definition 10.4.2 (Scheffé). *W is said to have structure S if for every point $E = (x_1, \ldots, x_N)$ with $x_i \neq x_j$ for all $i \neq j$, M points $(0 < M < N!)$ of the set $\{E'\}$ obtained by permuting the coordinates of E, are in W and the remaining $N! - M$ points are in $\bar{W}$.*

Theorem 10.4.2 (Scheffé, 1943a). *W has the property $\prod_2$ if it has the structure S.*

PROOF: Let $(i_1, \ldots, i_N)$ denote the ith permutation of $(1, \ldots, N)$ and (1, ..., N) denote the first of the $N!$ permutations of $(1, \ldots, N)$. Let A_i denote the region $\{x_{i_1} \leq \cdots \leq x_{i_N}\}$. $\{A_i\}$ is a collection of disjoint sets and covers all of R^N, the N-dimensional Euclidean space except the set H of points on the hyperplane $x_i = x_j$, $(i \neq j)$. The transformation $T_i : x_{i_1} \to x_1, \ldots, x_{i_N} \to x_N$ maps A_i into A_1 in such a way that $F^{(N)}$ remains invariant.

Now, assume that W has the structure S. Removal of $H \cap W$ from W does not affect $P(W|F)$ for every $F \in \Omega_2$. Hence,

$$P(W|F) = \sum_{i=1}^{N!} P(W \cap A_i|F) = \sum_{i=1}^{N!} \int_{W \cap A_i} dF^{(N)} = \sum_{i=1}^{N!} \int \chi_{W \cap Ai}(E) dF^{(N)}$$

where $\chi_B(E)$ denotes the characteristic function of the set B, that is, $\chi_B(E) = 1$ if $E \in B$, and zero otherwise. Now map each A_i onto A_1 via T_i. $F^{(N)}$ is invariant while, $\chi_{W \cap A_i}(E) \to h_i(E)$ such that $\sum_{i=1}^{N!} h_i(E) = M$ for $E \in A_i$. Thus,

$$P(W|F) = \sum_{i=1}^{N!} \int h_i(E) dF^{(N)} = \int_{A_1} \sum h_i(E) dF^{(N)} = M \int_{A_1} dF^{(N)} .$$

Also, we have $1 = P(R^N|F) = \sum_{i=1}^{N!} \int_{A_i} dF^{(N)}$ and by use of T_i we find

$$\int_{A_i} dF^{(N)} = \int_{A_1} dF^{(N)} = \frac{1}{N!} (i = 1, \ldots, N!) .$$

Hence,

$$P(W|F) = M/N! \text{ for all } F \in \Omega_2 .$$

Thus, W has the property $\prod_2$.

Corollary 10.4.2.1. *W has the property $\prod_2$, if it differs from a region with structure S by a region in ϕ_2 which denotes the class of regions constituting null sets for F (i.e., $B \in \phi_2 \Rightarrow P(B|F) = 0$). Scheffé (1943a) also obtains the following result which we shall state without proof.*

Theorem 10.4.3 (Scheffé, 1943a). *If the boundary of W is a region in ϕ_2, a necessary and sufficient condition that W have the property $\prod_2$ is that it has the structure S except on a subset of ϕ_2.*

PROOF: See Corollary 5 of Scheffé (1943a, p. 232).

10.5 Most Powerful Permutation Tests

Next we shall derive the most powerful permutation tests. Let $X_1, \ldots, X_m$ be a random sample from the pdf $f(x)$ and $Y_1, \ldots, Y_n$ be a random sample from the pdf $f(y - \Delta)$, where f is continuous almost everywhere. Now, let us write $X_i = Z_i$, $i = 1, \ldots, m$ and $Y_j = Z_{j+m}$, $j = 1, \ldots, n$. Then the joint pdf of $Z = (Z_1, \ldots, Z_N)$ where $N = m + n$ is

$$P_\Delta(\boldsymbol{z}) = f(z_1) \cdots f(z_m) f(z_{m+1} - \Delta) \cdots f(z_n - \Delta) = h(\boldsymbol{z}) \text{ (say)}. \quad (10.5.1)$$

Suppose we wish to test $H_0 : \Delta = 0$ against the alternative $H_1 : \Delta > 0$ or the pdf of $\boldsymbol{Z}$ is $h(\boldsymbol{z})$.

Then we have the following result of Lehmann (1959).

Theorem 10.5.1 (Lehmann, 1959). *The most powerful critical region for the permutation test of H_0 against an alternative with arbitrary pdf $h(\boldsymbol{z})$, is given by*

$$W = \{\boldsymbol{z} : h(\boldsymbol{z}) \geq C(T(\boldsymbol{z}))\} \quad (10.5.2)$$

such that $P(W|f) = \alpha$ and where $T(\boldsymbol{z})$ denotes the order statistic induced by $\boldsymbol{z}$.

PROOF: The conditional pdf of $\boldsymbol{Z}$ for given $T(\boldsymbol{Z})$ is given by

$$h(\boldsymbol{z}) / N! \sum_{i=1}^{N!} h(\boldsymbol{z}_i)$$

where $\boldsymbol{z}_i$ denotes the ith permutation of $\boldsymbol{z} = (z_1, \ldots, z_N)$. When H_0 is true, $P(\boldsymbol{Z} = \boldsymbol{z}_i) = 1/N!$. Thus by the Neyman-Pearson Lemma, the most powerful test of H_0 against the alternative is given by (10.5.2). in order to carry out the test, the $N!$ points obtained by permuting the coordinates of $\boldsymbol{z}$ are ordered according to the values of the density h. We certainly reject H_0 for the k largest values and for the $(k+1)^{\text{st}}$ value with probability γ such that

$$k + \gamma = \alpha(N!)\,. \tag{10.5.3}$$

It should be remarked that the most powerful permutation test against the alternative (10.5.1) does depend on Δ and f, and hence is not uniformly most powerful.

Example 10.5.1. Let $f(x)$ be the normal density having mean 0 and variance σ^2. Then (10.5.1) becomes

$$h(\boldsymbol{z}) = (\sqrt{2\pi}\sigma)^{-N} \exp\left[\sum_{i=1}^{m} -(z_i^2/2\sigma^2) - \sum_{i=m+1}^{N} (z_i - \Delta)^2/2\sigma^2\right]$$

since $\exp\left[\sum_{i=1}^{N} -z_i^2/2\sigma^2\right]$ is a function of the order statistic, the test rejects H_0 when

$$\exp\left(\Delta \sum_{m+1}^{N} z_i\right) > C\left(T(\boldsymbol{z})\right)$$

or when

$$\sum_{i=m+1}^{N} z_i > C\left(T(\boldsymbol{z})\right)\,. \tag{10.5.4}$$

where the constant C may depend on the order statistic $T(\mathbf{Z})$. Of the $N!$ values the statistic takes on only $\binom{N}{m}$ values will be distinct (because the value of the statistic does not change for permutations of $(x_1, \ldots, x_m)$ and for permutations of $(y_1, \ldots, y_n)$ among themselves).

Lehmann (1959, p. 187) shows the unbiasedness of the test given by (10.5.4) under the more general alternatives for which $X_i (i = 1, \ldots, m)$, $Y_j (j = 1, \ldots, n)$ are independent having cdfs F and G, respectively, such that Y_j is stochastically larger than X_i, that is, such that $G(x) \le F(x)$ for all x.

Multiplying both sides of (10.5.4) by $(1/m + 1/n)$ and subtracting $\left(\sum_1^m X_i + \sum_1^n Y_i\right)/m$ the rejection region is equivalent to

$$\left\{ \boldsymbol{z} : W = (\bar{Y} - \bar{X}) \Big/ \left\{ \sum_{i=1}^{N} (z_i - \bar{z})^2 \right\}^{1/2} > C\left(T(\boldsymbol{z})\right) \right\}$$

where $\bar{z} = (m\bar{X} + n\bar{Y})/(m+n)$. Numerator in W is

$$\left(\frac{1}{m} + \frac{1}{n}\right) n\bar{Y} - \frac{1}{m}(m\bar{X} + n\bar{Y}) = \bar{Y} - \bar{X}\,,$$

$$\begin{aligned}\sum_{i=1}^{N}(z_i - \bar{z})^2 &= \sum_{i=1}^{m}(X_i - \bar{X} + \bar{X} - \bar{z})^2 + \sum_{i=1}^{n}(Y_i - \bar{Y} + \bar{Y} - \bar{z})^2 \\ &= \sum_{i=1}^{m}(X_i - \bar{X})^2 + \sum_{i=1}^{n}(Y_i - \bar{Y})^2 + m(\bar{X} - \bar{z})^2 + n(\bar{Y} - \bar{z})^2\,.\end{aligned}$$

However,

$$m(\bar{X} - \bar{z})^2 + n(\bar{Y} - \bar{z}) = \frac{mn}{N}(\bar{Y} - \bar{X})^2\,.$$

Thus,

$$W = \frac{(\bar{Y} - \bar{X})}{\left\{(m+n-2)S_p^2 + \frac{mn}{N}(\bar{Y} - \bar{X})^2\right\}^{1/2}}$$

where

$$S_p^2 = \left\{\sum (X_i - \bar{X})^2 + \sum (Y_i - \bar{Y})^2\right\} / (m+n-2)\,.$$

Let

$$t = \frac{(\bar{Y} - \bar{X})}{(m^{-1} + n^{-1})^{1/2} S_p}\,. \tag{10.5.5}$$

Then

$$\begin{aligned} W &= \frac{(m^{-1} + n^{-1})^{1/2} t}{\left\{(m+n-2) + \frac{mn}{N}(m^{-1} + n^{-1})t^2\right\}^{1/2}} \\ &= \frac{(m^{-1} + n^{-1})^{1/2} t}{\left\{m+n-2+t^2\right\}^{1/2}}\,. \end{aligned} \tag{10.5.6}$$

Thus, W is an increasing function of t. Hence, we can write the rejection region as

$$\{\boldsymbol{z} : t \geq c\left(T(\boldsymbol{z})\right)\}\,.$$

Notice that t is the two-sample t-statistic. Now under certain regularity conditions on the underlying random variables and the sample sizes m and n (see Lehmann (1959, p. 189) or Hoeffding (1952, p. 181)), the difference between $C\left(T(z)\right)$ and $t_{N-2,1-\alpha}$ tends to zero in probability. Hence, for large samples, the permutation test is equivalent to the two-sample t-test. An analogous test and critical region holds for the alternative hypothesis: $\Delta < 0$. When $m = n$, Lehmann and Stein (1949) show that the test given by (10.5.4) is also most powerful against the alternative in which the generalized density w.r.t. the measure μ which is the $2n^{\text{th}}$ power of any one dimensional measure ν given by

$$P(\boldsymbol{Z} \in A) = \int_A C \, \exp\left\{\theta_1 \sum X_i + \theta_2 \sum Y_i + \sum r(X_i) + \sum r(Y_i)\right\} d\mu(\boldsymbol{z}) \tag{10.5.7}$$

where $\boldsymbol{z} = (X_1, \ldots, X_n, Y_1, \ldots, Y_n)$, the θ's are real numbers with $(\theta_1 < \theta_2)$ and r is any ν-measurable function. Notice that the generalized density can be specialized to the binomial, Poisson and other distributions.

10.6 One-sample Problems

Suppose we wish to test the null hypothesis that the components $Z_1, \ldots, Z_N$ of a random vector $\boldsymbol{Z}$ are independent and each Z_i is symmetric about the median 0. The hypothesis implies that the joint density of $\boldsymbol{Z}$ is invariant under the $M = 2^N$ transformations $g\boldsymbol{Z} = \left((-1)^{j_1} Z_1, \ldots, (-1)^{j_N} Z_N\right)$, $j_i = 0$ or 1, $i = 1, \ldots, N$. Then one can easily show that the most powerful similar test against the alternative that $Z_1, \ldots, Z_N$ are independent normal with a positive mean, is given by:

$$\text{Reject the null hypothesis if } \sum_{i=1}^{N} Z_i \text{ is too large.}$$

Since $\sum Z_i^2$ is invariant under the transformation g, the test statistic is equivalent to the one-sample t statistic, since

$$t(\boldsymbol{Z}) = \sum_{i=1}^{N} Z_i \Big/ \left(\sum_{i=1}^{N} Z_i^2\right)^{1/2} \tag{10.6.1}$$

is a monotonic increasing function of the t-statistic.

Test for Circular Serial Correlation

Suppose we wish to test the hypothesis that the joint density of $Z_1, \ldots, Z_N$ is symmetric in its N arguments against the alternative that the Z's are normally distributed with positive serial correlation. That is, $X_i - \xi = \delta(X_{i-1} - \xi) + \epsilon_i$ $(i = 1, 2, \ldots)$ with $X_0 - \xi = 0$. Then the joint density of $(Z_1, \ldots, Z_n)$ is

$$f(z) = C\left[\exp\left\{-(2\sigma^2)^{-1}\right\}\sum_{i=1}^{N}[(z_i - \xi) - \delta(z_{i+1} - \xi)]^2\right]$$

where $z_{N+1} = z_1$. The test criterion, $\sum_{i=1}^{N-1} z_i z_{i+1}$ proposed by Wald and Wolfowitz (1943) is most powerful similar (because of the theorem of Lehmann and Stein (1949)).

10.7 Tests in Randomized Blocks

Welch (1937) and Pitman (1938) have proposed permutation tests for the c-sample problem (one-way classification), Welch (1937) proposed to use the usual analysis of variance test criterion appropriate to testing for "no difference" of treatment effects. He transformed this into another statistic and computed its first two moments. The first moment agrees with that obtained under "normal theory". That is, for the case $X_{ij} = \mu_i + \epsilon_{ij}$ where the μ_i are constants and the ϵ_{ij} are i.i.d. normal (0,1) variables. However, the second moment depends on the subpopulation generated by the permutations of the observations. Welch fitted an incomplete beta distribution to the distribution of the test criterion. Pitman (1938), quite independently has obtained the same results and also computed the third and fourth moments of the Welch's statistic. Pitman (1938) surmises that when the number of treatments and the number of replications of each treatment, are both not too small, the usual normal-theory test may safely be applied. Pitman (1938) also suggests a method of testing the validity of the approximation and modifications to the procedure when necessary.

Wald and Wolfowitz (1944) derive a test criterion for the randomized blocks case and study its large sample distribution. The problem is as follows. Each of c different varieties of a plant is planted in one of the c cells which constitute a block. On the basis of the results from n blocks, it is desired to test the null hypothesis that there is no difference among the varieties. In order to eliminate the bias due to variations in fertility among the cells of a block, the varieties are assigned at random to the cells. That is, each permutation of the integers $(1, 2, \cdots, c)$ is allocated to the jth block by a chance mechanism and each permutation is equally likely, having the probability $(c!)^{-1}$.

Let X_{ijk} denote the yield of the ith variety in the kth cell of the jth block to which it was assigned by the randomization process. Assume that

$$X_{ijk} = \mu_{jk} + \alpha_i + \epsilon_{jk} \tag{10.7.1}$$

where μ_{jk} is the effect of the kth cell in the jth block, α_i is the effect of the ith variety and ϵ_{jk} are chance variables having an unknown distribution. We wish to test

$$H_0 : \alpha_1 = \cdots = \alpha_c = 0 . \tag{10.7.2}$$

Let a_{jk} be the yield in the kth cell of the jth block and X_{ij} be the yield of the ith variety in the jth block. If the null hypothesis is true, because of the randomization carried out within each block, the probability that $X_{1j}, \ldots, X_{cj}$ be any permutation of the elements $\{a_{jk}\}$ $(k = 1, \ldots, c)$ is $(c!)^{-1}$ for given $\{a_{jk}\}$. Now permuting in all the blocks simultaneously, for given $\{a_{jk}\}$, $j = 1, \ldots, r$ and $k = 1, \ldots, r$, one can infer that the conditional probability of any of the permutations is $(c!)^{-r}$. In order to test the null hypothesis of no varieties effects, the classical analysis of variance statistic which is employed in the normal-theory two-way classification is

$$F = \frac{r(r-1)\sum(X_{i\cdot} - X_{\cdot\cdot})^2}{\sum\sum(X_{ij} - X_{i\cdot} - X_{\cdot j} + X_{\cdot\cdot})^2} \tag{10.7.3}$$

where $X_{i\cdot} = r^{-1}\sum_j X_{ij}, X_{\cdot j} = c^{-1}\sum_i X_{ij}, X_{\cdot\cdot} = (rc)^{-1}\sum_i\sum_j X_{ij}$. The statistic proposed by Welch (1937) and Pitman (1938) is

$$W = F(r - 1 + F)^{-1} . \tag{10.7.4}$$

Since we use an upper tail with F and since W is a monotonic increasing function of F, the two tests are equivalent. The distribution of F or W is to be determined over the equally probable permutations of the values actually observed. Since X_{ij} takes on any one of the values $a_{j1}, \ldots, a_{jc}$ with probability c^{-1}, we have

$$\begin{aligned} E(X_{ij}) &= c^{-1}\sum_k a_{jk} = a_{j\cdot} \text{ (say)} \\ \text{var}(X_{ij}) &= c^{-1}\sum_k (a_{jk} - a_{j\cdot})^2 = b_{j\cdot} \text{ (say)} \end{aligned}$$

$$\begin{aligned}
\operatorname{cov}(X_{ij'}X_{l,j}) &= [c(c-1)]^{-1}\sum_{k\neq l} a_{jk}a_{jl} - a_{j\cdot}^2 \\
&= [c(c-1)]^{-1}\left[\left(\sum_k a_{jk}\right)^2 - \sum_k a_{jk}^2\right] - a_{j\cdot}^2 \\
&= \left[c^2a_{j\cdot}^2 - \sum_k a_{jk}^2\right][c(c-1)]^{-1} - a_{j\cdot}^2 \\
&= (c-1)^{-1}\left[a_{j\cdot}^2 - c^{-1}\sum_k a_{jk}^2\right] = -(c-1)^{-1}b_{j\cdot}\,.
\end{aligned}$$

Hence,

$$\begin{aligned}
E(X_{i\cdot}) &= r^{-1}\sum_j a_{j\cdot} = a_{\cdot\cdot}\ \text{(say)} \\
\operatorname{var}(X_{i\cdot}) &= r^{-2}\sum b_{j\cdot} = b\ \text{(say)} \\
\operatorname{cov}(X_{i\cdot}, X_{l\cdot}) &= \left[r^2(r-1)\right]^{-1}\sum b_{j\cdot} = d\ \text{(say) for } i \neq l\,.
\end{aligned}$$

Now, let $X_{ij}^* = \sum_{k=1}^{c} \lambda_{ik}X_{kj}$ $(i = 1, \ldots, c,\ j = 1, \ldots, r)$ where $((\lambda_{ik}))$ is an orthogonal matrix with

$$\lambda_{c1} = \lambda_{c2} = \cdots = \lambda_{cc} = c^{-1/2}\,.$$

Then

$$\begin{aligned}
E(X_{i\cdot}^*) = E\left(\frac{1}{r}\sum_j X_{ij}^*\right) &= E(\sum_k \lambda_{ik}X_{k\cdot}) \\
&= \sum_k \lambda_{ik}E(X_{k\cdot}) \\
&= \sum_k \lambda_{ik}\left(\frac{1}{r}\sum a_{j\cdot}\right) = 0\,,
\end{aligned}$$

$$\begin{aligned}
\mathrm{var}(X^*_{i\cdot}) &= \mathrm{var}(\sum \lambda_{ik} X_{k\cdot}) \\
&= \sum_k \lambda_{ik}^2 \mathrm{var}(X_{k\cdot}) + \sum_{k\neq l}\sum \lambda_{ik}\lambda_{il}\mathrm{cov}(X_{k\cdot}, X_{l\cdot}) \\
&= \sum_k \lambda_{ik}^2 b + \sum_{k\neq l}\sum \lambda_{ik}\lambda_{il} d \\
&= (b-d)\sum_k \lambda_{ik}^2 = (b-d),\ (i = 1, \ldots, c-1)
\end{aligned}$$

and

$$\mathrm{cov}(X^*_{i\cdot}, X^*_{l\cdot}) = 0,\ (i \neq l,\ i, l = 1, \ldots, c-1)\,.$$

Also,

$$\sum_{i=1}^{c-1} X^{*^2}_{i\cdot} = \sum_{i=1}^{c} (X_{i\cdot} - X_{\cdot\cdot})^2\,.$$

Applying the identity

$$\sum\sum (X_{ij} - X_{i\cdot} - X_{\cdot j} + X_{\cdot\cdot})^2 = \sum\sum (X_{ij} - X_{\cdot j})^2 - r\sum (X_{i\cdot} - X_{\cdot\cdot})^2$$

to the definitions of F and W we obtain

$$W = \frac{r\sum_i (X_{i\cdot} - X_{\cdot\cdot})^2}{\sum_i \sum_j (X_{ij} - X_{\cdot j})^2}\,. \tag{10.7.5}$$

The denominator in W is invariant under all permutations within each block and it equals

$$\sum_j \sum_k (a_{jk} - a_{j\cdot})^2 = (c-1)r^2(b-d)\,.$$

Hence, one can write

$$\begin{aligned}
W &= [r(c-1)(b-d)]^{-1}\sum_{i=1}^{c}(X_{i\cdot} - X_{\cdot\cdot})^2 \\
&= [r(c-1)(b-d)]^{-1}\sum_{i=1}^{c-1} X^{*^2}_{i\cdot}\,.
\end{aligned} \tag{10.7.6}$$

If the joint distribution of the $X^*_{i\cdot}$ $(i = 1, 2, \ldots, c-1)$ over the set of admissible permutations tends to a normal distribution having a non-singular correlation matrix as r, the number of blocks, becomes large, then $r(c-1)W$ tends in distribution to a chi-square variable having $(c-1)$ degrees of freedom. Thus it remains to find regularity conditions on the set $\{a_{jk}\}$ which would make the distribution of the $X_{i.}^*$ approach normality. Since each $X_{i.}$ is the mean of independent variables, these conditions need not be too restrictive. According to Cramér (1970, Theorem 21a, pp. 113–114) if the variances and covariances satisfy certain conditions (the limiting correlation matrix should also be non-singular) and if a generalized Lindeberg condition holds, asymptotic normality will follow. Somewhat more restrictive conditions that are easy to verify are that there exist positive constants δ_1 and δ_2 such that $0 < \delta_1 < b_{j.} < \delta_2$ for all j. Since the variance of X_{ij}^* is $c(c-1)^{-1}b_{j\cdot}$, it can be seen that the above inequalities imply the fulfillment of the conditions of Laplace-Liapunov theorem (see, for example, Uspensky (1937, p. 318)). Then it follows that the correlation matrix of $X_{i\cdot}$, is nonsingular.

Walsh (1959) proposes a class of nonparametric procedures for testing H: the hypothesis of no treatment effects in a randomized block experiment. The basic idea is to obtain from each block a statistic which is under H symmetrically distributed about zero and then to apply to the set of these statistics a nonparametric test of symmetry about zero.

10.8 Large-sample Power

Hoeffding (1952) has considered the large sample power properties of the permutations (one-sample two-sample, analysis of variance and independence cases).

Let $\mathcal{X}$ be the N-dimensional Euclidean space and let $\boldsymbol{X} = (Z_1, \ldots, Z_N)$ be a random variable taking values $\boldsymbol{z} \in \mathcal{X}$ let $\mathcal{G} = \{g\}$ be a finite group of transformations g of $\mathcal{X}$ onto itself. Let M be the number of elements in $\mathcal{G}$. Let H be the hypothesis that the distribution of X is invariant under the transformations in $\mathcal{G}$ so that for every $g \in \mathcal{G}$, $g\boldsymbol{X}$ has the same distribution as $\boldsymbol{Z}$. For example if each Z_i is symmetrically distributed about the median zero, then $g = \pm 1$ and $M = 2^N$. If Z_i, $i = 1, \ldots, m$ have a common d.f. $F(x)$ and Z_{m+i}, $i = 1, \ldots, N-m$ have a common d.f. $G(x)$, then $\boldsymbol{Z}$ is invariant under the $M = m!(N-m)!$ permutations which permit the first m and the last $N-m$ components among themselves. Here we shall use the randomized test $\phi(\boldsymbol{z})$, where $\phi(\boldsymbol{z})$ denotes the probability with which H is rejected when $\boldsymbol{Z} = \boldsymbol{z}$. The power of the test will be denoted by $E_P(\phi(\boldsymbol{z}))$ where P denotes the true distribution. We will be concerned with tests of the following type. Let $t(\boldsymbol{z})$ be a real-valued function on $\mathcal{X}$. For every $z \in \mathcal{X}$, let

$$t^{(1)}(\boldsymbol{z}) \le \cdots \le t^{(M)}(\boldsymbol{z}) \tag{10.8.1}$$

be the ordered values of $t(g\boldsymbol{z})$ for all $g \in \mathcal{G}$. For given α, $0 < \alpha < 1$, let k be defined by

$$k = M - [M\alpha]$$

where $[M\alpha]$ denotes the greatest integer smaller than or equal to $M\alpha$. Also, let

$$\begin{aligned} M^+(z) &= \text{number of } t^{(j)}(z) \text{ that are greater than } t^{(k)}(z) \\ M^o(z) &= \text{number of } t^{(j)}(z) \text{ that are equal to } t^{(k)}(z) \end{aligned}$$

Let

$$a(z) = \left\{M\alpha - M^+(z)\right\} / M^o(z) . \tag{10.8.2}$$

Since $M^+ \leq M - k \leq M\alpha$ and $M^+ + M^o \geq M - k + 1 > M\alpha$ we have $0 \leq a(z) < 1$. Now define $\phi(z)$ as

$$\phi(z) = \begin{cases} 1 & \text{if } t(z) > t^{(k)}(z) \\ a(z) & \text{if } t(z) = t^{(k)}(z) \\ 0 & \text{otherwise} . \end{cases} \tag{10.8.3}$$

Now, for every $z \in \mathcal{X}$,

$$\sum_{g \in \mathcal{G}} \phi(gz) = M^+ + a(z) \cdot M^o = M\alpha .$$

Hence,

$$M\alpha = E_P \sum_g \phi(gZ) = \sum_g E_P \phi(Z) = M E_P \phi(Z) .$$

That is, it is a similar test of size α for testing H. Most of the permutation tests encountered so far are of the form (10.8.3). Under certain regularity conditions, Hoeffding (1952) has shown that $t^{(k)}(Z)$ is close to a constant with high probability and that the power of the test can be approximated in terms of the distribution function of $t(Z)$.

Now assume that $\mathcal{X} = \mathcal{X}_N$, $\mathcal{G} = \mathcal{G}_N$, $t(z) = t_N(z)$, etc. are defined for an infinite sequence of positive integers N. Assume that for a given sequence $\{P_N\}$ of distributions of $Z = Z^{(N)}$ the following conditions are satisfied.

Condition A. There exists a constant λ such that $t_N^{(k)}(Z) \to \lambda$ in probability.

Condition B. There exists a function $H(x)$ continuous at $x = \lambda$ such that for every x at which $H(x)$ is continuous,

$$P(t_N(\boldsymbol{Z}) \leq x) \to H(x).$$

From (10.8.3) we have

$$P\left(t_N(\boldsymbol{Z}) > t_N^{(k)}(\boldsymbol{Z})\right) \leq E_{P_N}\phi_N(\boldsymbol{Z}) \leq P\left(t_N(\boldsymbol{Z}) \geq t_N^{(k)}(\boldsymbol{Z})\right). \quad (10.8.4)$$

Conditions A and B imply that

$$E_{P_N}\phi_N(Z) \to 1 - H(\lambda). \quad (10.8.5)$$

Conditions A and B may be satisfied by $t'(\boldsymbol{Z})$, but not by $t(\boldsymbol{Z})$ where

$$t'(\boldsymbol{z}) = c(\boldsymbol{z}) f\left(t(\boldsymbol{z})\right) + d(\boldsymbol{z})$$

where $f(y)$ is increasing, $c(\boldsymbol{z}) > 0$ and $c(\boldsymbol{z})$ and $d(\boldsymbol{z})$ are invariant under $\mathcal{G}$. Although λ and $H(y)$ will depend on the sequence $\{P_N\}$, Hoeffding (1952) shows that the dependence of λ on $\{P_N\}$ is much less pronounced than that of $H(x)$ in the sense that λ remains the same for a certain class of $\{P_N\}$ while $1 - H(x)$ varies from α to 1.

For every $\boldsymbol{z} \in \mathcal{X}$, let $F_N(y, \boldsymbol{z})$ denote the proportion of the number of elements g in $\mathcal{G}$ for which $t_N(g\boldsymbol{z}) \leq y$. Hoeffding (1952) provides sufficient conditions for the convergence in probability of $t_N^{(k)}(\boldsymbol{z})$. Let G be a random variable which takes each element g of $\mathcal{G}$ with probability M^{-1}. Then $F_N(y, \boldsymbol{z})$ is the distribution function of $t(G\boldsymbol{z})$. Let $m(\boldsymbol{z})$ and $v(\boldsymbol{z})$ denote the mean and variance of $t(G\boldsymbol{z})$ so that

$$m(\boldsymbol{z}) = M^{-1} \sum_g t(g\boldsymbol{z}), \; v(\boldsymbol{z}) = M^{-1} \sum_g \left[t(g\boldsymbol{z}) - m(\boldsymbol{z})\right]^2.$$

Let $t'(\boldsymbol{z}) = v(z)^{-1/2}\left[t(z) - m(\boldsymbol{z})\right]$ if $v(\boldsymbol{z}) > 0$ and $t'(\boldsymbol{z}) = 0$ if $v(\boldsymbol{z}) = 0$. Then the test $g(\boldsymbol{z})$ given by (10.8.3) is not changed if $t(z)$ is replaced by $t'(\boldsymbol{z})$. Thus, we may always assume that $F_N(y, \boldsymbol{z})$ has mean 0 and variance bounded by unity. If $F_N(y, \boldsymbol{z})$ tends to $F(y)$ in probability for all y, then $F(y)$ is a d.f. having the same properties. Furthermore, if $P\left(t(g\boldsymbol{Z}) = t(\boldsymbol{Z}) \text{ for all } g \in \mathcal{G}\right) \to 0$ as $N \to \infty$, then $P\left(v(\boldsymbol{Z}) = 0\right) \to 0$ and $F(y)$ has variance 1.

Next, let G' be i.i.d. as G and $\boldsymbol{Z}$ be independent of G and G'. Then

$$EF_N(y, \boldsymbol{Z}) = P(t_N(g\boldsymbol{Z}) \leq y) \quad (10.8.6)$$

$$E\{F_N(y,\boldsymbol{Z})\}^2 = P\left(t_N(G\boldsymbol{Z}) \le y, t_N(G'\boldsymbol{Z}) \le y\right) . \tag{10.8.7}$$

Notice that $t_N(G\boldsymbol{Z})$ and $t_N(G'\boldsymbol{Z})$ are identically distributed but not independent (except in the trivial case $F_N(y,\boldsymbol{Z})$ has variance 0).

Theorem 10.8.1 (Hoeffding, 1952). *Suppose that for some sequence $\{P_N\}$ of distributions $t_N(G\boldsymbol{Z})$ and $t_N(G'\boldsymbol{Z})$ have the limiting joint distribution function $F(y)F(y')$. Then for every continuity point y of $F(y)$,*

$$F_N(y,\boldsymbol{Z}) \to F(y) \text{ in probability}$$

and, if the equation $F(y) = 1-\alpha$ has a unique solution $y = \lambda$,

$$t_N^{(k)}(\boldsymbol{Z}) \to \lambda \text{ in probability} .$$

PROOF: See the proof of Theorem 3.2 of Hoeffding (1952, p. 174). The implication of this theorem is that if the permutation test criterion is equivalent to a normal theory test criterion and $t_N^{(k)}(\boldsymbol{Z}) \to \lambda$ in probability then the two tests are asymptotically equivalent. Let us verify the hypothesis of the theorem for some of the permutation tests.

Test for the median of a symmetrical distribution

The test is $\phi(\boldsymbol{z})$ with $t(\boldsymbol{z})$ given by (see Eq. (10.6.1))

$$t(\boldsymbol{z}) = \sum_1^N z_i \Big/ \left(\sum_1^N z_i^2\right)^{1/2}$$

or $t(\boldsymbol{z}) = 0$ if $\sum_1^N z_i^2 = 0$.

The random variable $G\boldsymbol{z}$ of $\boldsymbol{z}$ can be written as $G\boldsymbol{z} = (G_1 z_1, \dots, G_N z_N)$ where $G_1, \dots, G_N$ are independent, with $P(G_i = \pm 1) = 1/2$. Notice that $\sum z_i^2$ is invariant under g for all $g \in \mathcal{G}$ so that $g(G\boldsymbol{z})$ has mean 0 and variance 1 (unless $z_1 = z_2 = \cdots = z_N$). Let $Y_i = G_i Z_i$, $Y_i' = G_i' Z_i$ where all G_i, G_i' are independent, identically distributed and independent of the Z_i. Then $Y_i^2 = Y_i'^2 = Z_{i'}^2$,

$$\begin{aligned} t(G\boldsymbol{Z}) &= N^{-1/2}\sum_{i=1}^N Y_i (N^{-1}\sum_{i=1}^N Z_i^2)^{-1/2} \\ t(G'\boldsymbol{Z}) &= N^{-1/2}\sum_{i=1}^N Y_i' (N^{-1}\sum_{i=1}^N Z_i^2)^{-1/2} . \end{aligned}$$

Let $Z_1, \ldots, Z_N$ be i.i.d. having mean μ and variance σ^2. Then by Khintchine's theorem,

$$N^{-1} \sum_1^N Z_i^2 \to \sigma^2 + \mu^2 \text{ in probability}.$$

Hence,

$$\left(t(GZ), t(G'Z)\right)$$

is equivalent to

$$\left((\sigma^2+\mu^2)^{-1/2} N^{-1/2} \sum_1^N Y_i, \; (\sigma^2+\mu^2)^{-1/2} N^{-1/2} \sum Y_i'\right). \qquad (10.8.8)$$

The vectors $(Y_1, Y_1'), \ldots, (Y_N, Y_N')$ are i.i.d. with

$$\begin{aligned} EY_i = EY_i' = 0 \; EY_i^2 = EY_i'^2 &= \sigma^2 + \mu^2, \; EY_iY_i' = EG_iG_i'Z_i^2 \\ EG_iEG_i'EZ_i^2 &= 0. \end{aligned}$$

Hence, by the central limit theorem for i.i.d. vectors (for example, see Cramér (1946, p. 286), the random vector in (10.8.7) has the limiting distribution function $\Phi(y)\Phi(y')$ where Φ denotes the standard normal distribution function. The same is true of $(T(GZ)$ and $t(G'Z))$.

Then by Theorem 10.8.1, $t^{(k)}(z) \to \lambda$ in probability where $\Phi(\lambda) = 1 - \alpha$. Under the same conditions, we have, for every fixed y,

$$\lim_{N\to\infty} P\left\{t(Z) \le (y + N^{1/2}\mu/\sigma)\left(1 + (\mu/\sigma)^2\right)^{-1/2}\right\} = \Phi(y).$$

If μ/σ is free of N and is positive, the function $H(y)$ defined in Condition B is identically zero and the power of the test tends to unity. If the common d.f. of $Z_1, \ldots, Z_N$ depends on N, all the results remain true (because of Liapounov's form of the central limit theorem and its extension to vectors) provided $E|Z_1|^3\sigma^{-3} = o(N^{1/2})$. If $(\mu/\sigma)N^{1/2}$ converges to a constant c, then $H(y) = \Phi(y - c)$. Thus, the permutation test is asymptotically as powerful as the one-sided Student's t-test under the alternatives considered.

An analysis of variance test

Using Theorem 10.8.1, Hoeffding (1952, Section 5) shows that the test given by (10.7.5) is asymptotically equivalent to a chi-squared variable having $c-1$ degrees of freedom.

Two-sample test

Let $Z_i = X_i$, $i = 1, \ldots, m$ and $Z_{i+m} = Y_i$, $i = 1, \ldots, N - m$. Then under H, the distribution of $\boldsymbol{Z}$ is invariant under all $M = N!$ permutations of its components. Then $\phi(\boldsymbol{Z})$ is the test (10.8.3) with

$$t(\boldsymbol{z}) = \frac{\sum_1^N (a_i - \bar{a}) z_i}{\left\{\sum_1^N (a_i - \bar{a})^2 (N-1)^{-1} \sum_1^N (z_i - \bar{z})^2\right\}^{1/2}} \tag{10.8.9}$$

where

$$a_i = 1 \text{ for } i = 1, \ldots, m \text{ and } a_i = 0 \text{ for } i = m+1, \ldots, N\,. \tag{10.8.10}$$

Notice that the denominator of $t(\boldsymbol{z})$ in (10.8.9) is invariant under all permutations and is so chosen that $t(G\boldsymbol{Z})$ has mean 0 and variance unity (unless $z_1 = \cdots = z_N$).

Under certain regularity conditions on the $\{a_i\}$ when $E|Z_1|^3 < \infty$ and var $Z_1 > 0$, Hoeffding (1952, Theorems 6.1 and 6.2) shows that

$$F_N(y, \boldsymbol{z}) \to \Phi(y) \text{ in probability}$$

and

$$t_N^{(k)}(\boldsymbol{Z}) \to \lambda \text{ in probability where } \Phi(\lambda) = 1 - \alpha\,.$$

It should be noted that the regularity conditions imposed by Hoeffding (1952) are much less restrictive than those of Wald and Wolfowitz (1944) as well as those of Noether (1949). The sufficient conditions of Hoeffding have been relaxed by Dwass (1953, 1955) and Dwass's conditions have been relaxed by Hájek (1961). Thus, the two sample permutation test is asymptotically as powerful as the two-sample t-test. It should be noted that one can assume (without loss of generality) that $\bar{a} = 0$, $\sum a_i^2 = 1$, $EZ_1 = 0$, $EZ_1^2 = 1$. Then

$$t(\boldsymbol{z}) = \sum_1^N a_i z_i \left\{(N-1)^{-1} \sum_1^N (z_i - \bar{z})^2\right\}^{-1/2}.$$

Since $\bar{Z} \to 0$ and $N^{-1} \sum_1^N Z_i^2 \to 1$ in probability, we have

$$(N-1)^{-1} \sum_1^N (Z_i - \bar{Z})^2 \to 1$$

in probability. Hence, $(t(G(\boldsymbol{Z}), t(G'\boldsymbol{Z})))$ has the same limiting distribution as $(u(G\boldsymbol{Z}), u(G'\boldsymbol{Z}))$ where

$$u(\boldsymbol{z}) = \sum_{1}^{N} a_i z_i \,. \tag{10.8.11}$$

Hoeffding (1952, Section 7) studies the two-sample test when one of the subsample sizes is small. It is worthwhile to point out that under H, $t^{(N)}(\boldsymbol{Z}) \to \lambda$ in probability provided $E|Z_1|^3 < \infty$ and m/N is bounded away from zero and ∞. Hoeffding (1952) makes use of his Theorem 6.1.

Test for equality of medians

Let $X_1, \ldots, X_N$ be independent random variables such that each X_i has a continuous distribution function $F_i(x)$ that is symmetric about θ. We wish to test the hypothesis

$$H_o : \theta = \theta_0$$

against the alternative

$$H_1 : \theta > \theta_0 \,.$$

Let $Z_i = X_i - \theta_0$, $i = 1, \ldots, N$. Walsh (1949) has proposed the following permutation test procedure.

Consider the 2^N sets of values obtained by the transformations

$$z_i \to g_i z_i \ (i = 1, \ldots, N)$$

where $g_i = +1$ or -1. Form the mean of each of the 2^N sets of values. Then reject H if $\bar{z}$ exceeds the $(2^N - r)^{\text{th}}$ largest of the 2^N means, where the level of significance $\alpha = r/2^N$. One can also give the two-sided test.

10.9 Modified Permutation Tests

A practical short-coming of permutation tests is the great difficulty in enumerating the points $z^{(i)}$ and the evaluation of $U(\boldsymbol{z}^{(i)})$ (where $u(z)$ will be defined later). For instance in the two-sample case for $m = n = 5$, there are $\binom{10}{5} = 252$ permutations and for $m = n = 10$, there are $\binom{20}{10} = 184,765$ permutations to be examined. Hence Dwass (1957) proposes a procedure of examining a "random sample" of permutations and making the decision to accept or reject H on the basis of those permutations only. Bounds are determined for the ratio of the power of the original procedure to the modified one. Let us confine ourselves to the two-sample problem.

For any $\boldsymbol{z} = (z_1, \ldots, z_N)$ let $T(\boldsymbol{z})$ be the set of all points obtained from $\boldsymbol{z}$ by permuting its coordinates. With probability one, all sets $T(\boldsymbol{z})$ contain $M = N!$ points $z^{(1)}, \ldots, z^{(M)}$ which have been ordered so that $u(z^{(1)}) \geq \cdots \geq u(z^{(M)})$ where

$$u(z) = m^{-1} \sum_{i=1}^{m} z_i - n^{-1} \sum_{i=m+1}^{N} z_i = \bar{x} - \bar{y}\,. \tag{10.9.1}$$

Notice that since

$$\bar{x} - \bar{y} = m^{-1} \sum_{i=1}^{N} z_i - (N/m)\bar{y} = (N/n)\bar{x} - n^{-1} \sum_{i=1}^{N} z_i\,,$$

the same ordering can be induced by the functions $u(\boldsymbol{z}) = c\bar{x}$ or $u(\boldsymbol{z}) = -c\bar{y}$ for any $c > 0$.

Now, define $R^{(i)}$ to be the union over all sets $T(\boldsymbol{z})$ of the points $\boldsymbol{z}^{(i)}$ $(i = 1, \ldots, M)$. That is, $R^{(i)} = \left\{\boldsymbol{z}^{(i)} \in T(\boldsymbol{z}) : \boldsymbol{z} \in E_N\right\}$, E_N denoting the N-dimensional Euclidean space. Obviously, $R^{(1)}, \ldots, R^{(M)}$ are disjoint sets whose union is the whole sample space except for a set of probability zero. Let $p(i) = P(R^{(i)})$. Also let $Z_1 = X_1$ have d.f. $F(x)$ and $Z_{m+1} = Y_1$ have d.f. $F(x - \Delta)$, where $\Delta < 0$. Then the permutation test is given by $\phi(\boldsymbol{z})$ where

$$\phi(\boldsymbol{z}) = \begin{cases} 1 & \text{if } u(\boldsymbol{z}) \geq u(\boldsymbol{z}^{(k)}) \\ & \qquad\qquad\qquad (1 \leq k \leq M), \\ 0 & \text{if } u(\boldsymbol{z}) < u(\boldsymbol{z}^{(k)}) \end{cases}$$

and $z^{(1)}, \ldots, z^{(M)}$ are the points of $T(\boldsymbol{z})$ and $\phi(\boldsymbol{z})$ is the probability with which H is rejected when z is observed. Let $M^+ = M^+(\boldsymbol{z})$ =the number of $\boldsymbol{z}^{(i)}$ in $T(\boldsymbol{z})$ such that $U(\boldsymbol{z}^{(i)}) \geq u(\boldsymbol{z})$. Then $R^{(i)}$ is precisely the event $M^+(\boldsymbol{z}) = i$. Hence we reject H when $M^+(\boldsymbol{z}) \leq k$ and accept it otherwise.

The modified procedure is to make this decision on the basis of examining a random subset of $T(\boldsymbol{z})$. Specifically the modified test ϕ_L is as follows: Select at random M_o $(M_o < M)$ points of $T(\boldsymbol{z})$. For simplicity assume that the sampling from $T(\boldsymbol{z})$ is done with replacement. Let M'^+ be the number of points $\boldsymbol{z}^{(k)}$ in the sample for which $u(\boldsymbol{z}^{(k)}) \geq u(\boldsymbol{z})$. Then define

$$\phi_L(z) = \begin{cases} 1 & \text{if } M'^+ \leq d \\ 0 & \text{if } M'^+ > d \end{cases}, \tag{10.9.2}$$

where $d(0 \le d \le M_0)$ is a predetermined integer. Let

$$\psi(t) = \sum_{i=0}^{d} \binom{M_o}{i} t^i (1-t)^{M_o - i} \quad (0 \le t \le 1)\,.$$

Then we have the following easily verifiable proposition.

Proposition 10.9.1 (Dwass, 1957). *We have*

$$E\phi_L = \sum_{i=1}^{M} \psi\left(\frac{i}{M}\right) p(i)\,, \quad E\phi = \sum_{i=1}^{k} p(i)\,. \tag{10.9.3}$$

PROOF:

$$E\phi_L = E\left(E\phi_L | \boldsymbol{z} \in R^{(i)}\right) = EP\left(M'^{+} \le d | \boldsymbol{z} = \boldsymbol{z}^{(i)}\right)\,.$$

Also, $P\left(u(\boldsymbol{z}^{(k)}) \ge u(\boldsymbol{z}) | \boldsymbol{z} = z^{(i)}\right) = \frac{i}{M}$. Further, when H is true, $p(i) = 1/M$ and for large M,

$$E_H \phi_L \sim \int_0^1 \psi(t) dt\,. \tag{10.9.4}$$

Throughout we shall assume that $E_H\phi_L = E_H\phi = k/M = \alpha$ (say). Since

$$\psi(t) = \frac{M_o!}{d!(M_o - d - 1)!} \int_t^1 x^d (1-x)^{M_o - d - 1} dx$$

obviously, $\psi(t)$ is a non-increasing function of t in (0,1). Then we have

$$E\phi_L \ge \psi(\alpha) E\phi, \quad (\alpha = k/M)\,, \tag{10.9.5}$$

since

$$E\phi_L \ge \sum_{i=1}^{k} \psi\left(\frac{i}{M}\right) p(i) \ge \psi(\alpha) \sum_{i=1}^{k} p(i) = \psi(\alpha) E\phi\,.$$

The bound in (10.9.5) is quite weak and equality holds in (10.9.5) only when $p(i) = 0$ for $i \ne k$ and $p(k) = 1$. It is reasonable to assume that the alternatives against which ϕ is expected to be effective satisfy

$$p(1) \ge p(2) \ge \cdots \ge p(M)\,. \tag{10.9.6}$$

In particular, (10.9.6) is satisfied when the $p(i)$ are the probabilities induced by any simple alternative against which ϕ is most powerful. Lehmann and Stein (1949) confirm this assertion for the normal alternatives, uniformly for $\Delta < 0$. Hence we shall determine a lower bound for $E\phi_L/E_\phi$ over all $p(i)$, $i = 1, \ldots, M$ satisfying (10.9.6) and such that ϕ and ϕ_L have size $\alpha = k/M$.

Proposition 10.9.2 (Dwass, 1957). *Let the $p(i)$, $i = 1, \ldots, M$ satisfy (10.9.6). Then,*

$$E\phi_L \geq k^{-1} \sum_{i=1}^{k} \psi(i/M) E\phi \,. \tag{10.9.7}$$

PROOF: Consider

$$\begin{aligned} E\phi_L/E\phi &= \sum_{i=1}^{M} \psi(i/M)p(i)/\sum_{j=1}^{k} p(j) \\ &= \sum_{i=1}^{k} \psi(i/M)p(i)/\sum_{j=1}^{k} p(j) + \sum_{i=k+1}^{M} \psi(i/M)p(i)/\sum_{j=1}^{k} p(j) \,. \end{aligned}$$

Now, by replacing $p(i)$ by $p(i)/\sum_{j=1}^{k} p(i)$ for $i = 1, \ldots, k$ and with zero for $i = k+1, \ldots, M$, the value of $E\phi_L/E\phi$ is not increased. Hence, we may at the outset assume that $p(k+1) = \cdots = p(M) = 0$. Now, using the monotonicity of ψ, one can easily see that, subject to (10.9.6),

$$\sum_{i=1}^{k} \psi(i/M)p(i)/\sum_{j=1}^{k} p(i)$$

is minimized when $p(1) = \cdots = p(k) = 1/k$.

Remark 10.9.1. *It is obvious from the proof that (10.9.7) holds if (10.9.6) is replaced by*

$$p(1) \geq \cdots \geq p(k) \,. \tag{10.9.8}$$

Remark 10.9.2. *From (10.9.7) we have $E\phi_L/E\phi \geq (M/k)\sum_{i=1}^{k} \psi(i/M)/M$. For large M, $\sum_{i=1}^{k} \psi(i/M)/M$ is approximately equal to $\int_0^\alpha \psi(t)dt$.*

Hence, inf $E\phi_L/E\phi$ overall $p(i)$ satisfying (10.9.8) approximately equals

$$\alpha^{-1} \int_0^\alpha \psi(t)dt \,. \tag{10.9.9}$$

Let $B(M_o, t)$ denote the number of successes in M_o independent trials with t as the probability of success. Then

$$P\left(B(M_o, t) \leq d\right) = \psi(t) = 1 - \frac{M_o!}{d!(M_o - d - 1)!} \int_0^t u^d (1-u)^{M_o - d - 1} du \,.$$

Table 10.9.1: Giving the values of $\alpha^{-1}A(\alpha)$ defined by (10.9.11). Values in parentheses are based on a normal approximation. Computations are made only for those values of M_o such that $d+1=\alpha(M_o+1)$ is an integer.

	α					
M_o	.01		.05		.10	
19			.642		.743	
39			.736		.815	
49					.834	
59			.782		.848	
79			.810		.868	
99	.634	(.618)	.829	(.827)	.881	(.881)
119			.843	(.842)	.892	(.891)
149					.903	(.902)
199	.725		.877			(.915)
299	.774			(.900)		(.931)
499		(.824)		(.922)		(.946)
999		(.875)		(.945)		(.962)

Let $A(t)=\int_0^t \psi(u)du$. After integration by parts we get

$$A(t) = t\psi(t) + M_o\binom{M_o-1}{d}\int_0^t u^{d+1}(1-u)^{M_o-d-1}du$$
$$= t\psi(t) + \{(d+1)/(M_o+1)\}\,P\,(B(M_o+1,t)\geq d+2)\,. \tag{10.9.10}$$

By (10.9.4),

$$E_H\phi = E_H\phi_L = k/M \sim A(1) = (d+1)/(M_o+1)\,.$$

Let d and M_o be so chosen that $(d+1)/(M_o+1)=k/M=\alpha$. Then by (10.9.10), we have

$$\alpha^{-1}A(\alpha) = P\,(B(M_o,t)\leq d) + P\,(B(M_o+1,t)\geq d+2)\,. \tag{10.9.11}$$

Some of the values of $\alpha^{-1}A(\alpha)$ have been computed by Dwass and they are presented in Table 10.9.1.

The power of the modified test will be "close" to that of the most powerful permutation test. One may argue that Δ still has to be large. However, the optimum permutation test is almost impossible even for moderately large m

and n. Dwass (1957) remarks that if $m = n = 20$, then $\binom{m+n}{m} > 10^{11}$ and the job on a computer will take about 1000 years at the rate of checking 10 permutations per second. Thus, by resorting to the modified procedure, an impossible test can be made possible. For some alternatives the modified procedure may have better bound for the power ratio given by (10.9.11) since the sequence in (10.9.6) is often expected to be a strictly decreasing one.

The modified procedure can be extended to other permutation tests. Dwass (1957) generalizes the modified procedure in order to be applicable to randomized tests and to situations where the elements of $T(\boldsymbol{z})$ are selected without replacement and sequentially.

Problem 10.1

In the modified procedure let

$$\psi(t) = P(\text{deciding } M^+ \le k : M^+/M = t)\,.$$

Show that this $\psi(t)$ coincides with that defined in Proposition 10.9.1. Verify Propositions 10.9.1 and 10.9.2 and the bound (10.9.5).

10.10 Problems

10.2.1 John obtained the following scores (maximum score being 50) in two subjects in three different tests.

	Test 1	Test 2	Final
History:	40	43	41
Computer Science:	28	26	31

We wish to test the null hypothesis that the scores in the two subjects are independent. Use the permutation test based on T^* with $\alpha = \frac{1}{18}$.

10.3.1 Let (40, 43, 41) be a random sample from $F(x)$ and (38, 45, 42) be a random sample from $G(y)$ where F and G are both unknown. We wish to test

$$H_0\colon\ F(x) = G(x) \quad \text{for all } x$$
$$\text{against} \quad H_1\colon\ F(x) \ne G(x) \quad \text{for some } x$$

at $\alpha = 0.10$. Using Pitman's test criterion given by (10.3.2) carry out a permutation test of H_0.

10.3.2 Let (38, 36, 40, 43, 41) be a random sample from $F(x)$ and (32, 42, 39, 36, 44) be a random sample from $G(y)$. Test H_0: $F = G$ against $F \neq G$ using Dixon's criterion given by (10.3.3) at $\alpha = 0.05$.
(HINT: Use 0.833 as the critical value which is taken from Dixon (1940), Table 1 on p. 201).

10.5.1 To randomly selected two groups of students, a set of lectures were given live and via a closed video program respectively. A test based on the lectures was administered to the two groups and the following scores (maximum being 50) were obtained:

Live:	30	47	36	42	45	32	41	44	28	25	
TV	40	41	32	27	23	31	21	37	19	43	34.

Carry out a permutation test of the hypothesis, H_0: $\Delta = 0$ which is based on a students t-statistic with $\alpha = 0.10$.

10.6.1 Let us have the following sample of size 4 observations:

$$(z_1, z_2, z_3, z_4) = (-3, 1, 2, 4).$$

Suppose we wish to test the null hypothesis that the joint density of the above observations is symmetric in its four arguments against the alternative that the observations are normally distributed with positive serial correlation. Using $\sum_{i=1}^{3} z_i z_{i+1}$ as the test criterion, carry out a permutation test of the hypothesis with $\alpha = \frac{1}{12}$.

10.7.1 The mean length of developmental period (in days) for 3 strains of house flies at four densities is given below.[1] We wish to know whether these values differ in developmental period among strains? Use $\alpha = 0.05$.

	Strains		
	OL	BELL	BWB
Density Per Container			
60	9.6	9.3	9.3
80	10.6	9.1	9.2
160	9.8	9.3	9.5
320	10.7	9.1	10.0

(HINT: Use test criterion (10.7.6) and chi-squared approximation.)

[1]This data constitutes part of Problem 11.5 on p. 369 of Sokal, R.R. and F.J. Rohlf (1981). *Biometry.* 2nd Edition. W.H. Freeman and Co., San Francisco.

10.9.1 In the modified procedure let

$$\psi(t) = P(\text{deciding } M^+ \leq k : M^+/M = t)\,.$$

Show that this $\psi(t)$ coincides with that defined in Proposition 10.9.1. Verify Propositions 10.9.1 and 10.9.2 and the bound (10.9.5).

Chapter 11

Rank Order Tests

11.1 Introduction

The runs tests, although simple, will have low power. The permutation tests are time consuming, although they are most powerful or uniformly most powerful. Next we shall consider rank or rank order tests. We will distinguish between rank tests and rank order tests. Usually we reserve the term rank tests to mean linear rank tests. Those procedure that are based on the rank order will be called *rank order tests*.

Let $X_1, \ldots, X_N$ be independent random variables having distributions $F_1, \ldots, F_N$, respectively. Let R_i be the rank of X_i in the ordered $(X_1, \ldots, X_N)$ for $i = 1, \ldots, N$. Then $R = (R_1, \ldots, R_N)$ is called the rank order. Any procedure based on R is called a rank order procedure. However, a procedure based on $T = \sum_{i=1}^{N} b_i a_{R_i}$ will be called a *linear rank test* or simply a *rank test*, and T is called a *linear rank test statistic*. It is also worthwhile to point out that most of the distribution-free goodness of fit tests like the Kolmogorov-Smirnov test criterion, Cramér-von-Mises test criterion and Anderson and Darling test criterion (see Chapter 8) are nonlinear rank statistics. The die-hard advocates of normal theory statistics believe that there is a loss of information in replacing the observations by their associated ranks. In the following we shall show that the correlation between the observations and their ranks is very high and thus there is no need for alarm.

11.2 Correlation between Observations and Ranks

Stuart (1954) has evaluated the correlation between observations and their ranks. Let $(X_1, \ldots, X_N)$ be a random sample from a continuous distribution

$F(x)$, having mean μ and variance σ^2. Also, let $(R_1, \ldots, R_N)$ denote their rank vector. Then we have

Theorem 11.2.1 (Stuart, 1954). *The correlation between X_i and R_i, namely $\rho(X_i, R_i)$, is given by*

$$\rho(X_i, R_i) = E\left\{XF(X) - \frac{1}{2}\mu\right\}\left[\frac{12(N-1)}{(N+1)\sigma^2}\right]^{1/2} \tag{11.2.1}$$

where μ and σ^2 denote the mean and the variance of X_i.

PROOF: $\rho(X_i, R_i) = (EX_iR_i - \mu ER_i)/\sigma\sigma_{R_i}$ where $ER_i = \frac{1}{N}\sum_{k=1}^{N} k = (N+1)/2$, since R_i takes the value k with equal probability $1/N$,

$$ER_i^2 = \frac{1}{N}\sum_{k=1}^{N} k^2 = (N+1)(2N+1)/6\,.$$

Hence,

$$\sigma_{R_i}^2 = \text{var}(R_i) = \frac{(N+1)(2N+1)}{6} - \frac{(N+1)^2}{4} = \frac{N^2-1}{12}\,. \tag{11.2.2}$$

Next consider

$$\begin{aligned} E(X_iR_i) &= E\{E(X_iR_i|R_i = j\} \\ &= \frac{1}{n}\sum_{j=1}^{N} E\left(jX_{j,N}|R_i = j\right) \\ &= \frac{1}{n}\sum_{j=1}^{N} E\left(jX_{j,N}\right)\,, \end{aligned} \tag{11.2.3}$$

since the rank order and the order statistics are mutually independent (for example, see Hájek and Sidák (1967, p. 38) the proof of which we shall present later where $X_{j,N}$ denotes the jth smallest order statistic in a sample of size N drawn from $F(x)$. Hence,

$$E(X_iR_i) = \frac{1}{N}\sum_{j=1}^{N}\frac{N!j}{(j-1)!(N-j)!}\int_{-\infty}^{\infty} xF^{j-1}(1-F)^{N-j}dF\,.$$

Now writing $j = j-1+1$ and taking the summation underneath the integral sign we obtain

$$E(X_iR_i) = (N-1)\int_{-\infty}^{\infty} xF(x)dF(x) + \mu\,.$$

Thus,

$$\begin{aligned}\operatorname{cov}(X_i, R_i) &= (N-1)E\{XF(X)\} + \mu - \mu \cdot \frac{(N+1)}{2} \\ &= (N-1)\left[E\{XF(X)\} - \frac{\mu}{2}\right]. \end{aligned} \tag{11.2.4}$$

Now, using (11.2.2) in (11.2.4) completes the proof. Notice that $\rho(X_i, R_i)$ is free of i as is expected and

$$\lim_{N\to\infty} \rho(X_i, R_i) = \frac{2\sqrt{3}}{\sigma}\left[E\{XF(X)\} - \frac{\mu}{2}\right]. \tag{11.2.5}$$

In the following table we shall present a few values of $\rho(X_i, R_i)$ for some selected distributions.

Table 11.2.1: Giving the values of the correlation for some standard distributions

Distribution	μ	σ^2	$\int xF(x)dF(x)$	$\left(\frac{N+1}{N-1}\right)^{1/2}\rho$
Uniform (0,1)	1/2	1/12	1/3	1
Negative exponential	1	1	3/4	$\sqrt{3}/2$
Normal (0,1)	0	1	$1/2\sqrt{\pi}$	$(3/\pi)^{1/2}$

In general, let $\{a_{i,N}\}$ be a sequence of numbers and ask for what values of the $a_{i,N}$ the correlation between X_i and $a_{R_i,N}$ is maximum. The following result will provide the answer.

Theorem 11.2.2. *The correlation between X_i and $a_{R_i,N}$ is free of i and is maximum when*

$$a_{j,N} = \bar{a} - c\mu + c\mu_{jN}, \quad j = 1, \ldots, N \tag{11.2.6}$$

where

$$\bar{a} = N^{-1}\sum_{j=1}^{N} a_{j,N},\ \mu = N^{-1}\sum_{j=1}^{N}\mu_{jN} = EX_i\,,$$

c is some constant and R_i denotes the rank of X_i among $(X_1,\ldots,X_N)$.

PROOF: Let ρ_i denote the correlation between X_i and $a_{R_i,N}$. Then

$$\rho_i = \left\{N^{-1}E(X_i a_{R_i,N}) - \mu\bar{a}\right\}/\sigma\left\{N^{-1}\sum_{j=1}^{N}(a_{j,N}-\bar{a})^2\right\}^{1/2}$$

where σ^2 denotes the population variance and

$$\begin{aligned} E(X_i a_{R_i,N}) &= \sum_{j=1} E(X_i a_{R_i,N}|R_i = j)P(R_i = j) \\ &= N^{-1}\sum_{j=1}^{N} E(X_{j,N}a_{j,N}) \\ &= N^{-1}\sum_{j=1}^{N} a_{j,N}\mu_{j,N}\,. \end{aligned}$$

Thus,

$$\begin{aligned} \rho_i \equiv \rho &= N^{-1/2}\sum_{j=1}^{N}(\mu_{jN}-\mu)(a_{jN}-\bar{a})/\sigma\left\{\sum_{j=1}^{N}(a_{jN}-\bar{a})^2\right\}^{1/2} \\ &\le \frac{N^{-1/2}}{\sigma}\left\{\sum_{j=1}^{N}(\mu_{jN}-\mu)^2\right\}^{1/2}, \end{aligned}$$

after using the Cauchy-Schwarz inequality and equality holds when

$$a_{jN} - \bar{a} = c(\mu_{jN} - \mu) \tag{11.2.7}$$

for some constant c. This completes the proof of the assertion.

When (11.2.7) holds,

$$\rho = \left\{ N^{-1} \sum_{j=1}^{N} \mu_{jN}^2 - \mu^2 \right\}^{1/2} / \sigma \to 1 \text{ as } N \to \infty \tag{11.2.8}$$

where μ denotes the mean of the population, since $N^{-1} \sum_1^N \mu_{jN}^2 \to EX^2$ (see Hoeffding, 1953).

Thus, Theorem 11.2.2 provides a justification for employing expected values of order statistics as "scores" which generate a test statistic. Notice that Theorem 11.2.2 is a special case of the result of Brillinger (1966) which states that if (X, Y) is a bivariate random variable, then among all functions $g(Y)$ with $Eg^2(Y) < \infty$, $g(Y) = a + bE(X|Y)$ where a and b are constants, maximizes the square of the correlation between X and $g(Y)$. We set $Y =$ rank of X.

11.3 Properties of Rank Orders

Suppose we wish to test $H_o : F_1 = \cdots = F_N$ against the alternative $H_1 : F_1(x) \geq \cdots \geq F_N(x)$ Let G be the group of all transformations

$$x_i' = g(x_i),\ i = 1, \ldots, N$$

where g is continuous and strictly increasing. H_0 and H_1 remain invariant under G because the continuity of a distribution function and the property of two variables being either identically distributed or one being stochastically larger than the other are preserved by any member of G. Further (see Lehmann (1959, Example 3, p. 217)) that $R = (R_1, \ldots, R_N)$ is a maximal invariant under G.[1] That is, the maximal invariant is given by the equivalence

$$(x_1, \ldots, x_N) \approx (x_1', \ldots, x_N')$$

if and only if the two sets of numbers are in the same order. This relationship is invariant, since with respect to a strictly increasing function the x_i''s are in the same order relation as the x_i's. Conversely, given any two such sets of numbers there exists a strictly increasing function taking the first into the second.

[1]A function T is said to be a *maximal invariant* if it is invariant and $T(x_1) = T(x_2)$ implies that $x_2 = gx_1$ for some g belonging to G.

The maximal invariant classes in the space of distributions coincide with the classes $g_i(H)$ where g_i are strictly increasing with $g_i(0) = 0$, $g_i(1) = 1$, $i = 1, \ldots, N$. The classes $g_i(H)$ are called *Lehmann alternatives.* For a precise statement of this result and its proof the reader is referred to Lehmann (1953, pp. 41–42).

Next we shall give certain results which relate the order statistic and the rank order. Let $p(x_1, \ldots, x_N)$ denote the joint density of $\boldsymbol{X} = (X_1, \ldots, X_N)$ and let $\boldsymbol{X}_N = (X_{1N} < \cdots < X_{NN})$ denote the order statistic in $\boldsymbol{X}$. As before let $R = (R_1, \ldots, R_N)$ denote the rank order of $\boldsymbol{x}$. Also, let H denote the hypothesis that $p(x_1, \ldots, x_N)$ is invariant under permutations of its coordinates. That is $p(x_{r_1}, \ldots, x_{r_N}) = p(x_1, \ldots, x_N)$ for every permutation $(r_1, \ldots, r_N)$ of $(1, \ldots, N)$. For example, $p(x_1, \ldots, x_N) = \Pi_{i=1}^N f(x_i)$ where $f(x)$ is an arbitrary probability density defined on the real line.

Theorem 11.3.1. *With the above notation the joint probability density of* $\boldsymbol{X}_N$ *is given by*

$$\begin{aligned} \bar{p}(x_1, \ldots, x_N) &= \sum_{r \in \mathcal{R}} p(x_{r_1}, \ldots, x_{r_N}), \quad \textit{for } x_1 < \cdots < x_N \\ &= 0, \qquad\qquad\qquad \textit{elsewhere}, \end{aligned} \tag{11.3.1}$$

where $\mathcal{R}$ *denotes the* $N!$ *permutations of* $(1, \ldots, N)$. *Further,*

$$P(R = r | \boldsymbol{X}_N = x_N) = p(x_{r_1}, \ldots, x_{r_N}) / \bar{p}(x_1, \ldots, x_N) \tag{11.3.2}$$

for $x_1 < x_2 < \cdots x_N$ *and zero elsewhere.*

PROOF: The result is well known and we shall present the proof given by Hájek and Sídak (1967). Since $\mathcal{R}$ denotes the $N!$ permutations of $(1, \ldots, N)$, for any A belonging to the Borel field induced by $\boldsymbol{X}_N$,

$$\begin{aligned} P(\boldsymbol{X}_N \in A) &= \sum_{r \in \mathcal{R}} P(\boldsymbol{X}_N \in A, \ R = r) \\ &= \sum_{r \in \mathcal{R}} \int \cdots \int_{\boldsymbol{x}_N \in A} p(x_{r_1}, \ldots, x_{r_N}) \Pi_{i=1}^N dx_i \\ &= \int \cdots \int_A \bar{p}(x_1, \ldots, x_N) \Pi_{i=1}^N dx_i \end{aligned}$$

and thus the pdf of $\boldsymbol{X}_N$ is $\bar{p}(x_1, \ldots, x_N)$, since

$$\begin{aligned} P(\boldsymbol{X}_N \in A,\ R = r) &= \int_{R=r} \int_{\boldsymbol{y} \in A} p(y_1, \ldots, y_N) \Pi_{i=1}^N dy_i \\ &= \underset{\boldsymbol{x}_N \in A}{\int \cdots \int} p(x_{r_1}, \ldots, x_{r_N}) \Pi_{i=1}^N dx_i\,, \end{aligned}$$

where we have made a one-to-one linear transformation from $(y_1, \ldots, y_N)$ to $(x_1, \ldots, x_N)$ with the Jacobian equal to 1. Thus,

$$P(\boldsymbol{X}_N \in A,\ R = r) \underset{A}{\int \cdots \int} \frac{p(x_{r_1}, \ldots, x_{r_N})}{\bar{p}(x_1, \ldots, x_N)} \bar{p}(x_1, \ldots, x_N) \Pi_1^N\, dx_i \quad (11.3.3)$$

where we use the fact that $\bar{p}(x_1, \ldots, x_N) = 0$ implies that $p(x_{r_1}, \ldots, x_{r_N}) = 0$, for every $r \in \mathcal{R}$.

Thus, (11.3.3) and (11.3.1) prove (11.3.2).

Corollary 11.3.1.1 (Hájek and Sídak, 1967). *If $p \in H$, then R and $\boldsymbol{X}_N$ are mutually independent and*

$$P(R = r|H) = 1/N!\,, \quad r \in \mathcal{R}. \quad (11.3.4)$$

PROOF: If $p \in H$, then $\bar{p}(x_1, \ldots, x_N) = N!\; p(x_1, \ldots, x_N)$. Then the corollary readily follows from (11.3.2).

Corollary 11.3.1.2 (Hájek and Sídak, 1967 and Hoeffding, 1951). *Let $p_0(y_1, \ldots, y_N)$ denote the joint density of $\boldsymbol{X} = (X_1, \ldots, X_N)$ under H. Then, for any statistic $t(\boldsymbol{X})$*

$$E_H\left\{t(X_1, \ldots, X_N)|R = r\right\} = E_H\left\{t(X_{r_1,N}, \ldots, X_{r_N,N})\right\}, \quad (11.3.5)$$

and

$$P(R = r|p) = E_H\left\{\frac{p(X_{r_1,N}, \ldots, X_{r_N,N})}{p_0(X_{r_1,N}, \ldots, X_{r_N,N})}\right\} \cdot \frac{1}{N!}. \quad (11.3.6)$$

PROOF: Since $\boldsymbol{X}_N$ and R are independent under H,

$$\begin{aligned} E\left\{t(X_1, \ldots, X_N)|R = r\right\} &= E\left\{t(X_{r_1,N}, \ldots, X_{r_N,N})|R = r\right\} \\ &= E\left\{t(X_{r_1,N}, \ldots, X_{r_N,N})\right\}. \end{aligned}$$

$$
\begin{aligned}
P(R=r|p) &= \int_{\{R=r\}}\cdots\int p(y_1,\ldots,y_N)\Pi dy_i \\
&= \int_{x_1<\cdots<x_N}\cdots\int p(x_{r_1},\ldots,x_{r_N})\Pi_{i=1}^N dx_i \\
&= \int_{x_1<\cdots<x_N}\cdots\int \frac{1}{N!}\frac{p(X_{r_1},\ldots,X_{r_N})}{p_0(X_{r_1},\ldots,X_{r_N})}\cdot N!p_0(x_{r_1},\ldots,x_{r_N})\Pi dx_i \\
&= \int_{x_1<\cdots<x_N}\cdots\int \left\{\frac{p(x_{r_1},\ldots,x_{r_N})}{N!p_0(x_{r_1},\ldots,x_{r_N})}\right\}\cdot N!p_0(x_{r_1},\ldots,x_{r_N})\Pi_1^N dx_i \\
&= E_H\left\{\frac{p(X_{r_1,N},\ldots,X_{r_N,N})}{N!p_0(X_{r_1,N},\ldots,X_{r_N,N})}\right\}.
\end{aligned}
$$

Remark 11.3.1. *Equations (11.3.6) is due to Hoeffding (1951). Then as an immediate consequence of Neyman-Pearson Lemma, we have the following theorem.*

Theorem 11.3.2. *The most powerful rank order test of H against some simple alternative H_1 is given by*

$$
\begin{aligned}
\psi(r) &= 1 \quad \text{if } P(R=r|H_1) > k, \\
&= 0 \quad \text{if } P(R=r|H_1) < k,
\end{aligned}
$$

where $\psi(r) = P(\text{reject } H|R=r)$ and, k and $\psi(r)$ for r such that $P(R = r|H_1) = k$ can be determined so that $E[\psi(R)|H_1] = \alpha$.

In practice, it is not easy to evaluate explicitly the values of $P(R=r|H_1)$. However, for some alternatives called the Lehmann alternatives, one can do so.

11.4 Lehmann Alternatives

Consider two random variables X, Y having d.f.'s $F(x)$ and $G(y)$ respectively. If under the alternative, $G(x) < F(x)$ for some x, that is, Y is stochastically larger than X, then a subset of the alternative hypothesis can be stated as $H_{L,1}: F(x) = H(x)$, and $G(x) = H^\theta(x)$ where $\theta > 1$ and H is an unknown continuous d.f. If θ is a positive integer, one can interpret the sub-hypothesis as Y is distributed as the largest observation in a random sample of size θ drawn from $F(x)$. Analogously if, the alternative hypothesis is given by $G(x) > F(x)$ for some x, then the corresponding Lehmann alternative is $H_{L,2} : G(x) = 1 - [1 - F(x)]^{\theta}$, $\theta > 1$. The class of Lehmann alternatives is still composite and it is not vacuous. For example, let

$$\text{(i)}\ \ F(x) = 1 - \exp(-x), x > 0 \text{ and } G(x) = 1 - \exp(-\theta x),\ \theta > 1$$

or

$$\text{(i)}\ \ F(x) = x, 0 < x < 1 \text{ and } G(x) = x^{\theta},\ 0 < x < 1,\ \theta > 1\,.$$

Notice that $H_{L,2}$ can be obtained from $H_{L,1}$ by changing x to $-x$ and assuming H to be symmetric about zero.

Then, we have the following generalization of Lehmann's (1953) result.

Theorem 11.4.1 (Savage, 1956). *Let the random variables $X_1, \ldots, X_N$ be mutually independent such that X_i has the distribution function $F_i(x) = [H(x)]^{\theta_i}$, $i = 1, \ldots, N$, where $\theta_i > 0$, and $H(x)$ is an unknown continuous distribution function. Then*

$$P(X_{k_1} \le X_{k_2} \cdots \le X_{k_N} | H_L) = \left(\prod_{i=1}^{N} \theta_i\right) \Big/ \prod_{i=1}^{N} \left(\sum_{j=1}^{i} \theta_{k_j}\right) \tag{11.4.1}$$

where $(k_1, \ldots, k_N)$ is a permutation of $(1, \ldots, N)$ and H_L denotes the specified Lehmann alternative.

PROOF: Let $p = P(X_{k_1} < X_{k_2} < \cdots < X_{k_N} | H_L)$. Then

$$\begin{aligned} p &= \int\cdots\int_{-\infty<x_1<\cdots<x_N<\infty} \prod_{i=1}^{N} d\,(H(x_i)) \\ &= \int\cdots\int_{0<y_1<\cdots<y_N<1} \prod_{i=1}^{N} d(y_i^{\theta_{k_i}}) \\ &= \left(\prod_{i=1}^{N} \theta_{k_i}\right) \int\cdots\int_{0<y_1<\cdots<y_N<1} \prod_{i=1}^{N} (y_i^{\theta_{k_i}-1}) \prod_{i=1}^{N} dy_i\,. \end{aligned}$$

By performing repeated integration starting with y_1, then with y_2, etc., we obtain the result.

Corollary 11.4.1.1.

$$P(X_1 < X_2 < \cdots < X_N | H_L) = \left(\prod_{i=1}^{N} \theta_i\right) / \prod_{i=1}^{N} \left(\sum_{j=1}^{i} \theta_j\right).$$

Remark 11.4.1. *It should be noted here that $(k_1, \ldots, k_N)$ are subscripts of the X_j's that have the ranks $1, 2, \ldots, N$. These are called subscripts or anti-ranks. Since the correspondence is one to one, one can easily get the corresponding rank order. For example, let $N = 4$, and $(k_1, k_2, k_3, k_4) = (3, 2, 4, 1)$. That is, $X_3 = X_{14}$, $X_2 = X_{24}$, $X_4 = X_{34}$, $X_1 = X_{44}$. Thus, the rank order $(r_1, r_2, r_3, r_4) = (4, 2, 1, 3)$. The relation between r_i and k_j is $r_i = j$ if and only if $k_j = i$ for $i, j = 1, \ldots, N$. For instance suppose, we wish to find $P(R = r)$. Then*

$$\begin{aligned} P(R = r) &= \int_{\{R=r\}} \cdots \int \prod_{i=1}^{N} d\{H(y_i)\}^{\theta_i} \\ &= \int_{-\infty<x_1<\cdots<x_N<\infty} \cdots \int \prod_{i=1}^{N} d\{H(x_{r_i})\}^{\theta_i} \\ &= \int_{-\infty<x_1<\cdots<x_N<\infty} \cdots \int \prod_{i=1}^{N} d\{H(x_i)\}^{\theta_{k_i}} \end{aligned}$$

where $(k_1, \ldots, k_N)$ denotes the anti-rank vector of $(X_1, \ldots, X_N)$.

Trend Alternatives

Savage (1957) has considered the probabilities of rank orders under various "trend" alternatives. Here, we assume that $X_1, \ldots, X_N$ are independent random variables such that X_i has a d.f. of the form $F(x - \theta_i)$ $(i = 1, \ldots, N)$ where the θ_i form an increasing sequence. As before $r = (r_1, \ldots, r_N)$ denotes the rank order such that r_i is the rank of x_i, the observed value of X_i $(i = 1, \ldots, N)$.

Definition 11.4.1 (Savage, 1957b). *$r' L_{ij} r$ if $r'_k = r_k$ for $i \neq k \neq j$, $r'_i = r_j$, $r'_j = r_i$; and $(r_i - r_j)(i - j) > 0$. For example, if $r = (2, 3, 6, 5, 4, 1)$ and $r' = (2, 5, 6, 3, 4, 1)$, then $r' L_{24} r$. We shall write $r' L r$ as a short form of $r' L_{ij} r$ or to denote that there is a chain of rank orders $r^1, \ldots, r^t, \ldots, r^T$ such that*

$$r' L_{i_0, j_0} r^1, \ldots, r^t L_{i_t, j_t} r^{t+1}, \ldots, r^T L_{i_T j_T} r.$$

For example, if $r = (2,3,6,5,4,1)$ *and* $r' = (3,5,6,4,2,1)$, *then* $r'L_r$, $T = 2$ *and* $r'L_{15}(2,5,6,4,3,1)L_{45}(2,5,6,3,4,1)L_{24}r$. *For some of the hypotheses considered below,* $r'Lr$ *will imply that* r' *is less probable than* r.

Definition 11.4.2 (Savage, 1957b). r^*Cr *(rank order* r^* *is the complement of rank order* r*) if*

$$r_i^* = N + 1 - r_{N+1-i},\; i = 1, \ldots, N\,.$$

Note that r^*Cr implies rCr^* and that rCr if and only if $r_i + r_{N+1-i} = N+1$. Recall that we also defined the subscript rank order, $k = (k_1, \ldots, k_N)$ where $k_i = j$ when the ith smallest of the numbers $(x_1, \ldots, x_N)$ is x_j. Further it was pointed out that $r_i = j$ is equivalent to $k_j = i$ $(i, j = 1, \ldots, N)$. One can easily verify that $r'Lr$ is equivalent to $k'Lk$ and that r^*Cr is equivalent to k^*Ck.

Let us further assume that each X_i has the d.f. $F_i(x)$ and the pdf $f_i(x) = f(x;\theta_i)$ $(i = 1, \ldots, N)$. We will be interested in the following hypotheses.

- H_0: There exists a continuous d.f. $F(x)$ such that

$$F_i(x) = F(x), \text{ at all } x, \text{ for } i = 1, \ldots, N\,.$$

- H_1: The θ_i's are real-valued and the following conditions hold:
  1. $\theta_1 \le \theta_2 \le \cdots \le \theta_N$.
  2. If $\theta_i < \theta_j$ and $x < y$, then

  $$f(x;\theta_i)f(y;\theta_j) - f(x;\theta_j)f(y;\theta_i) \ge 0$$

  with strict inequality for some $x < y$.
  3. $f(x;\theta)$ is continuous in x for each θ.
  4. The set of points on which $f(x;\theta)$ is positive does not depend on θ.

- H_2: The θ_i's are real-valued such that
  1. $\theta_1 \le \theta_2 \le \cdots \le \theta_N$.
  2. $f(x;\theta_i) = g(\theta_i)h(x)\exp(\theta_i x)$ where g and h are nonnegative functions: (that is, f_i belongs to the exponential family).

- H_3: $F_i(x) = [H(x)]^{\theta_i}$, $0 < \theta_1 \le \cdots \le \theta_N$ and $H(x)$ is some continuous d.f.

- H_4: $\theta_i = i\theta > 0$ and $f(x;\theta_i) = f(x - i\theta) = f(i\theta - x)$.

One can notice the following relationship among the H_i, $i = 1, \ldots, 4$. H_3 (provided $H(x)$ has a density) $\Rightarrow H_2 \Rightarrow H_1$. H_4 is compatible with H_1 and H_2, but not H_3. The Cauchy density with translation parameter satisfies only H_4. The extreme value distribution with $\log\theta$ acting as the location parameter satisfies the assumption of H_3. Then we have the following results of Savage (1957b).

Theorem 11.4.2 (Savage, 1957b). *$r'Lr$ implies that*

$$P(R = r'|H_1) < P(R = r|H_1) \text{ when } \theta_i < \theta_j \text{ and } r_i < r_j .$$

PROOF: One can write

$$\begin{aligned} &P(R = r|H_1) - P(R = r'|H_1) \\ &= \int_{-\infty<x_1<\cdots<x_N<\infty} \cdots \int \prod_{\substack{k=1 \\ i \neq k \neq j}}^{N} f(x_{r_k};\theta_k) \\ &\quad \cdot \left[f(x_{r_i};\theta_i) f(x_{r_j};\theta_j) - f(x_{r_i},\theta_j) f(x_{r_j};\theta_i) \right] \cdot \prod_{i=1}^{N} dx_i . \end{aligned}$$

The first factor in the integrand is nonnegative since $f(x;\theta)$ is a density and the second factor is always nonncgativc and positive for some values of x_{r_i}, x_{r_j} say u ad v such that $u < v$. From assumptions (3) and (4) of H_1, the whole integrand can be made positive in a region of the following type: x_{r_k} is near u for $r_k < r_j$ and x_{r_k} is near v for $r_k \geq r_j$. Thus the integral is positive.

Corollary 11.4.2.1. *$r'Lr$ implies that $P(R = r|H_1) > P(R = r'|H_1)$ provided the θ_{r_i} corresponding to those i for which $r_i \neq r'_j$ are not all equal.*

Corollary 11.4.2.2. *When testing H_0 against H_1 with the restriction $\theta_1 < \theta_2 < \cdots < \theta_N$, an admissible rank order test must reject H_0 with probability one when $R = r$ provided $r'Lr$ and the probability of rejecting H_0 is positive when $R = r'$.*

Corollary 11.4.2.3. *If $\theta_1 < \cdots < \theta_N$, then $P(R_i = 1|H_1) > P(R_{i+1} = 1|H_1)$ for $i = 1, \ldots, N-1$.*

Theorem 11.4.3 (Savage, 1957b). *If (i) H_2 holds, (ii) the subscript orders k' and k are such that*

$$\sum_{j=1}^{i} k'_j \geq \sum_{j=1}^{i} k_j, \; i = 1, \ldots, N$$

and the inequality is strict for at most one value of i, and (iii) $\theta_i = i\theta$ where $\theta > 0$, then

$$P(K = k') < P(K = k).$$

PROOF: A direct computation yields

$$P(K = k') - P(K = k) = \left[\prod_{i=1}^{N} g(\theta_i)\right] \int_{-\infty<x_1<\cdots<x_N<\infty} \cdots \int \left[\prod_{i=1}^{N} h(x_i)\right] \cdot \left[\exp\left(\sum_{i=1}^{N} x_i \theta_{k_i}\right) - \exp\left(\sum_{i=1}^{N} x_i \theta_{k'_i}\right)\right] \prod_{i=1}^{N} dx_i .$$

Hence it suffices to show that

$$\sum_{i=1}^{N} x_i(\theta_{k_i} - \theta_{k'_i}) > 0 .$$

Towards this, since $\sum_{j=1}^{N}(\theta_{k_j} - \theta_{k'_j}) = 0$, one can write

$$\sum_{i=1}^{N} x_i(\theta_{k_i} - \theta_{k'_i}) = \sum_{i=1}^{N-1} (x_i - x_{i+1}) \left[\sum_{j=1}^{i} (\theta_{k_j} - \theta_{k'_j})\right]$$

(after summing on i first and j later) and $x_i - x_{i+1} < 0$ and $\sum_{j=1}^{i}(\theta_{k_j} - \theta'_{k_j}) = \theta \sum_{j=1}^{i}(k_j - k'_j) \geq 0$ with strict inequality for some i. This completes the proof.

Notice that assumption (ii) of Theorem 11.4.3 is equivalent to: $\sum_{j=i}^{N} k_j \geq \sum_{j=i}^{N} k'_j$, for $i = 1, \ldots, N$ and the inequality is strict for some i. Also, assumption (ii) of Theorem 11.4.3 does not imply $k'Lk$. This can be seen by examining $k = (2, 5, 1, 3, 4)$ and $k' = (3, 4, 2, 5, 1)$.

Corollary 11.4.3.1. *In testing H_0 against H_1, with $\theta_i = i\theta > 0$ an admissible rank order test must have probability of rejecting H_0 when k occurs equal to one provided the probability of rejecting H_0 when k' occurs is positive and assumption (ii) of Theorem 11.4.3 holds.*

From (11.4.1) we have

$$P(K=k|H_3)=\left(\prod_{i=1}^{N}\theta_i\right)\left[\prod_{i=1}^{N}\left(\sum_{j=1}^{i}\theta_{k_j}\right)\right]^{-1}.$$

Hence, a uniformly most powerful rank order test of H_0 against H_3 with $\theta_i = i\theta$ (here uniformity with respect to θ and $H(x)$) is to reject H_0 for large values of the statistic $T_3(k)$ where

$$T_3(k)=\prod_{i=1}^{N}\left(\sum_{j=1}^{i}k_j\right)^{-1}. \tag{11.4.2}$$

Theorem 11.4.4 (Savage, 1957b). *If $r^* cr$, then*

$$P(R=r^*|H_4)=P(R=r|H_4).$$

PROOF: Consider

$$P(R=r|H_4)=\int_{-\infty<x_1<\cdots<x_N<\infty}\cdots\int\prod_{i=1}^{N}[f(x_{r_i}-i\theta)dx_i]\,.$$

Now, let $x_i=(N+1)\theta-y_{N-i+1}$, $i=1,\ldots,N$, and obtain

$$\begin{aligned}P(R=r|H_4) &= \int_{-\infty<y_1<\cdots<y_N<\infty}\cdots\int\prod_{i=1}^{N}[f\left(-y_{N-r_i+1}+\theta(N+1-i)\right)dy_i]\\ &= \int_{-\infty<y_1<\cdots<y_N<\infty}\cdots\int\prod_{i=1}^{N}[fy_{r_i^*}-i\theta)dy_i]\\ &= P(R=r^*|H_4)\,.\end{aligned}$$

Savage (1957b) has presented a catalogue of rank order tests of trend and of their admissibility properties. In the following we shall list a few of these tests.

T_2 is the locally most powerful test against trend. For each statistic large values are in the rejection region. E_{iN} denotes the expected value of the ith smallest standard normal order statistic in a random sample of size N. If $r'Lr$ implies that $T(r) > T(r')$ then $+$ is recorded. If $r'Lr$

Table 11.4.1: Admissibility properties of rank order tests for trend

Hypothesis	H_1	H_2	H_4
Result	Cor. 11.4.2.2	Cor. 11.4.3.1	Theorem 11.4.4
Condition	$r'Lr$	$\sum_{j=1}^{i}(k'_j - k_j) \geq 0$	r^*Cr
Statistic			
$T_1 = \sum_{i=1}^N ik_i$	+	+	+
$T_2 = \sum_{i=1}^N k_i E_{i,N}$	+	+	+
$T_3 =$ (Eqn. 11.4.2)	+	+	-
$T_4 = \sum_{i=1}^N \sum_{j=1}^N d(r_i, r_j)$	+	-	+

where $d(x, y) = 1$ if $x < y$ and zero elsewhere.

implies that $T(r) < T(r')$ then $-$ is recorded. If $\sum_{j=1}^{i}(k'_i - k_i) \geq 0$ for all i and strict inequality for some i implies $T(r) > T(r')$ then $+$ is recorded; otherwise $-$ is recorded. The positive results for T_1 and T_2 are obtained in the same manner as the proof of Theorem 11.4.3. The negative results are obtained via counter examples. Thus, for T_4, consider $k = (1, 8, 2, 7, 6, 5, 4, 3)$ and $k' = (4, 5, 3, 6, 7, 8, 1, 2)$. The symbol $+$ is recorded if r^*Cr implies $T(r) = T(r^*)$ and the symbol - is recorded if for some r^* and r we have r^*Cr and $T(r) \neq T(r^*)$. Stuart (1954) has considered the asymptotic relative efficiencies of these and other tests of randomness against trend with normal alternatives. Tests T_1, and T_4 have been proposed as tests of independence in bivariate populations.

Independence of Tests of Randomness and other Hypotheses

Often we assume that we have a random sample at hand before we embark on any inference. It is worthwhile to test the hypothesis of randomness of the sample. Even if we accept the hypothesis of randomness, for subsequent tests are all conditional and it is of interest to evaluate the exact level of significance of the subsequent test that is carried out. If the test of randomness and the subsequent test are independent, then one can easily compute the levels of significance of the second test. Hence, it is of importance to investigate the class of tests that is independent of the test of randomness. Savage (1957a) has explored this class of statistics.

Theorem 11.4.5 (Savage, 1957a). *Let $X_1, \ldots, X_N$ be i.i.d. random variables having a continuous distribution function. Let R_i denote the rank of X_i $(i = 1, \ldots, N)$. Then the rank order statistic and symmetric statistics (that are symmetric functions of the observations) are independently distributed. In other words if $g(\boldsymbol{X})$ denotes a symmetric statistic, then*

$$P(R_i = r_i,\ i = 1, \ldots, N | g(\boldsymbol{X}) = c) = 1/N!$$

where $(r_1, \ldots, r_N)$ is a permutation of the first N integers.

PROOF: Tied observations will occur with zero probability since the underlying distribution is continuous. Given a value c for $g(\boldsymbol{x})$ the conditional probabilities of the rank orders are equal and this is true for those values of the symmetric statistic which do not imply tied observations. This completes the proof.

Remark 11.4.2. *An alternative proof of Theorem 11.4.5 is as follows. If $g(\boldsymbol{X})$ is symmetric, then it is a function of the order statistic. Also from Corollary 11.3.1.1 the order statistic and the rank order are mutually independent. Hence $g(\boldsymbol{X})$ and the rank order are also mutually independent. If α_1 and α_2 are the levels of significance of the randomness test and the subsequent test, then the level of significance of the second test is $(1 - \alpha_1)\alpha_2$ and the probability of rejecting the null hypothesis by either of the two tests is $1 - (1 - \alpha_1)(1 - \alpha_2) = \alpha_1 + \alpha_2 - \alpha_1\alpha_2$.*

Remark 11.4.3. *In Theorem 11.4.5, the X_i's need not be independent. It suffices if the joint distribution of $X_1, \ldots, X_N$ is symmetric in the N arguments. Then the conclusion of Theorem 11.4.5 is still valid.*

Applications

1. Suppose we wish to test the hypothesis that $X(t)$ is an observation on a Wiener Process. Under the null hypothesis, the quantities $X(i\delta) - X((i-1)\delta)$ $(i = 1, \ldots, N)$ constitute a random sample from a normal distribution with mean zero and variance proportional to δ. To test this hypothesis one must test for both randomness and normality. Suppose the test for randomness is carried out using one of the tests for 'trend' discussed earlier. The test for normality could be performed using one of the goodness of fit tests discussed in Chapter 8. Then the test for randomness and the goodness of fit test are independent.

2. Hotelling and Pabst (1936) consider the following example. kN observations are made and each is assigned to k categories. A test of the hypothesis that the categories have equal probability is based on the statistic $S = N^{-1} \sum_{i=1}^{k} (N_i - N)^2$ where N_i denotes the number of observations belonging to the ith category. This still will be approximately distributed (under the null hypothesis) as chi-square with $k-1$ degrees of freedom.

Let R_i denote the rank of N_i among $N_1, \ldots, N_k$ (tied N_i are assigned ranks at random.). Suppose we test the hypothesis of randomness of the categories against the alternative that the probabilities of the categories form an increasing sequence by using the statistic $T = \sum_{i=1}^{k} iR_i$. Then S is symmetric in the N_i, and T depends solely on the labelings of the categories. Thus the two statistics are independently distributed.

For other applications of Theorem 11.4.5, the reader is referred to Savage (1957a, pp. 55–56).

11.5 Two-sample Rank Orders

Let $X_1, \ldots, X_m$ be a random sample from $F(x)$ and $Y_1, \ldots, Y_n$ be a random sample from $G(x)$. Let $N = m + n$. Also, $W_{1N} \leq \cdots \leq W_{NN}$ denote the combined ordered X's and Y's. Define a new sequence $(Z_1, \ldots, Z_N)$ where $Z_{iN} = 0$ if W_{iN} is an X and $Z_{iN} = 1$ if W_{iN} is a Y, $(i = 1, \ldots, N)$. Further, let $v_i = \sum_{j=1}^{i} z_j$ and $u_i = i - v_i$ $(i = 1, \ldots, N)$. Let the ranks of the X's (Y's) be denoted by $(R_1, \ldots, R_m)$ $((S_1, \ldots, S_n))$. Also, let $(R_1' < \cdots < R_m')$ $((S_1' < \cdots < S_n'))$ be the ordered R's (S's). Obviously the R' sequence, S' sequence and the Z sequence are equivalent. Further, $P(R' = r') = m!.P(R = r)$ (where $r_1' < r_2', \cdots, r_m'$) and $P(S' = s') = n!\ P(S = s)$ (where

$s'_1 < s'_2 < \cdots < s'_n$), since $\{R = r\} \Rightarrow \{R' = r'\}$ and $\{S = s\} \Rightarrow \{S' = s'\}$ where $r'\{s'\}$ is ordered $r\{s\}$. Thus,

$$P(R' = r') = P(S' = s') = P(Z = z)\,. \tag{11.5.1}$$

Theorem 11.5.1 (Lehmann, 1953). *Under Lehmann alternatives, H_L : $F = H^{\Delta_1}$ and $G = H^{\Delta_2}$, we have*

$$P(Z = z) = m!n!\Delta_1^m \Delta_2^n / \prod_{i=1}^{N} \left\{ \sum_{j=1}^{i} [(1 - z_j)\Delta_1 + z_j \Delta_2] \right\}. \tag{11.5.2}$$

PROOF: In Corollary 11.4.1.1 set $\theta_j = (1 - z_j)\Delta_1 + z_j \Delta_2$.

One can simplify $P(Z = z)$ to

$$P(Z = z) = m!n!\,\delta^n / \prod_{i=1}^{N} (u_i + \delta v_i), \;\; \delta = \Delta_2/\Delta_1\,. \tag{11.5.3}$$

Example 11.5.1. Let $F(x) = x$, $0 < x < 1$ and $G(x) = x^2$, $0 \le x \le 1$. Then, set $\delta = 2$ in (11.5.3) and note that $\prod_{i=1}^{N}(u_i + 2v_i) = \prod_{i=1}^{N}(i + v_i)$, let $s'_1 \le \cdots < s'_n$ be the ordered ranks of the Y's. That is, $v_{s'_1} = 1, \ldots, v_{s'_n} = n$. Hence

$$P(S = s) = [(N+n)!]^{-1}\, m!\; 2^n s'_1 (s'_2 + 1) \cdots (s'_n + n - 1)\,.$$

Example 11.5.2 (Hájek and Sidak, 1967, pp. 53–54). Let $F(x) = x$, $0 \le x \le 1$ and $G(x) = 0$ for $x < \Delta$, $G(x) = x - \Delta$, $\Delta < x < 1 + \Delta$, and $G(x) = 1$ for $x \ge 1 + \Delta$. Then

$$P(S = s) = \sum_{j=0}^{a} \sum_{k=0}^{b} \Delta^{j+k} (1 - \Delta)^{N-j-k} / j!k!(N - j - k)! \tag{11.5.4}$$

where $a = N - \max(r_1, \cdots, r_m)$ = the number of Y's exceeding the largest X, and $b = \min(s_1, \ldots, s_n) - 1$ = the number of X's preceding the smallest Y. For the derivation of (11.5.4) see Hájek and Sidak (1967, pp. 53–54). Notice that $P(S = s)$ depends only on a and b and that it is increasing in both a and b. Thus if (a, b) corresponds to s and (a', b') to s', then, for $0 < \Delta < 1$, either $(a' \le a, b' < b)$ or $(a' < a$ and $b' \le b) \Rightarrow P(S = s) > P(S = s')$.

Even though we are able to obtain an explicit expression for $P(S = s)$ with Lehmann alternatives we cannot order the rank order probabilities unless δ is known. For instance, we cannot infer that rank order $z = (0, \cdots, 0, 1, 1, \cdots, 1)$ is more probable than the rank order $z' = (0, \ldots, 1, 0, 1, 1, \ldots, 1)$ for slippage alternatives. Consider the following counter example of Savage (1956).

Example 11.5.3. Let the two pdf's be as follows:

$$f(x) = \left\{ \begin{array}{ll} 0, & x < 0 \\ 1, & 0 \le x < \epsilon \\ 0, & \epsilon \le x < 2\epsilon \ (0 < \epsilon < 1) \\ 1, & 2\epsilon \le x < 1 + \epsilon \\ 0, & 1 + \epsilon \le x \end{array} \right\} \quad g(x) = \left\{ \begin{array}{ll} 0, & x < \epsilon \\ 1, & \epsilon \le x < 1 + \epsilon \\ 0, & 1 + \epsilon \le x \end{array} \right\}$$

Then, Savage (1956) shows that

$$P(Z = z') - P(Z = z) = \epsilon^m \{m(1-\epsilon)^n - 1\} > 0 \text{ provided } \epsilon < 1 - m^{-1/n}.$$

We have the following theorem of Savage (1956) showing that for the monotone likelihood ratio alternatives z is more probable than z'.

Theorem 11.5.2 (Savage, 1956, *Monotone likelihood ratio alternatives***).** *Let $f(x;\theta_1)$ be the common density of $X_1, \ldots, X_m$ and $f(x;\theta_2)$ denote the common density of $Y_1, \ldots, Y_n$, with $\theta_2 > \theta_1$. For $x_2 > x_1$ if $f(x_1;\theta_1)f(x_2;\theta_2) - f(x_1;\theta_2)f(x_2;\theta_1) \ge 0$, then the rank order z is more probable than the rank order z' where the two rank orders are identical except for their i^{th} and j^{th} elements ($i < j$) which are (0,1) for z and (1,0) for z'.*

PROOF: We have

$$P(Z = z) - P(Z = z') = \int_{-\infty < x_1 < \cdots < x_N < \infty} \cdots \int \prod_{\substack{k=1 \\ i \neq k \neq j}}^{N} f(x_k; \theta_{1+z_k})$$

$$\cdot \quad [f(x_i;\theta_1)f(x_j;\theta_j) - f(x_i,\theta_2)f(x_j;\theta_1)] \cdot \prod_{i=1}^{N} dx_i .$$

Since $x_i < x_j$ by assumption, the integrand is nonnegative and actually positive on a set of positive measure (except for the case $f(x;\theta_1) = f(x;\theta_2)$)

almost everywhere. Thus, when $m = n = 2$, the rank order $z = (0101)$ must be put into the critical region with probability one before the rank order $z' = (1001)$ is put into the critical region with nonzero probability. In the equal sample size case, the one- sided Smirnov test is based on large values of the statistic

$$\max_{1\le i\le N}(i - 2v_i), \text{ where } v_i = \sum_{j=1}^{i} z_j,$$

$$\begin{aligned}\Big(n \sup_x (F_n(x) - G_n(x)) &= n \max_i (F_n(W_i) - G_n(W_i)) \\ &= \max_i (u_i - v_i) \\ &= \max_i (i - 2v_i), \ u_i + v_i = i\Big) .\end{aligned}$$

However, for the two rank orders just mentioned, the Smirnov statistic has the same value, that is, 1. Thus, the Smirnov procedure could lead to the use of inadmissible tests of H_0 against H_M, the alternative that possesses the monotone likelihood ratio property.

Theorem 11.5.3 (Savage, 1956). *Let $T(z) = \sum_{i=1}^{N} v_i/i$ and let $\delta = \Delta_2/\Delta_i > 1$ under H_L. Then under H_L, if $T(z) < T(z')$, there exists a δ, say δ^*, such that $\delta^* > 1$ and for δ in the interval $(1, \delta^*)$ the probability of z is greater than the probability of z'. In fact, the δ^* may be chosen independently of z and z'.*

PROOF:

$$P(Z = z) = m!n!\delta^n / \prod_{i=1}^{N}(i - v_i + v_i\delta) = m!n!\delta^n/N! \prod_{1}^{N} \left(1 + (\delta - 1)\frac{v_i}{i}\right) .$$

Expanding $\prod_1^N \left(1 + (\delta - 1)\frac{v_i}{i}\right)^{-1}$ in powers of $(\delta - 1)$, we obtain

$$P(Z = z) = \frac{m!n!\delta^n}{N!} \left[1 - (\delta - 1)T(z) + O(\delta - 1)^2\right] .$$

Hence

$$P(Z = z) - P(Z = z') = \frac{m!n!\delta^n}{N!}(\delta - 1) \left[T(z') - T(z) + O\left((\delta - 1)\right)\right] .$$

Thus, for any z and z' such that $T(z) < T(z')$ there exists a $\delta^* > 1$ such that $P(Z = z) > P(Z = z')$ for $1 < \delta < \delta^*$; and since the number of rank orders if finite, δ^* can be chosen independently of z and z'.

Example 11.5.4. $m = 2, n = 2$

#	rank order	$T(z)$
1	0011	.8333
2	0101	1.3333
3	0110	1.6667
4	1001	2.3333
5	1010	2.6667
6	1100	3.1667

$1 \to 2 \to 3 \to 4 \to 5 \to 6$.

When the diagram corresponding to a particular combination of sample sizes is in the form of a simple chain, it is possible to construct a uniformly most powerful rank order test for every level of significance. When $m = 1$, or 2, and $n = 2, 3, 4$ or 5 UMP rank order tests of H_0 against H_L can be formed for every level of significance. For $m = n = 3$, there is not a simple ordering of the probabilities of the rank orders. Hence, it is not possible to construct UMP rank order tests for levels of significance in the intervals (.45, .55) and (.75, .85)

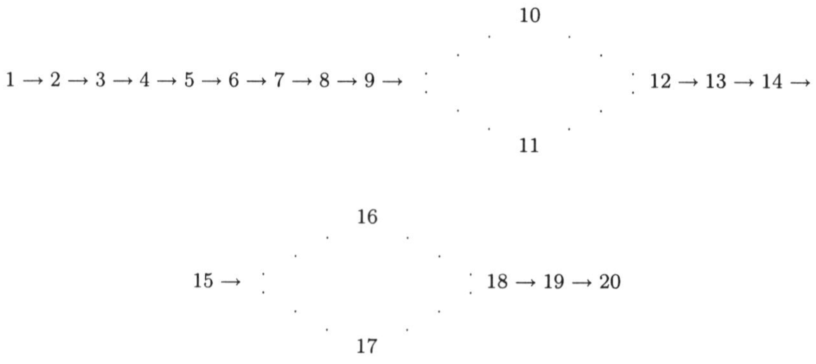

where the rank orders $1, 2, \cdots, 20$ are given below.

$m = n = 3$			
i	R.O	i	R.O
1	000111	11	011100
2	001011	12	100101
3	001101	13	100110
4	001110	14	101001
5	010011	15	101010
6	010101	16	101100
7	010110	17	110001
8	011001	18	110010
9	011010	19	110100
10	100011	20	111000

Recurrence relations for computing the probabilities of the rank orders under H_L

Let z be a rank order for sample sizes m and n, and let $z^o(z^1)$ be a rank order for sample sizes $m+1$ and n (m and $n+1$) such that the first $m+n$ elements of $z^o(z^1)$ are the same as the elements of z and the $(m+n+1)^{\text{th}}$ element of $z^o(z^1)$ is a 0(1). Then, if $h_z(\delta) = \prod_{i=1}^{N}(u_i + v_i\delta)$, we have

$$h_{z^o}(\delta) = [(m+1) + n\delta]\, h_z(\delta)$$

$$h_{z^1}(\delta) = [(m + (n+1)\delta]\, h_z(\delta)\,.$$

When two rank orders z and z' are identical except in their i^{th} and $(i+1)^{\text{th}}$ elements which are (0,1) for z and (1,0) for z', then we have the following relationship between their probabilities.

$$P(Z = z) = \frac{(u_i + \delta v_i + \delta - 1)}{(u_i + \delta v_i)} P(Z = z')$$

where u_i and v_i are computed for z. Also, the probability of I < II, all of the first sample less than the second, is

$$P(\text{I} < \text{II}) = n!\delta^n / \prod_{i=1}^{n}(m + i\delta) \text{ and } P(\text{I} > \text{II}) = m! \prod_{i=1}^{m}(i + n\delta)\,.$$

Rao, Savage and Sobel (1960) develop rank order theory for the two-sample problem in which censoring of the observations has occurred, that is, not all of the random variables are observed. For example, in life testing, the experiment is stopped before all of the lives of the experimental units are observed. In this case, the rank orders are not all equally likely when the null hypothesis is true and hence, it becomes important to work with the likelihood ratios of rank orders. The authors consider several censoring schemes;

1. continue experimentation until the N^* (non random) smallest random variables are observed $0 < N^* \leq N = m + n$.

2. continue experimentation until m^* random variables from $F(x)$ have been observed. If n^* is the number of random variables observed from $G(x)$, then n^* and $N^* = m^* + n^*$ are random.

3. continue experimentation until either the number of random variables from $F(x)$ is m^* or the number from $G(x)$ is n^*, where m^* and n^* are fixed integers.

4. continue experimentation until $(m^* - n^*)^2 \geq b_{m^*,n^*}$ where the b_{m^*,n^*} are preassigned numbers with $b_{m,n} = 0$.

5. continue experimentation until

$$\max_{-\infty < x < x_N^*} \left[F_{m^*}(x) - G_{n^*}(x)\right]^2 \geq a_{m^*,n^*}$$

where a_{m^*,n^*} are some preassigned numbers with $a_{m,n} = 0$.

11.6 One-sample Rank Orders

Savage (1959) has obtained certain results pertaining to the partial ordering of the one-sample rank order probabilities. If $X_1, \ldots, X_N$ is a random sample from $f(x;\theta)$ and $L_1, \ldots, L_m$ denotes the absolute values of the negative X's and $Y_1, \ldots, Y_n$ $(N = m + n)$ denotes the positive X's, define the rank order $Z = (Z_1, \ldots, Z_N)$ where $Z_i = 1$ if the ith smallest of the ordered values of the L's and Y's is a Y and zero otherwise. There are 2^N possible values of Z. For a fixed m, there are $\binom{N}{m}$ possible values of Z. For fixed m, the conditional distribution of Z is that of the two-sample problem where the first population has cdf:

$$P\left[|X| \leq x | X < 0\right] = F^-(x;\theta) = \begin{cases} \frac{F(0;\theta) - F(-x;\theta)}{F(0;\theta)}, & x \geq 0, \\ 0, & x < 0, \end{cases} \tag{11.6.1}$$

and the cdf underlying the Y's is

$$P\left[|X| \leq x | X > 0\right] = F^{+}(x;\theta) = \begin{cases} \frac{F(x;\theta)-F(0;\theta)}{1-F(0;\theta)}, & x \geq 0, \\ 0, & x < 0. \end{cases} \tag{11.6.2}$$

Thus for fixed m, the partial ordering problem is exactly the one treated by Savage (1956). However the previous results are not immediately applicable, since it is not clear how to impose conditions on $F(x;\theta)$ in order to get $F^{-}(x;\theta)$ and $F^{+}(x;\theta)$ to satisfy the conditions of Savage (1956).

Let $z'Lz$ denote the following relationship: $z'_k = z_k$ for all $k = 1, \ldots, N$ except i and j $(i < j)$ and $z_i = z'_j = 0$, $z_j = z'_i = 1$ Also

$$P(Z = z) = N! \int_{0<y_1<\cdots<y_N<\infty} \cdots \int \prod_{i=1}^{N} \left[f\left((2z_i - 1)y_i;\theta\right)\right] \prod_{i=1}^{N} dy_i\,. \tag{11.6.3}$$

The hypothesis of interest is $H_0 : F(-x;0) + F(x;0) = 1$, that is, symmetry about 0. $P(Z = z|H_0) = 2^{-N}$ for each z. An alternative of interest is $F(x;\theta) = \Phi(x - \theta)$ where Φ denotes the standard normal d.f. All of the following results apply to this alternative.

The case of fixed m

Theorem 11.6.1. *If*

1. $f(x;\theta) = u(x)v(\theta)\exp\{a(x)b(\theta)\}$,
2. $v(\theta) \geq 0$,
3. $u(x) = u(-x) > 0$,
4. $x < y$ *implies that* $a(x) < a(y)$,
5. $b(\theta) > 0$,

then $z'Lz$ *implies* $\Delta = P(Z = z) - P(Z = z') \geq 0$.

PROOF: Using (11.6.3) we have

$$\Delta = N! \int_{0<y_1<\cdots<y_N<\infty} \cdots \int A(y_i, y_j) \prod_{i=1}^{N} \left[f((2z_i - 1)y_i;\theta)\right] dy_i$$

where

$$\begin{aligned} A(y_i, y_j) &= 1 - \{f(y_i;\theta)f(-y_j;\theta)/f(-y_i;\theta)f(y_j;\theta)\} \\ &= 1 - \exp\{b(\theta)\,[a(y_i) - a(-y_i) + a(-y_j) - a(y_j)]\}\,. \end{aligned}$$

The theorem is proved by showing that $A(y_i, y_j) \geq 0$ which follows since the exponent is negative due to the monotonicity of $a(x)$.

Theorem 11.6.2. *If*

1. $f(x;\theta) = g(x-\theta) = g(\theta - x)$,

2. $x > y$ *and* $\theta > \delta$ *implies that*

$$f(x;\theta)f(y;\delta) - f(x;\delta)f(y;\theta) > 0$$

and

3. $\theta > 0$,

then $z'Lz$ *implies that* $\Delta = P(Z = z) - P(Z = z') > 0$.

PROOF:

$$\Delta = N! \underset{0<y_1<\cdots<y_N<\infty}{\int\cdots\int} B(y_i, y_j) \prod_{\substack{k=1 \\ i\neq k\neq j}}^{N} [f((2z_i - 1)y_i;\theta)] \prod_{k=1}^{N} dy_k$$

where $B(y_i, y_j) = f(y_j;\theta)f(-y_i;\theta) - f(-y_i;\theta)f(y_i;\theta)$ and the proof is complete if we can show that $B(y_i, y_j) > 0$. In (2), let $x = y_j$, $y = y_i$ and $\delta = -\theta$ so that

$$0 < f(y_j;\theta)f(y_i;-\theta) - f(y_i;\theta)f(y_j;-\theta)\,.$$

Now, using $g(x-\theta) = g(\theta - x)$ [assumption (1)], we have

$$\begin{aligned} 0 &< g(y_j - \theta)g(y_i + \theta) - g(y_j + \theta)g(y_i - \theta) \\ &= f(y_j;\theta)f(-y_i;\theta) - f(-y_j;\theta)f(y_i;\theta) \\ &= B(y_i, y_j)\,. \end{aligned}$$

The case of variable m

Let $z'Sz$ denote $z_k \geq z'_k$ for $k = 1, \ldots, N$ and $>$ holds for at least one value of k.

Theorem 11.6.3. *Under the assumptions of Theorem 11.6.1 if $z'Sz$, then $\Delta = P(Z = z) - P(Z = z') > 0$.*

PROOF: It is sufficient to consider only the special case $z'_k = z_k$ for all $k = 1, \ldots, N$ except $k = i$ where $z_i = 1$ and $z'_i = 0$. Then

$$\Delta = N! \int_{0<y_1<\cdots<y_N<\infty} \cdots \int c(y_i) \left\{ \prod_{k=1}^{N} f\left((2z_k - 1)y_k; \theta\right) dy_k \right\}$$

and the proof is completed by showing that $c(y_i) = 1 - f(-y_i, \theta)/f(y_i, \theta)$ is positive. Using the special form of $f(x;\theta)$, $c(y_i) = 1 - \exp\{b(\theta)\,[a(-y_i) - a(y_i)]\}$ and again the exponent is negative because of the monotonicity of $a(y)$.

Theorem 11.6.4. *1. $f(x;\theta) = g_\theta(x - \theta) = g_\theta(\theta - x)$,*

2. $x > y > 0$ implies that $g_\theta(y) > g_\theta(x)$, and

3. $\theta > 0$,
then $z'Sz$ implies that $\Delta = P(Z = z) - P(Z = z') > 0$.

PROOF: Consider

$$\Delta = N! \int_{0<y_1<\cdots<y_N<\infty} \cdots \int \int D(y_i) \prod_{\substack{k=1 \\ k \neq i}}^{N} f((2z_k - 1)y_k; \theta) \prod_{1}^{N} dy_k$$

and it is sufficient to show that $D(y_i) = f(y_i;\theta) - f(-y_i;\theta) > 0$. First, using assumption (1), $D(y_i) = g_\theta(y_i - \theta) - g_\theta(-y_i - \theta) = g_\theta(y_i - \theta) - g_\theta(y_i + \theta)$. Now, if $y_i > \theta$, the result follows from (2), since $y_i - \theta < y_i + \theta$. If $y_i < \theta$, the result follows from (2) when we write $D(y_i) = g_\theta(\theta - y_i) - g_\theta(y_i + \theta)$.

Remark 11.6.1. *In Theorem 11.6.4, writing $f(x;\theta) = g_\theta(x - \theta)$ allows $f(x;\theta)$ to incorporate not only translations of the H_0, but also other changes, such as changes in scale.*

Remark 11.6.2. *The assumptions of Theorem 11.6.2 imply those of Theorem 11.6.4, but not conversely. In (2) of Theorem 11.6.2, we set $\delta = 0$, $2\theta = x + y$, and $0 < y < x$, we obtain (2) of Theorem 11.6.4. The Cauchy density is a counter example of the converse.*

Some partial orderings. If the assumptions of Theorems 11.6.1 and/or 11.6.2 are satisfied, then the following orderings are obtained, where

$$P(Z = z) > P(Z = z') \Leftrightarrow z \to z' .$$

$N = 1:$ $1 \to 0$

$N = 2:$ $11 \to 01 \to 10 \to 00$

$$N = 3: \quad \begin{array}{ccccccccc} 111 & \to & 011 & \to & 101 & \to & 110 & & \\ & & & & \downarrow & & \downarrow & & \\ & & & & 001 & \to & 010 & \to & 100 \end{array}$$

$N = 4:$ See Savage (1959, p. 1021).

Next, consider the uniform distribution $f(x;\theta) = 1$ for $\theta - \frac{1}{2} \leq x \leq \theta + \frac{1}{2}$ and 0 otherwise ($0 \leq \theta \leq \frac{1}{2}$). If n' is the length of the last run of 1's in z or n' =the number of the positive observations greater than the maximum of the absolute values of the negative observations, then

$$P(Z = z) = \sum_{i=0}^{n'} \binom{N}{i} \left(\frac{1}{2} - \theta\right)^{N-i} (2\theta)^i .$$

To obtain the above expression, write

$$\begin{aligned} P(Z = z) &= \sum_{i=0}^{n'} P\left(Z = z | i \text{ observations are } > \frac{1}{2} - \theta\right) \\ &\quad \cdot P\left(i \text{ observations } > \frac{1}{2} - \theta\right) \end{aligned}$$

and use

$$P\left(Z = z | i \text{ observations } > \frac{1}{2} - \theta\right) = 2^{-(N-i)},$$

$$P\left(i \text{ observations } > \frac{1}{2} - \theta\right) = \binom{N}{i} (2\theta)^i (1 - 2\theta)^{N-i} .$$

Holding θ fixed, $P(Z = z)$ is an increasing function of n', and otherwise does not depend on z. Thus the most powerful rank order test depends only on n'.

We also have the following theorem for normal samples.

Theorem 11.6.5. *If $X_1, \ldots, X_N$ ($N \geq 3$) are independently and normally distributed, each with mean $\theta(> 0)$ and variance 1, then $\Delta = P(Z = z) - P(Z = z') > 0$ where z and z' are identical except $z_1 = z_2 = z_3' = 0$ and $z_1' = z_2' = z_3 = 1$.*

PROOF: See Savage (1959, pp. 1022–1023).

11.7 c-sample Rank Orders

Govindarajulu and Haller (1972) have extended the theory of rank orders to the c-sample case. Let $X_{i,1}, \ldots, X_{i,n_i}$ be a random sample from continuous d.f. $F_i(x)$ $(i = 1, \ldots, c)$. let $N = n_1 + \cdots + n_c$ and let $W_{1N} \leq \cdots \leq W_{NN}$ denote the combined ordered sample.

Definition 11.7.1. *The random vector $Z = (Z_1, \ldots, Z_N)$ is said to be a c-sample rank order where $Z_i = j$ if $W_{iN} = X_{jk}$ for some $k = 1, 2, \ldots, n_j$ $(i = 1, \ldots, N)$.*

The above definition is preferred since it is more compact than the definition due to Andrews and Truax (1964) who define Z_i as a c-dimensional vector having a 1 in j^{th} coordinate, if and only if $W_{iN} = X_{j,k}$ for some $k = 1, \ldots, n_j$ and zeros elsewhere. Let $\mathcal{Z} = \{z^i : i = 1, \ldots, M\}$ be the set of possible values of the rank order Z. In general we denote any $z \in \mathcal{Z}$. by $(z_1, \ldots, z_N)$. Obviously $M = N!/\prod_{i=1}^{c} n_i!$ and $\mathcal{Z}$ is a function of $n_1, \ldots, n_c$.

Suppose we are interested in testing $H_0 : F_1(x) = \cdots = F_c(x)$ for all x against some alternative H_1 at some level α. Then the test will consist of a sequence of real numbers $a_1, \ldots, a_M$ and a rule such that rank order z^i is observed, the null hypothesis is rejected with probability a_i $(i = 1, \ldots, M)$. Clearly since the rank orders are all equally likely under H_0, we have $M\alpha = a_1 + \cdot + a_M$. The following lemma gives a necessary condition for a rank order test to be most powerful (MP), which is a generalization to the c-sample case of a result due to Savage (1956).

Lemma 11.7.1. *Let H_1 be any alternative hypothesis and z^i, z^j be rank orders such that $P(Z = z^i|H_1) \geq P(Z = z^j|H_1)$. If $\phi = (a_1, \ldots, a_M)$ is a M.P. level α rank order test of H_0 against H_1, then $a_i \geq a_j$.*

PROOF: Assume that the set of probabilities $\phi = \{a_1, \ldots, a_M\}$ defines a MP level α test of H_0 against H_1. Then the set

$$\phi' = \{a_1', \ldots, a_M'\}$$

with $a_k' = a_k$ for all $k \neq i, j$, $a_j' = a_i$ and $a_i' = a_j$ also is a level α rank order test. Therefore, we have

$$P_\phi(\text{reject } H_0|H_1) \geq P_{\phi'}(\text{reject } H_0|H_1)\,. \tag{11.7.1}$$

Inequality (11.7.1) can be rewritten as

$$\sum_{k=1}^{M} a_k P(Z = z^k | H_1) \geq \sum_{k=1}^{M} a'_k P(Z = z^k | H_1) \tag{11.7.2}$$

which simplifies to

$$(a_i - a_j)\left[P(Z = z^i | H_1) - P(Z = z^j | H_i)\right] \geq 0\,.$$

Since by hypothesis of the lemma, $P(Z = z^i|H_1) \geq P(Z = z^j|H_1)$, we have $a_i \geq a_j$ and the proof is complete.

When the rank order probabilities are totally ordered, the most powerful rank order test exists and the form of the test is given by the following theorem.

Theorem 11.7.1. *Suppose that the rank order probabilities are totally ordered. For testing H_0 against H_1 at level α, the set $\phi = \{a_1, a_2, \ldots, a_M\}$ with*

$$a_i = \begin{cases} 1, & \text{if there are } M - K \text{ or more rank orders less probable than } z^i, \\ 0, & \text{if there are } K + 1 \text{ or more rank orders more probable than } z^i, \\ a = & (M\alpha - m_2)/(M - m_1 - m_2), \text{ otherwise.} \end{cases} \tag{11.7.3}$$

defines a MP level α rank order test where K is an integer such that $0 \leq K < M$ and $K \leq M\alpha < K + 1$, and m_1 [m_2] is the number of a_i's $= 0[1]$.

PROOF: It readily follows from the Neyman-Pearson lemma.

One can easily construct examples, such that for the rank order probabilities with respect to the class of slippage alternatives, namely: $H_S : F_1(x) \geq F_2(x) \geq \cdots \geq F_c(x)$ are not totally ordered. However, by restricting to certain subclasses of the class of slippage alternatives, a partial ordering of the rank order probabilities can be obtained. Let H_M denote the class of slippage alternatives, having probability density functions which satisfy the monotone likelihood ratio (MLR) property given below.

Definition 11.7.2. *Let $\mathcal{X}$ and Θ be the subsets of the real line. A pdf. $f(x;\theta)$ defined on $\mathcal{X} \times \Theta$ is said to have the MLR property if $x_1 > x_2$ and $\theta_1 > \theta_2$ imply that*

$$f(x_1;\theta_1) f(x_2;\theta_2) \geq f(x_1;\theta_2) f(x_2;\theta_1)\,.$$

Then we give a sufficient condition for ordering certain rank order probabilities.

Theorem 11.7.2. *Let $X_{i,1}, \ldots, X_{i,n_i}$ be a random sample from a population having $h(x_i; \theta_i)$ for its pdf, which satisfies the MLR property $(i = 1, \ldots, c)$. If $\theta_1 < \theta_2 < \cdots < \theta_c$, then the rank order z is more probable than the rank order z' whenever z and z' are identical except for the pth and q^{th} entries where $1 \le p < q \le N$,*

$$z_p = z'_q = i < j = z_q = z'_p, \text{ and } N = n_1 + \cdots + n_c .$$

PROOF: Let $A = \{x\colon \text{ rank order of } x \text{ is } z\}$, $A' = \{x\colon \text{ rank order of } x \text{ is } z'\}$. Then

$$P(Z = z|H_M) - P(Z = z'|H_M) = \int_A \prod_{i=1}^{c} \prod_{j=1}^{n_i} h(x_{ij}; \theta_i) dx_{ij}$$

$$- \int_{A'} \prod_{i=1}^{c} \prod_{j=1}^{n_i} h(x_{ij}; \theta_i) dx_{ij}$$

$$= \prod_{i=1}^{c} (n_i!) \quad \int \cdots \int_{-\infty < x_1 < \cdots < x_N < \infty} \prod_{i \ne p,q}^{N} h(x_i; \theta_{z_i})$$

$$\cdot \quad \left[h(x_p; \theta_{z_p}) h(x_q; \theta_{z_q}) - h(x_p; \theta_{z_q}) h(x_q; \theta_{z_q})\right] \prod_1^N dx_i \,. \quad (11.7.4)$$

By hypothesis, since $x_p < x_q$ and $\theta_i < \theta_j$, we have that the integrand is nonnegative and actually positive on a set of positive measure (except for the case $h(x; \theta_i) = h(x; \theta_j)$ almost everywhere). This completes the proof.

Under the assumptions of Theorem 11.7.2, the rank orders

$$\begin{aligned} z' &= (\underbrace{1, \ldots, 1}_{n_1}, \underbrace{2, \ldots, 2}_{n_2}, \ldots, \underbrace{3, \ldots, 3}_{n_3}) \\ z^M &= (\underbrace{c, \ldots, c}_{n_c}, \underbrace{2, \ldots, 2}_{n_2}, \ldots, \underbrace{1, \ldots, 1}_{n_1} \end{aligned} \qquad (11.7.5)$$

are most probable and least probable, respectively. Next, let us consider the

Lehmann alternatives

$$H_{L,1} = F_i(x) = [H(x)]^{\Delta_i}\,,\ i = 1, \dots, c,$$

$$H_{L,2} = F_i(x) = 1 - [1 - H(x)]^{\Delta_i}\,,\ i = 1, \dots, c,$$

where $H(x)$ is some unknown continuous d.f. Since $H_{L,2}$ can be gotten from H_{L1} by changing x to $-x$ and assuming that $H(x)$ is symmetric about zero, it suffices to consider $H_{L,1}$ which for brevity be denoted by H_L.

Theorem 11.7.3. *Let $z = (z_1, \dots, z_N) \in \mathcal{Z}$ and δ_{k,z_j} be the Kronecker's delta. If $F_i \in H_L$ $(i = 1, \dots, c)$, then*

$$P(Z = z|H_L) = \prod_{i=1}^{c} (n_i!\Delta_i^{n_i}) / \prod_{i=1}^{N} \left[\sum_{j=1}^{i} \sum_{k=1}^{c} \Delta_k \delta_{k,\, z_j} \right] . \tag{11.7.6}$$

PROOF: In Corollary 11.2.4.1 set $\theta_j = \sum_{k=1}^{c} \Delta_k \delta_{k,z_j}$, and the result readily follows. The factor $\prod_{i=1}^{c} n_i!$ accounts for the permutations within the subsamples which lead to the same rank order.

Let $\epsilon_k = (\Delta_k/\Delta_1) - 1$, for $k = 2, \dots, c$, and $v_{k,i}(z) = \sum_{j=1}^{i} \delta_{k,z_j}$. Then (11.7.6) takes the form of

$$P(Z = z|H_L) = M^{-1} \prod_{k=2}^{c} (1 + \epsilon_k)^{n_k} / \prod_{i=1}^{N} \left[1 + \left(\sum_{k=2}^{c} \epsilon_k v_{k,i}(z) \right) / i \right] \tag{11.7.7}$$

since

$$\sum_{k=1}^{c} v_{k,i}(z) = \sum_{k=1}^{c} \sum_{j=1}^{i} \delta_{k,z_j} = i\,.$$

The following theorem will enable one to obtain a partial ordering of the rank order probabilities under H_L.

Theorem 11.7.4. *Let z and $z' \in \mathcal{Z}$ be identical except for the pth and q^{th} entries, $1 < p < q < N$, where $z_p = z'_q = i < j = z_q = z'_p$. If $F_k(x) \in H_L$ $(k = 1, \dots, c)$ with $H(x)$ admitting a density, and $\Delta_i < \Delta_j$, then*

$$P(Z = z|H_L) > P(Z = z'|H_L)\,.$$

PROOF: When $F_k \in H_L$ and $H(x)$ has a density then the d.f.s F_k have the MLR property. Hence, the result follows from Theorem 11.7.2. By repeated application of Theorem 11.7.4 yields that rank orders z^1 and z^M defined by (11.7.5) have maximum and minimum probabilities under H_L provided $\Delta_1 < \Delta_2 < \cdots < \Delta_c$.

Let

$$T_k(z) = \sum_{i=1}^{N} v_{k,i}(z)/i\,. \tag{11.7.8}$$

Theorem 11.7.5. *For small* $\boldsymbol{\epsilon} = (\epsilon_2, \ldots, \epsilon_c)$,

$$P(Z = z|H_L) > P(Z = z'|H_L)$$

if

$$\sum_{k=2}^{c} \epsilon_k T_k(z') > \sum_{k=2}^{c} \epsilon_k T_k(z)\,. \tag{11.7.9}$$

PROOF: Let

$$\begin{aligned} H(\boldsymbol{\epsilon}) &= -\sum_{i=1}^{N} \log\left\{1 + \sum_{k=2}^{c} \epsilon_k v_{k,i}/i\right\} \\ &= -\sum_{i=1}^{N}\sum_{k=2}^{c} \epsilon_k v_{k,i}/i + o(\boldsymbol{\epsilon})\,, \end{aligned}$$

after expanding $H(\boldsymbol{\epsilon})$ in Taylor Series around $\boldsymbol{\epsilon} = 0$. Noting that

$$H(\boldsymbol{\epsilon}) = \log\left\{M \prod_{k=2}^{c} (\epsilon_k + 1)^{-n_k} P(Z = z|H_L)\right\},$$

one can easily infer that for all $\boldsymbol{\epsilon}$, z is more probable than z' if

$$\sum_{k=2}^{c} \epsilon_k T_k(z') > \sum_{k=2}^{c} \epsilon_k T_k(z)\,.$$

Next, let $F_k \in H_L^*$ if $F_k \in H_L$ and has parameter $\Delta_k = \Delta_1 + (k-1)\Delta$ with Δ_1, $\Delta > 0$. Clearly whenever $F_k \in H_L^*$ $(k = 1, \ldots, c)$ we have $F_1(x) \geq F_2(x) \geq \cdots \geq F_c(x)$. Then we have the following corollary.

Corollary 11.7.5.1. *If $F_k \in H_L^*$ $(k = 1, 2, \ldots, c)$, then for fixed n_k $(k = 1, \ldots, c)$ and Δ sufficiently small,*

$$P(Z = z|H_L^*) > P(Z = z'|H_L^*)$$

provided

$$\sum_{k=2}^{c}(k-1)T_k(z') > \sum_{k=2}^{c}(k-1)T_k(z),$$

or

$$\sum_{k=1}^{c} k\, T_k(z') > \sum_{k=1}^{c} k\, T_k(z), \tag{11.7.10}$$

since

$$\sum_{k=1}^{c} T_k(z) = \sum_{k=1}^{c}\sum_{i=1}^{N} v_{k,i}/i = \sum_{i=1}^{N}\frac{1}{i}\sum_{k=1}^{c} v_{k,i} = \sum_{i=1}^{N}\frac{1}{i}\cdot i = N\,.$$

It should be noted that $\sum_{k=1}^{c} T_k(z) = \sum_{k=1}^{c} T_k(z')$ does not necessarily imply that $z = z'$.

Example 11.7.1. Let four observations be drawn from each-of eight populations. Consider the rank orders $z = (1, 2, 2, 2, 1, 1, z_7, \ldots, z_{29}, 1, 2, z_{32})$ and $z' = (2, 1, 1, 1, 2, 2, z_7, \ldots, z_{29}, 2, 1, z_{32})$. Then since $T_2(z) = T_2(z')$, it follows that

$$\sum_{k=1}^{c} k\, T_k(z) = \sum_{k=1}^{c} k\, T_k(z')\,.$$

11.8 Locally Most Powerful (LMP) Rank Tests

In the previous sections we have defined a rank order and were able to derive explicit expressions for the probabilities of rank orders in special cases, especially for Lehmann alternatives. To order the rank order probabilities one ought to know the exponents occurring in the Lehmann alternatives. In this section we shall derive test criteria that are functions of the rank orders and which would enable us to order the rank order probabilities for local alternatives. Such tests are called locally most powerful rank tests. If a set of densities $\{q_\Delta\}$, $\Delta > 0$ is indexed, by a parameter Δ and if $q_0 \in H_0$, then a test is called locally most powerful (LMP) for H_0 against $\Delta > 0$ at some level α, if it is uniformly most powerful (UMP) at level α for H_0 against the alternative: $H_\delta : \{q_\Delta : 0 < \Delta < \delta\}$ for some $\delta > 0$. We can justify the LMP rank tests on two rounds:

1. if a test is sensitive to local alternatives we expect it to perform as well for global alternatives, and

2. it provides a method for constructing nonparametric tests whose performance with respect to other alternatives can be studied via asymptotic relative efficiencies.

Let $X_1, \ldots, X_N$ be N independent observations having joint density $f(\boldsymbol{x})$ under H_0 and $g(\boldsymbol{x})$ under the alternative H_1. Then from (11.3.6) we have

$$P(R = r|H_1) = \frac{1}{N!} E_{H_0} \{g(\boldsymbol{X}_r)/f(\boldsymbol{X}_r)\} , \tag{11.8.1}$$

where $\boldsymbol{X}_r = (X_{r_1,N}, \ldots, X_{r_N,N})$. If g is indexed by a parameter Δ and f corresponds to g when $\Delta = 0$, then one can expand $P(R = r|H_1)$ in powers of Δ and take the first nonvanishing coefficient of a power of Δ as the test criterion. For example, let us assume that

$$N!P(R = r|H_1) = 1 + \Delta T(r) + o(\Delta) \tag{11.8.2}$$

where $T(r) \neq 0$. Then we use $T(r)$ as the criterion that orders the rank order probabilities for some small values of Δ. Speaking geometrically, $T(r)$ maximizes the slope of the power function of the rank order test. Hoeffding (1951) and Terry (1952) were the first to propose the normal scores test as an LMP test criterion for detecting small shift in the location parameter with normal alternatives. An equivalent to normal scores test was proposed earlier by Fisher and Yates (1949). In Chapter 12 we shall derive LMP rank tests for shifts in location and changes in scale parameters in the two-sample case.

11.9 Problems

11.2.1 Evaluate the limit of the correlation coefficient between an observation and its rank for the following cases.

(i) when X has the double exponential distribution

(ii) when X has the Weibull distribution with shape parameter m.

11.4.1 For $N = 4$, using Theorem 11.4.1 evaluate $P(X_2 < X_4 < X_3 < X_1|H_L)$ when $\theta_i = i\theta (i = 1, \ldots, 4)$.

11.4.2 Suppose we have a sample of size N from a normal $(\theta, 1)$ population. First we wish to test whether the sample is random using T_2. Then we carry out a test based on $\overline{X} = (X_1 + \cdots + X_N)/N$ for $H_0 : \theta = 0$ versus $H_1 : \theta > 0$. Using Theorem 11.4.5 evaluate the conditional probability of rejecting H_0 when the hypothesis of randomness of the sample is accepted.

11.5.1 Suppose $m = n = 2$ and $z = (0011)$ and $z' = (0101)$. Evaluate $p(z')$ from $p(z)$ using the recursive relation.

11.6.1 For normal $(\theta, 1)$ sample of size $N = 4$, order (partially) the rank orders in terms of their probabilities.

Chapter 12

LMP Tests: Two-sample Case

12.1 Introduction

Let $X_1, \ldots, X_m$ be a random sample from $F(x)$ and $Y_1, \ldots, Y_n$ be a random sample from $G(y)$ where F and G are assumed to be continuous. We wish to test $H_0 : F(x) = G(x)$ for all x against the alternative $H_1 : F(x) \geq G(x)$ with strict inequality for some x. Let $N = m + n$ and $W_{1N} \leq \cdots \leq W_{NN}$ be the combined ordered X's and Y's.

Define the rank order statistic $\boldsymbol{Z} = (Z_1, \ldots, Z_N)$ where $Z_{N,i} = 1$ if W_{iN} is a Y and $Z_{N,i} = 0$ if W_{iN} is an X $(i = 1, \ldots, N)$. Then

$$P(\boldsymbol{Z} = \boldsymbol{z}|H_0) = 1/\binom{N}{m} \tag{12.1.1}$$

and

$$P(\boldsymbol{Z} = \boldsymbol{z}|H_1) = m!n! \int \cdots \int_{-\infty < w_1 < \cdots < w_N < \infty} \prod_{i=1}^{N} [f(w_i)]^{1-z_i} [g(w_i)]^{z_i} \, dw_i \, . \tag{12.1.2}$$

12.2 Location Parameter Case

Let us consider the location shift case. That is,

$$G(x) = F(x - \theta) \, .$$

Then H_0 and H_1 become

$$H_0 : \theta = 0 \text{ and } H_1 : \theta > 0\,.$$

Then we can express the probability of rank order as

$$P(\boldsymbol{Z} : \boldsymbol{z}|H_1) = m!n! \int \cdots \int_{-\infty<w_1<\cdots<w_N<\infty} \prod_{i=1}^{N} f(w_i - \theta z_i) dw_i\,. \tag{12.2.1}$$

The locally most powerful test of H_0 is to reject H_0 when T_N is large where

$$\begin{aligned} T_N &= \lim_{\theta\to 0} \left[\left\{P(\boldsymbol{Z} = \boldsymbol{z}|H_1) - P(\boldsymbol{Z} = \boldsymbol{z}|H_0)\right\}/\theta\right] \\ &= \lim_{\theta\to 0} m!n! \int \cdots \int_{-\infty<w_1<\cdots<w_N<\infty} \\ &\qquad \theta^{-1}\left\{\prod_{i=1}^{N} f(w_i - \theta z_i) - \prod_{i=1}^{N} f(w_i)\right\} \prod_{1}^{N} dw_l\,. \end{aligned} \tag{12.2.2}$$

Towards the evaluation of T_N, the following lemma can be used.

Lemma 12.2.1. *For any constants a_i and b_i $(i = 1, \ldots, N)$,*

$$\prod_{1}^{N} a_i - \prod_{1}^{N} b_i = \sum_{i=1}^{N} a_1 \cdots a_{i-1}(a_i - b_i)b_{i+1} \cdots b_N\,.$$

PROOF: Write

$$\begin{aligned} a_1 \cdots a_N - b_1 \cdots b_N &= a_1 \cdots a_{N-1}a_N - a_1 \cdots a_{N-1}b_N + a_1 \cdots a_{N-1}b_N - \\ &\quad a_1 \cdots a_{N-2}b_{N-1}b_N + \cdots + a_1b_2 \cdots b_N - b_1 \cdots b_N \\ &= a_1 \cdots a_{N-1}(a_N - b_N) + a_1 \cdots a_{N-2} \\ &\quad \cdot(a_{N-1} - b_{N-1})b_N + \cdots + (a_1 - b_1)b_2 \cdots b_N\,. \end{aligned}$$

So, using this lemma, we obtain

$$\begin{aligned} T(z) &= \sum_{i=1}^{N} \lim_{\theta\to 0} m!n! \int \cdots \int_{-\infty<w_1<w_2<w_N<\infty} \\ &\prod_{j=1}^{i-1} f(w_j - \theta z_j) \frac{[f(w_i - \theta z_i) - f(w_i)]}{\theta} \prod_{k=i+1}^{N} f(w_n) \prod_{1}^{N} dw_l\,. \end{aligned}$$

Suppose it is permissible to take the limit on θ underneath the multiple integral sign. Then we obtain

$$\begin{aligned}
T(z) &= \sum_{i=1}^{N} m!n! \int_{-\infty<w_1<\cdots<w_N<\infty} \cdots \int \left\{-z_i \frac{f'(w_i)}{f(w_i)}\right\} \prod_{i=1}^{N} f(w_i) dw_i \\
&= \sum_{i=1}^{N} E\left[-\frac{f'(W_{i,N})}{f(W_{i,N})}\right] z_i \\
&= \sum_{j=1}^{n} E\left[-\frac{f'(W_{S_j,N})}{f(W_{S_j,N})}\right], \qquad (12.2.3)
\end{aligned}$$

where $S_1, \ldots, S_n$ are the ranks associated with the Y-observations in the combined ordered sample.

In the following we show the interchange of limit on θ and the multiple integration is permissible provided $\int_{-\infty}^{\infty} |f'(x)|\, dx < \infty$. Towards this, consider

$$\begin{aligned}
&\int_{-\infty<w_1<\cdots<w_N<\infty} \cdots \int \prod_{j=1}^{i-1} f(w_j - \theta z_j) \left|\frac{f(w_i - \theta z_i) - f(w_i)}{\theta}\right| \prod_{k=i+1}^{N} f(w_i) \prod_{1}^{N} dw_l \\
&= \int_{-\infty}^{\infty} \cdots \int_{-\infty}^{\infty} \prod_{j=1}^{i-1} f(w_j - \theta z_j) \left|\frac{f(w_i - \theta z_i) - f(w_i)}{\theta}\right| \prod_{k=i+1}^{N} f(w_i) \prod_{1}^{N} dw_l \\
&= \int_{-\infty}^{\infty} \left|\frac{f(w_i - \theta z_i) - f(w_i)}{\theta}\right| dw_i \\
&= \int_{-\infty}^{\infty} \left|z_i f'(w_i - \theta^* z_i)\right| dw_i \le \int_{-\infty}^{\infty} |f'(y)|\, dy \,,
\end{aligned}$$

where $0 \le \theta^* \le \theta$.

Special Case 1

Let $f(x) = \phi(x)$. Then since $f'(x) = -xf(x)$, we obtain

$$T(z) = \sum_{i=1}^{N} z_i E(W_{iN}) \qquad (12.2.4)$$

where W_{iN} are the standard normal order statistics in a random sample of size N.

Special Case 2

Let $F(x) = (1 + e^{-x})^{-1}$. Then

$$f(x) = F(x)\,[1 - F(x)]$$

and

$$-f'(x)/f(x) = 2F(x) - 1\,.$$

Hence,

$$\begin{aligned} T(z) &= \sum_{i=1}^{N} z_i E\left\{2F(W_{iN}) - 1\right\} \\ &= \frac{2}{N+1}\sum_{1}^{N} i\, z_i - n \end{aligned} \tag{12.2.5}$$

which is equivalent to the Wilcoxon's rank sum test.

12.3 LMP Rank Tests for Scale Changes

Let X have d.f. $F(x)$ and Y have d.f. $F((1-\theta)x)$ where $\theta > 0$ and F is continuously twice differentiable. That is, Y has more spread than X. Let $X_1, \ldots, X_m$ be a random sample on X and $Y_1, \ldots, Y_n$ be a random sample on Y. Let $W_{1N} \le \cdots \le W_{NN}$ be the combined ordered observations with $N = m + n$ and the vector $Z = (Z_1, \ldots, Z_N)$ be as defined earlier. Then we wish to test

$$H_0 : \theta = 0 \text{ versus } H_\theta : \theta > 0\,.$$

Also, we can write

$$P(\boldsymbol{Z} = \boldsymbol{z}|H_\theta) = m!n! \int_{-\infty<w_1<\cdots<w_N<\infty} \cdots \int \prod_{i=1}^{N} (1-\theta z_i) f\left((1-\theta z_i)w_i\right) dw_i\,. \tag{12.3.1}$$

Consider

$$\begin{aligned} T(z) &= \lim_{\theta\to 0} \frac{[P(\boldsymbol{Z} = \boldsymbol{z}|H_\theta) - P(\boldsymbol{Z} = \boldsymbol{z}|H_0)]}{\theta} \\ &= \lim_{\theta\to 0} m!n! \int_{w_1<\cdots<w_N} \cdots \int \frac{1}{\theta}\left\{\prod_{i=1}^{N}(1-\theta z_i) f\left((1-\theta z_i)w_i\right) - \prod_{i=1}^{N} f(w_i)\right\} \prod_{j=1}^{N} dw_j \end{aligned}$$

$$= \lim_{\theta \to 0} m!n! \sum_{i=1}^{N} \int \cdots \int_{w_1 < \cdots < w_N} \prod_{j=1}^{i-1} (1 - \theta z_j) f((1 - \theta z_j) w_j)$$

$$\cdot \left\{ \frac{(1 - \theta z_i) f((1 - \theta z_i) w_i) - f(w_i)}{\theta} \right\} \cdot$$

$$\prod_{k=i+1}^{n} f(w_k) \cdot \prod_{l=1}^{N} dw_l \, .$$

Assume that the interchange of limit on θ and multiple integration is permissible. Then

$$T(\boldsymbol{z}) = m!n! \sum_{i=1}^{N} \int \cdots \int_{w_1 < \cdots < w_N} \lim_{\theta \to 0} \left\{ \frac{(1 - \theta z_i) f((1 - \theta z_i) w_i) - f(w_i)}{\theta f(w_i)} \right\} \cdot \prod_{1}^{N} f(w_i) dw_i \, . \tag{12.3.2}$$

Now writing

$$\begin{aligned}(1 - \theta z_i) f((1 - \theta z_i) w_i) - f(w_i) &= (1 - \theta z_i) f((1 - \theta z_i) w_i) - \\ &\quad (1 - \theta z_i) f(w_i) - \theta z_i f(w_i)\end{aligned}$$

and using the mean value theorem for the difference of the first two terms, we obtain

$$T(\boldsymbol{z}) = \sum_{i=1}^{N} -E_0 \left\{ 1 + \frac{f'(W_{iN})}{f(W_{iN})} W_{iN} \right\} z_i \tag{12.3.3}$$

and the rule is that we reject H_0 for large values of $T(\boldsymbol{z})$. Proceeding as in the location case, one can easily show that the sufficient condition for interchange of limit and integration is

$$\int_{-\infty}^{\infty} \left| x f'(x) \right| dx < \infty \, .$$

Special Case

Let $f(x) = \phi(x)$. Then

$$T(\boldsymbol{z}) = \sum_{i=1}^{N} \left\{ -1 + E_0(W_{iN}^2) \right\} z_i \, .$$

That is

$$T(z) + n = \sum_{i=1}^{N} E_0(W_{iN}^2) z_i \tag{12.3.4}$$

which is known as Capon's (1961) test.

Example 12.3.1. Let $g(x) = (1-\theta)e^{-x(1-\theta)}$, $x > 0$, $\theta > 0$, and

$$f(x) = e^{-x}, \ x > 0 .$$

Then

$$\frac{f'(x)}{f(x)} = -1 .$$

So

$$T(z) = \sum_{i=1}^{N} -E_0(1 + W_{iN}) z_i = -n + \sum_{i=1}^{N} E(W_{iN}) z_i$$

where $E(W_{iN})$ is the expected value of ith smallest order statistic in a random sample of size N drawn from $f(x)$. However,

$$E(W_{iN}) = \sum_{j=i}^{N} \frac{1}{j} .$$

Hence,

$$\begin{aligned}
\sum_{i=1}^{N} \sum_{j=i}^{N} \frac{1}{j} \cdot z_i &= \sum_{j=1}^{N} \frac{1}{j} \sum_{i=1}^{j} z_i \\
&= \sum_{j=1}^{N} \frac{1}{j} v_j, \quad v_j = \text{number of } Y\text{'s} \le W_{jN} \\
&= \text{Savage's test criterion} .
\end{aligned}$$

Alternatively, $n + T(z)$ is equivalent to

$$\begin{aligned}
\sum_{1}^{N} E(W_{iN}) z_i &= \sum_{1}^{N} \left(\sum_{j=i}^{N} \frac{1}{j} \right) z_i \\
&\doteq \sum_{1}^{N} \left(\log\left(\frac{N}{i}\right) \right) z_i \\
&= -\sum_{1}^{n} \log\left(\frac{s_i}{N}\right) .
\end{aligned}$$

12.4 Other Tests for Scale Alternatives

Suppose X is distributed normally with mean μ_1 and variance σ_1^2, Y is normal with mean μ_2 and variance σ_2^2 and X and Y are independent. If $(X_1, \ldots, X_m)\{(Y_1, \ldots, Y_n)\}$ is a random sample on $X\{Y\}$, then the parametric test of

$$H_0 : \sigma_1 = \sigma_2 \text{ versus } H_1 : \sigma_1 < \sigma_2$$

is based on the statistic

$$F_{n-1,m-1} = s_y^2/s_x^2 \tag{12.4.1}$$

and when H_0 is true, $F_{n-1,m-1}$ has Snedecor's F-distribution with $n-1$ and $m-1$ degrees of freedom, irrespective of the unknown values of μ_1 and μ_2. We reject H_0 for large values of $F_{n-1,m-1}$. Here the distributions of $Y - \mu_2$ and $X - \mu_1$ are related as

$$F_{Y-\mu_2}(y) = F_{X-\mu_1}(\theta y) \text{ for all } x, \text{ where } \theta = \sigma_1/\sigma_2\,. \tag{12.4.2}$$

The nonparametric analogue of (12.4.2) would be

$$F_{Y-M_2}(y) = F_{X-M_1}(\theta y) \text{ for all } y \tag{12.4.3}$$

where M_1 and M_2 denote the medians of X and Y respectively. When the medians are known, we can work with $X_i' = X_i - M_1$ $(i = 1, \ldots, m)$ and $Y_j' = Y_j - M_2$ $(j = 1, \ldots, n)$. Even if $M_2 - M_1$ is known, we can subtract this from each of the Y observations and then carry out a test for scale on the basis of X-sample, and Y-sample deviations from $(M_2 - M_1)$. If $M_1 = M_2 = M$ where M is unknown, we can represent the scale alternative as

$$F_Y(y) = F_X(\theta y), \text{ for all } y \text{ and } \theta > 0\,, \tag{12.4.4}$$

where $\theta < 1$ $(\theta > 1)$ if Y has more (less) spread than X. Then we can rewrite the hypotheses as

$$H_0 : \theta = 1 \text{ versus } H_1 : \theta < 1\,.$$

The well known tests for scale are Mood's test, Freund-Ansari-Bradley test, Siegel-Tukey's test, Capon's test, Klotz's normal scores test, the percentile modified rank tests and Sukhatme's (1958) test and Raghavachari's (1965) test. Duran (1976) provides a survey of nonparametric tests for scale.

Mood's Test

Mood's test is based on

$$M_N = \sum_{i-1}^{N} \left(i - \frac{N+1}{2} \right)^2 \tilde{z}_i \tag{12.4.5}$$

where $\tilde{z}_i = 1$ if i^{th} combined ordered X's and Y's is an X and zero otherwise $(i = 1, \ldots, N = m + n)$. A small value of M_N implies that Y has more spread than X. This belongs to the Chernoff-Savage class with $J_N\left(\frac{i}{N+1}\right) = \left(\frac{i}{N+1} - \frac{1}{2}\right)^2$. That is,

$$J_N(u) = \left(u - \frac{1}{2} \right)^2, \quad 0 < u < 1,$$

$$E(M_N|H_0) = \sum_{i=1}^{N} \left(i - \frac{N+1}{2} \right)^2 E(\tilde{Z}_i|H_0) = \frac{m}{N} \sum_{i=1}^{N} \left(i - \frac{N+1}{2} \right)^2,$$

which simplifies to $m(N^2 - 1)/12$.

$$\begin{aligned} \operatorname{Var}(M_N|H_0) &= \sum_{i=1}^{N} \left(i - \frac{N+1}{2} \right)^4 \operatorname{var}(Z_i|H_0) \\ &\quad + \sum\sum_{i \neq j} \left(i - \frac{N+1}{2} \right)^2 \left(j - \frac{N+1}{2} \right)^2 \operatorname{cov}(\tilde{Z}_i, \tilde{Z}_j|H_0). \end{aligned}$$

Notice that

$$\begin{aligned} \operatorname{Var}(\tilde{Z}_i|H_0) &= E(\tilde{Z}_i^2|H_0) - \left\{ E(\tilde{Z}_i|H_0) \right\}^2 \\ &= \frac{m}{N} - \left(\frac{m}{n} \right)^2 = \frac{m}{N} \left(1 - \frac{m}{N} \right), \end{aligned}$$

$$\begin{aligned}
\operatorname{Cov}(\tilde{Z}_i, \tilde{Z}_j | H_0) &= E(\tilde{Z}_i \tilde{Z}_j | H_0) - \left(\frac{m}{N}\right)^2 \\
&= P(\tilde{Z}_i = 1 \text{ and } \tilde{Z}_j = 1 | H_0) - \left(\frac{m}{N}\right)^2 \\
&= \frac{m}{N} \cdot \frac{m-1}{N-1} - \frac{m^2}{N^2} \\
&= \frac{m}{N^2(N-1)} \{N(m-1) - m(N-1)\} \\
&= \frac{-m(N-m)}{N^2(N-1)} .
\end{aligned}$$

Hence,

$$\begin{aligned}
\operatorname{Var}(M_N | H_0) &= \frac{m}{N^2}(N-m) \sum_{i=1}^{N} \left(i - \frac{N+1}{2}\right)^4 - \frac{m(N-m)}{N^2(N-1)} \\
&\quad \cdot \sum_{i \neq j}^{N} \sum^{N} \left(i - \frac{N+1}{2}\right)^2 \left(j - \frac{N+1}{2}\right)^2 \\
&= \left\{ \frac{m(N-m)}{N^2} - \frac{m(N-m)}{N^2(N-1)} \right\} \sum_{i=1}^{N} \left(i - \frac{N+1}{2}\right)^4 \\
&\quad - \frac{m(N-m)}{N^2(N-1)} \left[\sum_{i=1}^{N} \left(i - \frac{N+1}{2}\right)^2 \right]^2 \\
&= \frac{m(N-m)(N-2)}{N^2(N-1)} \sum_{i=1}^{N} \left(i - \frac{N+1}{2}\right)^4 \\
&\quad - \frac{m(N-m)}{N^2(N-1)} \left\{ \frac{N(N^2-1)}{12} \right\}^2 .
\end{aligned}$$

Using the facts $\sum_1^N i = N(N+1)/2$, $\sum_1^N i^2 = N(N+1)(2N+1)/6$,

$$\begin{aligned}
\sum_1^N i^3 &= [N(N+1)/2]^2 \\
\sum_1^N i^4 &= N(N+1)(2N+1)(3N^2+3N-1)/30 ,
\end{aligned}$$

we can simplify and obtain

$$\operatorname{var}(M_N|H_0) = mn(N+1)(N^2-4)/180\,. \tag{12.4.6}$$

Using the Chernoff-Savage theorem, one can evaluate the efficiency of the Mood's test relative to the F-test to be $15/2\pi^2 = 0.76$ when the underlying populations are normal. Laubsher, Steffens and De Lange (1968) tabulate the critical values of Mood's test. For large values of m and n, one can use the normal approximation provided by the Chernoff-Savage theorem. Let

$$\tilde{M}_N = (N+1)^2 \sum_{i=1}^{N} \left(\frac{i}{N+1} - \frac{1}{2}\right)^2 \tilde{Z}_i\,,$$

$$(N+1)^{-2} m^{-1} \tilde{M}_N = \int_{-\infty}^{\infty} \left(\frac{i}{N+1} - \frac{1}{2}\right)^2 dF_m(x)\,.$$

Asymptotic mean of $(N+1)^{-2}m^{-1}\tilde{M}_N$ is

$$\begin{aligned}\int_0^1 \left(u - \frac{1}{2}\right)^2 du &= \int_{-\frac{1}{2}}^{\frac{1}{2}} v^2 dv \\ &= \frac{1}{3}\left[\left(\frac{1}{2}\right)^3 - \left(-\frac{1}{2}\right)^3\right] = \frac{1}{12}\,,\end{aligned}$$

$$\therefore\ E(\tilde{M}_N|H_0) \sim (N+1)^2 m/12\,.$$

Asymptotic variance of $(N+1)^{-2}m^{-1}\sqrt{N}\,\tilde{M}_N$ under H_0 is

$$\begin{aligned}&\left(\frac{N-m}{m}\right)\left[\int_0^1 J^2(u)du - \left\{\int_0^1 J(u)du\right\}^2\right] \\ &= \frac{(N-m)}{m}\left[\int_0^1 \left(u-\frac{1}{2}\right)^4 du - \left(\frac{1}{12}\right)^2\right] \\ &= \frac{(N-m)}{m}\cdot\left[\frac{1}{5}\cdot\frac{1}{16} - \frac{1}{144}\right] \\ &= \frac{(N-m)}{m}\left(\frac{1}{180}\right)\,.\end{aligned}$$

$$\begin{aligned}\therefore\ \operatorname{var}\tilde{M}_N &\sim m^2\frac{(N+1)^4}{N}\cdot\frac{(N-m)}{m}\frac{1}{180} \\ &= mn\cdot(N+1)^4/180N\,.\end{aligned} \tag{12.4.7}$$

Freund-Ansari-Bradley Test

In Mood's test, the deviation of each rank from its null expected value was squared in order to eliminate the problem of keeping track of positive and negative deviations. On the other hand, if we keep absolute values of these deviations thereby giving equal weight to positive and negative deviation, the resultant linear rank statistic is given by

$$A_N = \sum_{i=1}^{N} \left| i - \frac{N+1}{2} \right| \tilde{Z}_i = (N+1) \sum_{i=1}^{N} \left| \frac{i}{N+1} - \frac{1}{2} \right| \tilde{Z}_i .$$

Then the Freund-Ansari-Bradley test criterion can be written as

$$F_N = \sum_{i=1}^{N} \left(\frac{N+1}{2} - \left| i - \frac{N+1}{2} \right| \right) \tilde{Z}_i = \frac{m(N+1)}{2} - A_N$$

or

$$\begin{aligned} F_N &= \sum_{i=1}^{(N+1)/2} i \cdot \tilde{Z}_i + \sum_{i=(N+1)/2+1}^{N} (N+1-i)\tilde{Z}_i \\ (N+1)^{-1} F_N &= \sum_{i=1}^{(N+1)/2} \left(\frac{i}{N+1} \right) \tilde{Z}_i + \sum_{i=(N+1)/2+1}^{N} \left(1 - \frac{i}{N+1} \right) \tilde{Z}_i \\ &= m \int_{-\infty}^{\infty} J_N \left(\frac{N}{N+1} H_N(x) \right) dF_m(x) , \end{aligned}$$

with

$$\begin{aligned} J_N(u) = J(u) &= u \qquad \text{for } 0 < u < \frac{1}{2}, \\ &= 1 - u \quad \text{for } \frac{1}{2} < u < 1 . \end{aligned}$$

Hence,

$$\{m(N+1)\}^{-1} E(F_N | H_0) = \int_0^{1/2} u\, du + \int_{1/2}^{1} (1-u) du = \frac{1}{4} .$$

$$\{m(N+1)\}^{-2} \operatorname{var}(F_N | H_0) = \frac{1}{N} \left(\frac{N-m}{m} \right) \left[\int_0^1 J^2(u) du - \frac{1}{16} \right]$$

where

$$\int_0^1 J^2(u)du = \int_0^{1/2} u^2 du + \int_{1/2}^1 (1-u)^2 du = \frac{1}{3}\left[\frac{1}{8} + \frac{1}{8}\right] = \frac{1}{12},$$

$$\therefore [m(N+1)]^{-2}\operatorname{var}(F_N|H_0) = \frac{n}{48mN}. \tag{12.4.8}$$

The Siegel-Tukey's Test

Their test is based on the following scores:

i	1,	2,	3,	4,	5,	...,	$\frac{N}{2}$	...	$N-3$,	$N-2$,	$N-1$,	N
a_i	1,	4,	5,	8,	9			...	7,	6,	3,	2.

The test criterion is

$$S_N = \sum_{i=1}^N a_i \tilde{Z}_i$$

where

$$\begin{aligned} a_i &= 2i && \text{for } i \text{ even}, \quad 1 < i \le N/2 \\ &= 2i-1 && \text{for } i \text{ odd}, \quad 1 \le i \le N/2 \\ &= 2(N-i)+2 && \text{for } i \text{ even}, \quad N/2 < i \le N \\ &= 2(N-i)+1 && \text{for } i \text{ odd}, \quad N/2 < i \le N\,. \end{aligned}$$

Since the probability distribution of S_N is the same as the Wilcoxon rank sum test, w_N.

$$E(S_N|H_0) = m(N+1)/2$$

and

$$\operatorname{var}(S_N|H_0) = mn(N+1)/12\,.$$

The critical values of w_N can be used in order to find the critical values of S_N. Further, we can write

$$\begin{aligned} \frac{1}{2(N+1)}a_i &\sim \frac{i}{N+1} && \text{for } 1 < i \le N/2 \\ &\sim 1 - \frac{i}{N+1} && \text{for } N/2 < i \le N\,. \end{aligned}$$

So, asymptotically, this is equivalent to the Freund-Ansari-Bradley test.

Klotz's (1962) Test for Scale and Its Modification

Let $X_1, \ldots, X_m$ be a random sample from $F((x-\mu)/\sigma)$ and $Y_1, \ldots, Y_n$ be a random sample from $F((x-\nu)/\tau)$ where μ and ν are the location parameters (for example, medians), σ and τ are the scale parameters and F is absolutely continuous. We are interested in testing $H_0 : \sigma = \tau$ against one- or two-sided alternative. Klotz (1962) proposed a normal scores test given by

$$mT(\boldsymbol{Z}) = \sum_{i=1}^{N} \left[\Phi^{-1}\left(\frac{i}{N+1}\right)\right]^2 \tilde{Z}_i \tag{12.4.9}$$

where $N = m+n$, $\tilde{Z}_i = 1 - Z_i$ and the Z_i are as defined in Section 12.3. When $\mu = \nu$ and the common value is unknown, any of the usual tests for scale can be applied. If μ and ν are not equal, but known, we can apply the existing tests for scale on the deviations of the X observations from μ and the deviations of the Y observations from ν. However, when μ and ν are unknown, the possibility of using the existing tests for scale to the deviations of the observations from certain consistent estimates of the unknown parameters has been noted by some workers in the field. For example, Sukhatme (1958) has constructed a test for this case which is asymptotically distribution-free under certain regularity assumptions on the underlying distributions. Crouse (1964) has also shown that the Mood's test when modified as suggested above, is asymptotically distribution-free under certain conditions. Raghavachari (1965) has modified Klotz's (1962) test so as to apply to the situation when the location parameters are completely unknown. This will be given in the following.

Let $X_1, \ldots, X_m$ be a random sample from $F(x-\mu)$ and $Y_1, \ldots, Y_n$ be a random sample from $G(y-\nu)$, where we assume that the distributions F and G have densities f and g respectively and that μ and ν are the medians of F and G respectively.

Let $N = m+n$ and $\hat{\mu}(X_1, \ldots, X_m)$ and $\hat{\nu}(Y_1, \ldots, Y_n)$ be some consistent estimates of μ and ν respectively such that $N^{1/2}(\hat{\mu}-\mu)$ and $N^{1/2}(\hat{\nu}-\nu)$ are bounded in probability. For instance, $\hat{\mu}$ and $\hat{\nu}$ could be the sample medians of the X and Y observations respectively.

Consider the combined sample $X_i - \mu$, $Y_j - \nu$ $(i = 1, \ldots, m;\ j = 1, \ldots, n)$ and define $\tilde{Z}_{N,i} = 1$ if the ith smallest in the combined sample is an X and 0 otherwise $(i = 1, \ldots, N)$. Define analogously the $\tilde{Z}^*_{N,i}$ for the combined sample $X_i - \hat{\mu}$, $Y_j - \hat{\nu}$ $(i = 1, \ldots, m;\ j = 1, \ldots, n)$. Define the statistics

(Klotz's and its modification)

$$\begin{aligned} nT_N &= \sum_{i=1}^{N} E_{N,i}\tilde{Z}_{N,i} \\ mT_N^* &= \sum_{i=1}^{N} E_{N,i}\tilde{Z}_{N,i}^* \end{aligned} \tag{12.4.10}$$

where

$$E_{N,i} = \left\{\Phi^{-1}\left(\frac{i}{N+1}\right)\right\}^2 .$$

Raghavachari (1965) shows that the statistics, T_N and T_N^* are asymptotically equivalent and T_N^*, when suitably standardized, has an asymptotically normal distribution (via the Chernoff-Savage theorem) provided

(1) f and g are symmetric about their respective location parameters,

and

(2) $f(x)/\phi\left[\Phi^{-1}(F(x))\right]$, $g(x)/\phi\left[\Phi^{-1}(G(x))\right]$ are bounded where $\phi(x)$ denotes the standard normal density function.

12.5 Chernoff-Savage (CS) Class of Statistics

Let $X_1, \ldots, X_m$ be a random sample from $F(x)$ and $Y_1, \ldots, Y_n$ be a random sample from $G(y)$. Let $F_m(x)$ and $G_n(y)$ denote the empirical distribution functions based on the X's and Y's respectively. Also, let $N = m + n$ and $\lambda_N = m/N$. Further, let

$$H_N(x) = \lambda_N F_m(x) + (1 - \lambda_N)G_n(x)$$

be the combined empirical d.f. If

$$\begin{aligned} W &= X \text{ with probability } \lambda_N \\ &= Y \text{ with probability } 1-\lambda_N\,, \end{aligned} \tag{12.5.1}$$

then $H(x) = \lambda_N F + (1-\lambda_N)G$ be the d.f. of W and $H_N(x)$ be the corresponding empirical d.f. Also let

$$W_{1N} \leq \cdots \leq W_{NN}$$

be the combined ordered sample. Consider the class of linear rank test statistics given by

$$T_N = \frac{1}{m}\sum_{i=1}^{N} a_{N,i}\tilde{Z}_i \tag{12.5.2}$$

where

$$\begin{aligned} \tilde{Z}_i &= 1 \text{ if } W_{iN} \text{ is an } X, \\ &= 0 \text{ if } W_{iN} \text{ is a } Y,\ i=1,\ldots,N\,. \end{aligned} \tag{12.5.3}$$

The following integral representation for T_N will be used:

$$T_N = \int_{-\infty}^{\infty} J_N\left(\frac{N}{N+1}H_N(x)\right)dF_m(x) \tag{12.5.4}$$

with

$$J_N\left(\frac{i}{N+1}\right) = a_{N,i}\,.$$

PROOF:

$$T_N = \frac{1}{m}\sum_{i=1}^{m} J_N\left(\frac{R_i}{N+1}\right) = \frac{1}{m}\sum_{i=1}^{N} J_N\left(\frac{i}{N+1}\right)\tilde{Z}_i\,,$$

where $R_1,\ldots,R_m$ denote the ranks associated with the X observations. Notice that $J_N(x)$ is a sequence of functions.

Examples of $J_N(x)$:

$$
\begin{aligned}
J_N\left(\frac{i}{N+1}\right) &= E(W_{iN}) \text{ yields the normal scores test,} \\
&= \frac{i}{N+1} \quad \text{yields the Wilcoxon rank sum test,} \\
&= \begin{cases} \dfrac{i}{N+1} & \text{for } i < \frac{N+1}{2} \\ 1 - \dfrac{i}{N+1} & \text{for } i > \frac{N+1}{2} \end{cases}
\end{aligned}
$$

yields the Ansari-Bradley test.

Assumptions on F and G

Chernoff and Savage (1958) assume that F and G are continuous. Then we can avoid ties among X's and among Y's and among X's and Y's. Since the ties among the X's and among the Y's do not cause any problem, there is no need for this strong restriction. All we need is that F and G do not have mutual discontinuities. Also, we make the following basic assumption on the score function $J_N(u)$, where we drop the superscript N without causing any confusion.

Basic Assumption. Let $J(u)$ be absolutely continuous with

$$
\begin{aligned}
& J' = J_1' + J_2' \\
& \left|J_2'(u)\right| \le f(u)g(u) \\
\text{and} \quad & \int \left|J_1'(u)\right| du \le b < \infty,
\end{aligned}
\tag{12.5.5}
$$

where $g(f) \ge 1$, U-shaped and square integrable (integrable).

Example 12.5.1. In particular, we can have

$$
\begin{aligned}
f(u) &= K\left[u(1-u)\right]^{-1+\delta}, && \text{for some } \delta > 0, \\
g(u) &= K\left[u(1-u)\right]^{-\frac{1}{2}+\delta'}, && \text{for some } \delta' > 0.
\end{aligned}
\tag{12.5.6}
$$

The main Chernoff-Savage Theorem can be stated as follows.

Theorem 12.5.1. *If $J(u)$ satisfies the basic condition and λ_N is bounded away from 0 and 1 (i.e., there exists a $\lambda_0 < \frac{1}{2}$ such that $0 < \lambda_0 < \lambda_N < 1 - \lambda_0 < 1$), then*

$$N^{1/2}\left[T_N - \int_{-\infty}^{\infty} J\left(H(x)\right) dF\right] \overset{d}{\approx} \text{normal}(0, \sigma^2)\,, \tag{12.5.7}$$

where

$$\begin{aligned}\sigma^2 &= 2(1-\lambda_N)^2 \\ &\cdot \left[(1-\lambda_N)^{-1} \iint_{x<y} G(x)\left[1-G(y)\right] J'\left(H(x)\right) J'\left(H(y)\right) dF(x)dF(y)\right. \\ &\left. + \lambda_N^{-1} \iint_{x<y} F(x)\left[1-F(y)\right] J'\left(H(x)\right) J'\left(H(y)\right) dG(x)dG(y)\right].\end{aligned} \tag{12.5.8}$$

When $F = G$

$$\sigma^2 = \frac{2(1-\lambda_N)}{\lambda_N} \iint_{0<u<v<1} u(1-v)J'(u)J'(v)du\,dv \tag{12.5.9}$$

$$= \frac{(1-\lambda_N)}{\lambda_N}\left[\int_0^1 J^2(u)du - \left\{\int_0^1 J(u)du\right\}^2\right]. \tag{12.5.10}$$

Example 12.5.2. Let us illustrate Theorem 12.5.1 by applying it to the Ansari-Bradley test for scale.

Let $X_1, \ldots, X_m$ be a random sample from $F(x)$ and $Y_1, \ldots, Y_n$ be a random sample from $F(x\theta)$, $0 < \theta$. We wish to test $H_0 : \theta = 1$ versus $H_1 : \theta < 1$. Ansari and Bradley (1960) propose a test procedure which is based on

$$\begin{aligned} T &= \sum_{i=1}^{m}\left[\frac{1}{2}(N+1) - \left(R_i - \frac{N+1}{2}\right)\right] \\ &= \sum_{j=1}^{N}\left[\frac{N+1}{2} - \left(j - \frac{N+1}{2}\right)\right]\tilde{Z}_j \end{aligned}$$

where $N = m + n$, $(R_1, \ldots, R_m)$ are the ranks of the X's in the combined ordered sample, and $\tilde{Z}_j = 1$ if W_{jN} is an X and zero if W_{jN} is a Y ($j = 1, \ldots, N$). Then

$$T = \sum_{j=1}^{(N+1)/2} j\tilde{Z}_j + \sum_{j=\frac{N+1}{2}+1}^{N} (N+1-j)\tilde{Z}_j\,.$$

Thus,

$$\frac{T}{N+1} = \sum_{j=1}^{\frac{N+1}{2}} \frac{j}{N+1}\tilde{Z}_j + \sum_{j\geq\frac{N+1}{2}+1}^{N} \left(1-\frac{j}{N+1}\right)\tilde{Z}_j .$$

So, in the Chernoff-Savage class,

$$\begin{aligned} J_N\left(\frac{j}{N+1}\right) &= \frac{j}{N+1} && \text{for } j \leq \frac{N+1}{2} \\ &= 1-\frac{j}{N+1} && \text{for } j > \frac{N+1}{2} . \end{aligned}$$

Thus, in order to assert the asymptotic normality of T when suitably standardized, we apply Theorem 12.5.1 with

$$\begin{aligned} J(u) &= u, && \text{for } 0 < u < \tfrac{1}{2}, \\ &= 1-u, && \text{for } \tfrac{1}{2} < u < 1, \end{aligned}$$

and infer that under H_0, $N^{\frac{1}{2}}\left(\frac{T}{N+1}-\frac{1}{4}\right)$ is asymptotically normal with mean 0 and variance σ^2 where $\sigma^2 = \frac{(1-\lambda_N)}{12\lambda_N}$.

12.6 Problems

12.2.1 For the two sample data in Problem 10.5.1, carry out a two sample Wilcoxon rank sum test (denote it by W) to test the hypothesis $H_0 : \Delta = 0$. Use $\alpha = 0.05$.
(HINT: Note that W under H_0 is asymptotically normal with mean $\frac{m(m+n+1)}{2}$ and variance $\frac{mn(m+n+1)}{12}$ where m and n denote the sizes of the two samples.)

12.2.2 The following data consists of observations on urinary concentration of cotanine, a major metabolite of ninotine on a set of infants that were exposed to household tobacco smoke and a set of infants that were not exposed.

Unexposed:	8, 11, 12, 14, 20, 43, 111
Exposed:	35, 56, 83, 92, 128, 150, 176, 208

Does the data suggest that the true average cotanine level is higher in exposed infants than in unexposed infants by more than 25?

[1]Data appears as Problem 15 on p. 115 of Devore, J.L. (2000). *Probability and statistics for engineering and the sciences.* Duxbury, Pacific Grove, CA.

(HINT: Use Wilcoxon rank sum test and its asymptotic normality.)

12.3.1 Let
3.219, 0.713, 1.427, 0.062, 0.288, 0.798, 0.151, 1.386, 2.302, 0.494
be a random sample from $F(x)$ and let
2.772, 4.604, 0.988, 0.082, 2.618, 0.144, 2.158, 0.862, 2.10, 0.684
be a random sample from $F(x/\theta)$ when F and θ are unkown. We wish to test $H_0 : \theta = 1$ against $H_1 : \theta > 1$ with $\alpha = 0.05$.
(HINT: Use Mood's test and its asymptotic normality.)

Problems 13.6.2, 13.6.3 and 13.6.4 which are placed in p. 323 should be placed under this chapter.

Chapter 13

One-sample Rank Order Tests

13.1 Introduction

Even though goodness of fit tests and the tests for randomness fall into the category of one-sample problems, here we will be concerned with tests for location of symmetry of a distribution. Let X be a random variable having a density $f(x-\theta)$ where $f(x)$ is symmetric about 0. We wish to test $H_0 : \theta = \theta_0$ versus $H_1 : \theta > \theta_0$ where, without loss of generality, we can set $\theta_0 = 0$. In a paired-comparison case, θ could be the difference between the means of two populations.

13.2 LMP Rank Order Test for Location

Let $X_1, X_2, \ldots, X_N$ denote a random sample from $f(x-\theta)$. Let $W_{1N} \leq \cdots \leq W_{NN}$ denote the ordered absolute X's. Let $\boldsymbol{Z} = (Z_1, \ldots, Z_N)$ where $Z_i = 1$ if W_{iN} came from a positive X, and 0 if W_{iN} arose from a negative observation ($i = 1, \ldots, N$). There are 2^N possible $\boldsymbol{Z}$ vectors:

$$P(\boldsymbol{Z} = \boldsymbol{z}|H_0) = 2^{-N} .$$

Savage (1959) gives

$$P(\boldsymbol{Z} = \boldsymbol{z}|H_0) = N! \int_{0<w_1<\cdots<w_N} \cdots \int \prod_{i=1}^{N} f^{z_i}(w_i - \theta) f^{1-z_i}(-w_i - \theta) dw_i$$

which can be simplified to [since $f(-u) = f(u)$]

$$P(\boldsymbol{Z} = \boldsymbol{z}|H_1) = N! \int\cdots\int_{0<w_1<\cdots<w_N<\infty} \prod_{i=1}^{N} f\left(w_i - (2z_i - 1)\theta\right) dw_i \,.$$

Theorem 13.2.1. *If* $\int_{-\infty}^{\infty} |f'(x)|\,dx < \infty$, *then the LMP test is to reject* H_0 *for large values of*

$$\sum_{i=1}^{N} z_i E\left[\frac{-f'(W_{iN})}{f(W_{iN})}\right] > K_\alpha \,. \tag{13.2.1}$$

PROOF:

$$\begin{aligned} &P(\boldsymbol{Z} = \boldsymbol{z}|H_1) - P(\boldsymbol{Z} = \boldsymbol{z}|H_0) \\ &= N! \int\cdots\int_{0<w_1<\cdots<w_N<\infty} \left[\prod_{i=1}^{N} f\left(w_i - (2z_i - 1)\theta\right) - \prod_{1}^{N} f(w_i)\right] \prod_{i}^{N} dw_i \\ &= N! \sum_{i=1}^{N} \int\cdots\int_{0<w_1<\cdots<w_N<\infty} \prod_{j=1}^{i-1} f\left(w_j - (2z_j - 1)\theta\right) \left[f\left(w_i - (2z_i - 1)\theta\right)\right. \\ &\quad \left. - f(w_i)\right] \prod_{k=i+1}^{N} f(w_k) \prod_{i=1}^{N} dw_i \,. \end{aligned}$$

Now, because of the assumption on f', one can take the limit underneath the integral sign. Thus,

$$\lim_{\theta\to 0} \left\{P\left[\boldsymbol{Z} = \boldsymbol{z}|H_\theta\right] - P\left[Z = z|H_\theta\right]\right\} \theta^{-1} = \sum_{i=1}^{N} (2z_i - 1) E\left[\frac{-f'(W_{iN})}{f(W_{iN})}\right] .$$

Since

$$\sum_{i=1}^{N} E\left\{\frac{f'(W_{iN})}{f(W_{iN})}\right\} = NE\left\{\frac{-f'(Y)}{f(Y)}\right\} = \text{a non-stochastic constant,}$$

where $Y = |X|$, we can use

$$T = \sum_{i=1}^{N} z_i E\left[\frac{-f'(W_{iN})}{f(W_{iN})}\right] \tag{13.2.2}$$

as the test criterion.

Special Cases

1. Let $f(x) = \phi(x)$, then

$$E\left\{\frac{-f'(W_{iN})}{f(W_{iN})}\right\} = E(W_{iN})\,.$$

So we get the absolute normal scores test.

2. Let $f(x) = e^x/(1+e^x)^2$, then

$$-f'(x)/f(x) = 2F(x) - 1 = G(x)$$

which is the d.f. of $|X|$. Hence, $W_{1N} < \cdots < W_{NN}$ are the O.S. in a random sample of size N drawn from $G(x)$.

$$EG(W_{iN}) = EU_{iN} = i/(N+1), \;\; i = 1, \ldots, N\,.$$

Thus we get the Wilcoxon signed rank test criterion.

3. Let $f(x) = \frac{1}{2}e^{-|x|}$, $-\infty < x < \infty$. Then

$$\frac{-f'(x)}{f(x)} = \begin{cases} 1 & \text{if } x > 0, \\ -1 & \text{if } x < 0, \end{cases}$$

implying

$$\frac{-f'(x)}{f(x)} = \text{Sgn } x\,.$$

Thus, the test criterion is

$$\begin{aligned} T &= \sum z_k E\,(\text{Sgn }(W_{kN})) = \sum z_k E(1) = \sum_1^N z_k \\ &= \text{sign test statistic}\,. \end{aligned}$$

Remark 13.2.1. *If $S_1, \ldots, S_n$ denote the ranks enjoyed by the positive observations in the sample where n denotes the number of positive X's, then one can write T as*

$$T = \sum_{j=1}^{n} E\left[-f'(W_{S_{j,N}})/f(W_{S_{j,N}})\right]\,. \tag{13.2.3}$$

13.3 Cases of Zero Observations

We have three possibilities for handling the zero observations.

1. If all the zeros are put in the negative category, we obtain a conservative test.

2. We could randomize on zero observations. That is, each zero is declared positive with probability $\frac{1}{2}$ and is considered negative with probability $\frac{1}{2}$.

3. We can reduce the sample by deleting the number of zeros.

If (3) is followed, notice that the test is conditional since the number of nonzero observations is random. Since the distribution is assumed to be continuous, zero observations should occur with probability zero and we should prefer option (1).

13.4 Tests for Randomness

The assumption of randomness is essential for almost all statistical procedures. In order to study this requirement, samples are randomly selected or treatments are randomly assigned to experimental units. This is, however, not always possible. The assumption is especially important when observations are made in a sequence, in which case we might suspect some kind of trend or dependency between successive observations. Lack of randomness is so broad that a variety of tests of randomness are available in the literature. Significant contributions have been made by Wald and Wolfowitz (1943), Mann (1943), Stuart (1954, 1956), Foster and Stuart (1954), Cox and Stuart (1955), Savage (1957), Gupta and Govindarajulu (1980) and Aiyar (1981). Most of the tests are sensitive to trend alternatives. First we derive an LMP test for trend alternatives.

Let $X_1, X_2, \ldots$ be an independent sequence of random variables having p.d.f.'s $f(x - c_i\theta)$, $i = 1, 2, \ldots$ where c_i are known. We wish to test $H_0 : \theta = 0$ vs. $H_1 : \theta > 0$.

Theorem 13.4.1. *If $\int |f'(x)|\,dx < \infty$, then the LMP rank test of H_0 vs. H_1 rejects H_0 for large values of*

$$T = \sum_{i=1}^{N} c_i E_0 \left\{ -f'(W_{R_{i,N}})/f(W_{R_{i,N}}) \right\} \qquad (13.4.1)$$

where $W_{1N} < \cdots < W_{NN}$ denote the ordered X's and $\boldsymbol{R} = (R_1, \ldots, R_N)$ denotes the rank vector of the X's.

PROOF: Proceed as in the two-sample case and note that

$$P(\boldsymbol{R} = \boldsymbol{r}|H_1) = \int_{-\infty<x_1<\cdots<x_N<\infty} \cdots \int \prod_{i=1}^{N} f(x_{r_i} - c_i\theta)dx_i$$

and $P(\boldsymbol{R} = \boldsymbol{r}|H_0) = 1/N!$.

13.5 LMP Rank Tests against Trend

Let $X_1, X_2, \ldots$ be an independent sequence of random variables having p.d.f.'s

$$(1 + c_i\theta)f\left(x(1 + c_i)\theta\right)\,,\quad i = 1, 2, \ldots$$

where the c_i are known. We wish to test

$$H_0 : \theta = 1 \text{ versus } H_1 : \theta < 1\,.$$

Then we have the following result.

Theorem 13.5.1. *If $\int |xf'(x)|\,dx < \infty$, the LMP rank test of H_0 versus H_1 rejects H_0 for large values of*

$$T = \sum_{i=1}^{N} c_i E_0 \left\{ -W_{R_i,N} \frac{f'(W_{R_i,N})}{f(W_{R_i,N})} \right\} \tag{13.5.1}$$

where $W_{1N} \le \cdots \le W_{NN}$ denote the ordered X's and $\boldsymbol{R} = (R_1, \ldots, R_N)$ denote the vector of ranks of the X's.

PROOF: Proceed as in the two-sample case and note that

$$P(\boldsymbol{R} = \boldsymbol{r}|H_1) = \int_{-\infty<x_1<\cdots<x_N<\infty} \cdots \int \prod_{i=1}^{N} (1 + c_i\theta)f\left(x_{r_i}(1 + c_i\theta)\right) dx_i$$

and

$$P(\boldsymbol{R} = \boldsymbol{r}|H_0) = \int_{-\infty<x_1<\cdots<x_N<\infty} \cdots \int \prod_{i=1}^{N} f(x_{r_i})dx_i\,.$$

Special Cases

1. Let $f(x) = \phi(x)$, then

$$T = \sum_{1}^{N} c_i E_0(W_{R_{i,N}}) \tag{13.5.2}$$

where W_{iN} are standard normal order statistics in a sample of size N.

2. Let $f(x) = e^x/(1+e^x)^2$. Then

$$T \approx \sum_{1}^{N} c_i R_i \,. \tag{13.5.3}$$

3. Let $f(x) = \frac{1}{2}\exp(-|x|)$. Then

$$\begin{aligned} T &= \sum c_i E\left\{\mathrm{Sgn}(W_{R_{i,N}})\right\} \\ &= \sum c_{k_i} E\left\{\mathrm{Sgn}(W_{i,N})\right\} \\ &\doteq \sum c_{k_i} \mathrm{Sgn}\left\{E(W_{i,N})\right\} \approx \sum_{i=[N/2]}^{N} c_{k_i} \end{aligned} \tag{13.5.4}$$

where $(k_1, \ldots, k_N)$ denotes the subscript vector. Notice that $R_i = j \Leftrightarrow k_j = i$.

4. **Lehmann Alternatives**. If X_i has d.f. $F_i(x) = [H(x)]^{\theta_i}$. Then Savage (1957) has obtained

$$P(\boldsymbol{K} = \boldsymbol{k}|H_\theta) = \left(\prod_{i=1}^{N} \theta_i\right) \Big/ \prod_{i=1}^{N} \left(\sum_{j=1}^{i} \theta_{k_j}\right) .$$

If $\theta_i = i\theta$, then

$$P(\boldsymbol{K} = \boldsymbol{k}|H_\theta) = N! \prod_{i=1}^{N} \left(\sum_{j=1}^{i} k_j\right)^{-1} .$$

The parametric test which is usually used is the serial correlation given by

$$
\begin{aligned}
S_N &= \left(\sum_{i=1}^{N-1}(X_iX_{i+1} - N\bar{X}^2)\right) / \left\{\sum_{i=1}^{N}(X_i - \bar{X})^2\right\} \\
&\approx \frac{\sum_1^{N-1}(X_i - \bar{X})(X_{i+1} - \bar{X})}{\sum(X_i - \bar{X})^2} . \qquad (13.5.5)
\end{aligned}
$$

Since $(N-1)^{-1}\sum(X_i - \bar{X})^2$ converges to the population variance and S_N is invariant with respect to location and scale parameters, we can take S_N to be

$$S_N = N^{-1}\sum_{i=1}^{N-1} X_iX_{i+1} . \qquad (13.5.6)$$

Wald and Wolfowitz (1943) proposed the distribution-free test given by

$$K_N = N^{-5/2}\sum_{1}^{N-1} R_iR_{i+1} \qquad (13.5.7)$$

where R_i denotes the rank of X_i $(i = 1, \ldots, N)$.

Gupta and Govindarajulu (1980) have derived locally most powerful tests for randomness against some dependence among the observations. This dependence can be one of the following.

1. The X's are multivariate normal, the variance-covariance matrix is a function of an unknown parameter ρ.

2. X's form a first order auto-regressive process defined by

   $$X_i = Z_i + \rho X_{i-1}, \quad i = 1, 2, \ldots,$$

   where the Z_i are i.i.d. normal $(0, \sigma^2)$.

3. X's follow a first order moving average process defined by

   $$X_i = Z_i + \rho Z_{i-1}, \quad i = 1, 2, \ldots,$$

   where the Z_i are i.i.d. normal $(0, \sigma^2)$.

Then the LMP test of $H_0 : \rho = 0$ against $H_1 : \rho > 0$ is given by:

reject H_0 for large values of

$$T_N = N^{-1/2}\sum_{i=1}^{N-1} E_0(Y_{R_{i,N}}, Y_{R_{i+1,N}}) \qquad (13.5.8)$$

where $Y_{1N} \leq \cdots \leq Y_{NN}$ constitute order statistics in a random sample of size N drawn from the standard normal distribution. The asymptotic normality of the preceding test under H_0 is established. Aiyar (1981) proposed an alternative test criterion given by

$$\tilde{T}_N = N^{-1/2} \sum_{i=1}^{N-1} \Phi^{-1}\left(\frac{R_i}{N+1}\right) \Phi^{-1}\left(\frac{R_{i+1}}{N+1}\right) \tag{13.5.9}$$

where Φ denotes the standard normal d.f. Aiyar (1981) establishes the asymptotic normality of $\tilde{T}_N$ under H_0 and studies Pitman efficiency of $\tilde{T}_N$ and K_N relative to S_N. Gupta and Govindarajulu (1980) tabulate the critical values of T_N and carry out Monte Carlo studies pertaining to power comparisons of T_N, K_N and S_N.

Remark 13.5.1. *Gupta and Govindarajulu (1980) also obtain LMP tests for j step auto-regressive and moving average alternatives.*

13.6 One-sample C-S Class of Statistics

In this section, we will formulate the one-sample Chernoff-Savage class of statistics and give the basic result pertaining to the asymptotic normality of the class of test statistics when suitably normalized. We defer the proof of the result to Chapter 25.

Notation and Assumptions

Let $X_1, \ldots, X_m$ $[Y_1, \ldots, Y_n]$ denote a random sample of size $m[n]$ drawn from a continuous distribution $F(x)$ $[G(x)]$ where m and n are random and $N = m + n$ is non-random. Continuity of F and G is not necessary provided F and G do not have common discontinuities. Then F and G can be made continuous by the continuization procedure described in Govindarajulu, Le Cam and Raghavachari (1967) [which will hereafter be abbreviated as GLR (1967)]. As a special case, let $V_1, \ldots, V_N$ be a random sample drawn from a distribution $L(x)$, $X_1, \ldots, X_m$ $[Y_1, \ldots, Y_n]$ are those V's or Borel measurable functions of V's which belong to Category I [Category II]. As a further special case, the X's can be the absolute values of the negative V's and the Y's be the positive V's. (Generally, values other than zero could have been used as the breaking point.)

Let $F_m(x)$ $[G_n(y)]$ denote the empirical distribution function (e.d.f.) based on $X_1, \ldots, X_m$ $[Y_1, \ldots, Y_n]$. Further, let

$$H_N(x) = \lambda_N F_m(x) + (1 - \lambda_N) G_n(x), \quad \lambda_N = m/N\,. \tag{13.6.1}$$

That is, $H_N(x)$ is the e.d.f. based on the combined sample of size N. Also, let

$$H^*(x) = p_N F(x) + (1 - p_N)G(x), \quad p_N = E(\boldsymbol{\lambda}_N). \tag{13.6.2}$$

Note that $L(x)$ and consequently F and G may depend on N and we suppress this fact for the sake of simplicity. Let m and n denote some specified values assumed by the random variables $\boldsymbol{m}$ and $\boldsymbol{n}$ respectively. Furthermore, let

$$\lambda_N = m/N, \; \mu_{2,N} = NE\left\{(\boldsymbol{\lambda}_N - p_N)\right\}^2, \quad s = N^{-1/2}(\boldsymbol{m} - Np_N)/\mu_{2,N}^{1/2}. \tag{13.6.3}$$

We also assume that there exists a p_0 ($p_0 \le 1/2$) such that

$$0 < p_0 \le p_N \le 1 - p_0 < 1. \tag{13.6.4}$$

We will be concerned with statistics of the form

$$T_N = \sum_{i=1}^{N} E_{N,i} Z_{N,i} \tag{13.6.5}$$

where $Z_{N,i} = 1$ if the ith smallest observation in the combined sample is an X and $Z_{N,i} = 0$, otherwise, and $E_N = (E_{N,1}, \ldots, E_{N,N})$ is a given vector of constants for each N. Notice that for the location problem with symmetry, $E_{N,i} - 1$ yields the sign test statistic, $E_{N,i} = i/(N+1)$ yields the Wilcoxon signed rank test, and the statistic becomes the absolute normal scores test when $E_{N,i}$ is the expected value of the ith smallest order statistic in a sample of size N drawn from the chi-population with one degree of freedom (or the folded normal distribution). We will use the following integral representation of T_N:

$$T_N = \boldsymbol{m} \int_{-\infty}^{\infty} J_N\left(\frac{N}{N+1} H_N(x)\right) dF_{\boldsymbol{m}}(x) \tag{13.6.6}$$

where representations (13.6.5) and (13.6.6) are equivalent when

$$E_{N,i} = J_N\left(\frac{i}{N+1}\right), \quad i = 1, \ldots, N.$$

We will be interested in the asymptotic behavior of the differences of the form

$$T_N^* = N^{1/2}\left[\boldsymbol{\lambda}_N \int J_N\left(\frac{N}{N+1} H_N(x)\right) dF_{\boldsymbol{m}}(x) - p_N \int J_N\left(H^*(x)\right) dF(x)\right]. \tag{13.6.7}$$

Hereafter, whenever there is no ambiguity, the subscript N in J_N, $\boldsymbol{\lambda}_N$, λ_N and p_N will be suppressed. Throughout K will be used as a generic finite constant which will not depend on F, G and N.

Although it suffices to define $J_N(u)$ at $1/(N+1), \ldots, N/(N+1)$, one can extend its domain of definition to $(0,1)$ by letting $J_N(u)$ be constant over $(i/(N+1), (i+1)/(N+1))$ $(i = 0, \ldots, N)$. Consider functions J defined by $J(x) = \int_{1/2}^{x} J'(v)dv + a$ where, without loss of generality, we can set $a = 0$ since this will not change the differences studied here. Then we have the main theorem of this section as given by the author (1985).

Theorem 13.6.1. *Let $N^{1/2}(\boldsymbol{\lambda}_N - p_N)$ be asymptotically normal with*

$$f(m) = P(\boldsymbol{m} = m) = P\left((\boldsymbol{\lambda} - p)/\sqrt{\mu_2} = s\right) = (N\mu_2)^{-1/2}\phi(s) + o(N^{-1/2}) \tag{13.6.8}$$

where $s = (m - Np)/\sqrt{N\mu_2}$ and ϕ denotes the standard normal density function. Let J be absolutely continuous with

$$J' = J_1' + J_2', \left|J_2'(u)\right| \le fg, \int \left|J_1'(u)\right| du \le b < \infty,$$

where f and g are defined as in Theorem 12.5.1. Then

$$\lim_{N\to\infty} P(T_N^* \le x) = \Phi(x/\sigma_N)$$

for all x, for every J and every triple $((F,G),p)$ provided σ_N^2 stays bounded away from zero where

$$\sigma_N^2 = \sigma_N^2\left((F,G),J,p\right) = p(1-p)^2 I_1 + p^2(1-p)I_2 + \mu_2 I_3^2 \tag{13.6.9}$$

where

$$I_1 = 2p(1-p)^2 \iint_{x<y} F(x)\left[1 - F(y)\right] J'\left(H^*(x)\right) J'\left(H^*(y)\right) dG(x)dG(y), \tag{13.6.10}$$

$$I_2 = 2p^2(1-p) \iint_{x<y} G(x)\left[1 - G(y)\right] J'\left(H^*(x)\right) J'\left(H^*(y)\right) dF(x)dF(y), \tag{13.6.11}$$

and

$$I_3 = \int \left[J(H^*) + p(F - G)J'(H^*)\right] dF\,. \tag{13.6.12}$$

Remark 13.6.1. *In the following case, the uniformity asserted in Theorem 13.6.1 is of much interest.*

$$F(x) = P\left(|V| \le x | V < 0\right) = \frac{\Psi(-\theta) - \Psi(-x-\theta)}{\Psi(-\theta)}$$

and

$$G(x) = P(|V| \le x | V > 0) = \frac{\Psi(x-\theta) - \Psi(-\theta)}{\Psi(\theta)}$$

where Ψ *is symmetric about zero and* θ *is in a bounded interval which depends on* N. *Furthermore, if* $\theta = \theta_N$ *where* $\theta_N \to 0$ *as* $N \to \infty$, *then the convergence of the distribution of* T_N^* *to normality is uniform for any* J *such that* $\int J^2(u)du \ge a > 0$. *Also notice that as* $\theta_N \to 0$, $\lim F(x) = \lim G(x) = 2\Psi(x) - 1$, $x \ge 0$, *and*

$$\sigma_N^2 [(F, G), J, p] \to \sigma_N^2 [(F, F), J, 1/2] = (1/4) \int_0^1 J^2(u)du\,. \tag{13.6.13}$$

Example 13.6.1. In the hypothesis-testing problem for location of symmetry, let $H_0 : \theta = 0$ vs. $H_1 : \theta > 0$. Let T_N be the Wilcoxon signed rank statistic with $J(u) = u$, $0 < u < 1$. Then under H_0,

$$\left(N^{-1}T_N - 1/2\right)(N/12)^{1/2}$$

is asymptotically normal $(0, 1)$. Under the alternative hypothesis, $\int J(H^*(x))\, dF(x)$ and σ_N^2 will take simpler forms when $J(u) = u$.

13.7 Application to Halperin's Statistic

Halperin (1960) proposed a Wilcoxon test procedure for censored samples and established its asymptotic normality under the null hypothesis. The present author (1985) has shown that the asymptotic normality of Halperin's statistic under all hypotheses follows from Theorem 13.6.1.

Let $X_1, \ldots, X_{m^*}$ $[Y_1, \ldots, Y_{n^*}]$ denote a random sample from a continuous distribution $F(x)$ $[G(y)]$. We assume that the X's and Y's are mutually independent. Let t_0 be a fixed point and we truncate the X's and Y's at the same point t_0. Let the uncensored samples be denoted by $X_1, \ldots, X_m$ and $Y_1, \ldots, Y_n$, where $m^* - m$ X-observations and $n^* - n$ Y-observations are censored. Halperin (1960) further assumes that $m^* + n^* - m - n$ is fixed. For instance, this can happen in life testing where one stops after observing a combined fixed number of failures of items from two populations. (See, for instance, Sobel, 1957.) Let

$$p_1 = P(X \le t_0) = F(t_0) \text{ and } q_1 = 1 - p_1\,, \tag{13.7.1}$$
$$p_2 = P(Y \le t_0) = G(t_0) \text{ and } q_2 = 1 - p_2 \tag{13.7.2}$$

and

$$N = m + n$$

which is nonrandom since $\boldsymbol{m} + \boldsymbol{n}$ is non-random. One can easily establish the following lemma.

Lemma 13.7.1. *The conditional distribution of* $\boldsymbol{m}$ *for given* $\boldsymbol{m} + \boldsymbol{n} = N$, *is asymptotically normal with*

$$mean = [m^* n^* p_1 p_2 (p_1 - p_2) + N m^* p_1 q_1] \, [m^* p_1 q_1 + n^* p_2 q_2]^{-1} \qquad (13.7.3)$$

and

$$variance = (m^* p_1 q_1)(n^* p_2 q_2)[m^* p_1 q_1 + n^* p_2 q_2]^{-1} \, . \qquad (13.7.4)$$

PROOF: See Lemma 5.2 of Govindarajulu (1985, p. 169).

Halperin's (1961) test statistic is an extension of the Mann-Whitney-Wilcoxon U-statistic and is given by

$$\sum_{1}^{\boldsymbol{m}} R_i - [\boldsymbol{m}(\boldsymbol{m}+1)/2] + \boldsymbol{n}(m^* - \boldsymbol{m}) \qquad (13.7.5)$$

where $R_1, \ldots, R_{\boldsymbol{m}}$ denote the ranks of the ordered uncensored X's in the combined uncensored sample.

After dividing by $N^{3/2}$ and considering only the random component, Halperin's (1960) statistic can be written as [with $p_N = E(\boldsymbol{\lambda}_N) = E(\boldsymbol{m}/N)$]

$$N^{-3/2} \sum_{i=1}^{\boldsymbol{m}} R_i - \left(N + m^* + \frac{1}{2} - Np \right) (\boldsymbol{\lambda} - p) N^{-1/2} + N^{1/2} (\boldsymbol{\lambda} - p)^2 / 2 \qquad (13.7.6)$$

where the first term belongs to the class of statistics T_N defined by (13.6.6). Also, one can easily see that $N^{1/2}(\boldsymbol{\lambda} - p)^2$ converges to zero in probability since $N^{1/2}(\boldsymbol{\lambda} - p)$ is bounded in probability, due to Lemma 13.7.1. Now, combining the $(\boldsymbol{\lambda} - p)$ term with the appropriate first order random term in the expansion of T_N, the asymptotic normality of Halperin's statistic under all the hypotheses follows from Theorem 13.6.1. The asymptotic mean of the statistic given by (13.7.6) is

$$N^{1/2} \int_{-\infty}^{t_0} \tilde{H}^* d\tilde{F}(x) \qquad (13.7.7)$$

and its variance

$$2p(1-p)^2 \int \cdots \int_{-\infty<x<y\le t_0} \tilde{F}(x)\left[1-\tilde{F}(y)\right] d\tilde{G}(x)d\tilde{G}(y)$$

$$+2p^2(1-p) \int \cdots \int_{-\infty<x<y\le t_0} \tilde{G}(x)\left[1-\tilde{G}(y)\right] d\tilde{F}(x)d\tilde{F}(y) \qquad (13.7.8)$$

$$+\mu_2\left\{\int_{-\infty}^{t_0}\left[\tilde{H}^* + p(\tilde{F}-\tilde{G})\right] d\tilde{F} - \left(N + m^* + \frac{1}{2} - Np\right)N^{-1}\right\}^2$$

where

$$\begin{aligned}\tilde{F}(x) &= F(x)\left[1-F(t_0)\right]^{-1}, \ -\infty < x < t_0, \\ \tilde{G}(x) &= G(x)\left[1-G(t_0)\right]^{-1}, \ -\infty < x < t_0, \\ \tilde{H}^* &= p\tilde{F} + (1-p)\tilde{G}\end{aligned}$$

and

$$\mu_2 = NE(\boldsymbol{\lambda}_N - p_N)^2 . \qquad (13.7.9)$$

Remark 13.7.1. *When H_0 is true, $\boldsymbol{m}$ has a hypergeometric distribution given by*

$$P(\boldsymbol{m} = m | N = \boldsymbol{m} + \boldsymbol{n}) = \binom{m^*}{m}\binom{n^*}{n} \Big/ \binom{N^*}{N}, \quad m = 0, \ldots, N \qquad (13.7.10)$$

where $N^* = m^* + n^*$.

Remark 13.7.2. *When $p_2 = p_1$ (i.e., H_0 holds) the asymptotic variance in (13.7.4) is $m^*n^*p_1q_1/N^*$. However, substituting $p_1 = N/N^*$ and $q_1 = 1-p_1$ since N/N^* converges to p_1 almost surely under H_0, this variance is equivalent to the conditional variance of $\boldsymbol{m}$ computed from (13.7.10). Thus, Halperin's test is distribution-free under H_0.*

Govindarajulu (1985, p. 169) derives the asymptotic efficiency of Halperin's test procedure relative to Student's t and F-test for location and scale changes respectively.

Remark 13.7.3. *Sobel (1957) proposed a different two-sample nonparametric test under censoring after a preselected total number of deaths have occurred in the two samples. However, Halperin's (1960) distribution results are conditional, whereas Sobel's (1947) results are unconditional.*

Savage (1959) obtained partial ordering of the one-sample rank order probabilities, which are discussed in Section 11.6. The contributions of Huskova (1970), Puri and Sen (1969), and Pyke and Shorack (1968) are briefly discussed in Remark 25.8.1.

13.8 Problems

13.2.1 Let $X_1, \cdots, X_N$ be a random sample from a distribution that is symmetric about θ. We wish to test $H_0 : \theta = 0$ against $H_1 : \theta > 0$. Specialize the LMP test statistic given by (13.2.2) for the triangular density function given by

$$\begin{aligned} f(x) &= x+1 && for\ -1 < x < 0 \\ &= x-1 && for\ 0 < x < 1 \\ &= 0 && elsewhere \end{aligned}$$

13.4.1 Evaluate the test criterion T when

(i) $f(x) = e^x(1+e^x)^{-2}, \quad -\infty < x < \infty$

and (ii) $f(x)$ is the triangular density given in Problem 13.2.1.

13.5.1 Specialize the test statistic T given by (13.5.1) to the case when $f(x)$ is the triangular density.

13.6.1 Let 0.40, -0.20, 1.28, 2.75, -0.39, 2.49, 0.61, 10.38, 0.56, 0.45 be a random sample from $F(x)$ which is symmetric about θ. Using Wilcoxon signed rank statistic, test $H_0 : \theta = 0.5$ against $H_1 : \theta > 0.5$ with $\alpha = 0.05$.

(HINT: Use the asymptotic normality of the test criterion under H_0 mentioned in Example 13.6.1.)

13.6.2[1] For the data in Problem 13.6.1, carry out the Freund-Ansari-Bradley test procedure to test H_0 at $\alpha = 0.05$.

(HINT: Use the asymptotic normality of the test statistic.)

13.6.3 For the data in Problem 13.6.1, carry out the Savage test procedure (using the version of the test statistic given towards the end of Section 12.3) in order to test H_0 at $\alpha = 0.05$.

(HINT: Note that Savage's statistic is under H_0, asymptotically normal with mean 1 and variance $2/N$.)

13.6.4 A random sample of 10 freshmen that were admitted to Mars University had the following SAT (scholastic assessment test) scores: 764, 650, 665, 680, 710, 784, 671, 682, 679, 662.

A random sample of 10 freshmen that were admitted into Jupiter University during the same year had the following SAT scores:

[1] Problems 13.6.2, 13.6.3 and 13.6.4 should be placed under Chapter 12.

660, 670, 710, 730, 740, 620, 570, 760, 780, 590.
The sample medians are very close to each other. Hence, assuming that the location parameters in the two populations are the same, use Ansari-Bradley test in order to test whether there is a difference in the variability of SAT scores of freshmen in the two Universities.

Chapter 14

Asymptotic Relative Efficiency

14.1 Introduction

For a typical problem there may be several applicable nonparametric procedures. We would like to know how to compare their efficiencies relative to the parametric competitor. Since most of the tests are consistent, the power of any test will be close to unity at global alternatives. So, the sensitivity of a test can be judged by considering local alternatives, i.e., alternatives that are close to the null hypotheses. These are called Pitman alternatives.

14.2 Pitman Efficiency

Definition 14.2.1. *Pitman alternatives is a sequence of alternatives converging to the null hypothesis.*

Consider the following example.

Example 14.2.1. Let $X_1, \ldots, X_n$ be a random sample from normal $(\theta, 1)$. Suppose we wish to test

$$H_0 : \theta = \theta_0 \text{ against } H_1 : \theta > \theta_0 \,. \tag{14.2.1}$$

Pitman alternatives are given by

$$H_n : \theta = \theta_n = \theta_0 + \xi / n^{\delta} \tag{14.2.2}$$

for some $\xi > 0$ and $\delta > 0$.

If the test is based on $\bar{X}_n$, we reject H_0 when

$$\bar{X}_n > k_\alpha = \theta_0 + z_\alpha/\sqrt{n},$$

where z_α is the $(1-\alpha)^{\text{th}}$ quantile of the standard normal d.f. Then the power at θ_n is obtained (after an elementary computation) as

$$\begin{aligned}
\pi(\theta_n) &= \Phi(n^{1/2-\delta}\xi - z_\alpha) \\
&\to 1 \text{ if } \delta < 1/2 \\
&= \Phi(\xi - z_\alpha) \text{ if } \delta = 1/2 \\
&\to \alpha \text{ if } \delta > 1/2\,.
\end{aligned}$$

So, choice of δ is important. Typically it is 1/2. However, there are some situations in which it may be 1 (exponential case) and may be 1/4 (in testing for association).

Definition 14.2.2. *Let T_n and T_n^* be two sequences of (one-sided) test statistics, all with the same significance level α. Let $\{n_i\}$ and $\{n_i^*\}$ be two monotonic increasing sequences of positive integers such that*

$$\lim_{i\to\infty} \pi(\theta_i, n_i) = \lim_{i\to\infty} \pi^*(\theta_i, n_i^*) = \gamma, \quad 0 < \gamma < 1\,, \tag{14.2.3}$$

where π and π^* denote the power functions of T_n and T_n^* respectively. Then the asymptotic efficiency of test T relative to T^* is

$$e(T, T^*) = \lim_{i\to\infty} \{n_i^*/n_i\}\,.$$

Assume the following regularity conditions. Let $\theta_n = \theta_0 + \xi/n^\delta$ for some $\xi, \delta > 0$,

$$A_1 : \lim_{n\to\infty} P_{\theta_n}\left[\frac{T_n - E_{\theta_n}(T_n)}{\sigma_{\theta_n}(T_n)} \le z\right] = \Phi(z) \text{ (asymptotic normality)},$$

$$A_2 : \lim_{n\to\infty} [\sigma_{\theta_n}(T_n)/\sigma_{\theta_0}(T_n)] = 1,$$

and

$$A_3 : \lim_{n\to\infty} [\{E_{\theta_n}(T_n) - E_{\theta_0}(T_n)\}/\xi\sigma_{\theta_0}(T_n)] = c\,.$$

Then we have the following theorem.

Theorem 14.2.1. *Under the regularity assumptions A_1 - A_3, the limiting power of the test T_n is*

$$\lim_{n\to\infty} \pi_n(\theta_n, n) = 1 - \Phi(z_\alpha - \xi c), \tag{14.2.4}$$

where z_α is such that $\Phi(z_\alpha) = 1 - \alpha$.

PROOF: $\pi(\theta_0, n) = \alpha = P(T_n > t_{n,\alpha} | \theta = \theta_0)$. That is,

$$\frac{t_{n,\alpha} - E_{\theta_0}(T_n)}{\sigma_{\theta_0}(T_n)} = z_\alpha . \tag{14.2.5}$$

Also,

$$\begin{aligned} \pi(\theta_n, n) &= P(T_n > t_{n,\alpha} | \theta = \theta_n) \\ &= P\left(\frac{T_n - E_{\theta_n}(T_n)}{\sigma_{\theta_n}(T_n)} > \frac{t_{n,\alpha} - E_{\theta_n}(T_n)}{\sigma_{\theta_n}(T_n)} | \theta = \theta_n \right) . \end{aligned} \tag{14.2.6}$$

Now consider

$$\begin{aligned} z_n = \frac{t_{n,\alpha} - E_{\theta_n}(T_n)}{\sigma_{\theta_n}(T_n)} &= \frac{\{t_{n,\alpha} - E_{\theta_0}(T_n)\}}{\sigma_{\theta_0}(T_n)} \cdot \frac{\sigma_{\theta_0}(T_n)}{\sigma_{\theta_n}(T_n)} \\ &- \frac{\{E_{\theta_n}(T_n) - E_{\theta_0}(T_n)\}}{\sigma_{\theta_0}(T_n)} \frac{\sigma_{\theta_0}(T_n)}{\sigma_{\theta_n}(T_n)} \\ &= z_\alpha - c\xi + o(1) . \end{aligned} \tag{14.2.7}$$

Hence using the continuity of Φ, we have

$$\lim_{n\to\infty} \pi(\theta_n, n) = 1 - \lim_{n\to\infty} \Phi(z_n) = 1 - \Phi(z_\alpha - c\xi) . \tag{14.2.8}$$

Theorem 14.2.2. *If T_n and T_n^* are two sequences of tests satisfying the regularity conditions A_1 - A_3, then*

$$e(T, T^*) = (c/c^*)^{1/\delta} ,$$

where c and c^* are quantities defined in A_3 for the statistics T_n and T_n^*.

PROOF: The limiting powers of T_n and T_n^* are equal if

$$1 - \Phi(z_\alpha - c\xi) = 1 - \Phi(z_\alpha - c^*\xi^*) ,$$

i.e., if

$$\xi c = \xi^* c^*$$

or

$$\frac{\xi^*}{\xi} = \frac{c}{c^*}.$$

Also,

$$\theta_n = \theta_0 + \frac{\xi}{n^\delta} = \theta_{n^*} = \theta_0 + \frac{\xi^*}{n^{*\delta}}.$$

That is,

$$\frac{\xi^*}{\xi} = \left(\frac{n^*}{n}\right)^\delta.$$

Consequently,

$$\frac{n^*}{n} = \left(\frac{\xi^*}{\xi}\right)^{1/\delta} = \left(\frac{c}{c^*}\right)^{1/\delta}. \tag{14.2.9}$$

Definition 14.2.3. $c^{1/\delta}$ *is also called the* **Pitman efficacy** *of test* T_n*. Thus, the asymptotic efficiency of* T *relative to* T^* *is the ratio of the efficacies.*

Remark 14.2.1. *One can replace* $E_{\theta_n}(T_n)$ *and* $\sigma_{\theta_n}(T_n)$ *by* $A(\theta_n)$ *and* $B(\theta_n)$ *provided* $\frac{T_n - A(\theta_n)}{B(\theta_n)} \to$ *normal* $(0,1)$ *in distribution.*

Example 14.2.2. Consider the sign test. Let $X_1, \ldots, X_n$ be a random sample from $F\left(\frac{x-\theta}{\sigma}\right)$, where the p.d.f. $f(x)$ is symmetric about zero. We wish to test

$$H_0 : \theta = \theta_0 \text{ against } H_1 : \theta > \theta_0 .$$

Let

$$Z_i = \begin{cases} 1 & \text{if } X_i > \theta_0 \\ 0 & \text{if } X_i \le \theta_0 ,\ i = 1, \ldots, n . \end{cases}$$

Then the sign test is based on the statistic

$$T = \sum_{i=1}^n Z_i ,$$

where

$$ET = np(\theta), \text{ var } T = np(\theta)\left[1 - p(\theta)\right],$$

$$p(\theta) = P(Z_1 = 1) = P(X > \theta_0 | \theta) = 1 - F\left(\frac{\theta_0 - \theta}{\sigma}\right) = F\left(\frac{\theta_0 - \theta}{\sigma}\right).$$

If $\theta_n = \theta_0 + \xi/n^{1/2}$, we have

$$\mathrm{var}(T|\theta_n)/\mathrm{var}(T|\theta_0) \to 1 \text{ as } n \to \infty\,,$$

and

$$\begin{aligned} E(T|\theta_n) - E(T|\theta_0) &= n\left[F\left(\xi/n^{1/2}\sigma\right) - F(0)\right] \\ &= nf(\tilde{\xi})(\xi/n^{1/2}\sigma)\,, \end{aligned}$$

for some $0 < \tilde{\xi} < \xi$. Hence,

$$\frac{E(T|\theta_n) - E(T|\theta_0)}{\xi\sigma_{\theta_0}(T)} = \frac{nf(\tilde{\xi})\cdot 2}{n^{1/2}\sigma\sqrt{n}} = \frac{2f(\tilde{\xi})}{\sigma}\,,$$

and

$$\lim_{n\to\infty}[\text{L.H.S.}] = 2f(0)/\sigma = c\,. \tag{14.2.10}$$

Also, the efficacy of Student's t test $= c^* = \sigma_x^2$ where σ_x^2 denotes the variance of X. Thus,

$$\begin{aligned} e(T, t^*) &= 4f^2(0)\sigma_x^2/\sigma^2\,, \\ &= 2/\pi \text{ (if } f \text{ is normal, } \sigma_x = \sigma) \\ &= 4\cdot\frac{1}{16}\cdot\frac{\pi^2}{3} = \frac{\pi^2}{12} \text{ (if } f \text{ is logistic)} \\ &= 2\left(\sigma_x^2 = 2\sigma^2,\ f(0) = \frac{1}{2} \text{ if } f \text{ is double exponential}\right) \\ &= 1/3\left(\sigma_x^2 = \frac{\sigma^2}{12},\ f(0) = 1\,, \text{ if } f \text{ is uniform}\right). \end{aligned}$$

Efficacy of Student's t test

The test statistic is

$$\begin{aligned}
T &= \sqrt{n}(\bar{X}-\theta_0)/s \sim \sqrt{n}(\bar{X}-\theta_0)/\sigma_x = \tilde{T} \text{ (say)},\\
E_\theta \tilde{T} &= \sqrt{n}(\theta-\theta_0)/\sigma_x\,,\\
E_\theta \tilde{T} - E_{\theta_0}(\tilde{T}) &= \sqrt{n}(\theta-\theta_0)/\sigma_x\,,\\
\theta-\theta_0 &= \xi/\sqrt{n}\,,\\
\operatorname{var}\tilde{T} &= 1\,,
\end{aligned}$$

and

$$c^2 = \lim_{n\to\infty}\left[\frac{E_\theta(\tilde{T})-E_{\theta_0}(\tilde{T})}{\xi\sigma_{\tilde{T}}(H_0)}\right] = 1/\sigma_x^2\,. \tag{14.2.11}$$

Example 14.2.3 (F-test statistic). Let $(X_1,\ldots,X_m)$ be a random sample from $F(x)$ and $(Y_1,\ldots,Y_n)$ be a random sample from $G(x)$. We wish to test $H_0 : F(x) = G(x)$ for all x against $H_1 : G(x) = F(x/\theta)$, $\theta > 1$. Then the F-statistic is given by

$$F = \frac{\sqrt{n}\sum_{j=1}^n (Y_j-\bar{Y})^2/(n-1)}{\sum(X_i-\bar{X})^2/(m-1)}\,. \tag{14.2.12}$$

Thus,

$$F \approx \sqrt{n}\sum_{j=1}^n (Y_j-\bar{Y})^2/(n-1)\sigma^2,$$

since $\sum_{j=1}^m (X_i-\bar{X})^2/(m-1) \to \sigma^2$ in probability and, without loss of generality, we can set $\sigma^2 = 1$.

Let $T = n^{-1/2}\sum_{j=1}^n (Y_j-\bar{Y})^2$. Then

$$ET \approx n^{1/2}\theta\,.$$

We can also write

$$T = n^{-1/2}\sum_{j=1}^n (Y_j-\mu)^2 - n^{1/2}(\bar{Y}-\mu)^2$$

where

$$\mu = EY_j\,.$$

Since $n^{1/2}(\bar{Y}-\mu)$ is bounded in probability,

$$T \approx n^{-1/2}\sum_{j=1}^{n}(Y_j-\mu)^2 \qquad (14.2.13)$$

and

$$ET^2 \approx n^{-1}\left[\sum_{1}^{n} E(Y_j-\mu)^4 + n(n-1)\theta^2\right].$$

Hence,

$$\begin{aligned}\text{var}\,T &\approx E(Y_1-\mu)^4 + (n-1)\theta^2 - n\theta^2\\ (\text{var}\,T|H_0) &\approx (\gamma_2+3)-1 = \gamma_2+2\end{aligned}$$

where γ_2 equals the kurtosis of the X-distribution. Consequently, the efficacy of F-test $= (\gamma_2+2)^{-1}$.

In the expansion of $E_{\theta_n}(T_n)=\Psi_n(\theta_n)$ about θ_0, if the first few derivatives of $\Psi_n(\theta)$ at θ_0 vanish, then the Pitman efficiency need to be modified accordingly. This has been considered by Noether (1955) which will be presented in the following. Let

$$E_{\theta_n}(T_n) = \Psi_n(\theta_n) = \Psi_n(\theta_0) + \cdots + \frac{(\theta_n-\theta_0)^m}{m!}\Psi^{(m)}(\hat{\theta})\,. \qquad (14.2.14)$$

If $\Psi^{(m)}(\theta_0)$ is the first non-vanishing derivative, then the power of T_n at θ_n will tend to

$$1-\Phi(z_\alpha - \xi^m c/m!) \qquad (14.2.15)$$

where

$$c = \lim_{n\to\infty}\left[\{\Psi_n(\theta_n)-\Psi_n(\theta_0)\}/\xi^m\sigma_{\theta_0}(T_n)\right]. \qquad (14.2.16)$$

If two tests T_n, T_n^* have identical powers at $\theta=\theta_n$, then we have

$$\xi^{m_1}c/m_1! = \xi^{*^{m_2}}c^*/m_2!\,. \qquad (14.2.17)$$

The alternatives are identical if

$$\xi/n^\delta = \xi^*/n^{*\delta}\,. \qquad (14.2.18)$$

If now $m_1 = m_2 = m$ as it should be if $m_i\delta = 1/2$ $(i = 1, 2)$, which is true in most cases, we have

$$\frac{n^*}{n} = \left(\frac{\xi}{\xi^*}\right) = \left(\frac{c^*}{c}\right)^{1/m\delta} = \lim_{n\to\infty} \frac{\left[\Psi_{n^*}^{(m)}(\theta_0)/\sigma_{n^*}^*(\theta_0)\right]^{1/m\delta}}{\left[\Psi_n^{(m)}(\theta_0)/\sigma_n(\theta_0)\right]^{1/m\delta}}. \tag{14.2.19}$$

If $m = 1$, we are back in Theorem 14.2.2.

14.3 Pitman Efficiency for C-S Class of Statistics

The Pitman efficacy of Chernoff-Savage class of statistics is given by

$$c = \lim_{N\to\infty} N^{1/2} \int_{-\infty}^{\infty} [J(H) - J(F)]\, dF/\xi \left\{2\left(\frac{1-\lambda_N}{\lambda_N}\right) I\right\}^{1/2} \tag{14.3.1}$$

where

$$H(x) = \lambda_N F(x) + (1 - \lambda_N)G(x)$$

and

$$2I = \int_0^1 J^2(u)du - \left[\int_0^1 J(u)du\right]^2.$$

Case 1 [location shift]

Let

$$H(x) = \lambda_N F(x) + (1-\lambda_N)G(x) = \lambda_N \tilde{F}(x-\theta_0) + (1-\lambda_N)\tilde{F}(x-\theta_N). \tag{14.3.2}$$

Then

$$\begin{aligned} H - F &= (1 - \lambda_N)\{G(x) - F(x)\} \\ &= (1 - \lambda_N)\left\{\tilde{F}(x - \theta_N) - \tilde{F}(x - \theta_0)\right\} \\ &= -(1 - \lambda_N)(\theta_N - \theta_0)\tilde{f}(x^*), \end{aligned} \tag{14.3.3}$$

where x^* lies between $x - \theta_0$ and $x - \theta_N$. Thus,

$$c = -\lim_{N\to\infty} N^{1/2-\delta}\xi\{\lambda_N(1-\lambda_N)\}^{1/2} \int J'(\tilde{F})\tilde{f}(x^*)d\tilde{F}/(2I)^{1/2}, \tag{14.3.4}$$

where

$$(\theta_N - \theta_0) = \xi N^{-\delta}.$$

So, here we choose $\delta = 1/2$, and hence,

$$c = -\lim_{N\to\infty} \{\lambda_N(1-\lambda_N)\}^{1/2} \cdot \int_{-\infty}^{\infty} J'(F)f(x)dF(x)/(2I)^{1/2}. \qquad (14.3.5)$$

Case 2 [scale change]

Let

$$H(x) = \lambda_N \tilde{F} + (1-\lambda_N)\tilde{F}\left(x(1+\xi/N^{\delta})\right). \qquad (14.3.6)$$

Hence,

$$H - \tilde{F} = (1-\lambda_N)x\xi N^{-\delta}\tilde{f}(x^*) \qquad (14.3.7)$$

where x^* lies between x and $x(1+\xi N^{-\delta})$. Here also we choose $\delta = 1/2$ and

$$c = \lim_{N\to\infty} [\lambda_N(1-\lambda_N)]^{1/2} \int xJ'(F)f(x)dF/(2I)^{1/2}. \qquad (14.3.8)$$

Pitman Efficacy of the Two-sample t-Test

The test statistic for testing the hypothesis of equality of the location parameters is given by

$$T = (\bar{X} - \bar{Y})/s\left(\frac{1}{m} + \frac{1}{n}\right)^{1/2} \qquad (14.3.9)$$

where $\bar{X}$ and $\bar{Y}$ denote the sample means based on sample sizes m and n respectively and s^2 denotes the pooled sample variance. Then T is asymptotically equivalent to

$$\tilde{T} = (\bar{X} - \bar{Y})/\sigma\left(\frac{1}{m} + \frac{1}{n}\right)^{1/2} \qquad (14.3.10)$$

and $\tilde{T}$ is asymptotically normally distributed with mean $(\mu_1 - \mu_2)/\sigma \cdot \left(\frac{1}{m} + \frac{1}{n}\right)^{1/2}$ and variance 1. Thus,

$$\begin{aligned} c = \lim_{N\to\infty} \frac{N^{-\delta}\xi}{\xi\sigma\left(\frac{1}{m}+\frac{1}{n}\right)^{1/2}} &= \lim_{N\to\infty} N^{-\delta}\frac{m^{1/2}n^{1/2}}{\sigma N^{1/2}} \\ &= \lim_{N\to\infty} [\lambda_N(1-\lambda_N)]^{1/2}/\sigma \qquad (14.3.11) \end{aligned}$$

upon setting $\delta = 1/2$.

14.4 Bahadur Efficiency

Introduction

In evaluating the Pitman efficiency of a test procedure one holds α (the level of significance) and the power fixed and lets the sequence of alternatives converge to the null hypothesized value The ratio of the sample sizes required by two procedures is used as the criterion for relative efficiency. However, Bahadur (1960a–c, 1967) says to hold the alternative hypothesized value θ fixed. The performance of test procedure A is superior to, or equivalent to, or inferior to that of test B according as the level of significance attained by A is less than, or equal to, or greater than the significance level attained by B. This idea seems to be reasonable and consistent with practice. In the following we will dwell on Bahadur efficiency.

14.4.1 Bahadur Efficiency: Limiting Case

Let T_0 and T_1 be two test procedures, let α $(0 < \alpha < 1)$ be a constant. Let $L_i(n)$ denote the observed level of significance for T_i $(i = 0, 1)$,

$$N_i^- = \begin{cases} \text{least integer} \quad m \text{ such that } L_i(m) \leq \alpha, \\ \qquad 1 \qquad\qquad \text{if no such } m \text{ exists,} \end{cases}$$

and

$$N_i^+ = \begin{cases} \text{least integer} \quad m \text{ such that } L_i(n) \leq \alpha \text{ for all } n \geq m, \\ \qquad \infty \qquad\qquad \text{if no such } m \text{ exists}. \end{cases}$$

The N_i^- and N_i^+ are respectively the lower and upper bounds to the number of observations required by the test T_i in order to achieve the level of significance α. Then, for the sample point $\boldsymbol{X}_\infty = (X_1, X_2, \ldots)$,

$$\frac{N_0^-}{N_1^+} \leq \frac{\text{sample size required by test } T_0}{\text{sample size required by test } T_1} \leq \frac{N_0^+}{N_1^-} \tag{14.4.1}$$

provided the extreme ratios in the above inequality are well-defined. Then we have the following definition of Bahadur efficiency.

Definition 14.4.1. *The Bahadur asymptotic efficiency of T_1 relative to T_0 is*

$$\psi = \lim_{\alpha \to 0} \{N_0^- / N_1^+\} = \lim_{\alpha \to 0} \{N_0^+ / N_1^-\}. \tag{14.4.2}$$

Consider the following example pertaining to samples from the normal distributions.

Example 14.4.1. Let $X_1, X_2, \ldots$ be a sequence of i.i.d. normal (μ, σ) variables. Suppose we wish to test $H_0 : \mu = 0$ versus $H_1 : \mu > 0$.

Case 1: σ is known

Let T_0 be based on

$$Y_n = n^{1/2}\bar{X}_n/\sigma\,, \quad \bar{X}_n = \sum_1^n X_i/n$$

and $S = T_1$ be based on

$$\begin{aligned} R_n &= (\#\text{ of positive } X_i\text{'s, } i = 1, \ldots, n) \\ &= \sum_{i=1}^n I\left(X_i > 0\right). \end{aligned}$$

Then if

$$L_i(n) = (\text{observed level of significance for } T_i,\ i = 0, 1),$$

then

$$L_0(n) = \Phi(-Y_n) \tag{14.4.3}$$

and

$$L_S(n) = \sum_{r=R_n}^n \binom{n}{r}\left(\frac{1}{2}\right)^n \tag{14.4.4}$$

where Φ denotes the standard normal distribution function. When H_0 is true, $L_0(n)$ is uniformly distributed on $(0,1)$ and

$$L_S(n) = P_0(S \geq R_n | \mu = 0) \doteq \Phi\left(\frac{(n/2) - R_n}{(n/4)^{1/2}}\right). \tag{14.4.5}$$

Thus, $L_S(n)$, for sufficiency large n, tends to be distributed uniformly on $(0,1)$.

The asymptotic joint distribution of $L_0(n)$ and $L_S(n)$ can be stated as

$$(L_0(n), L_S(n)) \to \{\Phi(U), \Phi(V)\}$$

in distribution where U and V are jointly normal with

$$EU = EV = 0, \text{ var } U = \text{var } V = 1 \tag{14.4.6}$$

and

$$\begin{aligned} \text{cov}(U,V) &= EUV \\ &= \left\{\sigma(n/4)^{1/2} n^{1/2}\right\}^{-1} E\left(\textstyle\sum_1^n X_i\right)\left(\textstyle\sum_1^n I(X_i > 0)\right) \\ &= 2E\left\{ZI(Z > 0)\right\} \end{aligned} \tag{14.4.7}$$

where Z is a standard normal variable. Hence,

$$E(UV) = 2 \cdot (2\pi)^{-1/2} \int_0^\infty z e^{-z^2/2} dz = (2/\pi)^{1/2} \doteq 0.8 \,. \tag{14.4.8}$$

Since the correlation between U and V is so high, it seems that, for sufficiently large n, the joint distribution of L_O and L_S is concentrated on the diagonal $L_O = L_S$ of the unit square. Let

$$\theta = \mu/\sigma, \;\; p(\theta) = 1 - \Phi(\theta), \;\; q(\theta) = 1 - p(\theta)$$

and

$$\tau = (2p^p q^q)\exp(-\theta^2/2) \,. \tag{14.4.9}$$

Also, let

$$S_n = S_n(\mu,\sigma) = n^{1/2}(\bar{X}_n - \mu)/\sigma$$

and

$$T_n = T_n(p) = (R_n - np)/(npq)^{1/2}, \;\; n = 1, 2, \ldots \,. \tag{14.4.10}$$

Then the joint distribution of L_O and L_S exists if the joint distribution of $\Phi(-S_n)$ and $\Phi(-T_n)$ exists and the two joint distributions will coincide since $|L_S(n) - \Phi(-T_n)| \to 0$ as $n \to \infty$ and Φ is monotone.

Theorem 14.4.1 (Bahadur, 1960a). *For any given μ and σ, the joint distribution of S_n and T_n tends to the bivariate normal with zero means, unit variances and correlation ρ where*

$$\rho(\theta) = \exp(-\theta^2/2)/(2\pi pq)^{1/2} \,.$$

Now, we can write

$$S_n = \sum_1^n Z_i/n^{1/2}, \;\; T_n = \sum_1^n W_i/n^{1/2}$$

where, for $i = 1, 2, \ldots,$

$$Z_i = Z(X_i) = (X_i - \mu)/\sigma, \quad W_i = W(X_i) = \begin{cases} (q/p)^{1/2} & \text{if } X_i > 0, \\ -(p/q)^{1/2} & \text{if } X_i \le 0. \end{cases} \tag{14.4.11}$$

Applying the central limit theorem for bivariate random variables (see Cramér, 1946, p. 285), we infer that (S_n, T_n) is asymptotically bivariate normal with zero means, unit variances and correlation ρ given by

$$\rho = \text{cov}(Z, W) = E(ZW)\,.$$

Note that W is a function of Z because,

$$(pq)^{1/2}W + p = \begin{cases} 0 & \text{if } Z \le -\theta, \\ 1 & \text{if } Z > -\theta. \end{cases}$$

Thus,

$$\begin{aligned} \rho(\theta) &= (pq)^{-1/2}E\left\{Z\left((pq)^{1/2}W + p\right)\right\} \\ &= (pq)^{-1/2}E\left\{ZI(Z > -\theta)\right\} \\ &= (pq)^{-1/2}\int_{-\theta}^{\infty} z\phi(z)dz = (pq)^{-1/2}\phi(-\theta)\,. \end{aligned} \tag{14.4.12}$$

One can easily see that when $\theta = 0$, $\rho = 0.8$ and $\rho \to 0$ as $\theta \to \infty$. Next consider

$$L_0(n) = \Phi(-Y_n) = \Phi(S_n + \sqrt{n}\,\theta)$$

where

$$S_n = \sqrt{n}\left(\frac{\bar{X}_n - \mu}{\sigma}\right) = \sqrt{2}\ \log\ \log(\sqrt{n}\,\sigma)\ \text{a.s.} \tag{14.4.13}$$

after using the law of the iterated logarithm.[1]

---

[1]For standardized sums of i.i.d. random variables, S_n, the law of the iterated logarithm says that

$$\overline{lim}_{n\to\infty}\,\frac{S_n}{2\ \log\ \log\ (\sqrt{n}\,\sigma)} = 1\ \text{a.s.}$$

Hence

$$\begin{aligned} L_0(n) &= 1-\Phi(S_n+\sqrt{n}\,\theta) \\ &\doteq 1-\Phi\left(2\log\log(n\sigma^2)+\sqrt{n}\,\theta\right) \\ &\doteq \frac{\phi\left(2\log\log(n\sigma^2)+\sqrt{n}\,\theta\right)}{2\log\log(n\sigma^2)+\sqrt{n}\,\theta} \end{aligned} \tag{14.4.14}$$

from which it follows that

$$\frac{1}{n}\log L_0(n) = \frac{-\theta^2}{2} + \epsilon_n \tag{14.4.15}$$

where

$$\epsilon_n = O\left(\frac{\log\log n}{n^{1/2}}\right) = o(1)\,.$$

By proceeding along similar lines, one can show that

$$\frac{1}{n}\log L_S(n) = -\log[2p^p q^q] + \epsilon_n' \tag{14.4.16}$$

where

$$\epsilon_n' = o(1)\,.$$

Now, by the definition of $N_0^-(\alpha)$ and $N_s^+(\alpha)$, we have

$$L_0\left(N_0^-(\alpha)-1\right) \le \alpha \le L_0\left(N_0^-(\alpha)\right)$$

and

$$L_S\left(N_S^+(\alpha)\right) \le \alpha < L_S\left(N_S^+(\alpha)-1\right)\,. \tag{14.4.17}$$

Hence, for any α,

$$L_0\left(N_0^-(\alpha)-1\right) > L_S\left(N_s^+(\alpha)\right) \text{ and } L_0\left(N_0^-(\alpha)\right) \le L_S\left(N_S^+(\alpha)-1\right)\,. \tag{14.4.18}$$

Since N_0^- and N_S^+ both tend to ∞ as $\alpha\to 0$, taking logarithms in (14.4.18) and applying (14.4.16) and (14.4.17), we obtain, since

$$\frac{N_0^+(\alpha)}{N_S^-(\alpha)}\cdot\frac{\log L_S(N_S^-)}{\log L_0(N_0^+)} \approx \frac{N_0^+(\alpha)}{N_S^-(\alpha)}\cdot\frac{\log\alpha}{\log\alpha} = \frac{N_0^+(\alpha)}{N_S^-(\alpha)} = 2\log\left\{(2p^p q^q)/\theta^2\right\}\,,$$

$$\psi(\theta) = \lim_{\alpha\to 0}\left\{N_0^+(\alpha)/N_S^-(\alpha)\right\} = 2\log\left\{(2p^p q^q)/\theta^2\right\}\,. \tag{14.4.19}$$

Bahadur (1960a) tabulated some of the values of $\psi(\theta)$ and are given by

θ	0	0.5	1	1.5	2	2.5	3	4	∞
$\psi(\theta)$	0.64	0.60	0.51	0.40	0.29	0.21	0.15	0.09	0

Case 2: σ is unknown

Then, we resort to the t-test procedure which is based on

$$Y_n^* = \sqrt{n}\,\bar{X}_n/V_n, \quad V_n^2 = \sum_1^n (X_i - \bar{X}_n)^2/(n-1)\,. \tag{14.4.20}$$

Thus,

$$L_0^*(n) = G_{n-1}(-Y_n^*) = \begin{cases} \text{the tail of the } t\text{-distribution with } (n-1) \text{ de-} \\ \text{grees of freedom.} \end{cases}$$

Bahadur (1960a) shows that

$$\psi^* = 2\,\log(2p^p q^q)/\log(1+\theta^2) \tag{14.4.21}$$

and computes some of the numerical values of ψ^* and are given by

θ	0	0.5	1.0	2.0	3.0	4.0	∞
ψ^*	0.64	0.67	0.74	0.73	0.59	0.26	0

Remark 14.4.1. *ψ^* exceeds ψ for each θ and this implies that the sign test is more efficient than the parametric test when σ is unknown, than when it is known. Hence, in using the t-test, the loss of information due to ignorance of σ is not asymptotically negligible (as is the case while computing Pitman efficiency).*

14.4.2 Bahadur Efficiency: General Setup

Let S be an abstract space of points s and that s has a distribution in S belonging to a given set of probability measures $\{P_\theta\}$, where θ is an abstract parameter taking values in Ω. We wish to test

$$H_0 : \theta \in \Omega_0 \text{ against } H_1 : \theta \in \Omega - \Omega_0\,.$$

Let T_n be a real-valued statistic defined on S. When n takes values $1, 2, \ldots$.

Definition 14.4.2. *T_n is said to be a* **standard sequence** *(for testing H_0) if*

1. *There exists a continuous d.f. F such that, for each $\theta \in \Omega_0$,*

$$\lim_{n\to\infty} P_\theta(T_n < x) = F(x) \text{ for every } x\,. \tag{14.4.22}$$

2. *There exists a constant a $(0 < a < \infty)$ such that*

$$\log[1 - F(x)] = -\frac{ax^2}{2}[1 + o(1)] \quad \text{as } x \to \infty. \tag{14.4.23}$$

3. *There exists a function b on $\Omega - \Omega_0$ with $0 < b < \infty$ such that, for each $\theta \in \Omega - \Omega_0$,*

$$n^{-1/2}T_n \text{ converges to } b(\theta) \text{ in probability as } n \to \infty. \tag{14.4.24}$$

Example 14.4.2. In order to fix ideas, let us consider the following special case. Let S denote the set of all sequences $s = (x_1, x_2, \ldots)$ with real coordinates x_n. Let Ω be a set of all d.f.'s $\theta(x)$ on the real line such that

$$\mu(\theta) = \int_{-\infty}^{\infty} x d\theta \geq 0 \text{ and } \int x^2 d\theta < \infty.$$

Let P_θ denote the product measure $\theta \times \theta \times \ldots$ on S. Let $H_0 : \mu = 0$. For each n, let T_n be the t-statistic based on the first n coordinates of s. Then I is satisfied with $F = \Phi(x)$ which in turn satisfies 2. with $a = 1$ (see Feller, 1968, p. 175) and 3. is satisfied with $b(\theta) = \mu(\theta)/\sigma(\theta)$ where $\sigma^2 = \int (x - \mu)^2 d\theta$ and n denotes the sample size.

In the general case, $1 - F(T_n(s))$ denotes the observed level of significance of T_n $(n = 1, 2, \ldots)$. Note that $1 - F(T_n(s))$ is only an approximate level which is of interest because the exact null distribution is too difficult to compute or n is large enough to justify the approximate distribution. (Note that in the case of the sign test, we were able to obtain the exact level of significance.) The approximate level of significance is also of interest when the exact level does not exist in the sense that the exact distribution of T_n varies with θ over Ω_0. Let

$$K_n(s) = -2 \log[1 - F(T_n(s))]. \tag{14.4.25}$$

Then, for each $\theta \in \Omega_0$,

$$\lim_{n\to\infty} P_\theta(K_n < v) = P(\chi_2^2 < v) = 1 - \exp(-v/2), \ v > 0 \tag{14.4.26}$$

where χ_2^2 is a chi-square variable with 2 degrees of freedom. Let

$$C(\theta) = \begin{cases} 0 & \text{for } \theta \in \Omega_0, \\ a[b(\theta)]^2 & \text{for } \theta \in \Omega_0. \end{cases} \tag{14.4.27}$$

Then, for any $\theta \in \Omega$, we have

$$K_n/n = C + \epsilon_n \tag{14.4.28}$$

where $\epsilon_n(s, \theta) \to 0$ in probability as $n \to \infty$.

In order to prove (14.4.28), let $\theta \in \Omega_0$ and let y, z be given constants such that $0 < y < z < 1$. Since F is continuous, there exist numbers a and b such that $F(a) = z - y$ and $F(b) = z$. Define the sets $A_n = \{s : T_n < a\}$, $B_n = \{s : F(T_n) < z\}$, $C_n = \{s : T_n < b\}$. Then $A_n \subset B_n \subseteq C_n\}$ for every n and hence

$$z - y \leq \text{lim inf } P_\theta(B_n) \leq \text{lim sup } P_\theta(B_n) \leq z \text{ by } (14.4.21)\,.$$

Since y and z are arbitrary, we infer that lim $P_\theta(B_n) = z$ for all z in $(0, 1)$, (14.4.25) follows from (14.4.24) and (14.4.27) is satisfied with $C = 0$.

Next, let $\theta \in \Omega - \Omega_0$. For any x let $h(x)$ be the $o(1)$ term on the right side of (14.4.23) where $-1 \leq h \leq \infty$. Then from (14.4.25), we have

$$K_n/n = a\left(\frac{1}{n}T_n^2\right)[1 + h(T_n)]\,.$$

Now use (14.4.24) in order to obtain (14.4.28).

Definition 14.4.3. *$C(\theta)$ is called the* **asymptotic slope** *of the test based on $\{T_n\}$ (or simply the slope of $\{T_n\}$) when θ is obtained.*

Definition 14.4.4. *$\{T_n\}$ is said to be strongly consistent if condition 3. is satisfied with (14.4.3) replaced by*

$$T_n/n^{1/2} \to b(\theta) \ a.s. \quad as\ n \to \infty\,. \tag{14.4.29}$$

If (14.4.28) is satisfied with $b = 0$ for every $\theta \in \Omega_0$, then ϵ_n in (14.4.28) tends to zero almost surely.

Remark 14.4.2. *Note that $\{K_n^{1/2}\}$ is a normalized version of $\{T_n\}$ and hence is equivalent to $\{T_n\}$ in the sense that the level obtained by $K_n^{1/2}$ equals the level attained by T_n.*

Comparison of Standard Sequences

Let $\{T_n^{(i)}\} = \{T_1^{(i)}, T_2^{(i)}, \ldots\}$, $i = 1, 2$ be two standard sequences defined on S. Let $F^{(i)}(x)$, a_i and $b_i(\theta)$ be the functions and constants prescribed by conditions 1, 2 and 3 for sequence i ($i = 1, 2$). Consider an arbitrary but fixed $\theta \in \Omega - \Omega_0$ and assume that s is distributed according to the probability measure P_θ. Then

$$\psi_{1,2}(\theta) = C_1(\theta)/C_2(\theta) \tag{14.4.30}$$

serves as the asymptotic efficiency of sequence 1 relative to sequence 2 where $C_i = a_i b_i^2$ $(i = 1, 2)$.

First fix n and compare the levels attained. We can reasonably say that the test procedure based on $T_n^{(i)}$ is less effective than the test based on $T_n^{(i)}$ if the level attained by $T_n^{(i)}$ exceeds the level attained by $T_n^{(j)}$, that is, $K_n^{(i)} < K_n^{(j)}$. Since $T_n^{(1)}$ and $T_n^{(2)}$ are standard sequences, it follows from (14.4.27) and (14.4.29) that

$$K_n^{(1)}/K_n^{(2)} \to \psi_{1,2} \text{ in probability as } n \to \infty\,. \tag{14.4.31}$$

Thus, with probability tending to unity, we say that $T_n^{(1)}$ is less, more or equally effective according as $\psi_{1,2} < 1$, $\psi_{1,2} > 1$ or $\psi_{12} = 1$, respectively. In order to compare the sample sizes required to attain the same level, for each i, let $m_1^{(i)}, m_2^{(i)}, \ldots$ be a sequence of positive integers such that

$$\lim_{r \to \infty} m_r^{(i)} = \infty \ (i = 1, 2)\,. \tag{14.4.32}$$

For simplicity, let $K_n^{(i)}(s) = K^{(i)}(n, s)$ and $T_n^{(i)}(s) = T^{(i)}(n, s)$. Then, the sample sizes $m_r^{(1)}$ and $m_r^{(2)}$ are said to be asymptotically equivalent for sequences 1 and 2, respectively, if

$$K^{(1)}(m_r^{(1)}, s)/K^{(2)}(m_r^{(2)}, s) \to 1 \text{ in probability as } r \to \infty\,. \tag{14.4.33}$$

Now, from (14.4.31) and (14.4.27) we obtain

$$\lim_{r \to \infty} \left\{ m_r^{(2)}/m_r^{(1)} \right\} = \psi_{1,2}\,. \tag{14.4.34}$$

It should be noted that asymptotically equivalent sample sizes always exist. For example, set $m_r^{(1)} = r$ and $m_r^{(2)} = [r\psi_{1,2} + 1]$ A stronger interpretation of ψ can be given if the sequences being compared are strongly consistent. For details, see Bahadur (1960b).

Discussion

1. There is a formal connection between the asymptotic slope of a standard sequence and the asymptotic power of the corresponding sequence

of tests. In Appendix I of Bahadur (1960a) it is shown that ψ can be regarded as the asymptotic relative efficiency when the power is held fixed (or at least bounded away from 0 and 1) and the resulting test sizes are compared.

2. Suppose that Ω is a metric space and $\Omega - \Omega_0$ is dense in Ω_θ. Let $\{T_n^{(1)}\}$ and $\{T_n^{(2)}\}$ be standard sequences and let $\psi_{1,2}(\theta)$ be defined by (14.2.30). Let

$$\psi_{1,2}(\theta_0) = \lim_{\theta \to \theta_0} \psi_{1,2}(\theta) \text{ for each } \theta_0 \in \Omega_0 .$$

Then, $\psi_{1,2}(\theta_0)$ coincides with Pitman's relative efficiency function in cases where Pitman's theory is also applicable.

3. Let $S^{(1)}$ and $S^{(2)}$ denote spaces of alternative experiments. In particular, suppose that $\left\{T_n^{(i)}\right\}$ is a natural or optimum sequence on $S^{(i)}(i = 1, 2)$. Then $\psi_{1,2}$ is, in a sense, the asymptotic efficiency of experiment 1 relative to experiment 2. This application is discussed in more detail in Bahadur (1960c). In this case, $\psi_{1,2}$ corresponds to the relative 'amount of information per observation' in the theory of estimation.

4. The concept of Bahadur efficiency can be extended to the situation where for each n, the level achieved by T_n is defined in terms of a distribution function depending on n. For details see Bahadur (1960b).

Examples of $F^{(i)}$ Which Satisfy Condition 2

1. $F^{(1)}(x) = \Phi(x)$ with $a = 1$,

2. $F^{(2)}(x) = P(\chi_k^2 < x)$ with $a = 1$,

3. $F^{(3)}(x) = 1 - 2\sum_{k=1}^{\infty}(-1)^{r-1}\exp(-2r^2x^2)$ with $a = 4$.

To show the property for $F^{(2)}$, choose an integer m such that $2m \geq k$. Then

$$\begin{aligned} 2\left[1 - F^{(1)}(x)\right] &= P(\chi_1^2 > x) \leq P(\chi_k^2 > x) \\ &= 1 - F^{(2)}(x) \leq P(\chi_{2m}^2 > x) = P(Y \leq m - 1) \end{aligned}$$

where Y is a Poisson variable with mean $x^2/2$. It follows that the lower and upper bounds for $1 - F^{(2)}(x)$ are of the form

$$\exp\left[-\frac{1}{2}x^2\left(1 + o(1)\right)\right].$$

The verification of the property for $F^{(3)}$ is clear from the first term in $1-F^{(3)}$.

Example 14.4.3. Let $S = (X_1, X_2, \ldots)$ where the X_i are i.i.d. normal (μ, σ) and, μ and σ are unknown. Let

$$T_n^{(1)} = |2U_n - n|/n^{1/2}$$

where U_n denotes the number of positive X's,

$$T_n^{(2)} = \text{ the } t\text{-statistic}.$$

We wish to test $H_0 : \mu = 0$ against $H_1 = \mu \neq 0$. Then, for $\theta = (\mu, \sigma)$,

$$\psi_{1,2}(\theta) = \left[2\Phi(\Delta) - 1\right]^2/\Delta^2, \quad \Delta = \mu/\sigma. \qquad (14.4.35)$$

This efficiency function is different from the one given in (14.4.19) and also from the expression of Hodges and Lehmann (1956). Also note that

$$\lim_{\mu \to 0} \psi_{1,2}(\theta) = 2/\pi$$

for every σ (coinciding with the Pitman efficiency of the sign test relative to the t-test) and that $\psi_{1,2}$ is a decreasing function of $|\Delta|$.

Assuming that σ is known, Bahadur (1960b) obtains the efficiency of Kolmogorov test statistic relative to the t-test and the sign test. Bahadur also provides the example where $S = (X_1, X_2, \ldots)$ where the X_i are i.i.d. normal $(0, \sigma)$ where σ is unknown. We wish to test $H_0 : \sigma = 1$, versus $H_1 : \sigma \neq 1$. He computes the efficiency of Kolmogorov's test relative to the chi-squared test and another test. For some examples of exact slopes, see Bahadur (1971, Section 8).

Abrahmson (1965) uses some large deviation results in order to compute the exact Bahadur efficiency for a number of classical statistics such as the F-statistic, studentized range, between-sample sum of squares, one and two sample Kolmogorov-Smirnov statistics. Also, Klotz (1965) computes the exact Bahadur efficiency of the one-sample Wilcoxon and normal scores tests. Hoadley (1967) computes the exact Bahadur efficiency of the Wilcoxon test relative to the t-test. Chernoff (1952) provides an alternative measure of asymptotic efficiency for tests of a hypothesis.

The concept of *deficiency* was introduced by Hodges and Lehmann (1970) to take care of cases in which Pitman efficiency of S relative to T (where S and T are two competing test procedures is equal to unity. They consider expansions (similar to those carried out in Pitman efficiency computations) to higher order terms in the sample sizes, say up to $O(n^{-1})$). Via this expansions, they examine the loss incurred by using the one-sample t-test instead of $\sqrt{n}\bar{X}/\sigma$ when σ is known. They show that, although there is a slight deficiency in using the t-test, this is not appreciable when compared with the protection provided by the t-test against possible errors in the assumed value of σ.

14.5 Problems

14.2.1 Obtain the ARE of Wilcoxon rank sum test relative to students t-test with respect to location alternatives. Hence, show that it becomes

$3/\pi$ when the underlying populations are normal,
and $3/2$ when the populations are double exponential.

14.2.2 Show that

(i) $ARE(T_1, T_3) = ARE(T_1, T_2) \cdot ARE(T_2, T_3)$
(ii) $ARE(T_1, T_3) = [ARE(T_3, T_1)]^{-1}$.

Now let T_1 be the sign test, T_2 be the Student's t-test and T_3 be the Wilcoxon signed-rank test (in the one-sample location case). Also, evaluate $ARE(T_1, T_3)$ for double exponential, normal and logistic cases.
(HINT: The answers are $4/3, 2/3, 3/4$ respectively.)

14.3.1 Obtain the ARE of Klotz's normal scores test for scale (see (12.4.9)) relative to the F-test when the underlying populations are normal. (HINT: Use Example 14.2.3 and Eq. (14.3.8).)

14.3.2 Obtain the ARE of Mood's test for scale (see (12.4.5)) relative to the F-test when the underlying populations are normal.

14.4.1 By looking at values of $\psi(\theta)$ (Bahadur efficiency of sign test relative to the Z-test when σ is known) and at values of $\psi^*(\theta)$ (Bahadur's efficiency of sign test relative to t-test when σ is unknown), what conclusion can be drawn in comparison with Pitman efficiency? How do you interpret $\psi(0) = \psi^*(0) = 0.64$?

Chapter 15

LMP Tests for Independence

15.1 Introduction

If X and Y denote random variables having a bivariate distribution, the covariance between these variables gives a measure of the association and its direction between these two variables. However, covariance is not a good measure of association because its value depends on the orders of magnitude and units of measurement of the variables X and Y. A unit-free relative measure of association such as the Pearson's product-moment correlation overcomes the above difficulty, which is given by

$$\rho(X,Y) = \frac{\text{cov}(X,Y)}{\sigma_X \sigma_Y}$$

where σ_X^2 and σ_Y^2 denote the variances of X and Y respectively. Note that $\rho(X,Y)$ lies between -1 and 1 and is invariant under location and scale changes in X and Y. If (X,Y) has a bivariate normal distribution, then ρ explicitly occurs in the bivariate density function and when $\rho = 0$, the variables become independent. Zero correlation does not always imply the independence of the variables except in the normal case. However, in non-parametric approach, we have to define dependence in a more general way other than correlation. Hence, we talk in terms of measures of association.

In the following we define a pair of dependent variables and derive LMP rank test criterion for independence along the lines of Hájek and Sidák (1967).

15.2 LMP Rank Tests

Let

$$X = X^* + Z \text{ and } Y = Y^* + Z \tag{15.2.1}$$

where X^*, Y^* and Z are independent with cdfs F, G and L which are known except for the distribution of Z. Assume that

$$EZ = \xi \text{ and } \sigma^2 = \text{var } Z < \infty .$$

We wish to test

$$H_0 : X, Y \text{ are independent}$$

vs.

$$H_1 : X, Y \text{ are positively associated.}$$

Let

$$H_\Delta : X = X^* + \Delta Z \text{ and } Y = Y^* + \Delta Z$$

where Δ is a positive constant and this formulation is similar to that of Bhuchongkul (1964). We have a random sample

$$(X_i, Y_i), \quad i = 1, \dots, n .$$

Assume that

$$\int_{-\infty}^{\infty} |f'(x)|\, dx < \infty \text{ and } \int_{-\infty}^{\infty} |g'(x)|\, dx < \infty$$

where f and g are the densities associated with X^* and Y^* and that the derivatives f' and g' are continuous almost everywhere. Let

$$\begin{aligned}
H_\Delta(x,y) &= P(X \le x,\ Y \le y | H_\Delta) = \text{the joint d.f. of } X \text{ and } Y, \\
h_\Delta(x,y) &= \frac{\partial^2}{\partial x \partial y} H_\Delta(x,y) = \text{the p.d.f. of } X \text{ and } Y, \\
f_\Delta(x) &= \int f(x - \Delta z) dL(z) = \text{the p.d.f. of } X, \\
g_\Delta(x) &= \int g(x - \Delta z) dL(z) = \text{the p.d.f. of } Y,
\end{aligned}$$

Let $(R_1, \dots, R_n)$ be the ranks of the X's and $(S_1, \dots, S_n)$ be the ranks of the Y's. Then Hájek and Sidák (1967, pp. 76–77) show that

$$\lim_{\Delta \to 0} \Delta^{-2} \left\{ (N!)^2 P_\Delta(R = r, S = s) - 1 \right\}$$

$$= \sigma^2 \sum_{i=1}^{N} \left\{ E \frac{f'(X_{r_i,N})}{f(X_{r_i,N})} \right\} \left\{ E \frac{g'(Y_{r_i,N})}{g(Y_{r_i,N})} \right\} . \tag{15.2.2}$$

15.3 Derivation of the LMP Rank Test

Lemma 15.3.1 (Hájek and Sídak, 1967, pp. 76–77). *We have*

$$\lim_{\Delta\to 0} \Delta^{-2}\left[h_\Delta(x,y) - f_\Delta(x)g_\Delta(y)\right] = f'(x)g'(y)\int_{-\infty}^{\infty}(z-\xi)^2 dL(z) \quad (15.3.1)$$

at every (x, y) *such that* x *is a continuity point of* $f'(\cdot)$ *and* y *is a continuity of* $g'(\cdot)$.

Recall that

$$\begin{aligned}
&\Delta^{-2}\{h_\Delta(x,y) - f_\Delta(x)g_\Delta(y)\} \\
&= \iint \{f(x-\Delta z)g(y-\Delta z) - f(x-\Delta z)g(y-\Delta z')\}dL(z)dL(z') \\
&= \frac{1}{2\Delta^2}\iint \{f(x-\Delta z) - f(x-\Delta z')\}\{g(x-\Delta z) \\
&\quad - g(x-\Delta z')\}dL(z)dL(z') \\
&= \frac{1}{2\Delta^2}\iint_{|\Delta z|\le\delta, |\Delta z'|\le\delta} + R_{\Delta\delta}\,, \qquad (15.3.2)
\end{aligned}$$

where

$$\begin{aligned}
R_{\Delta\delta} &\le 4\max_u f(u)\max_v g(v)\Delta^{-2}P\left(|\Delta z| > \delta\right) \\
&\le 4(\max f)(\max g)\delta^{-2}\int_{|z|>\delta/\Delta} z^2 dL(z)\,. \qquad (15.3.3)
\end{aligned}$$

Thus it follows that for each $\delta > 0$,

$$R_{\Delta\delta\to 0} \text{ as } \Delta \to 0\,.$$

Also, due to the continuity of $f'(x)$, for every $\epsilon > 0$, there exists a $\delta > 0$ such that for $|\Delta z| \le \delta$ and $|\Delta z'| \le \delta$,

$$\begin{aligned}
&\left|\frac{1}{\Delta(z-z')}\left[f(x-\Delta t) - f(x-\Delta z')\right] - f'(x)\right| \\
&= \frac{1}{\Delta(z-z')}\int_{x-\Delta z'}^{x-\Delta z}\left|f'(t) - f'(x)\right| dt < \epsilon\,.
\end{aligned}$$

An analogous inequality holds for g. Thus,

$$\begin{aligned}
&\lim_{\Delta\to 0}\Delta^{-2}\left[h_\Delta(x,y)-f_\Delta(x)g_\Delta(y)\right]\\
&= \left\{\frac{1}{2}\int_{-\infty}^{\infty}\int_{-\infty}^{\infty}(z-z')^2 dL(z)dL(z')\right\} f'(x)g'(y)\\
&= f'(x)g'(y)\text{var } Z\,. \qquad (15.3.4)
\end{aligned}$$

Also, writing

$$f(x-\Delta z)-f(x-\Delta z')=\int_{x-\Delta z'}^{x-\Delta z} f'(t)dt=\int_{\Delta z'}^{\Delta z} f'(x-t)dt\,,$$

one can easily show that

$$\begin{aligned}
&\Delta^{-2}\int_{-\infty}^{\infty}\int_{-\infty}^{\infty}\left|h_\Delta(x,y)-f_\Delta(x)g_\Delta(y)\right| dx\, dy\\
&\leq \frac{1}{2}\Delta^{-2}\iint(\Delta z-\Delta z')^2 dL(z)dL(z')\int_{-\infty}^{\infty}\int_{-\infty}^{\infty}\left|f'(x)g'(y)\right| dx\, dy\\
&= (\text{var } Z)\left\{\int_{-\infty}^{\infty}\left|f'(x)\right| dx\right\}\left\{\int_{-\infty}^{\infty}\left|g'(y)\right| dy\right\}. \qquad (15.3.5)
\end{aligned}$$

Next, consider

$$\begin{aligned}
&\lim_{\Delta\to 0}\Delta^{-2}\left[(N!)^2 P(R=r,S=s|H_\Delta)-1\right]\\
&= \lim_{\Delta\to 0}\Delta^{-2}(N!)^2\int\cdots\int_{\{R=r,S=s\}}\left\{\prod_{i=1}^{N}h_\Delta(x_i,y_i)-\prod_{i=1}^{N}f_\Delta(x_i)g_\Delta(y_i)\right\}\prod_{1}^{N}dx_i dy_i\\
&= \lim_{\Delta\to 0}(N!)^2\sum_{i=1}^{N}\int\cdots\int_{\{R=r,S=s\}}\prod_{k=1}^{i-1}h_\Delta(x_k,y_k)\,\cdot\\
&\quad\cdot\left[\frac{h_\Delta(x_i,y_i)-f_\Delta(x_i)g_\Delta(y_i)}{\Delta^2}\right]\prod_{k=i+1}^{N}f_\Delta(x_k)g_\Delta(y_k)\prod_{1}^{N}dx_k dy_k\,. \qquad (15.3.6)
\end{aligned}$$

Because of (15.3.5) we can take the limit underneath the integral sign and using Lemma 15.3.1 we obtain

$$\begin{aligned}\text{LHS} &= (\text{var } Z)(N!)^2 \sum_{i=1}^{N} \int \cdots \int_{\{R=r,S=s\}} \frac{f'(x_i)g'(y_i)}{f(x_i)g(y_i)} \prod_{k=1}^{N} f(x_k)g(y_k)dx_k dy_k \\ &= (\text{var } Z) \sum_{i=1}^{N} \left\{ E\frac{f'(X_{r_i,N})}{f(X_{r_i,N})} \right\} \left\{ E\frac{f'(Y_{s_i,N})}{f(Y_{s_i,N})} \right\} . \qquad (15.3.7)\end{aligned}$$

So the test procedure is to reject H_0 when

$$T = \sum_{i=1}^{n} \left\{ E\frac{f'(X_{r_i,N})}{f(X_{r_i,N})} \right\} \left\{ E\frac{g'(Y_{s_i,N})}{g(Y_{s_i,N})} \right\} > k_\alpha .$$

Now let $X_{1N} < \cdots < X_{NN}$ be the ordered X's and $s_1^*, s_2^*, \ldots, s_N^*$ be the ranks of the corresponding Y's. Then

$$T = \sum_{i=1}^{n} \left\{ E\frac{f'(X_{i,N})}{f(X_{i,N})} \right\} \left\{ E\frac{g'(Y_{s_i^*,N})}{g(Y_{s_i^*,N})} \right\} . \qquad (15.3.8)$$

Also, note that by replacing X's by their ranks and the Y's by their ranks in the product-moment correlation coefficient, one can obtain a statistic which is equivalent to the Spearman's test criterion.

Special Cases

1. Let f and g be standard logistic density functions. Then

$$T = \sum_{i=1}^{N} E\left\{2F(X_{i,N}) - 1\right\} E\left\{2G(Y_{s_i^*,N}) - 1\right\} \qquad (15.3.9)$$

   which is equivalent to

$$\sum_{i=1}^{N} i s_i^* , \qquad (15.3.10)$$

   namely, the Spearman test criterion.

2. **Fisher Yates Normal Scores Test** Let f and g be the standard normal density functions. Then

$$T = \sum_{i=1}^{N} a_N(i) a_N(s_i^*) , \qquad (15.3.11)$$

and

$$E(T|H_0) = \frac{1}{N}\left\{\sum_1^N a_N(i)\right\}^2 .$$

Since, for normal scores, $\sum_1^N a_N(i) = 0$, we have

$$E(T|H_0) = 0 .$$

Consider

$$\begin{aligned} \text{var}(T|H_0) = E_0(T^2) &= \sum a_N^2(i) E a_N^2(s_i^*) + \sum\sum_{i \neq j} \\ &= (N-1)^{-1}\left\{\sum_{i=1}^N (EV_{N,i}^2)\right\}^2 \\ &= N^2(N-1)^{-1} , \end{aligned}$$

a justification of which will be given in Section 15.4, when $V_{N,i}$ denote standard normal order statistics in a sample of size N.

3. **Van der Wearden Type of Test** is given by T where

$$T = \sum_{i=1}^N \Phi^{-1}\left(\frac{i}{N+1}\right)\Phi^{-1}\left(\frac{s_i^*}{N+1}\right) . \tag{15.3.12}$$

4. Let f and g be the standard double exponential density functions. Then

$$T = \sum_{i=1}^N E\ \text{Sgn}(X_{iN}) E\ \text{Sgn}(Y_{s_i^*,N}) \tag{15.3.13}$$

where $\text{Sgn}(x) = -1$ if $x < 0$, 0 if $x = 0$ and $+1$ if $x > 0$. Hájek and Sidák (1967, pp. 114 and 67) show that the test is asymptotically equivalent to the quadrant test of Elandt (1962).

Remark 15.3.1. *If one suspects that X and Y are negatively associated and wishes to test for independence of X and Y, then we write*

$$X = X^* + \Delta Z \ \text{and}\ Y = Y^* - \Delta Z$$

and derive the LMP rank test of $H_0 : \Delta = 0$ versus $H_1 = \Delta > 0$. Other forms of dependence are considered by Konijn (1956) and Farlie (1961). The former writes

$$X = \lambda_1 U + \lambda_2 V,\ \ Y = \lambda_3 U + \lambda_4 V \tag{15.3.14}$$

where U and V are independent. Farlie (1961) writes the joint distribution function of (X, Y) as

$$H(x,y) = F(x)G(y)\left[1 + \Delta A(x)B(y)\right], \ \Delta \geq 0\,. \tag{15.3.15}$$

15.4 The Variance of the Test Statistic under H_0

Consider

$$T = \sum_{1}^{N} a_N(i)a_N(s_i^*)\,.$$

Let

$$\bar{a} = \frac{1}{N}\sum_{1}^{N} a_N(i)\,.$$

Then,

$$\begin{aligned} E_0T = E(T|H_0) &= \sum a_n(i)\frac{1}{N}\sum a_N(j) \\ &= \frac{1}{N}\left(\sum a_N(i)\right)^2 = N\bar{a}^2\,, \end{aligned}$$

$$\begin{aligned} T - E_0T &= \sum a_N(i)a_N(s_i^*) - N\bar{a}^2 \\ &= \sum \{a_N(i) - \bar{a}\}\{a_N(s_i^*) - \bar{a}\}\ , \end{aligned}$$

$$\begin{aligned} &E_0(T - E_0T)^2 \\ &= E_0\left[\sum \tilde{a}_N(i)\tilde{a}_N(s_i^*)\right]^2, \ \tilde{a}_N(i) = a_N(i) - \tilde{a} \\ &= \left[\sum \tilde{a}_N^2(i)E_0\tilde{a}_N^2(s_i^*) + \sum\sum_{i\neq j} \tilde{a}_N(i)\tilde{a}_N(j)E_0\tilde{a}_N(s_i^*)\tilde{a}_N(s_j^*)\right] \\ &= \frac{1}{N}\left\{\sum \tilde{a}_N^2(i)\right\}^2 + \frac{1}{N(N-1)}\left[\sum\sum_{i\neq j} \tilde{a}_N(i)\tilde{a}_N(j)\right]^2 \end{aligned}$$

since under H_0, $P(s_i^* = k, s_j^* = l) = 1/N(N-1)$. Thus,

$$\begin{aligned}
\text{var}(T|H_0) &= \frac{1}{N}\left\{\sum \tilde{a}_N^2(i)\right\}^2 + \frac{1}{N(N-1)}\left[\left(\sum \tilde{a}_N(i)\right)^2 - \sum \tilde{a}_N^2(i)\right]^2 \\
&= \left\{\frac{1}{N} + \frac{1}{N(N-1)}\right\}\left\{\sum \tilde{a}_N^2(i)\right\}^2 \\
&= (N-1)^{-1}\left\{\sum \tilde{a}_N^2(i)\right\}^2 .
\end{aligned}$$

15.5 Other Rank Tests

A nonlinear rank test for testing independence was proposed by Kendall (see Kendall and Gibbons, 1992) which is based on concordant pairs of observations in the sample.

Definition 15.5.1. *A pair (X_i, Y_i) and (X_j, Y_j) is said to be a concordant pair if either $X_i < X_j$ and $Y_i < Y_j$ or $X_i > X_j$ and $Y_i > Y_j$. Otherwise the pair is said to be discordant. Then Kendall's tau is defined by*

$$\begin{aligned}
\tau &= \frac{1}{N(N-1)}\sum\sum_{i\neq j} Sgn(R_i - R_j)Sgn(S_i - S_j) \\
&= \frac{1}{N(N-1)}\sum\sum_{i\neq j} Sgn(i-j)Sgn(s_i^* - s_j^*) . \qquad (15.5.1)
\end{aligned}$$

If T denotes the number of pairs of indices (i, j), $i < j$ for which $s_i^* < s_j^*$ (i.e., T denotes the number of concordant pairs with $i < j$), then

$$\begin{aligned}
N(N-1)\tau &= -\sum\sum_{i<j} Sgn(s_i^* - s_j^*) + \sum\sum_{i>j} Sgn(s_i^* - s_j^*) \\
&= (\#\text{ of concordant pairs } - \#\text{ of discordant pairs})_{i<j} \\
&+ (\#\text{ of concordant pairs } - \#\text{ of discordant pairs})_{i>j} \\
&= \#\text{ of concordant pairs } - \#\text{ of discordant pairs} \\
&= 2(\#\text{ of concordant pairs}) - N(N-1) \\
&= 4T - N(N-1) .
\end{aligned}$$

That is,

$$\tau = \frac{4T}{N(N-1)} - 1 . \qquad (15.5.2)$$

Thus, for testing purposes, the simpler quantity T can be used.

Let p_c denote the probability that (X_i, Y_i) and (X_j, Y_j) are concordant.

If (X, Y) has any continuous bivariate distribution, $p_c + p_d = 1$ where p_d is the probability that the pair of random variables is discordant. Also, one can easily show that when X and Y are independent, $p_c = p_d = 1/2$. When (X, Y) has a bivariate normal distribution with correlation ρ, one can obtain an explicit expression for p_c. Towards this, let

$$V = (X_i - X_j)/\sqrt{2}\sigma_X \text{ and } W = (Y_i - Y_j)/\sqrt{2}\sigma_Y \,.$$

Then, since (V, W) is standard bivariate normal with correlation ρ, we have

$$\begin{aligned} p_c &= P(VW > 0) \\ &= 2\int_0^\infty \int_0^\infty \frac{1}{2\pi}(1-\rho^2)^{1/2}\, e^{-\frac{1}{2(1-\rho^2)}(x^2 - 2\rho xy + y^2)}\, dx\, dy \end{aligned}$$

upon noting that (V, W) is symmetric about $(0,0)$. Also, writing $(x^2 - 2\rho xy + y^2)$ as $(x-\rho y)^2 + (1-\rho^2)y^2$ and setting $z = (x-\rho y)/(1-\rho^2)^{1/2}$, we have

$$\begin{aligned} p_c &= 2\int_0^\infty \phi(y) \left[\int_{-\rho y/(1-p^2)^{1/2}}^\infty \phi(z)dz\right] dy \\ &= 2\int_0^\infty \phi(y)\Phi\left(\frac{\rho y}{(1-\rho^2)^{1/2}}\right) dy \,. \end{aligned}$$

Now consider

$$\begin{aligned} \frac{d}{d\rho}p_c &= \frac{(1-\rho^2)^{-3/2}}{\pi} \int_0^\infty y \exp\left(-\frac{y^2}{2(1-\rho^2)}\right) dy \\ &= \frac{(1-\rho^2)^{-1/2}}{\pi}, \text{ upon setting } y^2/2(1-\rho^2) = u \,. \end{aligned}$$

Thus,

$$p_c = \frac{1}{\pi}\int_a^\rho (1-t^2)^{-1/2} dt \,.$$

Now set $t = \sin\,\theta$ and integrating, we obtain

$$p_c = \frac{1}{\pi}(\text{Arc sin } \rho - \text{Arc sin } a) \,.$$

Now $p_c = 1/2$ when $\rho = 0$ implies that

$$p_c = \frac{1}{\pi} \text{ Arc sin } \rho + \frac{1}{2} \,.$$

So, in the case of the bivariate normal distribution,

$$E(\tau) = 2E\left(\frac{T}{N(N-1)/2}\right) - 1 = 2p_c - 1$$
$$= 2\left(\frac{1}{\pi}\text{ Arc sin }\rho + \frac{1}{2}\right) - 1$$
$$= \frac{2}{\pi}\text{ Arc sin }\rho\,. \qquad (15.5.3)$$

15.6 Variance of Kendall's Test

Let $(X_1, Y_1), \ldots, (X_N, Y_N)$ be the random sample drawn from the bivariate population. Let

$$W_{ij} = \text{Sgn}(X_i - X_j)\text{Sgn}(Y_i - Y_j)$$

where

$$\text{Sgn}(u) = \begin{cases} -1 \text{ if } u < 0 \\ 0 \text{ if } u = 0 \\ 1 \text{ if } u > 0\,. \end{cases} \qquad (15.6.1)$$

Then,

$$E(W_{ij}) = 1 \cdot p_c - 1 \cdot p_d = p_c - p_d\,. \qquad (15.6.2)$$

Recall that Kendall's tau is defined by

$$\frac{N(N-1)}{2}\tau = \sum\sum_{i<j} W_{ij}\,. \qquad (15.6.3)$$

Then,

$$N^2\frac{(N-1)^2}{4}\text{var } \tau$$
$$= \text{var}(\sum\sum_{i<j} W_{ij})$$
$$= \sum\sum_{i<j} \text{var } W_{ij} + \sum\sum_{\substack{i<j \\ i\neq h}} or \sum\sum_{\substack{h<k \\ j\neq k}} \text{cov}(W_{ij}, W_{h,k})\,. \qquad (15.6.4)$$

Note that when $h \neq i$ and $h \neq j$, $\text{cov}(W_{ij}, W_{hk}) = 0$ since we are dealing with independent pairs. The covariance terms can be partitioned as

1. with $i = h$, $j \neq k$, we obtain $\sum_{i=1}^{N-1} \sum_{j=i+1}^{N} \sum_{k=i+1, k\neq j}^{N} \text{cov}(W_{ij}, W_{ik})$,

2. with $i \neq h$, $j = k$, we obtain $\sum_{j=2}^{N} \sum_{i+1}^{j-1} \sum_{h=1,h\neq j}^{j-1} \text{cov}(W_{ij}, W_{hj})$,

3. with $i \neq k$, $j = h$, we obtain $\sum_{j=2}^{N} \sum_{i=1}^{j-1} \sum_{k=j+1,k\neq i}^{N} \text{cov}(W_{ij}, W_{jk})$,

4. with $i = k$, $j \neq k$, we obtain $\sum_{i=1}^{N-1} \sum_{j=i+1}^{N} \sum_{h=1,h\neq j}^{N} \text{cov}(W_{ij}, W_{hi})$.

Because of symmetry, all the covariances are equal. Within the first set we have two distinct permutations (W_{ij}, W_{ik}) and (W_{ik}, W_{ij}) for each of the $\binom{N}{j}$ choices of $i \neq j \neq k$ and similarly for the second set. However, the third and fourth sets do not allow the interchange of the W_{ij} and W_{jk} or W_{ij} and W_{hi} since this results in a different (X, Y) pair in common. Thus, the total number of distinct non-zero covariance terms in (15.6.4) is

$$(2+2+1+1)\binom{N}{3} = 6\binom{N}{3},$$

and consequently,

$$N^2(N-1)^2 \text{var } \tau = 4\left[\binom{N}{2} \text{var } W_{ij} + 6\binom{N}{3} \text{cov}(W_{ij}, W_{ik})\right], \quad (15.6.5)$$

with $i < j$, $i < k$, $j \neq k$. Hence,

$$N(N-1) \text{var } \tau = 2 \text{ var } W_{12} + 4(N-2) \text{cov}(W_{12}, W_{13}). \quad (15.6.6)$$

Also,

$$\begin{aligned} \text{var } W_{12} &= E(W_{12}^2) - (EW_{12})^2 \\ &= 1^2 p_c + (-1)^2 p_d - (p_c - p_d)^2 = (p_c + p_d)^2 - (p_c - p_d)^2, \end{aligned}$$

and

$$\text{cov}(W_{12}, W_{13}) = E(W_{12}W_{13}) - (p_c - p_d)^2$$

where

$$\begin{aligned} E(W_{12}W_{13}) &= 1 \cdot P(W_{12} = 1, W_{13} = 1) + 1 \cdot P(W_{12} = -1, W_{13} = -1) \\ &- 1 \cdot P(W_{12} = 1, W_{13} = -1) - 1 \cdot P(W_{12} = -1, W_{13} = 1) \\ &= p_{cc} + p_{dd} - 2p_{cd} \text{ (because of symmetry)}. \end{aligned} \quad (15.6.7)$$

When (X, Y) has a continuous bivariate distribution, $p_c + p_d = 1$ and $p_{cc} + p_{dd} - 2p_{cd} = 1$. Consequently,

$$\begin{aligned} N(N-1)\text{var}\,\tau &= 2\left\{1-(2p_c-1)^2\right\} + 4(N-2)\left\{1-4p_{cd}-(2p_c-1)^2\right\} \\ &= 2-2(2N-3)(2p_c-1)^2 + 4(N-2)(1-4p_{cd})\,. \end{aligned}$$

Now writing $(2p_c - 1)^2 = \{p_c - (1-p_c)\}^2 = 1 - 4p_c(1-p_c)$ and simplifying, we obtain

$$N(N-1)\text{var}\,\tau = 8(2N-3)p_c(1-p_c) - 16(N-2)p_{cd}\,. \tag{15.6.8}$$

Further, for continuous X and Y, we have

$$\begin{aligned} p_{cd} &= P(W_{12}=1 \text{ and } W_{13}=-1) \\ &= P(W_{12}=1) - P(W_{12}=1 \text{ and } W_{13}=1) \\ &= p_c - p_{cc}\,. \end{aligned} \tag{15.6.9}$$

Let (X, Y) have the df $F(x, y)$ and pdf $f(x, y)$. Then

$$\begin{aligned} p_c &= P(X_1 < X_2 \text{ and } Y_1 < Y_2) + P(X_1 > X_2 \text{ and } Y_1 > Y_2) \\ &= \int_{-\infty}^{\infty}\int_{-\infty}^{\infty} P(X_1 < x \text{ and } Y_1 < y) f(x,y)\,dx\,dy \\ &\quad + \int_{-\infty}^{\infty}\int_{-\infty}^{\infty} P(X_2 < x, Y_2 < y) f(x,y)\,dx\,dy \\ &= 2\int_{-\infty}^{\infty}\int_{-\infty}^{\infty} F(x,y) f(x,y)\,dx\,dy\,. \end{aligned} \tag{15.6.10}$$

Next consider

$$\begin{aligned} p_{cc} &= P(W_{12}=1 \text{ and } W_{13}=1) \\ &= P(X_1-X_2>0, Y_1-Y_2>0, X_1-X_3>0 \text{ and } Y_1-Y_3>0) \\ &\quad + P(X_1-X_2>0, Y_1-Y_2>0, X_1-X_3<0 \text{ and } Y_1-Y_3<0) \\ &\quad + P(X_1-X_2<0, Y_1-Y_2<0, X_1-X_3>0, Y_1-Y_3>0) \\ &\quad + P(X_1-X_2<0, Y_1-Y_2<0, X_1-X_3<0, Y_1-Y_3<0) \\ &= a_1+a_2+a_3+a_4 \text{ (respectively)}\,. \end{aligned} \tag{15.6.11}$$

Now,

$$\begin{aligned} a_1 &= \int_{-\infty}^{\infty}\int_{-\infty}^{\infty} P\left(X_2 < x, Y_2 < y, X_3 < x, Y_3 < y \mid \right. \\ &\quad \left. (X_1, Y_1) = (x, y)\right) f(x, y) dx\, dy \\ &= \int_{-\infty}^{\infty}\int_{-\infty}^{\infty} F^2(x, y) f(x, y) dx\, dy \,, \qquad (15.6.12) \\ a_2 &= \int\int P\left(X_2 < x, Y_2 < y, X_3 > x, Y_3 > y \mid \right. \\ &\quad \left. (X_1, Y_1) = (x, y)\right) f(x, y) dx\, dy \\ &= \int\int F(x, y) \cdot P(X > x, Y > y) f(x, y) dx\, dy \,. \qquad (15.6.13) \end{aligned}$$

It is easy to see that $a_3 = a_2$. Next consider

$$a_4 = \int\int P^2(X < x, Y < y) f(x, y) dx\, dy \,. \qquad (15.6.14)$$

Thus,

$$\begin{aligned} & p_{cc} \qquad (15.6.15) \\ &= \int_{-\infty}^{\infty}\int_{-\infty}^{\infty} \{F(x, y) + P(X > x, Y > y)\}^2 f(x, y) dx\, dy \\ &= \int_{-\infty}^{\infty}\int_{-\infty}^{\infty} \{1 - F_X(x) - F_Y(y) + 2F(x, y)\}^2 f(x, y) dx\, dy \,. \end{aligned}$$

Let us compute the variance of T when H_0 holds, i.e., when X and Y are independent. In this case, without loss of generality, we can assume that X and Y are standard uniform variables. Note that

$$p_c = 2\left\{\int_0^1 x\, dx\right\}^2 = 2 \cdot \frac{1}{4} = \frac{1}{2} \,. \qquad (15.6.16)$$

$$\begin{aligned} p_{cc} &= P(W_{12} = 1, W_{13} = 1) \\ &= P\left(\mathrm{Sgn}(X_1 - X_2)\mathrm{Sgn}(Y_1 - Y_2) = 1 \right. , \\ &\quad \left. \mathrm{Sgn}(X_1 - X_3)\mathrm{Sgn}(Y_1 - Y_3) = 1\right) . \end{aligned}$$

From (15.6.15) we have

$$\begin{aligned} p_{cc} &= \int_{-\infty}^{\infty}\int_{-\infty}^{\infty} \{1 - F_X(x) - F_Y(y) + 2F(x,y)\}^2 f(x,y)dx\,dy \\ &= \int_{-\infty}^{\infty}\int_{-\infty}^{\infty} \{F(x,y) + P(X > x, Y > y)\}^2 f(x,y)dy\,. \end{aligned}$$

If X and Y are independent standard uniform variables,

$$\begin{aligned} p_{cc} &= \int_0^1 \int_0^1 (1 - x - y + 2xy)^2 dx\,dy \\ &= \int_0^1 \int_0^1 \{xy + (1-x)(1-y)\}^2\,dx\,dy \\ &= \int \int \left\{x^2y^2 + (1-x)^2(1-y)^2 + 2x(1-x)y(1-y)\right\} dx\,dy \\ &= \left(\int_0^1 x^2 dx\right)\left(\int_0^1 y^2 dy\right) + \left\{\int_0^1 (1-x)^2 dx\right\}\left\{\int_0^1 (1-y)^2 dy\right\} \\ &\quad +2\left\{\int_0^1 x(1-x)dx\right\}^2 \\ &= \left(\frac{1}{3}\right)^2 + \left(\frac{1}{3}\right)^2 + 2\left(\frac{1}{2} - \frac{1}{3}\right)^2 \\ &= \frac{1}{9} + \frac{1}{9} + 2\frac{1}{36} = \frac{2}{9} + \frac{1}{18} = \frac{4+1}{18} = \frac{5}{18}\,. \end{aligned} \tag{15.6.17}$$

Hence,

$$\begin{aligned} N(N-1)\text{var } \tau &= 8(2N-3)p_c(1-p_c) - 16(N-2)(p_c - p_{cc}) \\ &= 8(2N-3)\frac{1}{4} - 16(N-2)\left(\frac{1}{2} - \frac{5}{18}\right) \\ &= 2(2N-3) - 16(N-2)\frac{2}{9} \\ &= \frac{2}{9}\left[-16(N-2) + 9(2N-3)\right] \\ &= \frac{2}{9}(2N-5) \end{aligned}$$

$$\therefore \operatorname{var}(\tau|H_0) = \frac{2}{9}\frac{(2N+5)}{N(N-1)}. \tag{15.6.18}$$

Kendall and Gibbons (1990, pp. 180–181) provide a proof for the variance of Kendall's tau when (X, Y) has a bivariate normal distribution with correlation ρ. The variance of tau is given by

$$\operatorname{var} \tau = \frac{2}{N(N-1)}\left\{1 - \left(\frac{2}{\pi} \text{ Arc sin } \rho\right)^2 + 2(N-2)\left[\frac{1}{9} - \left(\frac{2}{\pi} \text{ Arc sin } \frac{\rho}{2}\right)^2\right]\right\}. \tag{15.6.19}$$

Esscher (1924) originally obtained the above expression for the variance of tau. Setting $\rho = 0$, we obtain

$$\operatorname{var}(\tau|\rho = 0) = \frac{2}{N(N-1)}\left\{1 + \frac{2(N-2)}{9}\right\} = \frac{2(2N+5)}{9N(N-1)}$$

which coincides with the $\operatorname{var}(\tau|H_0)$ given by (15.6.18).

Remark 15.6.1. *By projecting Kendall's tau into the space of linear rank statistics, Hájek and Sidák (1967, p. 116) assert that the τ will be a constant multiple of the Spearman's correlation coefficient. The asymptotic normality of the Spearman correlation when suitably standardized follows, for instance, from Govindarajulu (1976, Theorem 6.1, p. 550 with $J(u) = u$). Hence, $3\tau\,[N(N-1)]^{1/2}\,/\,[2(2N+5)]^{1/2}$ is asymptotically standard normal under H_0 for sufficiently large N.*

15.7 Asymptotic Normality of a Class of Tests

Consider the LMP rank test statistics derived in Section 15.3 which can be embedded in the general class given by

$$T_N = N^{-1}\sum_{i=1}^{N} a_N(r_i) b_N(s_i) Z_{N,r_i} Z'_{N,s_i} \tag{15.7.1}$$

where $\{a_N(i)\}$, $\{b_N(i)\}$, $i = 1, \ldots, N$ are two sets of specified constants satisfying certain restrictions, and

$$Z_{N,r_i}(Z'_{N,s_i}) = \begin{cases} 1 & \text{when } X_1\ (Y_1) \text{ is the } r_i^{\text{th}}\ (s_i^{\text{th}}) \text{ smallest of the } X\text{'s } (Y\text{'s})), \\ 0 & \text{otherwise, for } i = 1, \ldots, N. \end{cases}$$

One can have the following integral representation of T_N:

$$T_N = \int_{-\infty}^{\infty}\int_{-\infty}^{\infty} J_N\left(\frac{N}{N+1}F_N(x)\right) L_N\left(\frac{N}{N+1}G_N(y)\right) dH_N(x,y) \tag{15.7.2}$$

where $J_N\left(\frac{i}{N+1}\right) = a_N(i)$ and $L_N\left(\frac{i}{N+1}\right) = b_N(i)$, F_N (G_N) is the empirical distribution function based on $X_1, \ldots, X_N$ $(Y_1, \ldots, Y_N)$ and $H_N(x,y)$ is the empirical distribution function based on (X_i, Y_i) $i = 1, \ldots, N$.

When $a_N(i) = b_N(i) =$ expected value of the ith smallest order statistic in a random sample of size N drawn from the standard normal distribution, T_N becomes the Fisher-Yates normal scores test. If $a_N(i) = b_N(i) = i/(N+1)$, T_N is equivalent to the Spearman test criterion. Bhuchongkul (1964) established the asymptotic normality of T_N under certain regularity assumptions on the J_N and L_N functions and their first two derivatives. Of course, it is assumed that the joint df of (X,Y), namely, $H(x,y)$ is continuous. These assumptions are satisfied by the normal scores and the Spearman's test statistic. In Chapter 26 these conditions will be relaxed and explicitly specified. That is, we have

$$\lim_{N\to\infty} P\left[\frac{T_N - \mu_N}{\sigma_N} \le x\right] = \Phi(x), \text{ for all } x\,, \tag{15.7.3}$$

uniformly with respect to F, G and H, provided $\sigma_N \neq 0$; where

$$\mu_N = \int_{-\infty}^{\infty}\int_{-\infty}^{\infty} J_N(F(x))\, L_N(G(x))\, dH(x,y) \tag{15.7.4}$$

and a complicated expression for σ_N^2. However, σ_N^2 takes a simpler form when H_0 is true.

$$\begin{aligned} N\operatorname{var}(T_N|H_0) &= \operatorname{var}_0\{J_N[F_0(x)]\,L_N[G_0(y)]\} \\ &\quad - E_0^2\{L_N[G_0(Y)]\}\operatorname{var}_0\{J_N[F_0(X)]\} \\ &\quad - E_0^2\{J_N[F_0(X)]\}\operatorname{var}_0\{L_N[G_0(Y)]\}\,. \end{aligned} \tag{15.7.5}$$

If you consider a sequence of alternatives that converge to zero such as considered by Bhuchongkul (1964, p. 141):

$$X = (1-\theta)V - \theta Z,\ Y = (1-\theta)W + \theta Z \tag{15.7.6}$$

where $0 \le \theta \le 1$ and V, W, Z are independent and H_0 now takes the form of $\theta = 0$, then one can show that

$$\lim_{N\to\infty} \operatorname{var}_N(T_N)/\operatorname{var}_0(T_N) = 1\,. \tag{15.7.7}$$

In the case of the normal scores test, $J_N = L_N = \Phi^{-1}$, $F = G = \Phi$ and hence

$$N \text{ var}(T_N|H_0) = \text{var}(XY) - E_0^2(Y)\text{var } X - E_0^2(X)\text{var } Y\,.$$

Since we assume that X and Y are standard normal,

$$\begin{aligned} N \text{ var}(T_N|H_0) &= E(X^2Y^2) - E^2(XY) - E_0^2(Y) - E_0^2(X) \\ &= E(X^2)E(Y^2) - 0 = 1\,. \end{aligned} \tag{15.7.8}$$

$$\mu_N = \int_{-\infty}^{\infty}\int_{-\infty}^{\infty} \Phi^{-1}\left[F(x)\right]\Phi^{-1}\left[F(y)\right]h(x,y)dx\,dy\,.$$

The local alternatives considered by Bhuchongkul (1964) which is similar to the ones given by Konijn (1956) is given by (15.7.6) where V, W and Z are independent having distributions F, G and F^* respectively. Further, without loss of generality, we assume that V, W and Z have zero means and unit variances and the density functions of V and W are twice differentiable. Thus,

$$H(x,y) = P[X \le x, Y \le y] = \int_{-\infty}^{\infty} F\left(\frac{x-\theta z}{1-\theta}\right) G\left(\frac{y-\theta z}{1-\theta}\right) dF^*(z)\,.$$

Consequently,

$$h(x,y) = (1-\theta)^2 \int_{-\infty}^{\infty} f\left(\frac{x-\theta z}{1-\theta}\right) g\left(\frac{y-\theta z}{1-\theta}\right) dF^*(z)\,. \tag{15.7.9}$$

Also,

$$\begin{aligned} F_\theta(x) &= \int_{-\infty}^{\infty} F\left(\frac{x-\theta z}{1-\theta}\right) dF^*(z), \\ f_\theta(x) &= \frac{1}{(1-\theta)} \int_{-\infty}^{\infty} f\left(\frac{x-\theta z}{1-\theta}\right) dF^*(z)\,, \\ G_\theta(y) &= \int_{-\infty}^{\infty} G\left(\frac{y-\theta z}{1-\theta}\right) dF^*(z), \\ g_\theta(y) &= \frac{1}{(1-\theta)} \int_{-\infty}^{\infty} g\left(\frac{y-\theta z}{1-\theta}\right) dF^*(z)\,. \end{aligned} \tag{15.7.10}$$

Straightforward computations yield

$$\rho = \rho(X,Y) = \frac{\theta^2 \text{var } Z}{(1-\theta)^2 - \theta^2 \text{var } Z} = \frac{\theta^2}{(1-\theta)^2 + \theta^2}\,. \tag{15.7.11}$$

$$\frac{d\rho}{d\theta} = 0, \; \frac{d^2\rho}{d\theta^2} = 2, \quad \text{when } \theta = 0$$

$$\frac{d}{d\theta}\mu_N(\theta)|_{\theta=0} = 0$$

and

$$\frac{d^2}{d\theta^2}\mu_N(\theta)|_{\theta=0} = \int_{\infty}^{\infty} J'\left[F(x)\right] f^2(x)dx \int_{\infty}^{\infty} J'\left[G(y)\right] g^2(y)dy \,. \quad (15.7.12)$$

By considering $\theta = \xi/N^{1/4}$ for some $\xi > 0$, and applying the extension of Pitman's efficiency due to Noether (1955), we obtain Efficacy of T_N is equal to

$$\left[\left\{\int J'\left(F(x)\right) f^2(x)dx\right\}\left\{\int J'\left(G(y)\right) g^2(y)dy\right\}\right]^2 / N \, \text{var}_0(T_N) \,. \quad (15.7.13)$$

Special Cases

1. **Normal scores.** $J = \Phi^{-1}$, $F(x) = G(x) = \Phi(x)$. Consequently, using (15.7.8), we obtain
$$\text{Efficacy of } T_N = 1 \,. \quad (15.7.14)$$
2. **Spearman test criterion.** Set $J(u) = u$. Then $N \, \text{var}_0(T_N) = \text{var}(UV) - (EV)^2 \text{var } U - (EU)^2 \text{var } V$, where U and V are independent standard uniform variables. Thus,
$$\text{Efficacy of } T_N = 144 \left[\int f^2(x)dx\right]^4 . \quad (15.7.15)$$

Efficacy of the Product-moment Correlation Coefficient

Since the correlation coefficient is invariant under location and scale changes, without loss of generality, we can assume that X and Y have zero means and unit variances. Then

$$r = (N-1)^{-1} \sum_1^N (X_i - \bar{X})(Y_i - \bar{Y})/s_X s_Y \quad (15.7.16)$$

where $\bar{X}$ and $\bar{Y}$ denote sample means and s_X and s_Y denote the sample deviations. Since $\bar{X}$ and $\bar{Y}$ converge to their means and s_X and s_Y converge to σ_X and σ_Y which we set equal to one. r is asymptotically equivalent to

$$r' = N^{-1} \sum_1^N X_i Y_i \,. \quad (15.7.17)$$

Thus,

$$\begin{aligned} Er' &= E(XY) = \rho \\ \text{var } r' &= Er'^2 - \rho^2 = N^{-2} E\left[\sum_1^N X_i^2 Y_i^2 + \sum_{i \neq j}\sum X_i Y_i X_j Y_j\right] - \rho^2 . \end{aligned}$$

Consequently,

$$\text{var}(r'|H_0) = N^{-1} E(X^2 Y^2 | H_0) = N^{-1} .$$

Let $\rho = \xi / N^{\delta}$ for some $\delta > 0$. Pitman efficacy of $r' = \lim_{N\to\infty}\left[\frac{(\xi N^{-\delta} - 0)}{\xi N^{-1/2}}\right]^2 =$ 1 provided $\delta = 1/2$. If we consider r itself, then note that Cramér (1947, Eq. (27.7.3)) obtained:

$$E(r) = \rho + O(N^{-1})$$

and

$$\text{var}(r|H_0) = \frac{1}{N} + O(N^{-1}). \tag{15.7.18}$$

From (15.7.12) and (15.7.13) we obtain the Pitman efficiency of the normal scores test and the Spearman's test criterion relative to Pearson's product-moment correlation coefficient is one and $\left[\int f^2(x)dx\right]^4$, respectively.

Bhuchongkul (1964, p. 146) tabulates the small sample null distribution of normal scores test for independence for $2 \leq N \leq 6$.

Remark 15.7.1. *Gokhale (1968), for testing*

H_0 : *X and Y are independent versus the alternative about the joint df:*

H_1 : $H(x,y) = (1-\theta)F(x)G(y) + \theta K(x,y)$,

where $K(x,y)$ has continuous marginal distributions F and G, proposes the test statistic

$$\underline{T}_N = \int J_N\left(\frac{N}{N+1}F_N(x)\right) J_N\left(\frac{N}{N+1}G_N(y)\right) dH_N(x,y) \tag{15.7.19}$$

where $J_N\left(\frac{i}{N+1}\right)$ denotes the expected value of the ith smallest order statistic in a sample of size N drawn from a population having df $J^{-1}(u)$, $0 < u < 1$ where $J(u) = \lim_{N\to\infty} J_N(u)$. Further, he shows that there exist alternatives for which the test based on the product-moment correlation coefficient is infinitely more efficient than $\underline{T}_N$. Also, for any members T_1 and T_2 belonging to the class in (15.7.19), he shows that there exist alternatives for which T_1 is infinitely more efficient than T_2 as well as the correlation r.

Some other test criterion. Olmstead and Tukey (1947) give a very simple test called the 'corner test' for bivariate independence. The statistic based on the difference $F_N(x,y) - F_N(x)G_N(y)$ where $F_N(x,y)$ denotes the two-dimensional empirical distribution function and $F_N(x)$ and $G_N(y)$ are the corresponding marginal empirical distribution functions based on the samples of sizes N has been proposed and investigated by Blum, Kiefer and Rosenblatt (1961). Hoeffding (1948) proposed an equivalent statistic. He also shows that there do not exist tests of independence based on ranks that are unbiased for any significance level with respect to all continuous alternatives. However, Lehmann (1966) has demonstrated that the one-sided test procedures based on Kendall's tau and Spearman's correlation coefficient are unbiased for all continuous (X,Y) such that Y is positively (negatively) regression dependent on X, that is, the conditional distribution of Y for given $X = x$, is nonincreasing (nondecreasing) in x. He also pointed out that positive (negative) regression dependence implies positive (negative) quadrant dependence. That is, $\text{cov}(X,Y)$ is nonnegative (nonpositive). Bell and Smith (1969) have characterized all distribution-free tests for independence and for the case of total independence, they have characterized the family of all rank tests.

15.8 Tests for Multi-variate Populations

If $(X_1, \ldots, X_p)$ has a continuous p-variate distribution, the hypothesis of complete independence is given by

$$H_0 : F(x_1, \ldots, x_p) = \prod_{i=1}^{p} F_i(x_i)$$

where the F_i $(i = 1, \ldots, p)$ denote the marginal distribution functions. In the parametric set-up, the test criterion used to test H_0 is a linear combination of all sample bivariate product-moment correlation coefficients. In the nonparametric case, Puri, Sen and Gokhale (1970) define a nonparametric association matrix whose elements are measures of bivariate association. For each null hypothesis H_0, a modified null hypothesis H_0^* in terms of the elements of the association matrix is set up and tested. In all cases considered by Puri et al. (1970), H_0 implies H_0^*, but the converse is not necessarily true. Thus, the procedures proposed by Puri et al. (1970) are appropriate for a subclass of the class of all alternatives. Further, their tests constitute nonparametric analogues of the parametric likelihood ratio tests.

Govindarajulu and Gore (1977) propose a statistic which is a linear combination of all sample bivariate Kendall's tau statistics, and two other test criteria that are based on certain multivariate analogues of bivariate rank correlations. Heuristically, the latter test procedures are directed to the problem of complete independence and may be looked upon as generalizations of Kendall's tau. Asymptotic distribution theory for the test statistics is developed and asymptotic relative efficiencies of the various tests are studied. Finally, a locally most powerful rank test for a class of mixture alternatives is derived and its asymptotic properties, especially those of the normal scores test for the p-variate case, are studied.

15.9 Problems

15.3.1 Evaluate the mean of the Spearman test criterion (see (15.3.10)) when H_0 is true (i.e. the variables X and Y are independent).

15.4.1 Evaluate the variance of the Spearman test criterion when H_0 is true.

15.4.2 The following constitute the scores of a random sample of 10 students in a statistics and a computer science undergraduate courses. (Maximum possible score in each test is 50).

statistics:	40 38 35 41 45 36 34 28 46 39
computer science:	45 42 40 38 48 38 32 30 45 42

Using the Spearman test criterion test

H_0: The scores in statistics and computer science courses are independent

against H_1: There is a positive association between these two scores.

Let $\alpha = 0.10$.

(HINT: Assume that the Spearman test criterion, when standardized, is asymptotically normal.)

15.4.3 A random sample of 10 cars from a used car lot yield the following data (in dollars):

year	: 1991 1992 1993 1994 1995 1996 1997 1998 1999 2000
sticker price	: 1400 1500 2000 1800 2200 2000 2500 3000 3500 4000

Using Spearman's test criterion test whether there is any trend in the price of the car according to its age. Use $\alpha = 0.05$.

15.5.1 For the data in Problem 15.4.2, compute Kendall's tau and its expected value under H_0.

15.6.1 For the data in Problem 15.4.2, compute the variance of Kendall's tau under H_0 and carry out the test for H_0 against H_1 with $\alpha = 0.10$. Do you reach the same conclusion as in Problem 15.4.2?
(HINT: Assume that Kendall's tau is asymptotically normal when suitably standardized. See Remark 15.6.1.)

15.6.2 For the data in Problem 15.4.3, carry out test based on Kendall's tau for the same hypotheses and level of significance.

15.7.1 Evaluate the Pitman efficiency of the Spearman test criterion when (i) f is normal, (ii) f is logistic, (iii) f is double exponential, (iv) f is uniform $(0, 1)$.

Chapter 16

c-sample Rank Order Tests

16.1 Introduction

Quite often one is faced with the problem of comparing more than two treatments, like methods of learning, varieties of corn, etc. In the parametric case there is an analysis of variance test based on the F-test. Several nonparametric test procedures are available for the c ($c \geq 3$) sample problem, for instance, the widely used Kruskal-Wallis test which is a heuristic test. In the following we derive a rank order (vector) statistic for the c-sample problem.

16.2 c-sample Rank Order Tests

Let X_i have a continuous d.f. $F_i(x)$ $(i = 1, \ldots, c)$. We will be interested in testing

$$H_0 : F_1(x) = \cdots = F_c(x) \text{ for all } x$$

versus

$$H_1 : F_i(x) \neq F_j(x) \text{ for some } x \text{ and some } i \neq j\,.$$

Suppose we have a random sample of size n_i denoted by $(X_{i,1}, \ldots, X_{i,n_i})$ from F_i $(i = 1, \ldots, c)$. Let $\boldsymbol{W}= (W_{1N}, \ldots, W_{NN})$ denote the combined ordered sample, where $N = n_1 + \cdots + n_c$. Then we have the following definition of c-sample rank order given by Govindarajulu and Haller (1972).

Definition 16.2.1. *The random vector $\boldsymbol{Z} = (Z_1, \ldots, Z_N)$ is said to be a c-sample rank order if $Z_i = j$ when $W_i = X_{j,k}$ for some $k = 1, \ldots, n_j$ $(i = 1, \ldots, N)$.*

Andrews and Truax (1964) defined Z_i as a c-dimensional vector having a one in the jth coordinate if and only if $W_i = X_{j,k}$ for some $k = 1, \ldots, n_j$ and zero elsewhere. The former notation is preferable because of its compactness. Let $\mathcal{Z} = \{\boldsymbol{Z}^i,\ i = 1, \ldots, M\}$ be the set of all possible values of the rank order Z. In general, we shall denote any $\boldsymbol{z} \in \mathcal{Z}$ by $(z_1, z_2, \ldots, z_N)$. Obviously,

$$M = N!/\prod_{i=1}^{c}(n_i!) \text{ and } \mathcal{Z} \text{ is a function of } (n_1, \ldots, n_c)\,.$$

One can construct a most powerful rank order test against any specified alternative provided the probabilities of the rank orders under the specified alternative can be ordered. Unfortunately, one cannot obtain a closed-form expression for the rank order probability under the alternative. Note that slippage alternatives are given by

$$H_L : F_1(x) \geq F_2(x) \cdots \geq F_c(x)$$

for all x, where each F_i is continuous. By restricting to a certain subclass of the class of slippage alternatives, namely H_M, which assumes the density functions indexed by a parameter θ satisfy the monotone likelihood ratio (MLR) property[1] a partial order of the rank order probabilities can be obtained. (See Theorem 2.2 of Govindarajulu and Haller, 1972.) Further, as noted in Chapter 11, if we consider the Lehmann class of alternatives, one can obtain an explicit expression for the rank order probabilities. This class of alternatives is given by

$$H_L : F_i(x) = 1 - \{1 - F(x)\}^{\theta_i} \quad (i = 1, \ldots, c) \text{ where } F \text{ is an unknown}\,,$$

continuous d.f. and $\theta_i > 0$ $(i = 1, \ldots, c)$.

One can readily obtain the following results (or see Govindarajulu and Haller, 1972, pp. 22, 23).

Theorem 16.2.1. *Let* $\boldsymbol{z} = (Z_1, \ldots, Z_N) \in \mathcal{Z}$ *and* $\delta_{k,z}$ *denote the Kronecker delta. If* $F_i \in H_L$ $(i = 1, \ldots, c)$*, then*

$$P(\boldsymbol{Z} = \boldsymbol{z}|H_L) = \prod_{i=1}^{c} [(n_i!)\theta_i^{n_i}] \Big/ \prod_{i=1}^{N} \left[\sum_{k=1}^{c} \theta_k u_{k,i}(\boldsymbol{z})\right] \tag{16.2.1}$$

where

$$u_{k,i}(\boldsymbol{z}) = \sum_{j=i}^{N} \delta_{k,z_j}\,. \tag{16.2.2}$$

[1]A probability density function $f(x;\theta)$ is said to have the MLR property if $x_1 < x_2$ and $\theta_1 < \theta_2$ imply that $f(x_1;\theta_1)f(x_2;\theta_2) \geq f(x_1;\theta_2)f(x_2;\theta_1)$.

Let $\Delta_k = (\theta_k/\theta_c) - 1$, *for* $k = 1, \ldots, c-1$. *Then (16.2.1) takes the form of*

$$P(\boldsymbol{Z} = \boldsymbol{z}|H_L) = M^{-1}\prod_{k=1}^{c-1}(\Delta_k+1)^{n_k}/\prod_{i=1}^{N}\left[1+\left(\sum_{k=1}^{c-1}\Delta_k u_{k,i}(\boldsymbol{z})\right)/(N-i+1)\right] \tag{16.2.3}$$

since $\sum_{k=1}^{c} u_{k,i}(\boldsymbol{z}) = N - i + 1$.

Let $\boldsymbol{Z} = (Z_1, \ldots, Z_N)$ *and* $\boldsymbol{Z}' = (Z_1', \ldots, Z_N')$ *be two rank vectors,* $u_{k,i} = \sum_{j=i}^{N}\delta_{k,z_j}$, $u'_{k,i} = \sum_{j=i}^{N}\delta_{k,z'_j}$. *Then from (16.2.3) we have*

$$\begin{aligned}
&M\prod_{k=1}^{c-1}(\Delta_k+1)^{-n_k}\left\{P(\boldsymbol{Z} = \boldsymbol{z}|H_L) - P(\boldsymbol{Z} = \boldsymbol{z}'|H_L)\right\}\\
&= \prod_{i=1}^{N}(1+A_i)^{-1} - \prod_{i=1}^{N}(1+A_i')^{-1}\\
&= \sum_{i=1}^{N}\prod_{j=1}^{i-1}(1+A_j)^{-1}\left[(1+A_i)^{-1} - (1+A_i')^{-1}\right]\prod_{j=i+1}^{N}(1+A_j')^{-1}
\end{aligned}$$

where $A_i = \sum_{k=1}^{c-1}\Delta_k u_{k,i}/(N-i+1)$ *and an analogous expression for* A_i'. *Hence,* $\boldsymbol{z}$ *is more probable than* $\boldsymbol{z}'$ *under* H_L *if* $A_i' > A_i$, *that is, if* $u'_{k,i} \geq u_{k,i}$ *for all* i *with at least one strict inequality.*

Further assuming the inequality $A_i' \geq A_i$, *we obtain that*

$$\sum_{k=1}^{c-1}\Delta_k S_k(\boldsymbol{z}') > \sum_{k=1}^{c-1}\Delta_k S_k(\boldsymbol{z})$$

implies that $\boldsymbol{z}$ *is more probable than* $\boldsymbol{z}'$ *under* H_L *where* $S_k(\boldsymbol{z}) = \sum_{i=1}^{N}\frac{u_{k,i}(\boldsymbol{z})}{N-i+1}$. *Thus, knowledge of* Δ_k $(k = 1, \ldots, c-1)$ *is required to order the rank order probabilities.*

Consider the following special case: Let $\theta_k = \theta_c + (c-k)\theta$ with $\theta, \theta_c > 0$, which will be denoted by H_L^*. Then the d.f.'s satisfy the stochastically increasing order, namely,

$$F_1(x) \geq F_2(x) \geq \cdots \geq F_c(x)\,, \text{ for all } x\,. \tag{16.2.4}$$

Hence, if $F_k \in H_L^*$ $(k = 1, \ldots, c)$, then for fixed n_i $(i = 1, \ldots, c)$ and sufficiently small θ,

$$P(\boldsymbol{Z} = \boldsymbol{z}|H_L^*) > P(Z = \boldsymbol{z}'|H_L^*) \tag{16.2.5}$$

provided

$$\sum_{k=1}^{c-1}(c-k)S_k(\boldsymbol{z}') > \sum_{k=1}^{c-1}(c-k)S_k(\boldsymbol{z}). \tag{16.2.6}$$

Note that the converse does not hold [i.e., (16.2.6) does not imply (16.2.5)]. For a counter example see Govindarajulu and Haller (1972, p. 26).

Recall that

$$\begin{aligned} S_k(\boldsymbol{z}) &= \sum_{i=1}^{N} u_{k,i}(z)/(N-i+1) = \sum_{i=1}^{N}\frac{1}{N-i+1}\sum_{j=i}^{N}\delta_{k,z_j} \\ &= \sum_{j=1}^{N}\delta_{k,z_j}\sum_{i=1}^{j}\frac{1}{N-i+1} = \sum_{j=1}^{N}E_{N,j}\delta_{k,z_j} \end{aligned} \tag{16.2.7}$$

where $E_{N,j} = \sum_{i=1}^{j}\frac{1}{N-i+1}$ is the expected value of the jth smallest order statistic in a sample of size N drawn from the standard exponential distribution given by $1-e^{-x}$ $(x \geq 0)$ and zero elsewhere. Thus, the statistics S_k $(k = 1, \ldots, c)$ belong to the class of Chernoff-Savage statistics with $J(u) = -\ln(1-u)$, $0 < u < 1$. Let

$$F_i(x) = \Psi\left((x-\theta_{N,i})/\beta_{N,i}\right),\ i = 1, \ldots, c$$

where Ψ admits a density ψ and as $N \to \infty$, $\theta_{N,i} \to 0$ and $\beta_{N,i} \to 1$. It follows from Theorem 7.4 and Corollary 5.2.1 of Govindarajulu, LeCam and Raghavachari (1967) that $N^{1/2}\left(\frac{S_1}{n_1} - \mu_{N,1} \cdots \frac{S_c}{n_c} - \mu_{N,c}\right)$ has asymptotically a c-variate normal distribution with zero mean vector and $\Sigma = ((\sigma_{N,i,j}))$ for the variance-covariance vector provided $|J'(u)| \leq K\left[u(1-u)\right]^{-3/2+\delta}$ for some $\delta < 1/2$ where

$$\mu_{N,j} = N^{1/2}\int_{-\infty}^{\infty} -\ln\left\{1 - \sum_{i=1}^{c}\lambda_i F_i(x)\right\} dF_j(x),\ (1 \leq j \leq c) \tag{16.2.8}$$

and

$$\lambda_j = n_j/N,\ \ N = n_1 + \cdots + n_c. \tag{16.2.9}$$

Furthermore, the asymptotic covariance matrix (i.e., $\lim_{N\to\infty}\Sigma$) is

$$\lim_{N\to\infty}\left\{\lambda_j\sigma_{N,j}^2/(1-\lambda_j)\right\} = \lim_{N\to\infty}(-\sigma_{N,i,j}) = 1. \tag{16.2.10}$$

Moreover, the asymptotic normality is uniform for $(\theta_{N,i}, \log\ \beta_{N,i})$ belonging to some bounded set of values and $\lambda_i \in [\lambda_0, 1-\lambda_0]$, $\lambda_0 > 0$. For tests of H_0

against location shift, consider the alternatives given by H_n^a which specify that for each n, $F_i(x) = F(x + \theta_i/n^{1/2})$ where F is a continuous distribution function. Assume that the number of observations n_i from F_i is a function of n such that

$$\lim_{n\to\infty} n_i(n)/n = s_i \ (i = 1, \dots, c)$$

exist and are positive. Govindarajulu and Haller (1972) propose the test statistic

$$L_S = \sum_{i=1}^{c} (S_i - n_i)^2/n_i \,. \tag{16.2.11}$$

Notice that the S_i are subject to one linear constraint,

$$\begin{aligned} \sum_{k=1}^{c} S_k(\boldsymbol{z}) &= \sum_{k=1}^{c}\sum_{j=1}^{N} E_{N,j}\delta_{k,z_j} \\ &= \sum_{j=1}^{N} E_{N,j} \sum_{k=1}^{c} \delta_{k,z_j} \\ &= \sum_{j=1}^{N} E_{N,j} = N \,. \end{aligned} \tag{16.2.12}$$

It follows that the limiting distribution of L_S under H_n^a is non-central chi-square with $c-1$ degrees of freedom and the non-centrality parameter is given by

$$\begin{aligned} \lambda_S(F) &= \sum_{j=1}^{c} \Bigg[\lim_{n\to\infty} \int_{-\infty}^{\infty} n_j^{1/2} \Bigg\{ -\ln\Bigg[1 - \sum_{i=1}^{c} \lambda_i F(x + (\theta_i - \theta_j)/n^{1/2})\Bigg] \\ &\quad + \ln[1 - F(x)] \Bigg\} dF(x) \Bigg]^2 \end{aligned} \tag{16.2.13}$$

provided the limit in (16.2.13) exists and is finite. Sufficient conditions for integerchange of the integration and limit operations in (16.2.13) are given by Govindarajulu (1980) which are less restrictive than those of Hodges and Lehmann (1961) and are easily verifiable for this case.

Hence, one obtains

$$\lambda_S(F) = \left\{ \int_{-\infty}^{\infty} \left[f^2(x)/\left(1 - F(x)\right)\right] dx \right\}^2 \cdot D \tag{16.2.14}$$

where

$$D = \sum_{j=1}^{c} s_j(\theta_j - \bar{\theta})^2, \ \ \bar{\theta} = \sum_{1}^{c} s_i\theta_i \Big/ \left(\sum_{1}^{c} s_i\right), \ \ s_i = n_i/n \,. \tag{16.2.15}$$

In particular, $n = N$ in which case $s_i \equiv \lambda_i$.

Remark 16.2.1. *If one considers the other Lehmann alternative*

$$\tilde{H}_L : F_i(x) = F^{\theta_i}(x) \text{ for all } x, \ (i = 1, \ldots, c),$$

then one can obtain

$$P(\boldsymbol{Z} = \boldsymbol{z}|\tilde{H}_L) = M^{-1} \prod_{k=2}^{c} (\tilde{\Delta}_k + 1)^{n_k} \Big/ \prod_{i=1}^{N} \left[1 + \left(\sum_{k=2}^{c} \tilde{\Delta}_k v_{k,i}(\boldsymbol{z})/i \right) \right] \tag{16.2.16}$$

where

$$\tilde{\Delta}_k = (\theta_k/\theta_1) - 1 \ (k = 2, \ldots, c)$$

and

$$v_{k,i}(\boldsymbol{z}) = \sum_{j=1}^{i} \delta_{k,z_j} \, .$$

Also, for testing $\tilde{H}_L$, we consider the statistic

$$\tilde{L}_T = \sum_{k=1}^{c} (T_k - n_k)^2/n_k \tag{16.2.17}$$

where

$$T_k(\boldsymbol{z}) = \sum_{i=1}^{N} v_{k,i}/i \, . \tag{16.2.18}$$

Then the limiting distribution of $\tilde{L}_T$ under H_n^a is non-central chi-square with $c - 1$ degrees of freedom and non-centrality parameter

$$\tilde{\lambda}_T(F) = \left\{ \int_{-\infty}^{\infty} \left[f^2(x)/F(x) \right] dx \right\}^2 \cdot D \tag{16.2.19}$$

where D is as defined in (16.2.15). For details, see Govindarajulu and Haller (1972, pp. 24–28).

The Parametric Test

In the parametric case, we assume that X_i is distributed normal (μ_i, σ^2) $(i = 1, \ldots, c)$ where the θ_i's and σ are unknown. We wish to test

$$H_0 : \mu_1 = \cdots = \mu_c \text{ against } H_1 : \mu_i \neq \mu_j \text{ for some } i \neq j \, .$$

The proposed statistic is the well-known F-statistic given by

$$\mathcal{F} = \frac{\sum_{i=1}^{c} n_i(\bar{X}_i - \bar{X})/(c-1)}{\sum_{i=1}^{c} \sum_{j=1}^{n_i}(X_{ij} - \bar{X}_i)^2/(N-c)} \tag{16.2.20}$$

where $(X_{i1}, \ldots, X_{i,n_i})$ constitutes a random sample of size n_i from normal (θ_i, σ^2),

$$\bar{X}_i = \sum_{j=1}^{n_i} X_{ij}/n_i, \; \bar{X} = \sum_{i=1}^{c} \sum_{j=1}^{n_i} X_{ij}/N \text{ and } N = n_1 + \cdots + n_c .$$

Note that the denominator in F converges to σ^2 in probability for large N and $\sum_{i=1}^{c} n_i(\bar{X}_i - \bar{X})^2/\sigma^2$ is distributed as chi-square with $c-1$ degrees of freedom and non-centrality parameter $\sum_1^c s_i(\theta_i - \tilde{\theta})^2/\sigma^2$ under $H_n^a : \mu_i = \theta_i/n^{1/2}$, $i = 1, \ldots, c$, where $\bar{\theta} = \sum n_i\theta_i/N = \sum_1^c s_i\theta_i/\left(\sum_1^c s_i\right)$. Thus, the asymptotic efficiency of S relative to F is equal to the ratio of the non-centrality parameters. Thus,

$$e(S, \mathcal{F}) = \sigma_X^2 \left(\int_{-\infty}^{\infty} \left[f^2(x)/\left\{1 - F(x)\right\}\right] dx \right)^2 . \tag{16.2.21}$$

Govindarajulu and Haller (1972) tabulate the values of $e(S, \mathcal{F})$ for various underlying distributions such as normal, logistic and exponential (see their Table 3.2).

16.3 Chernoff-Savage Class of Statistics

One can generalize the class of statistics given by (16.2.7) as follows. Let

$$T_{N,j} = n_j^{-1} \sum_{i=1}^{N} E_{N,i,j} Z_{N,i,j} \tag{16.3.1}$$

where $E_{N,i,j}$ are some specified scores and

$$Z_{N,i,j} = \left\{ \begin{array}{ll} 1 & \text{if the } i^{\text{th}} \text{ smallest in the combined sample, namely } W_{iN}, \\ & \text{comes from the } j^{\text{th}} \text{ sample,} \\ 0 & \text{otherwise.} \end{array} \right\}$$

One can have the following representation of $T_{N,j}$ in terms of an integral:

$$T_{N,j} = \int_{-\infty}^{\infty} J_{N,j}\left(\frac{N}{N+1} H_N(x)\right) dF_{j,n_j}(x) \tag{16.3.2}$$

where $J_{N,j}\left(\frac{i}{N+1}\right) = E_{N,i,j}$, $j = 1, \ldots, c$ and $i = 1, \ldots, N$. Puri (1964) has considered a subclass of the above class, namely when $E_{N,i,j} = E_{N,i}$, i.e., $J_{N,j}(u) = J_N(u)$, and established the joint asymptotic normality of $T_{N,j}$. Further, his regularity assumptions are somewhat restrictive. Here we will give the less restrictive conditions obtained in Govindarajulu, LeCam and Raghavachari (1967) that are specialized for the subclass. Hence, consider

$$T_{N,j} = \int_{-\infty}^{\infty} J_N\left(\frac{N}{N+1} H_n(x)\right) dF_{j,n_j}(x) \tag{16.3.3}$$

where $F_{j,n_j}(x)$ denotes the empirical distribution of $X_{j,k}$ $(k = 1, \ldots, n_j)$ and

$$H_N(x) = \sum_{j=1}^{c} \lambda_j F_{j,n_j}(x),\ \lambda_j = n_j/N\,. \tag{16.3.4}$$

If

1. there exists a $\lambda_0 \leq 1/c$ such that $0 < \lambda_0 < \lambda_j < 1 - \lambda_0 < 1$ $(j = 1, \ldots, c)$,
2. the $F_i(x)$ have no common discontinuities (continuity of the F_i will ensure this), and
3. $J_N(u)$ is absolutely continuous with $|J_N'(u)| \leq K\,[u(1-u)]^{-3/2+\delta}$ for some $0 < \delta < 1/2$,

then Govindarajulu, LeCam and Raghavachari (1967) assert that

$$\left\{N^{1/2}(T_{N,j} - \mu_{N,1}), \ldots, N^{1/2}(T_{N,c} - \mu_{N,c})\right\}$$

has a limiting joint c-variate normal distribution with

$$\mu_{N,j} = \int_{-\infty}^{\infty} J_N\left(H(x)\right) dF_j(x),\ j = 1, \ldots, c \tag{16.3.5}$$

and a certain variance-covariance matrix (which we will not display here) under all hypotheses. However, when H_0 holds, the variance-covariance matrix takes a much simpler form, namely

$$\sigma^2_{N,j} = \left(\frac{1-\lambda_j}{\lambda_j}\right) I^2,\ j = 1, \ldots, c \tag{16.3.6}$$

and

$$\sigma_{N,i,j} = -I^2 \text{ for } 1 \leq i \neq j \leq c \tag{16.3.7}$$

with

$$I^2 = \int_0^1 J_N^2(u)du - \left\{\int_0^1 J_N(u)du\right\}^2 = \text{var}\left\{J_N(U)\right\} \tag{16.3.8}$$

where U is uniformly distributed on $(0,1)$. That is, $\sqrt{n_j}(T_{N,j} - \mu_{N,j})$, $j = 1, \ldots, c$ has asymptotically *a* c-variate normal distribution with zero means and variance-covariance matrix given by

$$\begin{pmatrix} 1 & & 0 \\ & 1 & \\ & & \cdots \\ 0 & & 1 \end{pmatrix} - \begin{pmatrix} \sqrt{\lambda_1} \\ \vdots \\ \sqrt{\lambda_i} \end{pmatrix} (\sqrt{\lambda_1}, \ddots, \sqrt{\lambda_i}) . \tag{16.3.9}$$

Also note that $\sum_{j=1}^{c} \lambda_j T_{N,j} = N^{-1} \sum_{i=1}^{N} E_{N,i}$ = a non-stochastic constant.

Define the statistic

$$\mathcal{L}_T = N \sum_{j=1}^{c} \lambda_j (T_{N,j} - \mu_{N,j})^2 / (1 - \lambda_j) I^2 . \tag{16.3.10}$$

Consider the local shift alternatives

$$F_i(x) \equiv F(x + \theta_i/\sqrt{n}) \text{ for some } n \tag{16.3.11}$$

such that $n_j/n = s_j > 0$. Note that $\lambda_j = n_j/N = s_j / \left(\sum_{i=1}^{c} s_i\right)$.

By making the analysis of variance transformation

$$\begin{aligned} V_0 &= \sum_{j=1}^{c} \lambda_j^{1/2} W_j, \quad W_j = n_j^{1/2}(T_{N,j} - \mu_{N,j})/I \\ V_i &= \sum_{i'=1}^{c} a_{i,i'}, W_{i'}, \quad i = 1, 2, \ldots, c-1 \end{aligned} \tag{16.3.12}$$

where the $a_{i,i}$'s are so chosen that the transformation is orthogonal. Hence,

$$V_0^2 + \sum_{i=1}^{c-1} V_i = \sum_{1}^{c} W_i^2 .$$

Moreover, $V_0 \equiv 0$, we have $\sum_1^c W_i^2 = \mathcal{L}_T$ is asymptotically non-central chi-square with $c - 1$ degrees of freedom and non-centrality parameter

$$I^2 \lambda_{\mathcal{L}} = \sum_{j=1}^{c} \left[\lim_{n\to\infty} n_j^{1/2} \int_{-\infty}^{\infty} \left\{ J\left(\sum_{r=1}^{c} \lambda_r F\left(x + \frac{\theta_r - \theta_j}{n^{1/2}}\right)\right) - J\left(F(x)\right) \right\} \cdot dF(x) \right]^2 . \tag{16.3.13}$$

This result can be surmised by using a result of Rao (1965) or Hájek and Sidák (1967, p. 311). For these results, see Chapter 23.

Assuming that the limit on n can be taken underneath the integral sign (see, for instance, Govindarajulu, 1980, for sufficient conditions), we obtain

$$
\begin{aligned}
N^{1/2} & \int_{-\infty}^{\infty} \left[J\left\{ \sum_{r=1}^{c} \lambda_r F\left(x + (\theta_r - \theta_i)n^{-1/2}\right)\right\} - J\left(F(x)\right)\right] dF(x) \\
&= n^{-1/2} \sum_{r=1}^{c} \frac{s_r(\theta_r - \theta_j)}{(\sum_1^c s_r)} \int_{-\infty}^{\infty} J'(F) f^2(x) dx \\
&= n^{-1/2}(\bar{\theta} - \theta_j) \int_{-\infty}^{\infty} J'(F) f^2(x) dx, \ \bar{\theta} = \sum_1^c s_r \theta_r \Big/ \left(\sum_1^c s_r \right).
\end{aligned}
$$

Hence

$$
I^2 \lambda_{\mathcal{L}} = \sum_{j=1}^{c} s_j (\theta_j - \bar{\theta})^2 \left[\int_{-\infty}^{\infty} J'(F) f^2(x) dx \right]^2 . \qquad (16.3.14)
$$

Special Cases

1. If $E_{N,i}$ denote the expected value of the ith smallest order statistic in a random sample of size N drawn from the standard normal distribution, then we obtain the normal scores test. In this case, $J_N(u) = J(u) = \Phi^{-1}(u)$, $0 < u < 1$.

2. If $E_{N,i} = i/(N+1)$, i.e., $J_N(u) = u$, $0 < u < 1$, we obtain the Kruskal-Wallis test criterion. Also note that Kruskal-Wallis test is obtained by replacing the observations by their ranks in the $\mathcal{F}$ test criterion given by (16.2.9), the denominator of which becomes a non-stochastic constant. Also note that the test criterion of Terpestra (1952) is related to the Kruskal-Wallis test criterion given by

$$
H = \frac{12}{N(N+1)} \sum_{i=1}^{c} n_i \left(\bar{R}_i - \frac{N+1}{2} \right)^2 \qquad (16.3.15)
$$

where

$$
\bar{R}_i = n_j^{-1} \sum_{j=1}^{n_j} R_{ij}, \ R_{ij} = \text{ rank of } X_{ij} \, . \qquad (16.3.16)
$$

Andrews (1954) established the asymptotic chi-squareness of the Kruskal-Wallis and the Brown-Mood's median tests and evaluates their asymptotic efficiencies relative to the $\mathcal{F}$-test. The Brown-Mood test is based on the statistic

$$\begin{aligned} M &= \frac{N(N-1)}{b(N-b)} \sum_{i=1}^{c} \frac{1}{n_i} \left(m_i - \frac{bn_i}{N} \right)^2 \\ &\sim 4 \sum_{i=1}^{c} \frac{\left(m_i - \frac{n_i}{2}\right)^2}{n_i} = 4 \sum_{i=1}^{c} n_i \left(\frac{m_i}{n_i} - \frac{1}{2} \right)^2 \end{aligned}$$

where

$$b = \begin{cases} (N-1)/2 & \text{if } N \text{ is odd,} \\ N/2 & \text{if } N \text{ is even} \end{cases}$$

and m_i = number of observations in the ith sample which are less than the median of all the observations. In the next section we will provide a much simpler proof for the limiting distribution of the Brown-Mood's median test.

16.4 The Median Test

Mood-Brown median test (1950, 1951) which was originally proposed for the two-sample problem is equivalent (asymptotically) to the LMP rank test for the double exponential shift-alternatives. A c-sample analogue of the two-sample version has been proposed by Mood and Brown (1951) for the c-sample problem. Let, as before, $X_{ij} = j = 1, 2, \ldots, n_i$ denote a random sample from $F_i(x)$ $(i = 1, \ldots, c)$. Let $N = n_1 + \cdots + n_c$ and $\lambda_i = n_i/N$ $(i = 1, \ldots, c)$. Further, let $W_{1N} \leq \cdots \leq W_{NN}$ be the combined ordered sample of N observations. Without loss of generality, let N be even so that $\tilde{W}_N = \frac{1}{2}\left(W_{\frac{N}{2},N} + W_{\frac{N}{2}+1,N}\right)$ denote the combined sample median. Then let

$$\begin{aligned} T_i &= \left(\frac{m_i}{n_i} \right), \ i = 1, \ldots, c \qquad (16.4.1) \\ &= F_{i,n_i}(\tilde{W}_N) \,. \end{aligned}$$

Note that

$$\sum_{i=1}^{c} n_i T_i = N H_N(\tilde{W}_N) = \frac{N}{2}$$

where

$$H_N(x) = \sum_{1}^{c} \lambda_i F_{i,n_i}(x) \,.$$

Further, let ξ denote the median of $H(x) = \sum_1^c \lambda_i F_i(x)$. Consider

$$\begin{aligned} N^{1/2}(T_i - \mu_{iN}) &= N^{1/2}\left(F_{i,n_i}(\tilde{W}_N) - H(\xi)\right) \\ &= B_{N,i} + R_{N,i} \end{aligned}$$

where

$$B_{N,i} = N^{1/2}\left\{F_{i,n_i}(\xi) - H(\xi)\right\} \tag{16.4.2}$$

and

$$R_{N,i} = N^{1/2}\left\{F_{i,n_i}(\tilde{W}_N) - F_{i,n_i}(\xi)\right\}. \tag{16.4.3}$$

Since $\tilde{W}_N$ converges to ξ in probability, for sufficiently large N, $|R_{N,i}| \leq N^{1/2}/n_i$ in probability. So the remainder term can be neglected. So,

$$\sqrt{N}(T_i - \mu_{N,i}) \approx B_{N,i} \tag{16.4.4}$$

where $B_{N,i}$, $i = 1, \ldots, c$ are normalized sums of i.i.d. random variables. Hence, $\{B_{N,1}, \ldots, B_{n,c}\}$ has a c-variate normal distribution as N becomes large. Var $(B_{N,i}|H_0) = F(\xi)\left[1 - F(\xi)\right]/\lambda_i = 1/4\lambda_i$. By making the following analysis variance transformation

$$\begin{aligned} V_1 &= \sum \lambda_i W_{N,i} \\ V_2 &= \sum e_{i,2} W_{N,i} \\ &\vdots \\ V_c &= \sum_{i=1}^{c} e_{i,c} W_{N,i} \end{aligned}$$

where $W_{N,i} = 2N^{1/2}(T_i - \mu_{iN})\lambda_i^{1/2}$, $(i = 1, \ldots, c)$,

$$V_1^2 + \cdots + V_c^2 = \sum_1^c W_{N,i}^2 .$$

However, $V_1 = 0$. Hence, $\sum_1^c W_{N,i}^2$ is asymptotically non-central chi-square with $c - 1$ degrees of freedom and non-centrality parameter

$$\lambda^M = 4\sum_{i=1}^{c} n_i \left\{F_i(\xi) - H(\xi)\right\}^2 , \tag{16.4.5}$$

Now let $F_i(\xi) = F\left(\xi + \frac{\theta_i}{\sqrt{n}}\right)$, $i = 1, \ldots, c$ for some n getting large. Then

$$\begin{aligned} F_i(\xi) - H(\xi) &= \sum_{j=1}^{c} \lambda_j \left[F\left(\xi + \frac{\theta_i}{\sqrt{n}}\right) - F\left(\xi + \frac{\theta_j}{\sqrt{n}}\right) \right] \\ &\dot{\sim} \sum_{j=1}^{c} \lambda_j (\theta_i - \theta_j) n^{-1/2} F'(\xi), \end{aligned}$$

for sufficiently large n. Now let $s_i = n_i/n$ and hence

$$\sum \lambda_j \theta_j = \sum s_j \theta_j / \sum_1^c s_j = \bar{\theta}.$$

Thus,

$$F_i(\xi) - H(\xi) \sim n^{-1/2} F'(\xi)(\theta_i - \bar{\theta}).$$

Consequently, the non-centrality parameter becomes

$$\lambda^M = \left\{ \sum_{i=1}^{c} s_i (\theta_i - \bar{\theta})^2 \right\} \{2F'(\xi)\}^2. \tag{16.4.6}$$

Thus, the asymptotic efficiency of the median test relative to Kruskal-Wallis test is given by

$$e_{M,H} = \frac{1}{3} \left[f(\xi) / \int_{-\infty}^{\infty} f^2(x) dx \right]^2 \tag{16.4.7}$$

where

$$f(x) = F'(x).$$

Justification for Negligibility of the Remainder Term

Recall that the remainder term in the median test criterion is given by

$$R_{N,i} = N^{1/2} \left\{ F_{i,n_i}(\tilde{W}_N) - F_{i,n_i}(\xi) \right\}.$$

Now, for every ϵ $(0 < \epsilon < 1/n_i)$ there exists an $N_0(\epsilon)$ such that $|\tilde{W}_N - \xi| \le \epsilon$ in probability since the sample median converges to the population median. Hence,

$$\begin{aligned} R_{N,i} &\sim N^{1/2} \{F_{i,n_i}(\xi \pm \epsilon) - F_{i,n_i}(\xi)\} \\ &\le (n_i \lambda_i)^{-1/2} \{\text{number of } X_{ik}\text{'s in } [\xi - \epsilon, \xi + \epsilon]\} \\ &= (n_i \lambda_i)^{-1/2} S_{n_i}^* \text{ (say)}. \end{aligned}$$

Then,

$$ES^*_{n_i} = n_i \{F_i(\xi + \epsilon) - F_i(\xi - \epsilon)\} \approx 2\epsilon n_i f_i(\xi^*) \le 2f_i(\xi^*)$$

where ξ^* is in the interval $(\xi - \epsilon, \xi + \epsilon)$, and

$$\text{var}(S^*_{n_i}) = 2\epsilon n_i f_i(\xi^*)\,[1 - 2\epsilon f_i(\xi^*)] \le 2f_i(\xi^*)\,.$$

Now, letting $\epsilon \to 0$ (i.e., $n_i \to \infty$), we see that $R_{N,i}$ tends to zero in probability.

16.5 *U*-Statistics Approach

Bhapkar (1961) proposed a nonparametric test based on a generalized U-statistic given by

$$V = N(2c-1)\left[\sum_{i=1}^{c} \lambda_i (u^{(i)} - c^{-1})^2 - \left\{\sum_{1}^{c} \lambda_i (u^{(i)} - c^{-1})\right\}^2\right] \tag{16.5.1}$$

where

$$n_1 n_2 \cdots n_c u^{(i)} = v^{(i)}\,,$$

$$\begin{aligned} v^{(i)} &= \sum_{j=1}^{n_i} \prod_{r \ne i} \{\text{number of } X_{rs} > X_{ij},\ s = 1, \ldots, n_r\} \\ &= \sum_{t_1=1}^{n_1} \cdots \sum_{t_c=1}^{n_c} \psi^{(i)}(X_{1,t_1}, \ldots, X_{c,t_c})\,, \end{aligned} \tag{16.5.2}$$

$$\psi^{(i)}(x_{1,t_1}, \ldots, x_{c,t_c}) = \begin{cases} 1 & \text{if } x_{i,t_i} < x_{k,t_k} \text{ for all } k = 1, \ldots, c \text{ and } k \ne i\,, \\ 0 & \text{otherwise.} \end{cases}$$

In other words, $v^{(i)}$ is the number of c-plets that can be formed by selecting one observation from each sample such that the observation from the ith sample is the smallest. Since $E(v^{(i)}|H_0) = 1/c$, V may be viewed as a measure of departure from H_0. Using the asymptotic theory for U-statistics developed by Hoeffding (1948), Bhapkar (1961) was able to show that V has a limiting non-central chi-square distribution with $c-1$ degrees of freedom and non-centrality parameter

$$\lambda_V = (2c-1)c^2 \sum_{i=1}^{c} s_i(\theta_i - \bar{\theta})^2 \left[\int_{-\infty}^{\infty} [1 - F(y)]^{c-2} f^2(y)dy\right]^2, \tag{16.5.3}$$

under the local translation alternatives given by $F_i(x) = F(x - \theta_i n^{-1/2})$. Deshpande (1965) proposed a U-statistic based sum of $v^{(i)}$ and $v_1^{(i)}$ where $v_1^{(i)}$ is the number of c-plets that can be formed by selecting one observation from each sample such that the observation from the ith sample is the largest. One can develop a parallel asymptotic theory for the $v_1^{(i)}$ and propose a statistic similar to V. However, it is too complicated to be dealt with here. Bhapkar (1961) evaluates the asymptotic efficiency of V as

$$\begin{aligned} e_{V,H} &= 2(c-1)c^2\lambda_V^2/12\left\{\int_{-\infty}^{\infty} f^2(x)dx\right\}^2 \\ e_{V,M} &= 2(c-1)c^2\lambda_V^2/4\,[f(\xi)]^2 \\ &\text{and} \\ e_{V,\mathcal{F}} &= (2c-1)c^2\lambda_p^2\sigma_F^2 \end{aligned}$$

where λ_V is given by (16.4.3) and ξ denotes the median of F. For the uniform and exponential densities, $e_{V,H} \to \infty$ as $c \to \infty$ whereas for normal densities it goes to zero as $c \to \infty$.

16.6 Combining Two-sample Test Statistics

Dwass (1960) provides some new ideas on the use of rank order procedures for the problem of comparing several populations. These will be discussed in the following.

Let $\boldsymbol{X}_i = (X_{i,1}, \ldots, X_{i,n_i})$, $i = 1, \ldots, c$ be c independent vectors with i.i.d. components of $\boldsymbol{X}_i$ having continuous distribution $F_i(x)$. We wish to test the null hypothesis

$$H : F_1(x) = \cdots = F_c(x), \text{ for all } x\,.$$

Dwass (1960) provides a general method of combining two-sample tests in order to provide a c-sample test. Let $\boldsymbol{R}_j = (R_{j1}, \ldots, R_{j,n_j})$, $j = 2, \ldots, c$ be the rank order associated with observational vector $\boldsymbol{X}_j$ $(j = 2, \ldots, c)$ such that

$$R_{j,i} = \text{number of } X_{rs} \le X_{ji},\ r = 1, \ldots, j,\ s = 1, \ldots, n_r\,.$$

That is, $\boldsymbol{R}_j$ is the vector of the rank-order positions of the elements of $\boldsymbol{X}_j$ in the combined sample of the $n_1 + \cdots + n_j$ observations $\boldsymbol{X}_1, \ldots, \boldsymbol{X}_j$. $\boldsymbol{R}_j$'s are also known as the sequential ranks. Then the following result of Dwass (1960) is of independent interest, the proof of which is somewhat elementary.

Theorem 16.6.1. *The $c-1$ vectors $\boldsymbol{R}_2, \boldsymbol{R}_3, \ldots, \boldsymbol{R}_c$ defined above are statistically independent if the null hypothesis H is true.*

Let us follow the following convention: $T(\boldsymbol{X}_1, \ldots, \boldsymbol{X}_{j-1}; \boldsymbol{X}_j)$ denotes some two-sample test statistic where the first sample consists of the pooled sample of the $n_1 + \cdots + n_{j-1}$ observations which are the components of $\boldsymbol{X}_1, \ldots, \boldsymbol{X}_{j-1}$ and the second sample consists of the n_j observations which are the components of $\boldsymbol{X}_j$. Then the following corollary readily follows.

Corollary 16.6.1.1. *Let the statistic T be invariant under permutations within each of the two samples (In other words, $T(\boldsymbol{X}_1, \ldots, \boldsymbol{X}_{j-1}; \boldsymbol{X}_j)$ is a function of $\boldsymbol{R}_j$ only.) Then the $c - 1$ random variables $T(\boldsymbol{X}_1; \boldsymbol{X}_2)$, $T(\boldsymbol{X}_1, \boldsymbol{X}_2; \boldsymbol{X}_3)$, ..., $T(\boldsymbol{X}_1, \ldots, \boldsymbol{X}_{c-1}; \boldsymbol{X}_c)$ are statistically independent when H is true.*

Example 16.6.1. Let $T(\boldsymbol{X}_i, \boldsymbol{X}_j) = U_{ij}$, the Mann-Whitney statistic (i.e., the number of times components of $\boldsymbol{X}_j$ are greater than or equal to components of $\boldsymbol{X}_i$). It is easy to verify that in this special case $T(\boldsymbol{X}_1, \ldots, \boldsymbol{X}_{j-1}; \boldsymbol{X}_j) = U_{1j} + U_{2j} + \cdots + U_{j-1,j}$. Hence the following linear combinations of Mann-Whitney statistics are statistically independent under H:

$$\begin{array}{lll} U_{12} & & \\ U_{13} & + & U_{23} \\ \vdots & & \\ U_{1c} & + & U_{2c} + \cdots + U_{c-1,c} \,. \end{array}$$

Now a test of H can be based on the $c - 1$ random variables and the distribution of the two-sample statistics is required for the determination of the level of significance. For instance, we accept H if $V < K$ where

$$V = \max\left\{D_2 \left|U_{12} - \frac{n_1 n_2}{2}\right|, \ldots, D_k \left|\left(U_{1k} - \frac{n_1 n_k}{2}\right) + \cdots + \left(U_{c-1,c} - \frac{n_{c-1} n_c}{2}\right)\right|\right\}$$

and $D_2, \ldots, D_c$ and K are suitably chosen constants.

A general analogue of the above test is that we accept H when

$$\max\left\{D_2 T(\boldsymbol{X}_1; \boldsymbol{X}_2), \ldots, D_k T(\boldsymbol{X}_1, \ldots, \boldsymbol{X}_{c-1}; \boldsymbol{X}_c)\right\} < K \,.$$

The following will serve as a heuristic rationale of such a test: Each of the $T(\boldsymbol{X}_1, \ldots, \boldsymbol{X}_{j-1}; \boldsymbol{X}_j)$ statistics plays the role of the estimate of a "contrast" among the c-populations. There are only $c-1$ independent contrasts and the test is based on one convenient choice of estimates of such a set of contrasts.

Procedure II. Let

$$W = \max_{i,j \text{ in } A} d_{ij} T(\boldsymbol{X}_i; \boldsymbol{X}_j)$$

where A is a set of pairs of indices from $(1, \ldots, c)$ and d_{ij} are suitably chosen constants. For instance, A can consist of all $\binom{c}{2}$ pairs of indices. Then the test procedure is to accept H only if $W < K$. However, this procedure is having some difficulties in finding the distribution of W since here W is not a maximum of independent variables. A general inequality which may be useful here is the easily verifiable one:

$$P(W < K) \geq 1 - \sum_{i,j \text{ in } A} P\left(d_{ij} T(\boldsymbol{X}_i; \boldsymbol{X}_j) \geq K\right) .$$

An example in which Dwass (1960) was able to deal with the distributional problem, at least asymptotically, is when $n_1 = n_2 = \cdots = n_c = n$ and $T(\boldsymbol{X}_i; \boldsymbol{X}_j) = |U_{ij} - n^2/2|$. Then Dwass (1960) obtains the following theorem.

Theorem 16.6.2. *When H is valid,*

$$\lim_{n \to \infty} P\left\{\sqrt{\frac{12}{n^2(2n+1)}} \max_{i,j=1,\ldots,c} \left|U_{ij} - \frac{n^2}{2}\right| \leq t\right\} = G(t)$$

where G denotes the distribution function of the range of c independent standard normal variables.

PROOF: If $Z_1, \ldots, Z_c$ are independent standard normal variables and Y is the random vector with $\binom{c}{2}$ components $(Z_i - Z_j)$, $i, j = 1, \ldots, c (i < j)$, let Y_n be the random vector with $\binom{c}{2}$ components $\left\{\frac{12}{n^2(2n+1)}\right\}^{1/2} (U_{ij} - n^2/2)$. Then Dwass (1961) claims that the limiting distribution of Y_n, as n gets large, is the same as Y.

16.7 Kolmogorov-Smirnov Type of Statistics

Dwass (1960) and Kiefer (1959) propose Kolmogorov-Smirnov type of test statistics for testing the null hypothesis of equality of k-distribution functions. Here we will implement one of two procedures proposed by Kiefer

(1959). Let X_{ij}, $j = 1, \ldots, n_i$ be a random sample from $F_i(x)$ $(i = 1, \ldots, c)$. We wish to test

$$H_0 : F_1(x) = \cdots = F_c(x) \text{ for all } x\,.$$

Let $F_{i,n_i}(x)$ be the e.d.f. based on X_{ij} $(j = 1, \ldots, n_i)$ for $i = 1, \ldots, c$. Let $N = n_1 + \cdots + n_c$ and $H_N(x) = \sum_{i=1}^{c} \lambda_i F_{i,n_i}(x)$, $\lambda_i = n_i/N$. Then Kiefer proposes test procedures based on

$$T_N = \sup_x \sum_{i=1}^{c} n_i \left[F_{i,n_i}(x) - H_N(x)\right]^2 \tag{16.7.1}$$

and

$$W_N = \int_{-\infty}^{\infty} \sum_1^c n_i \left[F_{i,n_i}(x) - H_N(x)\right]^2 dH_N(x)\,. \tag{16.7.2}$$

Notice that the latter statistic is of Cramér-Von-Mises type. Kiefer (1959) then strives to obtain the limiting distribution of T_N and W_N when H_0 holds. When H_0 is true, without loss of generality, one can set F_1 to be uniform on $(0, 1)$ for $i = 1, \ldots, c$. Kiefer shows that

$$\lim_{\text{all } n_i \to \infty} P[T_N \le a] = A_{c-1}(a) \tag{16.7.3}$$

where

$$A_h(a) = P\left[\max_{0 \le t \le 1} \sum_{i=1}^{h} |Y_i(t)|^2 \le a\right], \tag{16.7.4}$$

and the $Y_i(t)$ are assumed to be as follows: $Y_1(t), Y_2(t), \ldots, Y_h(t)$ are independent separable Gaussian processes whose sample functions are functions of the same "time" parameter t, $0 \le t \le 1$ and such that $EY_i(t) = 0$, and $EY_i(t)Y_i(s) = \min(s, t) - st$ for each i. Thus the Y_i are independent "tied down Wiener processes" which can be written as

$$Y_i(t) = (1 - t)^{-1} W_i\left(t/(1 - t)\right)$$

where the W_i are independent Wiener processes such that $EW_i(t) = 0$, $EW_i(t)W_i(s) = \min(s, t)$, for $0 \le t, s < \infty$. Kiefer obtains explicit expressions for $A_i(a)$ for $i = 1, 3$ and tabulates, numerically, the values of $A_i(x^2)$ for $1 \le i \le 5$ and $x = 0.37(0.01)2.56$.

Similarly, if we define

$$B_h(a) = P\left\{\int_0^1 \sum_{i=1}^{h} [Y_i(t)]^2\, dt \le a\right\}, \tag{16.7.5}$$

then Kiefer (1959) obtains

$$\lim_{\text{all } n_j \to \infty} P(W_N \le a) = B_{c-1}(a) \tag{16.7.6}$$

and tabulates, numerically, the values of $B_i(x)$ for $1 \le i \le 5$ and $x = 0.01(0.01)2.42$.

If

$$T_{ij} = \max_{-\infty<x<\infty} \left|F_{i,n_i}(x) - F_{j,n_j}(x)\right| ,$$

one can use

$$\max_{i,j(i\neq j)} T_{ij}$$

to test

$$H_0 : F_1(x) = \cdots = F_c(x) \text{ for all } x .$$

Such statistics have been proposed by David (1958) and Fisz (1960) for $c = 3$. Dwass (1960) proposed the following test criterion for testing H_0:

$$\max_{2\le j\le c} T_j = \max_{2\le j\le c} \max_{-\infty<x<\infty} \left| F_{j,n_j}(x) - \sum_{i=1}^{j-1} \lambda_i^* F_{i,n_i}(x) \right| \tag{16.7.7}$$

where $\lambda_i^* = n_i / \sum_{k=1}^{j-1} n_k$. Note that from Theorem 16.6.1, the statistics $T_2, T_3, \ldots, T_c$ are statistically independent when H_0 is true.

16.8 Problems

16.2.1 A plant ecologist wishes to know whether the height of plant species depend on the type of soil it grows in. He measured the height of three plants in each of 4 plots representing different soil types, all 4 plots being contained in an area of two square miles. Does the data[2] support the hypothesis? (Height is measured in centimeters) Let $\alpha = 0.05$.

	plots			
replications	1	2	3	4
1	10	15	17	25
2	13	9	23	21
3	16	14	20	19

[2]For data see Sokal, R.R. and F.J. Rohlf (1981). *Biometry* Second Edition W.H. Freeman and company, San Francisco, CA, pp. 264–265.

(HINT: Assume that the alternative hypothesis is of the form H_L : $F_i(x) = 1 - [1 - F(x)]^{\theta_i}, i = 1, \cdots, 4$. Test H_0 using L_S given by (16.2.11) as the test criterion.)

16.2.2 Evaluate $e(S, \mathcal{F})$ given by (16.2.21) when $\mathcal{F}$ is (i) normal, (ii) logistic and (iii) negative exponential.

16.3.1 For the data in Problem 16.2.1, carry out a Kruskal-Wallis test H given by (16.3.15) and refer to the chi-square tables for the critical value.

16.4.1 For the data in Problem 10.7.1, carry out the Brown-Mood Test defined towards the end of Section 16.3 with $\alpha = 0.05$.
(HINT: See Equ (16.4.5).)

16.4.2 Evaluate the ARE of Brown-Mood's test relative to Kruskal-Wallis test when f is (i) normal, (ii) logistic, and (iii) double exponential.

16.5.1 For the data in Problem 16.2.1, carry out the test procedure based on the V statistic given by (16.5.1).

Chapter 17

c-sample Tests for Scale

17.1 Introduction

In this chapter we will give some nonparametric as well as parametric tests of homogeneity of scale parameters. First, we will give the parametric test procedures that are not so well-known.

17.2 Parametric Procedure

Let X_{ij}, $j = 1, \ldots, n_i$ $(i = 1, \ldots, c)$ denote a random sample from a normal population with mean μ_i and variance σ_i^2 $(i = 1, \ldots, c)$. We wish to test

$$H_0 : \sigma_1^2 = \cdots = \sigma_c^2$$

against the alternative

$$H_1 : \sigma_i^2 \neq \sigma_j^2$$

for some j. Let σ^2 denote the common value of the variances. Then the unconditional maximum of the likelihood function is obtained by setting

$$\hat{\mu}_i = \bar{X}_i \text{ and } \hat{\sigma}_i^2 = n_i^{-1} \sum_{j=1}^{n_i} (X_{ij} - \bar{X}_i)^2 = S_i^2$$

giving

$$\sup_{H_1} L(\mu_1, \ldots, \mu_c, \sigma_1^2, \ldots, \sigma_c^2 | \underline{X}) = (2\pi)^{-N/2} \prod_{i=1}^{c} (S_i^2)^{-n_i/2} e^{-N/2} \qquad (17.2.1)$$

where

$$N = \sum_{i=1}^{c} n_i .$$

Under H_0, the maximum likelihood estimates of the μ_i and σ^2 are

$$\hat{\mu}_2 = \bar{X}_i \text{ and } \hat{\sigma}^2 = \frac{1}{N} \sum_{1}^{c} n_i S_i^2 = S^2$$

so that

$$\sup_{H_0} L(\mu_1, \ldots, \mu_c, \sigma^2 | \underline{X}) = (2\pi)^{-N/2} (S^2)^{-N/2} e^{-N/2} . \tag{17.2.2}$$

Thus, the likelihood ratio criterion is $\wedge = \prod_{i=1}^{c} (S_i^2/S^2)^{n_i/2}$ so that

$$-2 \log \wedge = N \log(S^2) - \sum_{i=1}^{c} n_i \log(S_i^2) \tag{17.2.3}$$

where

$$\wedge = (\sup_{H_1} L)/(\sup_{H_0} L) .$$

Bartlett's (1937) modification of the LR statistic is obtained by replacing n_i by $\nu_i = n_i - 1$ (the degrees of freedom) so that N is replaced by $\nu = \sum_1^c (n_i - 1) = N - k$. Thus,

$$\wedge^* = \prod_{i=1}^{c} (S_i^2/S^2)^{\nu_i/2} \tag{17.2.4}$$

where now

$$\begin{aligned} S_i^2 &= \frac{1}{\nu_i} \sum_{j=1}^{n_i} (X_{ij} - \bar{X}_i)^2 \\ S^2 &= \frac{1}{\nu} \sum_{1}^{c} \nu_i S_i^2 . \end{aligned}$$

Thus,

$$-2 \log \wedge^* = \nu \log S^2 - \sum_{1}^{c} \nu_i \log S_i^2 . \tag{17.2.5}$$

$\wedge^*$ has the advantage over $\wedge$ of being an unbiased test for any value of the n_i. Kendall (1960, p. 236) shows that when you are sampling from the normal populations,

$$E(-2 \log \wedge^*) = (c-1) + \frac{1}{3}\left(\sum_{i=1}^{c} \frac{1}{\nu_i} - \frac{1}{\nu}\right) + O(\nu_i^3).$$

Hence,

$$\frac{-2 \log \wedge^*}{1 + \frac{1}{3(c-1)}\left(\sum_1^c \nu_i^{-1} - \nu^{-1}\right)} \overset{D}{\approx} \chi^2_{c-1} \text{ under } H_0. \tag{17.2.6}$$

However, when one is sampling from non-normal populations, the correction factor might be different which will be explored in the following. First we need the following lemma.

Lemma 17.2.1. *Let $X_1, \ldots, X_n$ be a random sample from a distribution having unknown mean μ and unknown variance σ^2 and a finite fourth moment. Let*

$$s^2 = (n-1)^{-1} \sum_1^n (X_i - \bar{X}^2).$$

Then,

$$var\ s^2 = \frac{\sigma^4}{n}\left\{\gamma_2 + 2 + \frac{2}{n-1}\right\} = \frac{\sigma^4}{n}\left(\gamma_2 + \frac{2n}{n-1}\right) \tag{17.2.7}$$

where $\gamma_2 = \frac{\mu_4}{\sigma^4} - 3 =$ the kurtosis of the distribution.

PROOF: Since s^2 is a function of the deviations of the observations from the sample mean, in order to compute the variance of s^2, we can, without loss of generality, set $\mu = 0$.

Since $\bar{X} = n^{-1} \sum_1^n X_i$,

$$n\bar{X}^2 = n^{-1}\left(\sum X_i\right)^2 = n^{-1}\left(\sum X_j^2 + \sum\sum_{i \neq j} X_i X_j\right).$$

Hence,

$$(n-1)s^2 = \sum X_i^2 - n\bar{X}^2 = \left(\frac{n-1}{n}\right)\sum X_i^2 + \frac{1}{n}\sum\sum_{i \neq j} X_i X_j$$

and

$$\begin{aligned}
&(n-1)^2 \text{var } s^2 \\
&= \text{var}\left\{\left(\frac{n-1}{n}\right)\sum X_i^2 + \frac{1}{n}\sum_{i\neq j}\sum X_i X_j\right\} \\
&= n^{-2}\left[(n-1)^2 \text{var}\left(\sum X_i^2\right) + \text{var}\left(\sum_{i\neq j}\sum X_i X_j\right) + 2(n-1)\right. \\
&\qquad \left.\cdot \text{cov}\left(\sum X_k^2, \sum_{i\neq j}\sum X_i X_j\right)\right] \\
&= n^{-2}\left[(n-1)^2 n \text{ var } X_1^2 + E\left\{2\sum_{i\neq j}\sum X_i^2 X_j^2 + \sum_{i\neq j}\sum \sum_{k\neq l}\sum\right.\right. \\
&\qquad \left.\left. X_i X_j X_k X_l\right\}\right]
\end{aligned}$$

since the covariance term is zero. Thus,

$$\begin{aligned}
(n-1)^2 \text{var } s^2 &= n^{-2}\left[n(n-1)^2(\mu_4 - \sigma^4) + 2n(n-1)\sigma^4\right] \\
&= \left(\frac{n-1}{n}\right)\left[(n-1)\mu^4 + 2\sigma^4\right].
\end{aligned}$$

That is,

$$\text{var } s^2 = \frac{1}{n(n-1)}\left[(n-1)(\gamma_2+2)\sigma^4 + 2\sigma^4\right] = \frac{\sigma^4}{n}\left(\gamma_2 + 2 + \frac{2}{n-1}\right).$$

Lemma 17.2.2. *Let*

$$-2 \log \wedge = N \log S^2 - \sum n_i \log S_i^2 \ (\textit{see } (17.2.3))$$

where

$$S^2 = \sum_1^c \lambda_i S_i^2, \quad \lambda_i = n_i/N.$$

Then

$$E(-2 \log \wedge | H_0) = (2+\gamma_2)(c-1)/2 \tag{17.2.8}$$

when the sample sizes are large where it is assumed that all the c populations have the same kurtosis.

PROOF: By a second order Taylor's expansion, we have

$$\log\, S_i^2 \doteq \log\, \sigma_i^2 + (S_i^2 - \sigma_i^2)\sigma_i^{-2} - \frac{1}{2}(S_i^2 - \sigma_i^2)^2\sigma_i^{-4}\,.$$

Hence,

$$\begin{aligned} E(\log\, S_i^2) &= \log\, \sigma_i^2 - \frac{1}{2}\sigma_i^{-4}\text{var}(S_i^2) \\ &\doteq \log\, \sigma_i^2 - \frac{1}{2n_i}(2+\gamma_2) \end{aligned} \tag{17.2.9}$$

after using Lemma 17.2.1. Similarly, expanding $\log\, S^2$ as

$$\log\, S^2 = \log\, \sigma^2 + (S^2 - \sigma^2)\sigma^{-2} - \frac{1}{2}(S^2 - \sigma^2)^2\sigma^{-4}$$

we obtain

$$\begin{aligned} E(\log\, S^2) &= \log\, \sigma^2 - \frac{1}{2}\sigma^{-4}\sum \lambda_i^2 \text{var}\, S_i^2 \\ &= \log\, \sigma^2 - \frac{1}{2}\sigma^{-4}\sum \sigma_i^4\lambda_i^2(2+\gamma_2)/n_i \\ &= \log\, \sigma^2 - \frac{1}{2N}\sigma^{-4}(2+\gamma_2)\sum \lambda_i\sigma_i^4\,. \end{aligned} \tag{17.2.10}$$

Using (17.2.9) and (17.2.10) we have

$$\begin{aligned} E(-2\, \log\, \wedge) &= NE(\log\, S^2) - \sum n_i E(\log\, S_i^2) \\ &= N\, \log\, \sigma^2 - \sum n_i\, \log\, \sigma_i^2 + \\ &\quad \frac{(2+\gamma_2)}{2}\left\{c - \sum \lambda_i\sigma_i^4\sigma^{-4}\right\}. \end{aligned} \tag{17.2.11}$$

In particular, we have

$$\begin{aligned} E(-2\, \log\, \wedge|H_0) &= (2+\gamma_2)(c-1)/2 \\ &= c-1 \text{ if the distributions are normal}. \end{aligned}$$

Thus, in order to test the equality of variances, using the asymptotic property of the likelihood ratio test criterion, it seems to be more precise if we say that $2(-2\, \log\, \wedge)/(2+\gamma_2) \overset{d}{\approx} \chi_{c-1}^2$ when H_0 holds. Hence, using the large-sample property of the likelihood ratio test criterion under the alternative, we infer that

$$2(-2\, \log\, \wedge)/(2+\gamma_2) \overset{d}{\approx} \chi_{c-1}^2(\lambda_{\mathcal{F}})$$

when the alternative is true where

$$\lambda_{\mathcal{F}} = 2(N \log \sigma^2 - \sum n_i \log \sigma_i^2)/(2+\gamma_2)\,. \tag{17.2.12}$$

Now, let us evaluate the non-centrality parameter $\lambda_{\mathcal{F}}$ under Pitman alternatives given by

$$\sigma_i^2 = \theta_0 + \theta_i/\sqrt{n},\ \sigma^2 = \theta_0 + \sum n_i\theta_i/N\sqrt{n} \tag{17.2.13}$$

where n is getting large. Letting $s_i = n_i/n$ and consequently $\lambda_i = n_i/N = s_i/\sum_1^c s_i$, and setting $\bar\theta = \sum s_i\theta_i/\sum s_i$, we obtain

$$N \log \sigma^2 - \sum n_i \log \sigma_i^2 = \sum n_i \left\{\log\left(1+\frac{\bar\theta}{\sqrt{n}\,\theta_0}\right) - \log\left(1+\frac{\theta_i}{\sqrt{n}\,\theta_0}\right)\right\}. \tag{17.2.14}$$

Now, using the expansion

$$\log(1+a_i) = \log(1+\bar a) + \frac{(a_i-\bar a)}{(1+\bar a)} - \frac{1}{2}\frac{(a_i-\bar a)^2}{(1+\bar a)^2}\,,$$

with $a_i = \theta_i/\sqrt{n}\,\theta_0$ and $\bar a = \bar\theta/\sqrt{n}\,\theta_0$, we have

$$\lim_{n\to\infty}\left\{N \log \sigma^2 - \sum n_i \log \sigma_i^2\right\} = \frac{1}{2}\sum_1^c s_i(\theta_i-\bar\theta)^2/\theta_0^2\,. \tag{17.2.15}$$

Thus,

$$\lambda_{\mathcal{F}} = \sum_1^c s_i(\theta_i-\bar\theta)^2/\theta_0^2(2+\gamma_2)\,. \tag{17.2.16}$$

Alternate Parametric Test

The likelihood ratio test and its modification due to Bartlett (1937) may be sensitive to kurtosis of the populations if they are non-normal. Hence, Scheffé (1959, pp. 83–87) has proposed an alternative test which also has been mentioned by Puri (1965). This will be described below. Let $\eta_i = \log \sigma_i^2$ $(i = 1,\ldots,c)$. Then H_0 can be written as

$$H_0 : \eta_1 = \eta_2 = \cdots = \eta_c$$

and

$$H_1 : \eta_i \neq \eta_j \text{ for some } i \neq j\,.$$

Randomly divide the sample from the ith population, namely, $X_{i1}, \ldots, X_{i,n_i}$ into J_i subsamples, the jth subsample size being n_{ij}. Assume that not all J_i are equal to 1. Let S_{ij}^2 denote the sample variance (unbiased version) of the jth subsample of the ith original sample. Let

$$Y_{ij} = \log\ S_{ij}^2 .$$

Then,

$$E(Y_{ij}) \sim \eta_i \ \ (j = 1, \ldots, J_i)$$

and due to Lemma 17.2.1 and the delta method,

$$\operatorname{var}(Y_{ij}) \sim \frac{1}{n_{ij}}\left(\gamma_{2i} + 2 + \frac{2}{n_{ij}-1}\right) = \frac{\gamma_{2i}}{n_{ij}} + \frac{2}{n_{ij}-1} \tag{17.2.17}$$

where γ_{2i} denotes the kurtosis of the ith population. Hereafter, we assume that $\gamma_{2i} \equiv \gamma_2$ for all $i = 1, \ldots, c$. In order to test H_0, Scheffé (1959) proposes the usual F-statistic given by

$$F = \frac{\sum_1^c \nu_i (y_{i\cdot} - \bar{y})^2/(c-1)}{\sum_{i=1}^c \sum_{j=1}^{J_i} \nu_{ij}(y_{ij} - y_{i\cdot})^2 / \sum_{i=1}^c (J_i - 1)} \tag{17.2.18}$$

where $y_{i\cdot} = \sum \nu_{ij} y_{ij}/\nu_i$, $\bar{y} = \sum \nu_i y_{i\cdot}/\nu$, $\nu_{ij} = n_{ij} - 1$, $\nu_i = \sum_{j=1}^{J_i} \nu_{ij}$, $\nu = \sum_1^c \nu_i$. Now, F has under H_0 (approximately) the F-distribution with $c-1$ and $\nu_e = \sum_i (J_i - 1)$ degrees of freedom. The use of the F statistic is justified since

$$\operatorname{var}(Y_{ij}) = \frac{1}{\nu_{ij}}\left[2 + \frac{\gamma_2}{1+\nu_{ij}^{-1}}\right] \approx \frac{2+\gamma_2}{\nu_{ij}} \text{ for large } \nu_{ij} .$$

Further, since the denominator converges in probability to $2+\gamma_2$, $(c-1)$ times (the statistic) behaves (asymptotically) like

$$\sum_{i=1}^{c} \nu_i (y_{i\cdot} - \bar{y})^2/(2+\gamma_2)$$

which has a non-central chi-square distribution with $c-1$ degrees of freedom and non-centrality parameter

$$\sum_{i=1}^{c} \nu_i (\eta_i - \bar{\eta})^2/(2+\gamma_2) \tag{17.2.19}$$

where

$$\eta_i = \log\ \sigma_i^2 \text{ and } \bar{\eta} = \sum_1^c \nu_i \eta_i/\nu . \tag{17.2.20}$$

For local alternatives, keeping only the first two terms in the expansion of $\log(1 + a_i)$, (17.2.19) simplifies to (17.2.16). Puri (1965) gives the non-centrality parameter as

$$4 \sum_{1}^{c} \nu_i(\theta_i - \bar{\theta})^2/(2 + \gamma_2)\,.$$

In other words, we do not get the factor 4 and we will have an additional factor θ_0^2 in the denominator (see 17.2.19).

17.3 Rank Order Tests

We have a random sample of size n_i, namely X_{ij} $j = 1, \ldots, n_i$ from the distribution $F_i(x) = F\left(x(1 + \theta_i)\right)$, $i = 1, \ldots, c$. We wish to test

$$H_0 : F_1(x) = \cdots = F_c(x) \text{ for all } x$$

against the alternative hypothesis

$$H_1 : F_i(x) \neq F_j(x) \text{ for some } i \neq j\,.$$

Let $N = n_1 + \cdots + n_c$ and $W_{1N} \leq \cdots \leq W_{NN}$ denote the combined ordered sample. As in Section 16.3, define the Chernoff-Savage class of statistics:

$$T_{N,j} = n_j^{-1} \sum_{i=1}^{N} E_{N,i} Z_{N,i,j} \tag{17.3.1}$$

where $E_{N,i}$ are some specified scores and

$$Z_{N,i,j} = \begin{cases} 1 & \text{if } W_{i,N} \text{ comes from the } j^{\text{th}} \text{ sample,} \\ 0 & \text{otherwise}\,. \end{cases}$$

We can have the following integral representation of $T_{N,j}$,

$$T_{N,j} = \int_{-\infty}^{\infty} J_N\left(\frac{N}{N+1} H_N(x)\right) dF_{j,n_j}(x) \tag{17.3.2}$$

where

$$J_N\left(\frac{i}{N+1}\right) = E_{N,i}, \text{ for } j = 1, \ldots, c, \text{ and } i = 1, \ldots, N\,,$$

$$H_N(x) = \sum_{j=1}^{c} \lambda_j F_{j,n_j}(x), \ \lambda_j = n_j/N\,.$$

As special cases we have

$$\begin{aligned} E_{N,i} &= \left|\frac{i}{N+1} - \frac{1}{2}\right| && \text{for Ansari-Bradley or Siegel-Tukey's scores;} \\ &= \left[\Phi^{-1}\left(\frac{1}{N+1}\right)\right]^2 && \text{for Klotz's scores;} \\ &= \left(\frac{i}{N+1} - \frac{1}{2}\right)^2 && \text{for Mood's scores;} \\ &= -\log\left(1 - \frac{i}{N+1}\right) && \text{for Savage's scores.} \end{aligned} \tag{17.3.3}$$

Define the statistic

$$\tilde{\mathcal{L}}_T = N \sum_{j=1}^{c} \lambda_j (T_{N,j} - \mu_{N,j})^2/(1-\lambda_j) I^2 \tag{17.3.4}$$

where

$$I^2 = \int_0^1 J_N^2(u)du - \left\{\int_0^1 J_N(u)du\right\}^2 \tag{17.3.5}$$

and

$$\mu_{N,j} = \lambda_j \int_{-\infty}^{\infty} J_N\left(H(x)\right) dF_j(x), \ \ H(x) = \sum_1^c \lambda_i F_i(x)\,.$$

Consider the local alternatives

$$F_i(x) \doteq F\left(x(1+\theta_i/\sqrt{n})\right), \ \ i = 1, \ldots, c \text{ for some } n \text{ large}$$

such that $n_j/n = s_j > 0$. Note that

$$\lambda_j = n_j/N = s_j/\left(\sum_{i=1}^{c} s_i\right).$$

As in Section 16.3, one can show that $\tilde{\mathcal{L}}_T$ is asymptotically non-central chi-square with $c-1$ degrees of freedom and non-centrality parameter $\lambda_{\tilde{\mathcal{L}}}$ given by

$$\begin{aligned} I^2\lambda_{\tilde{\mathcal{L}}} = \sum_{j=1}^{c}\Bigg[\lim_{n\to\infty} n_j^{1/2}\int_{-\infty}^{\infty}\Bigg\{J\Bigg(\sum_{r=1}^{c}\lambda_r F\left(\left(\frac{1+\theta_r/\sqrt{n}}{1+\theta_j/\sqrt{n}}\right)x\right) \\ -J(F(x))\Bigg)\Bigg\}\cdot dF(x)\Bigg]^2. \end{aligned} \tag{17.3.6}$$

Assuming that the limit on n can be taken underneath the integral sign (see, for instance, Govindarajulu, 1980, for sufficient conditions),

$$
\begin{aligned}
N^{1/2} & \int_{-\infty}^{\infty} \left[J\left\{ \sum_{r=1}^{c} \lambda_r F\left(\frac{1+\theta_r/\sqrt{n}}{1+\theta_j/\sqrt{n}} \right) x \right\} - J\left(F(x)\right) \right] dF(x) \\
&\approx \; n^{-1/2} \sum_{r=1}^{c} s_r \frac{(\theta_r - \theta_j)}{(\sum_1^c s_r)} \int_{-\infty}^{\infty} x J'(F) f^2(x) dx \\
&\approx \; n^{-1/2}(\bar{\theta} - \theta_j) \int_{-\infty}^{\infty} x J'(F) f^2(x) dx \, .
\end{aligned}
$$

Hence,

$$
I^2 \lambda_{\tilde{\mathcal{L}}} = \sum_{j=1}^{c} s_j (\theta_j - \bar{\theta})^2 \left[\int_{-\infty}^{\infty} x J'(F) f^2(x) dx \right]^2 . \tag{17.3.7}
$$

Now, one can readily compute the asymptotic efficiency of one test procedure relative to the other by taking the ratios of their non-centrality parameters.

17.4 A Class of Nonparametric Tests

Govindarajulu (1976) proposed a class of nonparametric tests which in particular includes the class studied by Sen (1963) for the two-sample case and also the class of c-sample test statistics proposed by Sen and Govindarajulu (1966). This class of statistics is useful to detect location shifts or changes of scale in the populations. In the class of Chernoff-Savage statistics, the ranks of one ordered sample in the combined ordered sample have been weighted with functions of the combined sample distribution function (or the empirical distribution function). However, one might weight the ranks of one ordered sample with functions of the other sample distribution function, which might be easier from the computational point of view. Further the Chernoff-Savage theorems or their generalizations are applicable at least if the distribution functions do not have mutual discontinuities. The class of statistics by the author (1976) include those arising from discrete distributions having the same support (i.e., having common discontinuities).

Let $X_{i,j}$, $j = 1, \ldots, n_i$ be a random sample drawn from a population having $F_i(x)$ for its distribution function ($i = 1, \ldots, c$). Assume that the c-samples are mutually independent. Define $\eta_i = n_i/n_1$ and assume that

$$
0 \le \eta_0 \le \eta_i \;\; (i = 1, \ldots, c) \tag{17.4.1}
$$

for some fixed η_0. Further, let $N = n_1 + \cdots + n_c$ and $\lambda_i = n_i/N$ $(i = 1, \ldots, c)$ and $F_{i,n_i}(x)$ denote the empirical d.f. based on $X_{i,1}, \ldots, X_{i,n_i}$. Now define the set of statistics by

$$\begin{aligned} T_{n_1,i} &= n_1^{-1} \sum_{j=1}^{n_1} E(j, n_1) F_{i,n_i}(X_{1,j}) \\ &= \int_{-\infty}^{\infty} F_{i,n_i}(x) J_{n_1}\left(\frac{n_1}{n_1+1} F_{1,n_1}(x)\right) dF_{1,n_1}(x), \quad i = 1, \ldots, c. \end{aligned} \tag{17.4.2}$$

Let

$$\mu_{n_{1,i}} = \int F_i(x) J\left(F_1(x)\right) dF_1(x), \quad i = 1, \ldots, c. \tag{17.4.3}$$

From Theorem 6.1 of the author (1976), one can easily establish that

$$(T_{n_{1,i}} - \mu_{n_1,i}), \quad i = 1, \ldots, c$$

have a degenerate c-variate normal distribution having mean zero and a certain variance-covariance matrix which we do not spell out here because the expressions are too long. In order to compute the variance-covariance matrix, it suffices to consider only the first-order random component in the decomposition of $(T_{n_{1,i}} - \mu_{n_{1,i}})$ $i = 1, \ldots, c$. Also, assume that $\left|J'_{n_1}(u)\right| \le K\left[u(1-u)\right]^{-3/2+\delta}$ for some $K > 0$ and $0 < \delta < 1/2$.

Now, in order to test $H_0 : F_1(x) = \cdots = F_c(x)$ for all x against the alternative $H_1 : F_i(x) \neq F_k(x)$ for some x and for some pair (j, k), let

$$\bar{T}_N = \sum_{i=1}^{c} \lambda_i T_{n_{1,i}}, \quad \bar{\mu}_N = \sum_{1}^{c} \lambda_i \mu_{n_{1,i}}\,. \tag{17.4.4}$$

Consider the test statistic

$$S_N = \sum_{i=1}^{c} n_i (T_{n_{1,i}} - \bar{T}_N)^2 / I \tag{17.4.5}$$

where

$$I = \int_0^1 J^2(u) du - \left(\int_0^1 J(u) du\right)^2 \tag{17.4.6}$$

and the subscript n_1 on J is suppressed for the sake of simplicity.

Straightforward computations yield

$$\operatorname{var}\left\{\sqrt{n_i}(T_{n_{1,i}} - \bar{T}_N) | H_0\right\} = (1 - \lambda_i) I \tag{17.4.7}$$

and

$$\mathrm{cov}\left\{\sqrt{n_i}(T_{n_{1,i}} - \bar{T}_N), \sqrt{n_k}(T_{n_{1,k}} - \bar{T}_N)\right\} = -\sqrt{\lambda_i \lambda_k}\, I \text{ for } 1 \le i, k \le c\,. \tag{17.4.8}$$

Now, applying Corollary 23.6.1.1, one can infer that S_N under H_0 is distributed as chi-square with $c - 1$ degrees of freedom. Also, one can show that if

$$F_i(x) = F(x - \theta_i/N^{-1/2}) \left[F\left(x(1 + \theta_i N^{-1/2})\right)\right]$$

for $i = 1, \ldots, c$, S_N is approximately distributed as non-central chi-square with $c - 1$ degrees of freedom and non-centrality parameter equal to

$$\sum_{i=1}^{c} \lambda_i \left(\theta_i - \sum_{k=1}^{c} \lambda_k \theta_k\right)^2 \left\{\int_{-\infty}^{\infty} f^2(x) J(F) dx\right\}^2 / I\,,$$

$$\left[\sum_{i=1}^{c} \left(\theta_i - \sum_{k=1}^{c} \lambda_k \theta_k\right)^2 \left\{\int_{-\infty}^{\infty} x f^2(x) J(F) dx\right\}^2 / I\right]. \tag{17.4.9}$$

Also, computations yield that the Pitman efficiency of S_N relative to the classical F-test is

$$\left\{\sigma^2 \int f^2(x) J(F) du\right\}^2 / I \tag{17.4.10}$$

where σ denotes the common scale parameter of $F_1, \ldots, F_c$, and the Pitman efficiency of S_N relative to the parametric test discussed in Section 17.2 is given by

$$\sigma^4 \left\{\int_{-\infty}^{\infty} x f^2(x) J(f) dx\right\}^2 (2 + \gamma_2)/I\,, \tag{17.4.11}$$

where γ_2 denotes the common kurtosis of the distributions.

17.5 Problems

17.2.1 Consider the following artificially generated data with 3 treatments.

Treatment 1:	3.22, 0.71, 1.43, 0.06, 0.29, 0.80, 0.15, 1.39, 2.30, 0.49
Treatment 2:	2.77, 4.60, 0.99, 0.08, 2.62, 0.14, 2.16, 0.86, 2.10, 0.68
Treatment 3:	2.14, 0.49, 1.89, 0.02, 0.41, 0.98, 0.67, 4.49, 2.57, 5.87

We wish to test

H_0: the distributions are the same
against H_1: the distributions differ in scale

Carry out (i) Bartlett's test, and (ii) Scheffé's modified test at $\alpha = 0.05$.

17.3.1 For the data in Problem 17.2.1 carry out the c-sample Chernoff-Savage test given by (17.3.4) with $J_N(u) = -\ln(1-u)$ and compute its ARE relative to Bartlett's and Scheffé's test by taking the ratios of the noncentrality parameters.

17.3.2 Consider the following artificially generated data with three treatments.

Treatment 1:	-1.75, -0.02, -0.38, -0.71, 1.55, 0.67, 0.33, -0.31, -0.10, 1.80
Treatment 2:	-0.91, -1.80, 0.42, 2.63, -0.92, 2.23, -0.58, 0.56, -0.66, 0.83
Treatment 3:	-1.41, 1.17, -6.8, 0.20, 4.67, 1.41, 0.40, 0.72, -3.29, -1.83

Carry out Bartlett's test for homogeneity of scale parameters and also Chernoff-Savage test given by (17.3.4) with $J_N(u) = u$ and compute its ARE relative to Barlett's test. Use $\alpha = 0.05$.

17.4.1 For the data in Problem 17.2.1, carry out the S_N test (see Equ. (17.4.5)) with $J_N(u) = u$ and compare it with its parametric competitors.

17.4.2 For the data in Problem 17.3.2, carry out the S_N test (see Equ. (17.4.5)) with $J_N(u) = u$ and compare it with the test procedures carried out in Problem 17.3.2.

Chapter 18

c-sample Tests for Ordered Alternatives

18.1 Introduction

In certain situations, more information is available regarding the alternative hypotheses. For example, Jonckheere (1945) considered a test for the null hypothesis that stress has no effect on the task of manual dexterity against the alternative hypothesis that increasing stress produces an increasing effect on manual dexterity. Alternatives of this type are referred to as *ordered alternatives*. In this chapter we will consider parametric as well as nonparametric test procedures for the ordered alternatives case.

18.2 Parametric Test Procedures

Parametric test procedures have been proposed for this problem by Bartholomew (1959a, 1959b, 1961a, 1961b), Chacko (1963) and Kudo (1963). To fix ideas on parametric tests for ordered alternatives, consider the following case of Bartholomew (1959a). Let $X_1, \ldots, X_c$ be independent normal with means $\mu_1, \ldots, \mu_c$ and variances σ_i^2 $(i = 1, \ldots, c)$. We wish to test

$$H_0 : \mu_1 = \cdots = \mu_c$$

vs.

$$H_1 : \mu_1 \leq \cdots \leq \mu_c$$

with at least one strict inequality. Assume that the σ_i are known. Later on we will consider the case $\sigma_i = \sigma/\sqrt{n_i}$ $(i = 1, \ldots, c)$ and n_i are sample sizes, the x_i denote the sample means and σ is estimated from the data. The likelihood ratio test yields the following test criterion:

$$\bar{\chi}_c^2 = \sum_1^c a_i(X_i - \bar{X})^2 - \sum_1^c a_i(X_i - \hat{\mu}_i)^2 \tag{18.2.1}$$

where $a_i = \sigma_i^{-2}$, $\bar{X} = \sum_1^c a_i X_i / \sum_1^c a_i$, and $\hat{\mu}_1, \ldots, \hat{\mu}_c$ are the values of $\mu_1, \ldots, \mu_c$ which maximize (18.2.1) subject to the restriction $\hat{\mu}_1 \le \cdots \le \hat{\mu}_c$. Thus, we want to find the values of $\mu_1, \ldots, \mu_c$ which minimize

$$U_c = \sum_1^c a_i(X_i - \mu_i)^2 .$$

If the absolute minimum of U_c lies in the parametric space specified by H_1, then the restricted and unrestricted problems have the same solution which, in this case, is $\hat{\mu}_i = X_i$ for $i = 1, \ldots, c$. This solution lies in H_1 if $X_1 \le X_2 \le \cdots \le X_c$. If this is not the case, then there is some j such that $X_j > X_{j+1}$. The absolute minimum would require $\hat{\mu}_j > \hat{\mu}_{j+1}$ which is outside of H_1. In order to avoid this, we must introduce the restriction $\mu_j \le \mu_{j+1}$ before minimizing U_c as follows:

$$\begin{aligned} U_c &= \sum_{i=1, i\ne j,j+1}^{c} a_i(X_i - \mu_i)^2 + \left\{a_j(X_j - \mu_j)^2 + a_{j+1}(X_{j+1} - \mu_{j+1})^2\right\} \\ &= R_1 + R_2 \text{ (say)}, \end{aligned}$$

where the minimum of R_1 is not affected by the restriction and hence is zero when $\hat{\mu}_i = X_i$. Next consider the minimum of R_2. The point $(\mu_j, \mu_{j+1}) = (X_j, X_{j+1})$ falls outside the region: $\mu_j \le \mu_{j+1}$ in the (μ_j, μ_{j+1}) plane. Hence, the minimum of R_2 occurs on the boundary of the region, namely when $\mu_j = \mu_{j+1}$.

So, setting the derivative of R_2 with respect to μ_j equal to zero, we obtain

$$\hat{\mu}_j = \hat{\mu}_{j+1} = (a_j X_j + a_{j+1} X_{j+1})/(a_j + a_{j+1}) . \tag{18.2.2}$$

For the sake of convenience, let us introduce the following further notation:

$$X(r, s) = \sum_{i=r}^{s} a_i X_i / \sum_{i=r}^{s} a_i . \tag{18.2.3}$$

Thus, u_c is minimized subject to the restriction $\mu_j \le \mu_{j+1}$ at

$$\begin{aligned} \hat{\mu}_j &= \hat{\mu}_{j+1} = X(j, j+1) \\ \hat{\mu}_i &= X_i, \text{ for } i \ne j, j+1 . \end{aligned} \tag{18.2.4}$$

Now, the problem has been reduced to one involving $c-1$ variables. If

$$X_1 \le \cdots \le X_{j-1} \le X(j, j+1) \le X_{j+2} \le \cdots \le X_k\,,$$

we have achieved the final solution. If not, the problem is the same as the original one except that we have reduced the number of variables to $c-1$. We can treat the reduced problem in the same way by considering $(X_{j,j+1})$ as a single observation with weight $(a_j + a_{j+1})$. Thus, the procedure for obtaining $\hat{\mu}_1, \ldots, \hat{\mu}_c$ is as follows.

Arrange the observations in the order predicted by the alternative (i.e., as $X_1, \ldots, X_c$). If any consecutive pair (X_j, X_{j+1}) is not in the expected order (i.e., $X_{j+1} < X_j$) collapse the pair to an average given by (18.2.2). Now there are $c-1$ members of which $c-2$ are unchanged and one is the average of two original observations. Go to the next pair which does not conform to the order specified by the alternative treating $(X_{j,j+1})$ as a single observation having weight $(a_j + a_{j+1})$. Proceed in this manner until the resulting values are in the order specified by H_1. Thus, for each i, the MLE $\hat{\mu}_i$ of μ_i is equal to that one of the final quantities to which the original X_i contributed. Let us illustrate the method by the following example.

Example 18.2.1. Let $X_1 = 40$, $X_2 = -32$, $X_3 = -10$, $X_4 = 5$ with $a_1 = a_2 = a_3 = a_4$. Since $X_1 > X_2$, we form the average $X(1,2) = (40-32)/2 = 8$. Now $X(1,2)$ is greater than X_3, we form $X(1,3) = \{2(X_{1,2}) + X_3\}/3 = (16-10)/3 = 2$. Since $X_4 > X(1,3)$, the solution is

$$\hat{\mu}_1 = \hat{\mu}_2 = \hat{\mu}_3 = 2,\ \hat{\mu}_4 = 5\,.$$

After computing $\hat{\mu}_1, \ldots, \hat{\mu}_c$ via the above described procedure, $\bar{\chi}_c^2$ can be obtained by substituting these estimates in (18.2.1). Now, since

$$\sum a_i(X_i - \bar{X})^2 = \sum a_i(X_i - \hat{\mu}_i)^2 + \sum a_i(\hat{\mu}_i - \bar{X})^2$$

$$\bar{\chi}_c^2 = \sum a_i(X_i - \bar{X})^2 - \sum a_i(X_i - \hat{\mu}_i)^2 = \sum_1^c a_i(\hat{\mu}_i - \bar{X})^2\,. \qquad (18.2.5)$$

Recall if one is testing H_0 against the unrestricted alternative, namely H_1' : $\mu_i \neq \mu_j$ (for some $i \neq j$), the likelihood ratio test criterion is

$$\chi_c^2 = \sum_1^c a_i(X_i - \bar{X})^2\,, \qquad (18.2.6)$$

the difference being that we use the estimates $\hat{\mu}_1, \ldots, \hat{\mu}_c$ in (18.2.5) instead of the original observations.

When the averaging process is completed, we have l means,

$$X(1, c_1), X(c_1+1, c_1+c_2), \ldots, X(c-c_l+1, c)$$

with weights

$$A(1, c_1), A(c_1+1, c_1+c_2), \ldots, A(c-c_l+1, c)$$

where

$$A(r,s) = \sum_{i=r}^{s} a_i, \quad \sum_{i=1}^{l} c_i = c\,.$$

Then, $\bar{\chi}_c^2$ is the sum of squares of the deviations of these means from $\bar{X}$. We call this the reduced form of the problem.

Remark 18.2.1. *Brunk (1955, 1958) and van Eaden (1957) were the first to consider MLE's of restricted parameters and they have shown that the MLE's of μ's under H_1 are unique and can be formally represented as*

$$\hat{\mu}_i = \max_{1\le r\le i}\ \min_{i\le s\le c}\ X_{[r,s]}\,. \tag{18.2.7}$$

Brunk (1955) also gave the preceding averaging process in order to obtain MLE's of the μ_i's under H_1.

Case when $\sigma_i = \sigma/\sqrt{n_i}$, $X_i = \bar{X}_i = n_i^{-1}\sum_{j=1}^{n_i} X_{ij}$ $(i = 1, \ldots, c)$ and σ is Estimated from the Data

Let $\bar{X} = \sum_1^c n_i\bar{X}_i / \sum_1^c n_i$, here $a_i = n_i/\sigma^2$. Then the likelihood ratio test needs to be modified by replacing the unknown variance σ^2 by its residual mean square, namely S_e^2 obtained from the analysis of variance table. Thus, the test criterion takes the form of

$$\sum_1^c n_i(\hat{\mu}_i - \bar{X}_i)^2/S_e^2\,. \tag{18.2.8}$$

Bartholomew points out that a slight improvement can be obtained as follows. If $\bar{X}_i > \bar{X}_{i+1}$, then we form the weighted average

$$X(i, i+1) = (n_i\bar{X}_i + n_{i+1}\bar{X}_{i+1})/(n_i + n_{i+1})\,.$$

Then the residual mean square calculated from the new ANOVA table will have one more degree of freedom than that from the original table since the 'between groups' degrees of freedom have been reduced by one. In general, if the reduced form of the problem consists of l groups, then the increase in the residual degrees of freedom will be $c - l$. The modified test criterion is defined as

$$\bar{F}_c = \sum_{1}^{k} n_i(\hat{\mu}_i - \bar{X})^2/(l-1)S_e'^2 \tag{18.2.9}$$

where $S_e'^2$ is the residual mean square of the ANOVA table obtained from the reduced form of the problem having $N - l$ degrees of freedom ($N = \sum_1^c n_i$), one can show that $\bar{F}_c$ is the test criterion produced by the likelihood ratio method.

The Distribution of $\bar{\chi}_c^2$ and $\bar{F}_c$

First consider the case of $c = 2$. Then

$$\begin{aligned} \bar{\chi}_2^2 &= \sum_{1}^{2} a_i(X_i - \bar{X})^2 \qquad \text{if } X_1 < X_2\,, \\ &= 0 \qquad \text{if } X_1 \geq X_2\,. \end{aligned}$$

Hence, we need

$$P(\bar{\chi}_2^2 \geq \gamma | H_0) = P\left(X_1 < X_2 \text{ and } \sum_{1}^{2} a_i(X_i - \bar{X})^2 \geq \gamma | H_0\right).$$

Now the value of $\sum_1^2 a_i(X_i - \bar{X})^2$ does not depend on the order of X_1 and X_2, but only on their absolute difference. Hence,

$$\begin{aligned} P(\bar{\chi}_2^2 \geq \gamma | H_0) &= P(X_1 < X_2 | H_0) P\left(\sum_{1}^{2} a_i(X_i - \bar{X})^2 \geq \gamma | H_0\right) \\ &= \frac{1}{2} P(\chi_1^2 \geq \gamma) \end{aligned}$$

irrespective of the values of a_1 and a_2.

For general c, let $P(l, c; a_1, \ldots, a_c)$ denote the probability that the reduced method of the problem consists of l means when the original observations $X_1, \ldots, X_c$ had weights $a_1, \ldots, a_c$. Without causing any confusion, for the sake of simplicity, we will denote the preceding probability by $P(l, c)$. Hence,

$$\begin{aligned} P(\bar{\chi}_c^2 \geq \gamma | H_0) &= \sum_{l=2}^{c} P(l, c) P(\bar{\chi}_c^2 \geq \gamma | H_0) \\ &= \sum_{l=2}^{c} P(l, c) P(\chi_{l-1}^2 \geq \gamma) \end{aligned} \tag{18.2.10}$$

with $P(\bar{\chi}_c^2 = 0 | H_0) = P(1, c)$.

The following is justification for (18.2.10). Recall that

$$\bar{\chi}_c^2 = \sum_1^l A_i(X_i - \bar{X})^2$$

where $\bar{X} = \sum_1^l A_i X_i / \sum_1^l A_i$ and var $X_i = 1/A_i$ $(i = j, \ldots, l)$. Then $\bar{\chi}_c^2 \stackrel{d}{=} \chi_{l-1}^2$ under H_0 provided:

(i) each X_i $(i = 1, \ldots, l)$ is normally distributed about a common mean;

(ii) the distribution of $\bar{\chi}_c^2$ is independent of the restriction $X_1 \leq \cdots \leq X_c$ (i.e., the X_i's may be treated as if they are independent).

(i) is satisfied under H_0. In order to show (ii), we can write $\bar{\chi}_c^2$ in an algebraically equivalent form

$$\bar{\chi}_c^2 = \sum_{i<j}\sum A_i A_j (X_j - X_i)^2 / \sum_{i=1}^l A_i \,. \tag{18.2.11}$$

If the X's were independent, $(X_j - X_i)^2$ would be the square of a normal variate with zero mean and hence $\bar{\chi}_c^2$ would be a linear function of such variables. When the X's are ordered, $X_j - X_i$ is distributed like the modulus of a normal variable (since $X_i \leq X_j$ when $i < j$). However, the square of the modulus has the same distribution as the square of the difference when the sign is not restricted. Thus, the distribution of $\bar{\chi}_c^2$ is not affected by the dependence among the X's from the restriction $X_1 \leq \cdots \leq X_c$. Thus, the distributional problem is reduced to that of finding the probabilities $P(l, c)$.

Bartholomew (1959a) obtains the exact expressions for $P(l, c)$ for $c = 3$ and 4 and a partial result for $k = 5$ (i.e., when the weights are equal).

Let $Z_i = X_{i+1} - X_i$ $(i = 1, \ldots, c-1)$ and ρ_{ij} denote the product-moment correlation coefficient between Z_i and Z_j. Then

$$\rho_{i,i+1} = -\left\{\frac{a_i a_{i+2}}{(a_i + a_{i+1})(a_{i+1} + a_{i+2})}\right\}^{1/2} \quad (i = 1, \ldots, c-1)\,,$$

$$\begin{aligned} \rho_{ij} &= 1 \quad \text{if } i = j\,, \\ &= 0 \quad \text{if } |i - j| > 1\,. \end{aligned}$$

Then the probabilities depend only on the correlation coefficients ρ_{ij} which are independent of σ^2. Hence,

$$P(\bar{F}_c \geq \gamma | H_0) = \sum_{l=2}^{c} P(l,c) P(F_{l-1,N-l} \geq \gamma) , \tag{18.2.12}$$

$$P(\bar{F}_c = 0 | H_0) = P(1,c) .$$

Also note that $P(F_{l-1,N-l} \geq \gamma) = I_{1-\eta}\left(\frac{1}{2}(N-l), \frac{1}{2}(l-1)\right)$ with $\eta = (l-1)\gamma / \{N - l + (l-1)\gamma\}$.

For equal weight case, the following recurrence relation was established by Niles (1959):

$$P(l,c) = \left[P(l-1,c-1) + (c-1)P(l,c-1)\right]/c . \tag{18.2.13}$$

Thus, one can compute the values of $P(l,c)$ from its values for $c-1$ where the sample sizes are equal (i.e., equal weights case). Chacko (1963) derived explicit expressions for $P(l,c)$ when the sample sizes are equal and calculated the 5% and 1% critical values of T_c for $c = 3$, 4, 5 and 6 and $n_i = 2(1)16$ where

$$T_c = \sum_{1}^{c} n_i(\hat{\mu}_i - \bar{X})^2 / s_0^2 , \tag{18.2.14}$$

$$s_0^2 = \sum_{1}^{c} n_i(\bar{X}_i - \bar{X})^2 + \sum_{1}^{c} n_i s_i^2 \tag{18.2.15}$$

and

$$s_i^2 = n_i^{-1} \sum_{j=1}^{n_i} (X_{ij} - \bar{X}_i)^2 .$$

Asymptotic Distribution of $\bar{F}_c$ under the Alternative

Since S_e^2 and $S_e'^2$ converge to σ^2 in probability, the distribution of $\bar{F}_c$ for given l and for large n_i will be the same as $\bar{\chi}_c^2$ which is non-central chi-square with $l-1$ degrees of freedom and non-centrality parameter

$$\sum_{1}^{c} n_i(\mu_i - \bar{\mu})^2 / \sigma^2$$

where $\bar{\mu} = \sum_1^c n_i \mu_i / N$.

Kudo (1963) proposed a multi-variate parametric test which, when specialized to the univariate case is to reject H_0 if

$$N^{1/2} \sum_{i<j}^{c} \sum^{c} (\bar{X}_j - \bar{X}_i)/c^2 \tag{18.2.16}$$

is large. One can establish that the Pitman efficacy of Kudo's statistic (see, for instance, Puri, 1965, p. 59) for the equally spaced alternatives is

$$(c^2 - 1)/12\sigma^2 .$$

Regression Alternatives

In some situations we may have prior knowledge that the means constitute known multiples of an unknown parameter where the multiplying constants are ordered. In this case, we can obtain a reasonable test criterion.

Let X_{ij} $(j = 1, \ldots, n_i)$ denote a random sample from normal (μ_i, σ^2) for $i = 1, \ldots, c$ where $\mu_i = \theta_0 + \theta b_i$ $(i = 1, \ldots, c)$ where b_i are specified constants and θ_0, θ and σ are unknown. We wish to test

$$H_0 : \theta = 0 \text{ versus } H_1 : \theta > 0 . \tag{18.2.17}$$

The log likelihood of θ_0, θ and σ^2 is

$$l(\theta_0, \theta, \sigma^2) = -\frac{N}{2}\{\log(2\pi) + \log\ \sigma^2\} - \frac{1}{2\sigma^2}\sum_{1}^{c}\sum_{1}^{n_i}(X_{ij} - \theta_0 - \theta b_i)^2 \tag{18.2.18}$$

and the MLE's of θ_0 and σ^2 under H_0 are

$$\hat{\theta}_0 = \bar{X} = \frac{1}{N}\sum\sum X_{ij} = \sum_{1}^{c} \frac{n_i}{N}\bar{X}_i = \sum_{1}^{c} \lambda_i X_i$$

where $\bar{X}_i = \sum_{j=1}^{n_i} X_{ij}/n_i$ and $\lambda_i = n_i/N$ $(i = 1, \ldots, c)$.

The MLE's of θ_0, θ and σ^2 under $H_0 \cup H_1$ are:

$$\begin{aligned} \hat{\hat{\theta}} &= \sum \lambda_i b_i(\bar{X}_i - \bar{X}) / \sum \lambda_i (b_i - \bar{b})^2, \ \bar{b} = \sum_{1}^{c} \lambda_i b_i , \\ &= \sum \lambda_i (b_i - \bar{b}) X_i / \sum \lambda_i (b_i - \bar{b})^2 , \end{aligned}$$

$$\hat{\hat{\theta}}_0 = \bar{X} - \bar{b}\hat{\hat{\theta}}$$

and

$$\hat{\hat{\sigma}}^2 = N^{-1} \sum\sum (X_{ij} - \hat{\hat{\theta}}_0 - \hat{\hat{\theta}} b_i)^2 .$$

Hence the likelihood ratio test criterion takes the form of

$$\Lambda_N = (\hat{\hat{\sigma}}^2/\hat{\sigma}^2)^{N/2} . \tag{18.2.19}$$

Consider

$$\begin{aligned} \frac{\hat{\hat{\sigma}}^2}{\hat{\sigma}^2} &= \frac{\sum\sum \left\{X_{ij} - \bar{X} - \hat{\hat{\theta}}(b_i - \bar{b})\right\}^2}{\sum\sum (X_{ij} - \bar{X})^2} \\ &= 1 - \frac{\hat{\hat{\theta}}^2 \sum \lambda_i (b_i - \bar{b})^2}{N^{-1}\sum\sum (X_{ij} - \bar{X})^2} . \end{aligned}$$

So, rejecting H_0 for small values of Λ_N is equivalent to rejecting H_0 for large values of $\hat{\hat{\theta}}^2 \sum \lambda_i (b_i - \bar{b})^2 / N^{-1} \sum\sum (X_{ij} - \bar{X})^2 = S_N^*$ (say). Under H_0, $N^{-1}\sum\sum (X_{ij} - \bar{X})^2$ tends to σ^2 in probability, and $\hat{\hat{\theta}}^2 \sum \lambda_i (b_i - \bar{b})^2 = \sum \lambda_i (b_i - \bar{b})(\bar{X}_i - \theta_0)$, since $\sum \lambda_i (b_i - \bar{b}) = 0$. Hence,

$$S_N^* \overset{H_0}{\approx} \left[\sum \lambda_i (b_i - \bar{b})(\bar{X}_i - \theta_0)\right]^2 / \sigma^2 \sum \lambda_i (b_i - \bar{b})^2 .$$

So,

$$N S_N^* \overset{d}{\approx} \chi_1^2 .$$

Since

$$\sum\sum (X_{ij} - \theta_0 - \bar{b})^2 = \sum\sum (X_{ij} - \bar{X})^2 + N(\bar{X} - \theta_0 - \theta\bar{b})^2 ,$$

$$\sum\sum (X_{ij} - \bar{X})^2/\sigma^2 \overset{d}{=} \chi_{N-1}^2 .$$

Hence, $N^{-1}\sum\sum (X_{ij} - \bar{X})^2 \to \sigma^2$ in probability under all hypotheses. Thus,

$$N S_N^* \approx \chi_{1,\lambda_S}^2$$

where the non-centrality parameter

$$\lambda_S = N\theta^2 \left[\sum \lambda_i (b_i - \bar{b}) b_i\right]^2 / \sigma^2 \sum \lambda_i (b_i - \bar{b})^2 .$$

18.3 Nonparametric Test Procedures

In several experiments, there seems to be no suitable metric by which the different treatments under investigation may be characterized. The only thing an experimenter can assert, especially in psychology, is that the treatments may be ranked in some order such as increasing stress. In this case, it is no longer possible to use any form of regression analysis since the independent variable is not adequately quantified. Jonckheere (1954) provides some practical situations where one would be interested in testing against the ordered alternative hypothesis.

Example 18.3.1. Suppose on each of N successive occasions, any one of c events may occur. Then we can test the hypothesis that the c-events occur randomly in the series of N occasions, against the alternative that they tend to occur in a particular ordered time sequence.

Example 18.3.2. Births of siblings might be of three types: normal, abnormal (with respect to some characteristic) and still births. The effect of birth rank on the type of birth could be tested when, for instance, the alternative to random occurring is that the earlier births tend to be normal, the later births abnormal, with finally, the appearance of still births.

Let X_{ij}, $j = 1, \ldots, n_i$ be a random sample from $F_i(x)$ $(i = 1, \ldots, c)$ and we are interested in testing

$$H_0 : F_1(x) = \cdots = F_c(x) \text{ for all } x$$

against

$$H_1 : F_1(x) < F_2(x) < \cdots F_c(x) \text{ for some } x\,.$$

Let

$$\begin{aligned} a_{i,j,k,l} &= 1 \text{ if } X_{i,j} < X_{k,l} \\ &= 0 \text{ if } X_{i,j} > X_{k,l} \end{aligned}$$

for $j = 1, \ldots, n_i$ and $l = 1, \ldots, n_k$. Let

$$\tilde{T}_{ik} = \sum_{j=1}^{n_i} \sum_{l=1}^{n_k} a_{i,j,k,l} \tag{18.3.1}$$

and

$$S = 2\sum_{i=1}^{c-1} \sum_{j=i+1}^{c} \left(\tilde{T}_{ik} - \frac{n_i n_k}{2}\right). \tag{18.3.2}$$

Jonckheere (1954) proposed S as the test criterion for testing H_0 vs. H_1 where we reject H_0 for large values of S. Notice that $\tilde{T}_{ik}$ is the Mann-Whitney statistic for samples from i^{th} and k^{th} populations. Jonckheere (1954) provides tables of $P(S \geq s_0|H_0)$ for c samples of equal size n for $c = 3$, $n = 2, 3, 4, 5$; $c = 4$, $n = 2, 3, 4$, $c = 5$, $n = 2, 3$ and $c = 6$, $n = 2$. He also derives the first four cumulants, the skewness and kurtosis of S. (The first and third cumulants of S are zero). Since S is a linear sum of several Mann-Whitney statistics which are asymptotically normal, one can easily assert that S is asymptotically normal when the n_i become large. Jonckheere (1954) obtains the variance of S to be

$$\text{var}(S|H_0) = \frac{1}{18}\left\{N^2(2N+3) - \sum_{i=1}^{c} n_i^2(2n_i+3)\right\} \tag{18.3.3}$$

and hence

$$\frac{S}{\{\text{var}(S|H_0)\}^{1/2}} \overset{d}{\approx} \text{normal}(0,1)\,.$$

Further, the author remarks that since the interval between all possible adjacent values of S is always two, an improvement in the normal approximation to the true distribution will be obtained if unity is subtracted from an observed value of S prior to its division by its standard deviation.

Chacko (1963), by replacing the observations by the ranks in the test statistic given by (18.2.14), obtains a nonparametric test which is similar to the Kruskal-Wallis test and shows that its Pitman efficiency relative to the parametric test is

$$12\sigma^2\left[\int_{-\infty}^{\infty} f^2(x)dx\right]^2$$

which coincides with the Pitman efficiency of Kruskal-Wallis test relative to the classical F-test.

Next, Puri (1965) generalized Jonckheere's test to

$$V = N^{1/2}\sum_{i<j}^{c}\sum^{c} \tilde{h}_{ij} \tag{18.3.4}$$

where

$$\tilde{h}_{ij} = \int_{-\infty}^{\infty} J_N\left[\frac{n_iF_{i,n_i}(x) + n_jF_{j,n_j}(x)}{n_i+n_j}\right] d\left(F_{i,n_i}(x) - F_{j,n_j}(x)\right)\,. \tag{18.3.5}$$

Note that the statistic $\tilde{h}_{ij}$ is equivalent to $\tilde{T}_{ij}$ of Jonckheere when $J_N(u) \equiv u$.

Since Bartholomew's likelihood ratio test has a different limiting distribution than Puri's (1965) test criterion V, asymptotic relative efficiency criterion cannot be used to compare those two tests. So, Puri (1965, Table 1) carries out asymptotic power comparisons of his test with $J = \Phi^{-1}$, Jonckheere's test and Bartholomew's test when the sample sizes are equal and the μ's are equally spaced and the samples are arising from normal populations for $c = 3, 4, 8, 12$, and for $\Delta = 0(1)4$ where $\Delta - \left\{\sum_1^c(\mu_i - \bar{\mu})^2\right\}^{1/2}$, $\bar{\mu} = \sum_1^c \mu_i/c$. He surmises that his test is superior to Jonckheere's and Bartholomew's for equally spaced μ_i and inferior to Bartholomew's when all but one of the μ_i's are equal.

The class of nonparametric test statistics proposed by Puri (1965) is equivalent to the sum of all possible pairs of two-sample Chernoff-Savage statistics because (with

$$H_{N_{ij}}(x) = \tilde{\lambda}_i F_{i,n_i}(x) + (1 - \tilde{\lambda}_i) F_{j,n_j}(x) ,$$

and

$$\tilde{\lambda}_i = \frac{n_i}{N_{ij}} = \frac{n_i}{n_i + n_j}) ,$$

$$\begin{aligned}
&\int J_N\left(H_{N_{ij}}(x)\right) d\left[F_{i,n_i}(x) - dF_{j,n_j}(x)\right] \\
&= \int J_N\left(H_{N_{ij}}(x)\right) \left\{ dF_{i,n_i}(x) - d\frac{\left(H_{N_{ij}} - \hat{\lambda}_i F_{i,n_i}(x)\right)}{1 - \hat{\lambda}_i} \right\} \\
&= -(1 - \tilde{\lambda}_i)^{-1} \int J_N(H_{N_{ij}})|dH_{N_{ij}}(x) \\
&\quad + (1 - \tilde{\lambda}_i)^{-1} \int J_N\left(H_{N_{ij}}(x)\right) dF_{in_i}(x) \\
&= -(1 - \tilde{\lambda}_i)^{-1} \frac{1}{N_{ij}} \sum_{k=1}^{N_{ij}} J_N\left(\frac{k}{N_{ij}}\right) \\
&\quad + (1 - \tilde{\lambda}_i)^{-1} \int J_N\left(H_{N_{ij}}(x)\right) dF_{in_i}(x) ,
\end{aligned} \qquad (18.3.6)$$

where the first term is non-stochastic and the second term, except for a scalar multiplier, is the two-sample Chernoff-Savage statistic based on the ith and j^{th} samples. Thus, the statistic proposed by Puri (1965) is equivalent to

$$T_N = \sum_{i=1}^{c-1} \sum_{j=i+1}^{c} a_{ij} T_{ij}, \quad a_{ij} \geq 0 \qquad (18.3.7)$$

where $T_{i,j,N}$ is (i, j) samples Chernoff-Savage statistic.

Let Γ denote the class of statistics defined by T_N. Let Γ^* be the subclass of Γ consisting of linear combinations of $(T_{12}, T_{23}, \ldots, T_{c-1,c})$. Then, assuming that $n_1 = n_2 = \cdots = n_c$, Tryon and Hettmansperger (1973) prove that for each T_N in Γ, there corresponds an equivalent statistic T_N^* in Γ^* such that

$$T_N^* = \sum_{k=1}^{c-1} a_k T_{k,k+1,N} \tag{18.3.8}$$

where

$$a_k = \sum_{i=1}^{k} \sum_{j=k+1}^{c} a_{ij}, \; k = 1, \ldots, c-1 \,. \tag{18.3.9}$$

Note that the "equivalence" mentioned above is meant in the sense that the difference of T_N and T_N^*, when suitably standardized, converges in probability to zero under H_0 and that for testing H_0 against H_1 (the ordered alternative), the Pitman efficiency of T_N^* relative to T_N is one. Besides, T_N^* requires only the computation of $c-1$ instead of $c(c-1)/2$ two-sample test statistics required by T_N.

Tryon and Hettmansperger (1973) obtain optimal weight a_k in T_N^* for which the Pitman efficiency of T_N^* is maximized when the alternatives governing the location shift parameter are equally spaced. They also show that the statistics proposed by Jonckheere and Puri have maximum Pitman efficacy when shift alternatives are equally spaced. The assumption of equal spacing is of much practical interest which is implicitly assumed in applying Jonckheere's statistic.

First let us give the LMP rank test derived by Haller (1968) or see Govindarajulu and Haller (1977) for the Lehmann alternatives. As before, let $X_{i,j}$ $(j = 1, \ldots, n_i)$ be a random sample from $F_i(x)$, $i = 1, \ldots, c$. Let $N = n_1 + \cdots + n_c$. We will obtain an LMP rank test for H_0 vs. H_1. We could get analogous results for the alternative, $H_1' : F_1(x) \geq \cdots \geq F_c(x)$ for all x with at least one strict inequality. Consider $H_{L,1} : F_i(x) = \{F(x)\}^{\theta_i}$, $\theta_i = \theta_0 + \theta b_i$, θ_0, $\theta > 0$ and $0 \leq b_1 \leq \cdots \leq b_c$ where we assume that $F(x)$ is continuous and unknown. Let $(W_1, W_2, \ldots, W_N)$ denote the combined ordered sample, and

$$Z = (Z_1, \ldots, Z_N)$$

where $Z_i = k$ when $W_i = X_{k,j}$ for some $j = 1, \ldots, n_k$. Further, let

$$T_k(Z) = \sum_{i=1}^{N} \left(\sum_{j=1}^{i} \delta_{k,Z_j}/i \right) \tag{18.3.10}$$

where δ_{k,Z_j} denotes the Kronecker's delta. When $c = 2$, $T_k(Z)$ is known as Savage's test. Let $\mathcal{F}$ denote the collection of all $M = N!/\left(\prod_{i=1}^{c} n_i!\right)$ possible rank orders. If $\theta = \Delta\theta_0$, one can easily show that

$$P(Z = z|H_{L,1}) = \prod_{i=1}^{c}(b_i\Delta + 1)^{n_i}/M \left\{ \prod_{i=1}^{c} \left[1 + \Delta \left(\sum_{k=2}^{c} b_k \sum_{j=1}^{i} \delta_{k,z_j} \right) \right] \right\} . \tag{18.3.11}$$

The derivative of the above rank order probability with respect to θ evaluated at $\theta = 0$ is

$$(M\theta_0)^{-1} \left[\sum_{k=2}^{c} b_k n_k - \sum_{k=2}^{c} b_k T(z) \right] . \tag{18.3.12}$$

Hence, it follows from the Neyman-Pearson lemma that a LMP rank order test of H_0 against $H_{L,1}$ is to reject H_0 for small values of $\sum_2^c b_k T(z)$.

A Larger Class of Nonparametric Tests

Earlier we saw that a LMP rank order test against ordered Lehmann alternatives is a weighted sum of c-sample Savage type of statistics. Hence, it is of interest to consider a wider class of c-sample tests against ordered alternatives which, in particular, include the above statistics. In order to assess the merits of the various tests, the asymptotic efficiency (in the sense of Pitman) will be computed relative to other parametric and nonparametric competitors for location shift as well as change of scale. Hereafter, assume that $n_1 = n_2 = \cdots = n_c = n$. Define the statistic

$$\psi_i(z)/n = \int_{-\infty}^{\infty} J_{N,i}\left[F_{1,n}(x), \ldots, F_{c,n}(x)\right] dF_{i,n}(x) \tag{18.3.13}$$

for $i = 1, \ldots, c$.

Consider two sequences of classes of "near" alternatives $\{\mathcal{F}_N^*\}$ and $\{\mathcal{F}_N^{**}\}$ where

$$\begin{aligned} \mathcal{F}_N^* &= \{(F_1, \ldots, F_c) : F_i(x) = F(x - b_i\theta N^{-1/2}), \\ &\qquad F \text{ is absolutely continuous with } 0 \le b_1 \le \cdots \le b_c; \theta > 0\}, \end{aligned} \tag{18.3.14}$$

$$\begin{aligned} \mathcal{F}_N^{**} &= \{F_1, \ldots, F_c) : F_i(x) = F\left(x(1 - b_i\theta N^{-1/2})\right), \\ &\qquad F \text{ is absolutely continuous with } 0 \le b_1 \le \cdots \le b_c, \theta > 0\} \end{aligned} \tag{18.3.15}$$

and not all the b_i's are equal. Notice that $N = cn$. Assume that for $0 < u_i < 1$ $(i = 1, \ldots, c)$, the function $J_{N,i}(u_1, \ldots, u_c)$ converges in

Lebesgue measure to an absolutely continuous function $J_i(u_1,\ldots,u_c)$ and $\lim_{n\to\infty} J_i(u_1,\ldots,u_c) = J(u,\ldots,u)$. Further assume that $\sum_1^c \psi_i(z)$ is equal to a non-stochastic constant. Also assume that

$$N^{1/2}\left\{\psi_i(z)/n - \int_{-\infty}^{\infty} J_i\left(F_1(x),\ldots,F_c(x)\right)\right\} dF_i(x),\ i = 1,\ldots,c$$

is asymptotically distributed as a c-variate normal having zero mean vector and $(c-1)A^2$ for the variances and $-A^2$ for the covariances where A^2 is a function of c.

Let

$$T = N^{1/2}\sum_{i=1}^{c} d_i\left[\psi_i(z)/n - \int_{-\infty}^{\infty} J_i\left(F_1(x),\ldots,F_c(x)\right)\right] dF_i(x)\,. \tag{18.3.16}$$

Then the following theorem of Govindarajulu and Haller (1977) gives the optimal weight such that the Pitman efficacy of T is maximized.

Theorem 18.3.1. *Under the preceding assumptions about the statistic $\psi_i(z)/n$, defined in (18.3.13), the efficacy of T defined by (18.3.16) is maximized if $d_i = \beta\sum_{j=1}^{c}(b_i - b_j)$ where β is a real constant.*

PROOF: It follows that T is asymptotically normal with mean zero and variance

$$A^2\left[c\sum_{i=1}^{c} d_i^2 - \left(\sum_{j=1}^{c} d_j\right)^2\right]. \tag{18.3.17}$$

In order to compute the efficacy of T, consider

$$\lim_{N\to\infty}\frac{N^{1/2}}{\theta}\left[\int J_i(F_1,\ldots,F_c)dF_i - \int J(F,\ldots,F)dF\right]$$

$$= \lim_{\xi\to 0}\left[\int_{-\infty}^{\infty}\{J_i(F(y+(b_i-b_j)\xi),\ldots,F(y+(b_i-b_j)\xi)\right.$$

$$\left. - J(F(y),\ldots,F(y)))\}dF(y)\right] \qquad \text{if } (F_1,\ldots,F_c)\in\mathcal{F}_N^*$$

$$= \lim_{\xi\to 0}\left[\int_{-\infty}^{\infty}\left\{J_i\left(F\left(\frac{t_1}{t_i}y\right),\ldots,F\left(\frac{t_c}{t_i}y\right) - J\left(F(y),\ldots,F(y)\right)\right)\right\}dF(y)\right]$$

where $t_j = 1 - b_j\xi$ if $(F_1,\ldots,F_c)\in\mathcal{F}_N^{**}$.

Thus,

$$\text{LHS} = \sum_{j=1}^{c}(b_i - b_j)B$$

where

$$\begin{aligned} B &= \int \frac{\partial}{\partial u_j} J_i(u_1, \ldots, u_c)|_{u_1=\cdots=u_c=F(y)} \cdot f(y)dF(y) \text{ if } (F_1 \cdots F_c) \in \mathcal{F}_N^* \\ &= \int \frac{\partial}{\partial u_j} J_c(u_1, \ldots, u_c)|_{u_1=\cdots=u_c=F(y)} \cdot yf(y)dF(y) \text{ if } (F_1 \cdots F_c) \in \mathcal{F}_N^{**}. \end{aligned} \tag{18.3.18}$$

Thus the efficacy of T is given by

$$(B/A)^2 \left[\sum_{i=1}^{c} d_i \sum_{j=1}^{c} (b_i - b_j)\right]^2 / \left[c\sum_{i=1}^{c} d_i^2 - \left(\sum_{j=1}^{c} d_i\right)^2\right] \tag{18.3.19}$$

where B is as defined in (18.3.18). Now one can see that the numerator and the denominator remains the same if the values of d are shifted by a constant. Hence, without loss of generality, we can set $\sum_1^c d_i = 0$. Hence, the efficacy of T simplifies to

$$(B^2/cA^2) \left[\sum_{i=1}^{c} d_i \sum_{j=1}^{c} (b_i - b_j)\right]^2 / \sum_{i=1}^{c} d_i^2 . \tag{18.3.20}$$

However, from the Cauchy-Schwarz inequality, we have

$$\left\{\sum_{i=1}^{c} d_i \sum_{j=1}^{c} (b_i - b_j)\right\}^2 / \sum_{i=1}^{c} d_i^2 \le \sum_{i=1}^{c} \left(\sum_{j=1}^{c} (b_i - b_j)\right)^2$$

with equality when $d_i = \beta \sum_{j=1}^{c} (b_i - b_j)$ $(i = 1, \ldots, c)$ where β is an arbitrary real constant which, without loss of generality, can be set equal to unity.

We consider three special cases of Theorem 18.3.1.

Case (i). Let $J_{N,i}(F_{1,n}(x), \ldots, F_{c,n}(x)) = \tilde{J}_N\left(\frac{N}{N+1} H_N(x)\right)$ for some $J_N(u)$ and $i = 1, \ldots, c$, where $H_N(x) = \sum_{i=1}^{c} \frac{1}{c} F_{i,n}(x)$. In this case, T is a weighted sum of c-sample Chernoff-Savage statistics. From Govindarajulu, Le Cam and Raghavachari (1967), it follow that if

$$\left|\tilde{J}_N'(u)\right| \le K\, [u(1-u)]^{-\frac{3}{2}+\delta}$$

for some $0 < \delta < \frac{1}{2}$, $0 < u < 1$, where K is a fixed constant and the subscript N on $\tilde{J}$ is suppressed for the sake of simplicity,

$$\left[N^{1/2}\left\{\psi_i(z)/n - \int_{-\infty}^{\infty} \tilde{J}\left[\sum_{j=1}^{c} F_j(x)/c\right] dF_i(x)\right\},\ i = 1, \ldots, c\right]$$

is asymptotically c-variate normal having zero mean vector, $(c-1)A^2$ for the variances and $-A^2$ for the covariances where

$$A^2 = \int_0^1 \tilde{J}^2(u)du - \left(\int_0^1 \tilde{J}(u)du\right)^2 \tag{18.3.21}$$

and

$$\begin{aligned} B &= \frac{1}{c}\int_{-\infty}^{\infty}\left\{\frac{d}{dF}\tilde{J}\left(F(x)\right)\right\} f^2(x)dx \text{ if } (F_1 \cdots F_c) \in \mathcal{F}_N^* \\ &= \frac{1}{c}\int_{-\infty}^{\infty} x\left\{\frac{\partial}{\partial F}\tilde{J}\left(F(x)\right)\right\} f^2(x)dx \text{ if } (F_1 \cdots F_c) \in \mathcal{F}_N^{**}. \end{aligned} \tag{18.3.22}$$

Case (ii). Let

$$J_{N,i}\left(F_{1,n}(x), \ldots, F_{c,n}(x)\right) = -\prod_{\substack{k=1 \\ k\neq i}}^{c} \{1 - F_{k,n}(x)\},\ i = 1, \ldots, c.$$

In this case, the statistic T is of the type considered by Bhapkar (1961). Let

$$P_{N,i} = \int_{-\infty}^{\infty} J_{N,i}\left(F_{1,n}(x), \ldots, F_{c,n}(x)\right) dF_{i,n}(x)$$

and

$$p_i = -\int_{-\infty}^{\infty}\left[\prod_{\substack{k=1 \\ k\neq i}} \{1 - F_k(x)\}\right] dF_i(x),\ i = 1, \ldots, c.$$

Bhapkar (1961) showed that asymptotically $N^{1/2}(P_{N,1} - p_1, \ldots, P_{N,c} - p_c)$ has a c-variate normal distribution. Using this result we obtain

$$A^2 = (2c - 1)^{-1}$$

and

$$B = \begin{cases} \displaystyle\int_{-\infty}^{\infty} \{1 - F(x)\}^{c-2} f^2(x)dx, & \text{for "near" location alternatives,} \\ \displaystyle\int_{-\infty}^{\infty} \{1 - F(x)\}^{c-2} xf^2(x)dx, & \text{for "near" scale alternatives.} \end{cases}$$

Case (iii). Let

$$J_{N,i}\left(F_{1,n}(x), \ldots, F_{c,n}(x)\right) = \prod_{\substack{k=1 \\ k \neq i}}^{c} F_{k,n}(x) - \prod_{\substack{k=1 \\ k \neq i}}^{c} \left[1 - F_{k,n}(x)\right], \; i = 1, \ldots, c.$$

Despande (1965) considered this type of scores for a location problem. For further details on this case, the reader is referred to Govindarajulu and Haller (1977, pp. 97–98).

Application of Asymptotic Results

Consider the case of equally spaced alternatives. That is, set $b_i = i$ ($i = 1, \ldots, c$) and let $d_i^* = i - (c+1)/2$. It has been shown in Govindarajulu and Haller (1977, Theorem 4.1) for location shift alternatives with $b_i = i$ ($i = 1, \ldots, c$) and $J = \Phi^{-1}$ that their test and Puri's test have the same Pitman efficacy given by

$$(c^2 - 1)\left[\int_{-\infty}^{\infty} \left\{\frac{d}{dx} J\left(F(x)\right)\right\} f(x)dx\right]^2 / 12A^2 \tag{18.3.23}$$

where A^2 is given by (18.3.21). They list the Pitman efficacies of Kudo's, Jonckheere ($J(u) = u$), normal scores $\left(J(u) = \Phi^{-1}(u)\right)$, Savage ($J(u) = -\log(1 - u)$), Bhapkar and Despande's tests for equally spaced location and scale alternatives. In particular, they numerically evaluate these for exponential, logistic and normal shift alternatives, and exponential and normal scale alternatives. The test procedures of Bhapkar and Deshpande are not efficient (especially when c is large) for scale alternatives since they are designed primarily for location alternatives. For location alternatives, they are not as efficient as the normal scores test or the Jonckheere's test. Notice that the test statistics of Bhapkar and Deshpande belong to the class of U-statistics. If the reader is interested in U-statistics, he/she is referred to Madhava Rao (1982) who proposed nonparametric tests for homogeneity of scale against ordered alternatives.

For testing the homogeneity of scale against ordered alternatives, Govindarajulu and Gupta (1978) have derived a locally most powerful rank test (LMPRT) assuming that the location parameters of the populations are all equal, but unknown. They also derive a parametric test based on the likelihood derivative method for the ordered scale alternatives. Asymptotic distributions of these statistics are studied and are compared via the Pitman efficiency criterion. A heuristic class of rank tests is also proposed for the above hypothesis-testing problem, which can accommodate unequal sub-sample sizes (which, for equal sample sizes, coincides with the class of statistics proposed by Govindarajulu and Haller, 1977). It is shown that the weighting constants can be chosen optimally so as to maximize the Pitman efficacy. Further, an asymptotically distribution-free test is also proposed for the case when the location parameters are unequal and unknown.

Since the *Transaction of the* 8th *Prague Conference* is not readily available, in the following, the results of Govindarajulu and Gupta (1978) will be summarized without proofs.

Let X_{ij} $(j = 1, \ldots, c)$ be a random sample from $F_i(x)$ where

$$F_i(x) = F\left((x - \mu_i)\theta_i\right)$$

where μ_i is the location parameter and $\theta_i > 0$ is the scale parameter $(i = 1, \ldots, c)$. We are interested in testing

$$H_0 : \theta_1 = \cdots = \theta_c \text{ vs. } H_1 : \theta_1 \leq \cdots \leq \theta_c$$

with strict inequality for at least one pair of θ's. Let R_{ij} denote the rank of X_{ij} among the combined sample of size $N = n_1 + \cdots + n_c$. Consider the local alternatives given by $H_\Delta : \theta_i = (1 + b_i\Delta)\theta_0$ with normal alternatives and $\mu_i = \mu$ which we can set it equal to zero. Then the LMP rank test of H_0 vs. H_Δ is to reject H_0 for large values of

$$S_{1N} = N^{-1/2} \sum_{i=1}^{c} \sum_{j=1}^{n_i} b_i E(Z^2_{R_{ij},N}) \tag{18.3.24}$$

where $Z_{u,N}$ is the uth smallest standard normal order statistic in a sample of size N. It is shown that if the $\lambda_i = n_i/N$ $(i = 1, \ldots, c)$ are bounded away from 0 and 1, for sufficiently large N, S_{1N} is normally distributed with mean

$$\mu(S_{1N}) = N^{-1/2} \sum_{i=1}^{c} n_i b_i \int_{-\infty}^{\infty} \left[\Phi^{-1}\left(\sum_{1}^{c} \lambda_i F_i(x)\right)\right]^2 dF_i(x)$$

and

$$\mathrm{var}(S_{1N}|H_0) = \sigma_0^2(S_{1N}) = 2\left[\sum \lambda_i b_i^2 - \left(\sum \lambda_i b_i\right)^2\right].$$

The Pitman efficacy of S_{1N} for the sequence of 'near' alternatives $\theta_i = 1 - \Delta b_i N^{-1/2}$ is given by

$$e(S_{1N}) = 2\left[\sum \lambda_i b_i^2 - \left(\sum \lambda_i b_i\right)^2\right] I^2 \tag{18.3.25}$$

where

$$I = \int_{-\infty}^{\infty} y f^2(y) \Phi^{-1}(F(y)) \left[\phi\left(\Phi^{-1}(F(y))\right)\right]^{-1} dy.$$

The likelihood derivative test proposed by Rao (1948) and showed by Neyman (1959) to be locally asymptotically most powerful is given by

$$\tilde{S}_{2N} = N^{-1/2} \sum_{1}^{c} b_i \sum_{j=1}^{n_i} (X_{ij} - \bar{X})^2 \tag{18.3.26}$$

which is shown to be asymptotically equivalent to

$$S_{2N} = N^{-1/2} \sum_{i=1}^{c} b_i \sum_{j=1}^{n_i} (X_{ij} - \mu)^2.$$

The Pitman efficacy of S_{2N} for the near alternatives given earlier is

$$e(S_{2N}) = \frac{1}{c}\left(\sum b_i^2\right) / \left[E(Y^4) - 1\right]$$

where Y denotes the parameter-free random variable having distribution $F(y)$. Further, Govindarajulu and Gupta (1978) define (for unequal sample sizes) the weighted sum of Chernoff-Savage type of statistics given by

$$S_{3N} = N^{-1/2} \sum_{i=1}^{c} d_i n_i \psi_i \tag{18.3.27}$$

where

$$\psi_i = \int_{-\infty}^{\infty} J_N\left(\sum_{k=1}^{c} \lambda_k F_{k,n_k}(x)\right) dF_{i,n_i}(x),\ i = 1, \ldots, c. \tag{18.3.28}$$

Note that this reduces to the class of statistics considered by Govindarajulu and Haller (1977) when $\lambda_i \equiv 1/c$ (i.e., $n_1 = \cdots = n_c = n$). It is shown that S_{3N} is asymptotically normal (under some regularity assumptions on J such as J is absolutely continuous with $|J'(u)| \le K\,(u(1-u))^{-\frac{3}{2}+\delta}$ for $0 < \delta < \frac{1}{2}$) with mean

$$\mu(S_{3N}) = N^{-1/2} \sum_{i=1}^{c} n_i d_i \int_{-\infty}^{\infty} J\left(\sum_{k=1}^{c} \lambda_k F_k(x)\right) dF_i(x)$$

and

$$\operatorname{var}(S_{3N}|H_0) = \sigma_0^2(S_{3N}) = \left[\sum \lambda_i d_i^2 - \left(\sum \lambda_i d_i\right)^2\right] A^2$$

where

$$A^2 = \int_0^1 J^2(u)du - \left[\int_0^1 J(u)du\right]^2 .$$

Computations yield the Pitman efficacy of S_{3N} to be

$$e(S_{3N}) = \left[\sum \lambda_i d_i \sum_k \lambda_k (b_i - b_k)\right]^2 I_1^2/\sigma_0^2(S_{3N}) . \tag{18.3.29}$$

Then via the Cauchy-Schwarz inequality, $e(S_{3N})$ is maximized when $d_i = \beta b_i^*$, where $b_i^* = \sum_k \lambda_k(b_i - b_k)$, $i = 1, \ldots, c$ and, without loss of generality, we can set $\beta = 1$.

With the above choice of d_i, one can show that $\sum \lambda_i d_i = 0$, and

$$e(S_{3N}) = \left(\sum \lambda_i b_i^{*2}\right) I_1^2/A^2 . \tag{18.3.30}$$

In the special case of $J(u) = \left[\Phi^{-1}(u)\right]^2$, $I_1^2 = 4I^2$ and $A^2 = 2$ and hence,

$$e(S_{3N}) = 2\left[\sum \lambda_i b_i^2 - \left(\sum \lambda_i b_i\right)^2\right] I^2 = e(S_{1N}) .$$

Further, substituting $d_i = \sum_k \lambda_k(b_i - b_k) = b_i - \sum \lambda_k b_k$ in the expressions for S_{3N}, we obtain

$$S_{3N} = S_{1N} - (\text{a non-stochastic constant}) .$$

Case when Location Parameters are Unknown

When the μ_i are unknown, one can estimate them by some consistent estimates (such as sample medians) and let S_N^* denote the test statistic based on the deviations of observations, namely $X_{ij} - \hat{\mu}_i$ ($j = 1, \ldots, n_i$ and $\lambda = 1, \ldots, c$). Further, it is assumed that $N^{1/2}(\hat{\mu}_i - \mu_i)$ is bounded in probability ($i = 1, \ldots, c$). The modified test is asymptotically distribution-free for a fairly general class of alternatives provided the $F_i(x)$ are symmetric about their respective location parameters, and the $f_i(x)/\phi\left[\Phi^{-1}(F_i)\right]$ are bounded for all x and $i = 1, \ldots, c$.

All the previous test procedures assume that the underlying observations are mutually independent. Shetty and Govindarajulu (1988) assume that $\boldsymbol{X} = (X_1, \ldots, X_n)$ has a multivariate density given by

$$f_1(x_1, \ldots, x_n) = \prod_{i=1}^{n} \sigma_i^{-1} f(x_i/\sigma_i) \left[1 + \lambda \prod_{j=1}^{n} (1 - 2F(x_j/\sigma_j))\right]$$

where $-1 < \lambda < 1$ and f and F respectively denote some specified univariate density and distribution function.

Assuming that $\sigma_i = (1 + i\Delta)\sigma_0$ where σ_0 is known, for testing $H_0 : \Delta = 0$ vs. $H_1 : \Delta > 0$, they derive LMP rank tests and study the asymptotic distribution of the test criterion for the special case $f(x) = \exp(-x)$ and $F(x) = 1 - \exp(-x)$ for $x > 0$.

18.4 Problems

18.2.1 Consider the following data
$(X_1, X_2, X_3, X_4, X_5) = (31, 29, 40, 37, 45)$
Assume that their true means are ordered as $\mu_1 < \mu_2 < \cdots < \mu_5$. Using (18.2.7) estimate the μ_i by the method of maximum likelihood.
(Hint: Assume that $a_1 = \cdots = a_5$.)

18.2.2 Consider the following data with $c = 3$ samples

Sample 1:	-0.75, 0.98, 0.62, 0.29, 2.55, 1.67, 1.33, 0.69, 0.90, 2.8
Sample 2:	1.39, 0.79, 2.28, 3.75, 1.39, 3.48, 1.62, 2.38, 1.56, 2.55
Sample 3:	2.29, 3.58, -0.40, 3.10, 5.33, 3.71, 3.1, 3.36, 1.36, 1.08

Assume the model X_{ij} ($j = 1, \ldots, 10$) is a random sample from normal (μ_i, σ^2) for $i = 1, 2, 3$, where $\mu_i = \theta_0 + i\theta$ ($i = 1, \ldots, 3$) where θ_0, θ and σ are unknown. Test $H_0 : \theta = 1$ versus $H_1 : \theta > 1$ using the test criterion NS_N^* (here $N = 30$) and $\alpha = 0.10$. Find the explicit distribution of NS_N^* when H_1 is true.

18.3.1 The World Almanac and Book of Facts (2003, p. 73) gives the following data pertaining to live births per 1000 women by certain age groups.

year	Age groups (in years) 25–29	30–34	35–39	> 40
1995	112.2	82.5	34.4	6.9
1996	113.1	83.9	35.3	7.1
1997	113.8	85.3	36.1	7.5
1998	115.9	87.4	37.4	7.7
1999	117.8	89.6	38.3	7.8
2000	121.4	94.1	40.4	8.4

Let $F_i(x)$ denote the distribution of live births per 1000 women in i^{th} age group ($i = 1, \ldots, 4$). We wish to test

$$H_0\colon F_1(x) = F_2(x) = F_3(x) = F_4(x) \quad \text{for all } x$$

against $H_1\colon F_1(x) < F_2(x) < F_3(x) < F_4(x)$.

Carry out Jonckheere's test based on S defined in (18.3.2) (using its large sample property) with $\alpha = 0.05$.

18.3.2 Evaluate explicitly the expressions for a_{ij} given in (18.3.7) in terms of $\tilde{\lambda}_i$ and hence the a_k defined in (18.3.9).

18.3.3 Evaluate T_N given in (18.3.7) and T_N^* given in (18.3.8) with $J_N(u) = u$, for the data in Problem 18.3.1, and using their asymptotic normalities, test H_0 versus H_1 specified in Problem 18.3.1.

18.3.4 Using (18.3.23) evaluate the Pitman efficacy of Jonckheere's test procedure (for which $J(u) = u$) when F is normal.

Chapter 19

Tests in Two-way Layouts

19.1 Introduction

In earlier chapters we have analyzed data arising from observations which were classified according to a single criterion. Of course, such a simple state of affairs is not often the case. In most scientific research, there are several factors to be considered and the data are typically classified according to several different criteria. Thus, it is necessary to provide methods by which we may handle such data.

19.2 Randomized Block Design

The first design we will study is one in which the observations are classified according to two different criteria referred to as *blocks* and *treatments*. Usually we use the word *treatment* in a general sense of a combination of factors, often called a *treatment combination* to which an experimental unit is subjected to. If we have various groups of experimental units, we should separate the group effect from the treatment effect. So we arrange the experimental units in such a way that the units within each group are fairly homogeneous. Then we assign the various treatments at random to the units within each group. For example, if the fertility trend of the ground is running from north to south, the plots in north being more fertile, we should treat fertility of ground as the blocking factor. When the number of units in each block is equal to the number of treatments, the layout is called a *randomized complete block design* and the word 'complete' is typically omitted. The model can be expressed as

$$X_{ij} = \mu + \beta_i + \tau_j + e_{ij},\ i = 1, \ldots, b,\ j = 1, \ldots, c \qquad (19.2.1)$$

where we assumed that $\sum_{i=1}^{b} \beta_i = \sum_{j=1}^{c} \tau_j = 0$,

$$E(e_{ij}) = 0 \text{ and var } e_{ij} = \sigma^2 .$$

Let $\bar{X}_{i\cdot} = \frac{1}{c}\sum_{j=1}^{c} X_{ij} = \mu + \beta_i + e_{i\cdot}$ etc. In the parametric setup, we further assume that e_{ij} are independently distributed as normal $(0, \sigma^2)$. Typically we are interested in testing

$$H_0 : \tau_1 = \cdots = \tau_c$$

versus

$$H_1 : \tau_j \neq \tau_l$$

at least for some pair (j, l), $j \neq l$. The test criterion is

$$\begin{aligned} F &= \frac{b\sum_{j=1}^{c}(\bar{X}_{\cdot j} - \bar{X}_{\cdot\cdot})^2/(c-1)}{\sum_{i=1}^{b}\sum_{j=1}^{c}(X_{ij} - \bar{X}_{i\cdot} - \bar{X}_{\cdot j} + \bar{X}_{\cdot\cdot})^2/(b-1)(c-1)} \\ &= \frac{\text{mean square for treatments}}{\text{mean square for error}} . \end{aligned} \tag{19.2.2}$$

When H_0 is true, F has the Snedecor's F distribution with $c-1$ and $(b-1)(c-1)$ degrees of freedom. When b becomes large, the denominator converges to σ^2 in probability, and hence,

$$\begin{aligned} (c-1)F &\approx \frac{b}{\sigma^2}\sum_{j=1}^{c}(\bar{X}_{\cdot j} - \bar{X}_{\cdot\cdot})^2 \stackrel{d}{=} \chi^2_{c-1} \text{ under } H_0, \\ (c-1)F &\stackrel{d}{\approx} \chi^2_{c-1,\lambda_F} \text{ under } H_1 , \end{aligned}$$

where the noncentrality parameter λ_F is given by

$$\lambda_F = \frac{b}{\sigma^2}\sum_{j=1}^{c}(E\bar{X}_{\cdot j} - E\bar{X}_{\cdot\cdot})^2 = \frac{b}{\sigma^2}\sum_{j=1}^{c}\tau_j^2 = \sigma^{-2}\sum_{j=1}^{c}\tau_j^{*2} , \tag{19.2.3}$$

where $\tau_j^* = \tau_j b^{1/2}$ $(j = 1, \ldots, c)$.

19.3 Nonparametric Test Procedures

Let $P(X_{ij} \leq x) = F_{ij}(x) = F(x - \mu - \beta_i - \tau_j)$ where we can set, without loss of generality, $\mu = 0$. We wish to test

$$H_0 : \tau_1 = \tau_2 = \cdots = \tau_c$$

against the alternative

$$H_1 : \tau_j \neq \tau_l \text{ for at least some pair } (j,l),\ j \neq l\,.$$

Let R_{ij} denote the rank of X_{ij} among $(X_{i1}, \ldots, X_{ic})$, $i = 1, \ldots, b$. That is, $(R_{i1}, \ldots, R_{ic})$ is a permutation of $(1, \ldots, c)$ for each $i = 1, \ldots, b$.

Notice [or one can easily establish, since R_{ij} are discrete uniform on $(1, \ldots, c)$] that

$$E(R_{ij}|H_0) = (c+1)/2, \quad \text{var}(R_{ij}|H_0) = (c^2-1)/12$$

and for $j \neq l$,

$$\text{cov}(R_{ij}, R_{i,l}) = -(c+1)/12\,. \tag{19.3.1}$$

Then the Friedman's (1937) test criterion is F^* where

$$(c-1)F^* = b^2(c-1)\sum_{j=1}^{c}\left(\bar{R}_{\cdot j} - \frac{c+1}{2}\right)^2 \Big/ \sum_{i=1}^{b}\sum_{j=1}^{c}\left(R_{ij} - \frac{c+1}{2}\right)^2. \tag{19.3.2}$$

However,

$$\begin{aligned}
\sum_{i=1}^{b}\sum_{j=1}^{c}\left(R_{ij} - \frac{c+1}{2}\right)^2 &= \sum_{i-1}^{b}\sum_{j=1}^{c}\left\{R_{ij}^2 - (c+1)R_{ij} + \left(\frac{c+1}{2}\right)^2\right\} \\
&= b\sum_{l=1}^{c} l^2 - (c+1)b\sum_{l=1}^{c} l + bc\frac{(c+1)^2}{4} \\
&= b\frac{c(c+1)(2c+1)}{6} - \frac{bc(c+1)^2}{2} + bc\frac{(c+1)^2}{4} \\
&= \frac{bc(c+1)}{12}\{2(2c+1) - 6(c+1) + 3(c+1)\} \\
&= bc(c^2-1)/12\,.
\end{aligned}$$

Hence,

$$(c-1)F^* = \frac{12b}{c(c+1)}\sum_{j=1}^{c}\left(\bar{R}_{\cdot j} - \frac{c+1}{2}\right)^2. \tag{19.3.3}$$

First we will show that $E\{(c-1)F^*|H_0\} = (c-1)$ and $\text{var}\{(c-1)F^*|H_0\} \approx 2(c-1)$ for sufficiently large b,

$$E\{(c-1)F^*|H_0\} = \frac{12b}{c(c+1)}\sum_{j=1}^{c} E\left\{\left(\bar{R}_{\cdot j} - \frac{c+1}{2}\right)^2 |H_0\right\}. \tag{19.3.4}$$

Since $\bar{R}_{\cdot j} = \frac{1}{b}\sum_{i=1}^{b} R_{ij}$ = the average of i.i.d. random variables having mean $(c+1)/2$ and variance $(c^2-1)/12$, we have,

$$\mathrm{var}(\bar{R}_{\cdot j}|H_0) = \frac{1}{b}\,\mathrm{var}(R_{ij}|H_0) = (c^2-1)/12b\,. \qquad (19.3.5)$$

Using (19.3.5) in (19.3.4), we obtain

$$E\{(c-1)F^*|H_0\} = (c-1)\,.$$

We can also write

$$(c-1)F^* = \frac{12}{bc(c+1)}S \qquad (19.3.6)$$

where

$$S = \sum_{j=1}^{c}\left(\sum_{i=1}^{b} Q_{ij}\right)^2 = \sum_{j=1}^{c}\left[\sum_{i=1}^{b} Q_{ij}^2 + 2\sum^{b}\sum_{i<k}^{b} Q_{ij}Q_{kj}\right] \qquad (19.3.7)$$

and

$$Q_{ij} = R_{ij} - (c+1)/2\,.$$

Then

$$\begin{aligned} S &= \sum_{i=1}^{b}\sum_{l=1}^{c}\left(l - \frac{c+1}{2}\right)^2 + 2\sum_{j=1}^{c}\sum^{b}\sum_{i<k}^{b} Q_{ij}Q_{kj} \\ &= \frac{bc(c^2-1)}{12} + 2\tilde{S} \end{aligned} \qquad (19.3.8)$$

where

$$\tilde{S} = \sum_{j=1}^{c}\sum^{b}\sum_{i<k}^{b} Q_{ij}Q_{kj}\,.$$

Then

$$E(\tilde{S}|H_0) = \sum_{j=1}^{c}\sum^{b}\sum_{i<k}^{b} E(Q_{ij}|H_0)E(Q_{kj}|H_0) = 0\,. \qquad (19.3.9)$$

Next,

$$\mathrm{var}(S|H_0) = 4\,\mathrm{var}(\tilde{S}|H_0) = 4E(\tilde{S}^2|H_0)\,,$$

$$
\begin{aligned}
E(\tilde{S}^2|H_0) &= E\left[\sum_{j=1}^{c}\sum_{l=1}^{c}\sum\sum_{i<k}\sum\sum_{i'<k'} Q_{ij}\,Q_{kj}Q_{i',l}\,Q_{k',l}|H_0\right] \\
&= \sum_{j=1}^{c}\sum^{b}\sum_{i<k}^{b} E(Q_{ij}^2|H_0)\cdot E(Q_{kj}^2|H_0) \\
&\quad +\sum^{c}\sum_{j\neq l}^{c}\sum^{c}\sum_{i<k}^{b} E(Q_{ij}\,Q_{il}|H_0)\cdot E(Q_{kj}\,Q_{kl}|H_0)\,,
\end{aligned}
$$

and all other terms will be equal to zero since $E(Q_{ij}|H_0)=0$. Notice that we also used the fact that Q_{ij} and Q_{kj} are independent for $i\neq k$. Thus,

$$
\begin{aligned}
E(\tilde{S}^2|H_0) &= \frac{b(b-1)}{2}\cdot c\left(\frac{c^2-1}{12}\right)^2\cdot\frac{b(b-1)}{2} \\
&\quad +c(c-1)\frac{b(b-1)}{2}\cdot\left\{-\frac{(c+1)}{12}\right\}^2
\end{aligned}
$$

since

$$
E(Q_{ij}\,Q_{ik}|H_0)=\operatorname{cov}(R_{ij},R_{ik}|H_0)=\frac{-(c+1)}{12}\,.
$$

Hence,

$$
\begin{aligned}
\operatorname{var}(S|H_0) &= 4E(\tilde{S}^2|H_0) \\
&= 4\left[c\left(\frac{c^2-1}{12}\right)^2\cdot\frac{b(b-1)}{2}+c(c-1)\frac{b(b-1)}{2}\left(\frac{c+1}{12}\right)^2\right] \\
&= c^2(c-1)(c+1)(c+1)^2b(b-1)/72\,. \qquad (19.3.10)
\end{aligned}
$$

Thus,

$$
\operatorname{var}\left((c-1)F^*|H_0\right)=2(c-1)(b-1)/b\approx 2(c-1)
$$

when b is large.

Since $\bar{R}_{.j}$ is the average of b i.i.d. random variables, under H_0 as b becomes large, $(c-1)F^*$ is asymptotically distributed as central chi-square with $(c-1)$ degrees of freedom since the $\bar{R}_{.j}$ are subject to the linear constraint, namely, $\sum_1^c \bar{R}_{.j}$ is a non-stochastic constant. Under H_1, $(c-1)F^*$ is approximately distributed as noncentral chi-square with $c-1$ degrees of freedom and noncentrality parameter

$$\begin{aligned}\lambda_{F^*} &= \frac{12b}{c(c+1)}\sum_{j=1}^{c}\left(E\bar{R}_{\cdot j}-\frac{c+1}{2}\right)^2 \\ &= \frac{12}{bc(c+1)}\sum_{j=1}^{c}\left\{\sum_{i=1}^{b}\left(ER_{ij}-\frac{c+1}{2}\right)\right\}^2 . \qquad (19.3.11)\end{aligned}$$

Now, since

$$R_{ij}=\sum_{\substack{l=1\\ l\neq j}}^{c} I(X_{ij}>X_{ik})+1\,,$$

$$E(R_{ij}|H_1)=\sum_{\substack{l=1\\ l\neq j}}^{c} P(X_{ij}>X_{il}|H_1)+1\,.$$

Next, for $l \neq j$,

$$\begin{aligned}P(X_{ij}>X_{il}) &= \int_{-\infty}^{\infty} F(x+\mu+\beta_i+\tau_l)dF(x+\mu+\beta_i+\tau_j) \\ &= \int_{-\infty}^{\infty} F(y+\tau_l-\tau_j)dF(y) \\ &= \int_{-\infty}^{\infty} [F(y)+(\tau_l-\tau_j)f(y)+o(\tau_l-\tau_j)]\,dF(y) \\ &= \frac{1}{2}+(\tau_l-\tau_j)\int_{-\infty}^{\infty} f^2(y)dy+o(\tau_l-\tau_j)\,.\end{aligned}$$

Thus,

$$\begin{aligned}E(R_{ij}|H_1) &= \sum_{\substack{l=1\\ l\neq j}}\left[\frac{1}{2}+(\tau_l-\tau_j)\int f^2(y)dy+o(\tau_l-\tau_j)\right]+1 \\ &= \frac{(c+1)}{2}+c\left[\int f^2(y)dy\right]\tau_j+o(\tau_j)\,.\end{aligned}$$

If $\tau = b^{-1/2}\tau_j^*$ $(j = 1, \ldots, c)$, we have from (19.3.11)

$$\lambda_{F^*} \to \frac{12c}{(c+1)} \left[\int_{-\infty}^{\infty} f^2(y)dy\right]^2 \sum_{j=1}^{c} \tau_j^{*2} \text{ as } b \to \infty \,. \tag{19.3.12}$$

Thus, the asymptotic efficiency of Friedman's test relative to the classical F-test is given by

$$e(F^*, F) = \frac{12c\sigma^2}{(c+1)} \left[\int_{-\infty}^{\infty} f^2(y)dy\right]^2 . \tag{19.3.13}$$

Elteren and Noether (1959) compute the asymptotic efficiency of Friedman's test extended to balanced incomplete block designs by Durbin (1951). One can obtain (19.3.13) by setting $k = c$ in their expression given in equation (7). When f is normal,

$$e(F^*, F) = \frac{3c}{(c+1)\pi} \,. \tag{19.3.14}$$

Remark 19.3.1. *If one can look upon the blocks as judges, then we can interpret the hypotheses as*

H_0: *There is concordance among the judges*

and

H_1: *There is discordance between at least one pair of judges.*

19.4 Nonparametric Tests for Ordered Alternatives

The contributions to this problem can be categorized as (i) with one observation per cell, (ii) with more than one observation in each cell, (iii) distribution-free tests, and (iv) asymptotically distribution-free tests. Govindarajulu and Mansouri-Ghiassi (1986) – which will be abbreviated as GMG (1986) – have provided an excellent survey of these contributions and they will be summarized below.

Let

$$X_{ijk} \left(1 \le i \le b, 1 \le j \le c, 1 \le k \le n_{ij}; n = \sum_{i}^{b} \sum_{j}^{c} n_{ij}\right)$$

be independent random variables having continuous cumulative distribution functions (cdf's)

$$P(X_{ijk} \leq x) = F(x - \mu - \beta_i - \tau_j) \tag{19.4.1}$$

where μ and $(\beta_1, \ldots, \beta_b)$ are nuisance parameters and $(\tau_1, \ldots, \tau_c)$ are treatment effects. We wish to test

$$H_0 : \tau_1 = \cdots = \tau_c \text{ versus } H_1 : \tau_j = \tau a_j \ (1 \leq j \leq c) \tag{19.4.2}$$

where $a_1 \leq \cdots \leq a_c$, not all a_j's are equal, $\tau > 0$ is an unknown parameter and F is also unknown.

In randomized block design with $n_{ij} = 1$, some nonparametric test procedures have been proposed by Jonckheere (1954), Page (1963), Hollander (1967), Doksum (1967), Puri and Sen (1967) and Pirie and Hollander (1972).

Let ξ_i be the Kendall's rank correlation coefficient between the postulated order and the observation order in the ith block. Then Jonckheere's (1954) procedure is to reject H_0 for large values of

$$\xi = \sum_{i=1}^{b} \xi_i \,. \tag{19.4.3}$$

Page's (1963) procedure rejects H_0 for large values of

$$\rho = \sum_{i=1}^{b} \rho_i \,, \tag{19.4.4}$$

where ρ_i denotes the Spearman's correlation coefficient between the postulated order and the observation order in the ith block.

Let R_i (as in Friedman's case) denote the sum of ranks for the ith treatment ($i = 1, \ldots, c$). Then the Page's test criterion can be written equivalently as P where

$$P = \sum_{j=1}^{c} jR_j = \sum_{j=1}^{c} j \sum_{i=1}^{b} R_{ij} \,. \tag{19.4.5}$$

Straightforward computations yield

$$E_0(P) = \frac{bc(c+1)^2}{4} \,,$$

$$\text{var}_0(P) = bc^2(c+1)(c^2-1)/144 \,.$$

Since we can also write

$$P = \sum_{i=1}^{b} W_i$$

where the W_i are i.i.d. under all the hypotheses. Hence, applying the usual central limit theorem, we can assert that

$$\tilde{P}^* = \frac{P - EP}{(\text{var } P)^{1/2}} = \frac{P - bE(W_1)}{[b \text{ var } W_1]^{1/2}}$$

is distributed as normal $(0, 1)$ for sufficiently large b. Hollander and Wolfe (1999) tabulate the upper tail probabilities of P under H_0 for $c = 3(1)8$ and values of $b \leq 10$.

Pirie and Hollander (1972) proposed a normal scores test for ordered alternatives among the c-treatments in the randomized block design. They reject H_0 for large values of B where

$$B = \sum_{i=1}^{b} \sum_{j=1}^{c} jE_{R_{ij},c}\,, \tag{19.4.6}$$

where R_{ij} denotes the rank of X_{ij} $(j = 1, \ldots, c)$ (that is, observations are ranked within each block) and $E_{k,c}$ denotes the expected value of the kth smallest standard normal order statistic in a random sample of size c; the latter are known as *normal scores*.

For $c = 2$, $E_{1,2} = -E_{22}$ and for $c = 3$, $E_{1,3} = -E_{33}$ and $E_{2,3} = 0$, one can easily show that P and B are linearly related when $c = 2$ or 3. Hence, P and B test procedures are equivalent. Further, when $c = 2$, both the test procedures are equivalent to the sign test. Both P and B are distribution-free under H_0 because the within block rank matrix $R = ((R_{ij}))$ is distribution-free under H_0. Also, when

$$P(X_{ij} \leq x) = F(x - \mu - \beta_i - \tau_j) = \Phi(x - \mu - \beta_i - j\tau) \tag{19.4.7}$$

where Φ is the standard normal cdf and τ is some unknown positive constant, the locally most powerful test of H_0 against normal ordered alternatives is given by B. Straightforward computations yield

$$E_0(B) = 0$$

and

$$\text{var}_0(B) = \{bc(c+1)/12\} \sum_{j=1}^{c} E_{j,c}^2\,.$$

The above can be obtained upon using the following facts:

$$\sum_1^c E_{k,c} = 0\,,$$

and for $j \neq j'$,

$$P(R_{ij} = k|H_0) = \frac{1}{c} \text{ and } P(R_{ij} = k, R_{ij'} = l|H_0) = \frac{1}{c(c-1)}\,.$$

Consequently, one can carry out the test using normal tables when b is large using the fact that $B/\left\{\mathrm{var}_0(B)\right\}^{1/2}$ is asymptotically standard normal under H_0.

Using the asymptotic normality of B (since it is a sum of i.i.d. random variables under H_1), Pirie and Hollander (1972) were able to show the consistency of B provided

$$\mu = E\left(\sum_{j=1}^{c} E_{R_{ij},c}\right) > 0\,. \tag{19.4.8}$$

They also evaluate the Pitman efficacy of the B and P test procedures. Based on numerical computations, the asymptotic efficiency of B relative to P tends to infinity as $c \to \infty$ for uniform and exponential populations. For normal, uniform and exponential populations, this efficiency exceeds unit for $c \geq 4$. They also provided a table of critical values of the B test for $c = 4$ and 5 and small values of b (say ≤ 5). Hollander (1967) proposed an asymptotically distribution-free criterion based on the sum of Wilcoxon signed rank statistics (the sum being over all distinct pairs) with $n_{ij} \equiv 1$. Let

$$Y_{u,v}^{(i)} = |X_{iu} - X_{iv}|$$

and $R_{uv}^{(i)}$ be the rank of $Y_{u,v}^{(i)}$ in the ranking of $Y_{u,v}^{(i)}$ $(i = 1, \ldots, b)$. Then let

$$T_{u,v} = \sum_{i=1}^{b} R_{u,v}^{(i)}\,\psi_{u,v}^{(i)} \tag{19.4.9}$$

with

$$\psi_{u,v}^{(i)} = \begin{cases} 1 & \text{if } X_{iu} < X_{iv} \\ 0 & \text{otherwise, } i = 1, \ldots, b\,. \end{cases} \tag{19.4.10}$$

Towards the test criterion, let

$$T = \sum_{u<v}^{c} \sum^{c} T_{u,v} \, . \tag{19.4.11}$$

Notice that $T_{u,v}$ represents a measure of the difference between the uth and v^{th} treatments and the summation over $1 \leq u < v \leq c$ takes into account the prior ordering of the treatments. H_0 will be rejected for large values of T. However, T is not distribution-free because $\text{var}(T|H_0)$ depends on F defined in (19.4.1). Since one can obtain a consistent estimate of the variance of T under H_0, T will be asymptotically distribution-free.

Doksum (1967) proposed a test procedure based on a modified T statistic. Let

$$U_{u,v} = T_{u,v} - \sum_{i=1}^{b} \psi_{u,v}^{(i)} \tag{19.4.12}$$

and consider the statistic

$$U = \sum_{u<v}^{c} \sum^{c} (U_{u\cdot} - U_{v\cdot}) \text{ with } U_{u\cdot} = \frac{1}{c} \sum_{s=1}^{c} U_{u,s} \, . \tag{19.4.13}$$

An asymptotically equivalent test statistic is

$$T' = \sum_{u<v}^{c} \sum^{c} (T_{u\cdot} - T_{v\cdot}) \text{ where } T_{u\cdot} = \frac{1}{c} \sum_{s=1}^{c} T_{u,s} \, . \tag{19.4.14}$$

Doksum (1967) shows that the asymptotic efficiency of T' relative to T is at least unity for all F and c.

Puri and Sen (1968) generalize Hollander's (1967) criterion to Chernoff-Savage class of statistics. Let

$$Y_{i,u,v} = X_{i,u} - X_{i,v}, \; u < v = 1, \ldots, c \text{ and } i = 1, \ldots, b,$$

and let $G_{u,v}(x)$ denote the cdf of $Y_{i,u,v}$.

Further, define the random variable

$$S_{u,v} = \frac{1}{b} \sum_{i=1}^{b} E_{b,i} Z_{u,v,i}, \; u < v = 1, \ldots, c, \tag{19.4.15}$$

where

$$Z_{u,v,i} = \begin{cases} 1 & \text{if the } i^{\text{th}} \text{ smallest among } |Y_{i,u,v}|, \, i = 1, \ldots, b \text{ is from a positive } Y, \\ 0 & \text{otherwise}, \end{cases} \tag{19.4.16}$$

and $E_{b,i}$ denotes the expected value of the ith smallest order statistic in a sample of size b drawn from the distribution

$$\begin{aligned} \Psi^*(x) &= \Psi(x) - \Psi(-x) \quad \text{for } x \geq 0 \\ &= 0 \quad \text{for } x < 0 . \end{aligned} \tag{19.4.17}$$

Let

$$V = \sum\sum_{u<v} S_{u,v} . \tag{19.4.18}$$

Special Cases of V. If Ψ is uniform on $(-1, 1)$, then V reduces to T given by (19.4.11). If Ψ is standard normal, V reduces to normal scores type of a statistic. Let

$$\begin{aligned} S_{u\cdot} &= \frac{1}{c} \sum_{v=1}^{c} S_{u,v}, \quad 1 \leq u \leq c, \\ S^*_{u,v} &= S_{u\cdot} - S_{v\cdot}, \quad \text{for } 1 \leq u \leq v \leq c, \end{aligned}$$

and

$$V^* = \sum\sum_{u<v} S^*_{u,v} . \tag{19.4.19}$$

Note that V and V^* are not distribution-free, although they can be made asymptotically distribution-free by employing consistent estimators for the variances of V and V^* under H_0. Puri and Sen (1968) show that the asymptotic efficiency of V^* relative to V is at least one for all $G_{u,v}$ and all Ψ^{-1} satisfying certain regularity assumptions.

Shorack (1967) obtained a parametric test procedure when the model is given by

$$X_{ij} = \mu + \beta_i + \tau_j + e_{ij}, \; i = 1, \ldots, b \text{ and } j = 1, \ldots, c,$$

where $\bar{\beta}. = \frac{1}{b} \sum_1^b \beta_i = 0$ and $\bar{\tau}. = \frac{1}{c} \sum_{j=1}^{c} \tau_j = 0$ (without loss of generality) and e_{ij} are independent normal $(0, \sigma^2)$ variables. Then we can write

$$\begin{aligned} \sum\sum e_{ij}^2 &= \sum\sum (X_{ij} - \mu - \beta_i - \tau_j)^2 \\ &= bc(\bar{X}_{\cdot\cdot} - \mu)^2 + c\sum_{i=1}^{b} (\bar{X}_{i\cdot} - \bar{X}_{\cdot\cdot} - \beta_i)^2 + b\sum_{j=1}^{c} (\bar{X}_{\cdot j} - \bar{X}_{\cdot\cdot} - \tau_j)^2 \\ &\quad + \sum_{i=1}^{b} \sum_{j=1}^{c} (\bar{X}_{ij} - \bar{X}_{i\cdot} - \bar{X}_{\cdot j} + \bar{X}_{\cdot\cdot})^2 . \end{aligned}$$

Let

$$\begin{aligned}
SS_B &= c\sum_{i=1}^{b}(\bar{X}_{i\cdot} - \bar{X}_{\cdot\cdot})^2, \\
SS_\tau &= b\sum_{j=1}^{c}(\bar{X}_{\cdot j} - \bar{X}_{\cdot\cdot})^2 \\
SS_E &= \text{Error Sum of Squares} = \sum_{i=1}^{b}\sum_{j=1}^{c}(X_{ij} - \bar{X}_{i\cdot} - \bar{X}_{\cdot j} + \bar{X}_{\cdot\cdot})^2, \\
SAS_\tau &= \inf_{\tau_1 < \cdots < \tau_c} b\sum_{j=1}^{c}(X_{ij} - \bar{X}_{\cdot\cdot} - \tau_j)^2
\end{aligned}$$

where SAS_τ is an abbreviation for the sum of amalgamated squares for the τ-effect. This sum can be computed as in the one-way layout model described in chapter 18, section 2. We wish to test

$$H_0\colon \tau_1 = \cdots = \tau_c$$

versus

$$H_1\colon \tau_1 \le \cdots \le \tau_c \text{ with at least one strict inequality.}$$

The likelihood ratio (LR) test criterion for H_0 against H_1 is given by

$$\bar{B} = (SS_\tau - SAS_\tau)/(SS_E + SS_\tau)\,.$$

If σ^2 is known, the LR-test criterion for H_0 versus H_1 becomes

$$\bar{D} = (SS_\tau - SAS_\tau)/\sigma^2\,.$$

Under H_0, $\bar{D}$ has the distribution of a linear combination of central chi-square variables, whereas $\bar{B}$ has the distribution of a linear combination of beta variables. Shorack (1967, Table 1) tabulates the upper 5% and 1% values of $\bar{B}(b, bc)$ for $b = 3, 4, 5, 6$ and $bc = 4(1)20(2)24, 27, 30, 40, 50, \infty$.

Shorack (1967) also proposed a modified Friedman's chi-square test procedure using the 'amalgamation procedure' of Bartholomew (1959) (see, also, Barlow *et al.*, 1972, Section 1.2).

Let

$$\bar{R}_j = \frac{1}{b} \sum_{i=1}^{b} R_{ij} \,. \tag{19.4.20}$$

Now apply the amalgamation procedure to the $\bar{R}_j$'s and obtain m distinct integers $t_1, \ldots, t_m$ such that $t_1 + \cdots + t_m = c$ and m distinct quantities $\bar{R}_{[t_1]}, \ldots, \bar{R}_{[t_m]}$. Note that the amalgamation procedure establishes the same order among the $\bar{R}_{[t_j]}$, $(j = 1, \ldots, m)$ as was postulated among the τ_j in H_1. Then we reject H_0 for large values of

$$\bar{\chi}^2 = \frac{12b}{c(c+1)} \sum_{j=1}^{m} t_j \left(\bar{R}_{[t_j]} - \frac{c+1}{2} \right)^2 . \tag{19.4.21}$$

Shorack (1967, p. 1748 and 1742) shows that under H_0, $\bar{\chi}^2$ is a mixture of central chi-square distribution functions when $b \to \infty$. He also computed the asymptotic efficiency of $\bar{\chi}^2$ test relative to his parametric competitor $\bar{B}$ as

$$\left(\frac{c}{c+1} \right) 12\sigma_F^2 \left[\int_{-\infty}^{\infty} F'(t) dF(t) \right]^2 \tag{19.4.22}$$

where σ_F^2 denotes the variance of F.

Case of $n_{ij} > 1$. So far we have stated the results with $n_{ij} \equiv 1.$. (i.e., a single observation in each cell). However, the experimenter may have replication in each cell. Hettmansperger (1975) has extended Page's (1963) test criterion to this case. Let X_{ijk} be as defined at the beginning of Section 19.4. For each i, we rank the X_{ijk}. Let $R_{ij\cdot}$ denote the sum of the ranks of the observations in the $(i,j)^{\text{th}}$ cell. Then $\bar{R}_{ij} = R_{ij\cdot}/n_{ij}$ is the average rank in the $(i,j)^{\text{th}}$ cell and $\sum_{i=1}^{b} \bar{R}_{ij}$ is the total corresponding to treatment j. Hettmansperger (1975) proposes the test statistic

$$\tilde{T} = \sum_{j=1}^{c} j \sum_{i=1}^{b} \bar{R}_{ij} = \sum_{i=1}^{b} \sum_{j=1}^{c} j \bar{R}_{ij} \,. \tag{19.4.23}$$

When the alternative hypothesis H_1 holds, a high degree of agreement is expected between the hypothesized ordering and the ordering exhibited by the treatment totals. Hence, H_0 will be rejected for large values of $\tilde{T}$. When $n_{ij} \equiv 1$, $\tilde{T}$ reduces to Page's (1963) statistic.

Mansouri-Ghiassi and Govindarajulu (1986) – to be abbreviated as M-GG (1986) – have proposed an asymptotically distribution-free test which is obtained by replacing the nuisance parameters by some consistent estimators and then using a Kruskal-Wallis type of test procedure on the residuals. In other words, the approach is based on converting the two-way layout to a one-way layout.

Let $\{\hat{\beta}_i\}_{i=1}^b$ be some $n^{1/2}$-consistent estimators of $\{\beta_i\}_{i=1}^b$. That is, given $\epsilon_i > 0$, there exist δ_i such that, for n sufficiently large,

$$P\left(n^{1/2}|\hat{\beta}_i - \beta_i| \geq \delta_i\right) \leq \epsilon_i,\ i = 1, \ldots, b,\ n = \sum_i \sum_j n_{ij}\,.$$

(See also Lehmann, 1963, for some robust estimates.)

Let $R_{ijk} =$ rank of $(X_{ijk} - \hat{\beta}_i)$ in the overall ranking of n observations. Since rank $(X_{ijk} - \hat{\mu} - \hat{\beta}_i) =$ rank $(X_{ijk} - \hat{\beta}_i)$, without loss of generality, we can assume that $\mu = 0$. Let

$$\hat{R}_{\cdot j\cdot} = \frac{1}{n_{\cdot j}} \sum_{i=1}^{b} \sum_{k=1}^{n_{ij}} R_{ijk}\ (1 \leq j \leq c)\,, \tag{19.4.24}$$

where $n_{\cdot j} = \sum_{i=1}^b n_{ij}$. That is, $\bar{R}_{\cdot j\cdot}$ denotes the average of ranks assigned to j^{th} treatment. The proposed test criterion is

$$M = \sum_{j=1}^{c} d_j \bar{R}_{\cdot j\cdot} \tag{19.4.25}$$

where $d_1 \leq d_1 \cdots \leq d_c$ are some real-valued constants and not all the d's are equal. One rejects H_0 for large values of M. In particular, the $\hat{\beta}_i$ could be the least squares estimators of the block-effects or the Lehmann (1963) robust estimators of the block effects. The latter ones are given by

$$\hat{\beta}_i = (bc^2)^{-1} \sum_{i'=1}^{b} \sum_{j=1}^{c} \sum_{j'=1}^{c} Y_{i,j,i',j'}\,, i = 1, \ldots, b \tag{19.4.26}$$

where

$$Y_{i,j,i',j'} = \text{median}\{X_{ijk} - X_{i'j'k'},\ 1 \leq k,\ k' \leq n_{ij},\ 1 \leq i,\ i' \leq b,\ 1 \leq j,\ j' \leq c\}$$

Asymptotic Distribution of M

If $\beta_1, \beta_2, \ldots, \beta_b$ are known, let

$$Q_{ijk} = \text{rank}(X_{ijk} - \beta_i),\ 1 \leq i \leq b,\ 1 \leq j \leq c \text{ and } k = 1, \ldots, n_{ij}\,, \tag{19.4.27}$$

and define

$$\underline{Q}' = \left(\frac{12}{n}\right)^{1/2} (\bar{Q}_{\cdot 1 \cdot}, \ldots, \bar{Q}_{\cdot c \cdot})$$

$$\bar{Q}_{\cdot j \cdot} = \frac{1}{n_{\cdot j}} \sum_{i=1}^{b} \sum_{k=1}^{n_{ij}} Q_{ijk} \,.$$

It is well-known that when $\lambda_{\cdot j} = n_{\cdot j}/n$ are bounded away from 0 and 1 and F is continuous, $\underline{Q}$ is asymptotically distributed as multivariate normal with mean $E(\underline{Q})$ and a certain variance-covariance matrix. Under H_0, the variance-covariance matrix takes the form of

$$\Sigma = ((\sigma_{j,j'}))$$

with

$$\sigma_{j,j'} = -1 + \delta_{jj'}(\lambda_{\cdot j}\lambda_{\cdot j'})^{-1/2}$$

where δ_{j,j^*} is the Kronecker's delta function.

M-GG (1986), after going through several asymptotic equivalences, show that

$$L^* = \left\{ 12 \Big/ \left[\sum_{j=1}^{c} n_{\cdot j}^{-1} (nd_j - cn_{\cdot j}\bar{d})^2 \right] \right\}^{1/2} \left[L - \left(\frac{n}{12}\right)^{1/2} \underline{d}' E_0(\underline{Q}) \right] \tag{19.4.28}$$

is asymptotically standard normal where

$$\bar{d} = \frac{1}{c} \sum_{1}^{c} d_j \text{ and } E_0(Q_{\cdot j \cdot}) = (n+1)\left(\frac{3}{n}\right)^{1/2} .$$

Hence we reject H_0 when $L^* > z_\alpha$ where z_α denotes the $(1-\alpha)^{\text{th}}$ quantile of the standard normal distribution and α denotes the level of significance. It also follows that the test based on L^* is consistent.

Pitman Efficacy of L

Consider a sequence of "near" alternatives of the form

$$H_n : \tau_{jn} = a_j n^{-1/2} \tau, \; 1 \le j \le c \,.$$

Then the Pitman efficacy of L is given by

$$\text{eff}(L) = \frac{12 \left[\sum_{j=1}^{c} \lambda_{\cdot j} (a_j - \bar{a})(\lambda_{\cdot j}^{-1} d_j - c\bar{d}) \right]^2}{\sum_{j=1}^{c} \lambda_{\cdot j} (\lambda_{\cdot j}^{-1} d_j - c\bar{d})^2} \left[\int f^2(x)dx \right]^2 \tag{19.4.29}$$

where f is the density of F which is assumed to be continuous and $\lambda_{\cdot j} = n_{\cdot j}/n$ and $\bar{a} = \sum_{j=1}^{c} \lambda_{\cdot j} a_j$. From Cauchy-Schwarz inequality, eff(L) is maximized when

$$d_j = \lambda_{\cdot j}(a_j - \bar{a}),\ 1 \le j \le c\,.$$

Notice that, without loss of generality, we can set $\bar{d} = 0$. Thus, with the above optimal choice of the d_j's, we have

$$\text{eff}(L) = 12 \left[\sum_{j=1}^{c} \lambda_{\cdot j}(a_j - \bar{a})^2\right] \left[\int f^2(x)dx\right]^2 . \tag{19.4.30}$$

Also note that (19.4.30) with $a_j = j$ coincides with the Pitman efficacy of Hettmansperger's (1975) test when $n_{ij} = n/bc$ and $n \to \infty$.

Comparison of L with the Parametric Competitor

Assuming normality for F, one can derive the likelihood derivative test (see also Knoke, 1975, for a reference on this) for testing $H_0 : \tau = 0$. The likelihood derivative test criterion, originally proposed by C.R. Rao, is the derivative of the likelihood function with respect to the parameter of interest evaluated at the null hypothesized value of the parameter and the maximum likelihood estimators of the nuisance parameters, if any, evaluated at the null hypothesis.

In the following, we will derive the likelihood derivative test for $H_0 : \tau_1 = \cdots = \tau_c = 0$ against $H_1 : \tau_j = a_j\tau$, not all a_j are the same and $\tau > 0$.

Let

$$X_{ijk} = \beta_i + a_j\tau + \epsilon_{ijk},\ k = 1, \ldots, n_{ij},\ i = 1, \ldots, b \text{ and } j = 1, \ldots, c \tag{19.4.31}$$

where ϵ_{ijk} are i.i.d. normal $(0, \sigma^2)$. Let

$$n = \sum_{i=1}^{b}\sum_{j=1}^{c} n_{ij}\,.$$

Then,

$$\begin{aligned} X_{ij\cdot} &= \beta_i + \alpha_j\tau + \epsilon_{ij\cdot}\,, \\ X_{i\cdot\cdot} &= \beta_i + \bar{a}\tau + \epsilon_{i\cdot\cdot}\,, \\ X_{\cdot j\cdot} &= \bar{\beta} + a_j\tau + \epsilon_{\cdot j\cdot}\,, \\ X_{\cdot\cdot\cdot} &= \bar{\beta} + \bar{a}.\tau + \bar{\epsilon}_{\cdot\cdot\cdot} \end{aligned}$$

where

$$\bar{a} = \sum_{1}^{c} n_{\cdot j} a_j / n$$

and

$$\epsilon_{\cdots} = \sum_{j=1}^{c} n_{\cdot j} \epsilon_{\cdot j \cdot} / n \,.$$

Then the likelihood of $\boldsymbol{\beta}$, τ and σ^2 is given by

$$L(\boldsymbol{\beta}, \tau, \sigma^2) = (2\pi\sigma^2)^{-n/2} \exp\left\{ -\frac{1}{2\sigma^2} \sum_{i=1}^{b} \sum_{j=1}^{c} \sum_{k=1}^{n_{ij}} (X_{ijk} - \beta_i - a_j \tau)^2 \right\},$$

$$l(\boldsymbol{\beta}, \tau, \sigma^2) = \log L = \frac{n}{2} \log(2\pi) - \frac{n}{2} \log \sigma^2 - \frac{1}{2\sigma^2} \sum_{1}^{b} \sum_{1}^{c} n_{ij} (X_{ij\cdot} - \beta_i - a_j \tau)^2 \,, \tag{19.4.32}$$

$$\begin{aligned} \frac{\partial l}{\partial \tau}\Big|_{\tau=0} &= \sum_{1}^{b} \sum_{1}^{c} (X_{ij\cdot} - \beta_i) a_j n_{ij} \\ &= \sum_{j=1}^{c} (X_{\cdot j \cdot} - \bar{\beta}) a_j n_{\cdot j} \,. \end{aligned}$$

Also,

$$\begin{aligned} \frac{\partial l}{\partial \beta_i}\Big|_{\tau=0} &= \sum_{i=1}^{b} \sum_{j=1}^{c} (X_{ij\cdot} - \beta_i) n_{ij} \\ &= \sum_{j=1}^{c} (X_{ij\cdot} - \beta_i) n_{\cdot j} = 0 \,. \end{aligned}$$

Thus,

$$\begin{aligned} \hat{\beta}_i &= \sum_{j=1}^{c} X_{\cdot j \cdot} n_{\cdot j} \Big/ \sum_{j=1}^{c} n_{\cdot j} \\ &= X_{\cdots} \,. \end{aligned} \tag{19.4.33}$$

Hence,

$$\begin{aligned}\frac{\partial l}{\partial \tau}|_{\tau=0} &= t = \sum_{j=1}^{c}(X_{.j.} - X_{...})n_{.j}a_j \\ &= \sum_{j=1}^{c}(X_{.j.} - X_{...})n_{.j}(a_j - \bar{a}) \end{aligned} \tag{19.4.34}$$

since $\sum n_{.j}(X_{.j.} - X_{...}) = 0$.

Efficacy of t

For evaluating the efficacy of t, let us assume that

$$\tau = n^{-1/2}\tau^* .$$

We can write t as

$$t = \sum_{j=1}^{c}(a_j - \bar{a})n_{.j}\left\{(\epsilon_{.j.} - \epsilon_{...}) + (a_j - \bar{a})\tau\right\} ,$$

$$E(t|H_0) = 0 \text{ and } E(t|H_1) = \sum_{j=1}^{c}(a_j - \bar{a})^2 n^{-1/2}\tau^* \tag{19.4.35}$$

and

$$\begin{aligned}\mathrm{var}(t|H_0) &= \sum_{1}^{c}(a_j - \bar{a})^2 n_{.j}^2 \mathrm{var}(\epsilon_{.j.} - \epsilon_{...}) + \sum_{j\neq k}^{c}\sum^{c}(a_j - \bar{a})(a_k - \bar{a})n_{.j}n_{.k} \\ &\quad \cdot \mathrm{cov}(\epsilon_{.j.} - \epsilon_{...}, \epsilon_{.k.} - \epsilon_{...}) .\end{aligned} \tag{19.4.36}$$

Now,

$$\begin{aligned}\mathrm{var}(\epsilon_{.j.} - \epsilon_{...}) &= \mathrm{var}(\epsilon_{.j.}) + \mathrm{var}(\epsilon_{...}) - 2\mathrm{cov}(\epsilon_{.j.}, \epsilon_{...}) \\ &= \frac{\sigma^2}{n_{.j}} + \frac{\sigma^2}{n} - \frac{2\sigma^2}{n} = \frac{\sigma^2}{n}\left(\frac{1-\lambda_{.j}}{\lambda_{.j}}\right) ,\end{aligned} \tag{19.4.37}$$

since $\epsilon_{...} = \sum_1^c n_{.j}\epsilon_{.j.}/n$.

$$\begin{aligned}\mathrm{cov}(\epsilon_{.j.} - \epsilon_{..}, \epsilon_{.k.} - \epsilon_{...}) &= \mathrm{cov}(\epsilon_{.j.}, \epsilon_{.k.}) - \mathrm{cov}(\epsilon_{.j.}, \epsilon_{...}) \\ &\quad - \mathrm{cov}(\epsilon_{.k.}, \epsilon_{...}) + \mathrm{var}\ \epsilon_{...} \\ &= 0 - \frac{\sigma^2}{n} - \frac{\sigma^2}{n} + \frac{\sigma^2}{n} = \frac{-\sigma^2}{n} .\end{aligned} \tag{19.4.38}$$

Consequently,

$$\begin{aligned}\operatorname{var}(t|H_0) &= \sum_1^c (a_j-\bar{a})^2 n_{\cdot j}^2\left(\frac{1-\lambda_{\cdot j}}{\lambda_{\cdot j}}\right)\frac{\sigma^2}{n} \\ &\quad -\sum_{j\neq k}^{c}\sum^{c}(a_j-\bar{a})(a_k-\bar{a})n_{\cdot j}n_{\cdot k}\frac{\sigma^2}{n} \\ &= n\sigma^2\left\{\sum(a_j-\bar{a})^2(1-\lambda_{\cdot j})\lambda_{\cdot j}-\sum_{j\neq k}\sum(a_j-\bar{a})(a_k-\bar{a})\lambda_{\cdot j}\lambda_{\cdot k}\right\} \\ &= n\sigma^2\left[\sum(a_j-\bar{a})^2\lambda_{\cdot j}-\left\{\sum(a_j-\bar{a})\lambda_{\cdot j}\right\}^2\right] \\ &= n\sigma^2\sum(a_j-\bar{a})^2\lambda_{\cdot j}\,. \end{aligned} \tag{19.4.39}$$

Thus, the efficacy of t is

$$\text{eff } t = \frac{\left\{\sum(a_j-\bar{a})^2\lambda_{\cdot j}\sqrt{n}\right\}^2}{n\sigma^2\left[\sum(a_j-\bar{a})^2\lambda_{\cdot j}\right]} = \sum(a_j-\bar{a})^2\lambda_{\cdot j}/\sigma^2\,, \tag{19.4.40}$$

where σ^2 denotes the variance of F. From (19.4.30) and (19.4.40), we obtain the asymptotic efficiency of L^* relative to t to be

$$ARE(L^*,t) = 12\sigma^2\left[\int_{-\infty}^{\infty} f^2(x)dx\right]^2 \tag{19.4.41}$$

which has been shown by Hodges and Lehmann (1956) to have a lower bound of 0.846.

In order to compare L with other nonparametric tests in randomized blocks with one observation per cell, we assume that $n_{ij}\equiv 1$ and set

$$P(X_{ij}\le a) = F(x-\beta_i-\tau_j)$$

where $\tau_j = j\tau$ and we let b, the number of blocks become large. Then the test statistic L takes the form of

$$L = \sum_{j=1}^{c} j\bar{R}_{\cdot j}\,.$$

Now, considering alternatives of the form $\tau_j = j\tau b^{-1/2}$, it can be shown that

$$\text{eff}(L) = c(c+1)(c-1)\left[\int_{-\infty}^{\infty} f^2(x)dx\right]. \tag{19.4.42}$$

Denoting the tests proposed by Jonckheere (1954), Page (1963) and Hollander (1967) by ξ, P and T, respectively, we find that

$$ARE(L,\xi) = (2c+5)/2(c+1)\,, \tag{19.4.43}$$
$$ARE(L,P) = (c+1)/c\,,$$

and

$$ARE(L,T) = \frac{\{3+2(c-2)\rho^*(F)\}\left[\int f^2(x)dx\right]^2}{2(c+1)\int g^2(x)dx}\,, \tag{19.4.44}$$

where

$$g(x) = \frac{d}{dx}G(x)\,,$$
$$G(x) = P(X_1 - X_2 \le x)\,,$$
$$\rho^*(F) = 12\eta(F) - 3\,,$$
$$\eta(F) = P(X_1 - X_2 < X_3 - X_4 \text{ and } X_1 - X_5 < X_6 - X_7)$$

and where $X_1, X_2, \ldots, X_7$ are i.i.d. random variables with continuous cdf $F(x)$. From (19.4.43) one can easily see that

$$1 \le \{ARE(L,\xi),\ ARE(L,P)\} \le 1.5\,.$$

M-GG (1986) tabulate the numerical values of $ARE(L,T)$ for certain well-known forms of F. From this, we infer that L performs favorably relative to Hollander's (1967) test. Perhaps this may be attributed to the fact that the ratios of the τ_j are known constants in the case of L.

19.5 Problems

19.2.1 For the data in 18.3.1, treating age groups as treatments and the years as blocks, carry out an F test for the equality of the treatment effects. Use $\alpha = 0.05$.

19.2.2 The World Almanac and the Fact Book (2003, p. 80) gives the deaths in the U.S. involving firearms by age in 1999.

	Age groups						
Cause	< 14	15–19	20–24	25–44	45–64	65–74	> 75
Unintentional	88	336	403	1824	1020	248	208
Suicides	103	975	1340	5718	4537	1791	2135
Homicides	282	1708	2330	4916	1266	193	133
Undetermined	15	68	59	108	42	15	17

Treat the age groups as blocks, and using an F test, test H_0 that there is no differences among the causes of deaths involving firearms. Use $\alpha = 0.05$.

19.3.1 For the data in Problem 19.2.2, carry out the Friedman's Test of H_0.

19.4.1 For the data in Problem 18.3.1, carry out the Page's test given by (19.4.5) with $\alpha = 0.05$. (HINT: Use the asymptotic normality of the test criterion.)

19.4.2 The World Almanac and Book of Facts (2003, p. 78) gives the following data pertaining to drug use in U.S. of high school seniors.

	1998	1999	2000	2001
Marijuana	49.1	49.7	48.8	49.0
Stimulants	16.4	16.8	15.6	16.2
Barbituates	8.7	8.9	9.2	8.7
Tranquilizers	8.5	9.3	8.9	9.2

Treating different drugs as treatments and the years as blocks, carry out Friedman's as well as Page's test using $\alpha = 0.05$.

Chapter 20

Rank Tests for Random Effects

20.1 Introduction

The random effects model naturally arises in several situations, especially when only certain levels of a treatment such as levels of a drug or amounts of a fertilizer are considered. The random effects model for the analysis of variance with the assumption of normality was considered by Scheffé (1959, Chapter 9). In this chapter, we will provide LMP rank tests for random effects in one-factor, two-factor (mixed) and balanced incomplete block designs. For one-factor experiments, the model is given by

$$X_{jk} = \mu + Y_j + \epsilon_{ik}, \; k = 1, \ldots, n_j, \; j = 1, \ldots, c \tag{20.1.1}$$

where Y_j and ϵ_{ij} are mutually independent random variables. Assuming the existence of the variance, the null hypothesis of interest is

$$H_0 : \text{var}(X_{ij}) = \text{var}(\epsilon_{ij}) \text{ for every } j \tag{20.1.2}$$

or equivalently, $\text{var}(Y_j) = 0$ for every j.

Without loss of generality, we can assume that Y_j and ϵ_{ij} have zero means. In the parametric case, normality of Y_j and ϵ_{ij} is assumed. Neyman (1967) considered the asymptotically 'optimal' $c(\alpha)$ tests for the one-sample problem (that is, when $n_j \equiv 1$ and $c = n$). Kulkarni (1969, 1970) obtained asymptotically 'optimal' $c(\alpha)$ tests for the c-sample problem when the underlying model is linear or nonlinear. However, the above mentioned $c(\alpha)$ tests assume specific form of the density function of the underlying variables. In the following, we say a little more about $c(\alpha)$ tests.

Typically, the hypotheses to be tested in applied research are composite, involving one or more nuisance parameters. Hence, the existing most powerful tests cannot readily be applied in such case for the following reasons. The distribution of the observable variables either does not conform to the form for which optimal tests can be found or the hypothesis to be tested and/or the alternatives are non-standard. In dealing with a null hypothesis that is composite, one of the main difficulties is finding a class of tests that would maintain, at least approximately, a specified level of significance, say α. This constitutes the problem of obtaining approximate similar regions. If such a class of tests, say $k(\alpha)$ is found, the next problem is to compute at least approximately, the power functions of each member of this class and having these functions, to determine the particular member of this class that is optimal in some sense. Neyman (1959) was able to construct a class of asymptotically optimal tests called $c(\alpha)$ tests which are applicable in a wide variety of situations. The rigorous asymptotic justification was given by Neyman (1959). A recipe for constructing such $c(\alpha)$ tests with several applications is given by Neyman and Scott (1965). The form of the underlying density is assumed to be known except for the unknown parameters. We wish to test a hypothesis about a particular parameter of interest and the rest of the parameters are considered to be nuisance parameters. Cramér's type of conditions on maximum likelihood estimation, square root of sample size consistency-estimates for the nuisance parameters and the stochastic independence of the logarithmic derivatives of the likelihood functions with respect to the nuisance parameters and the chosen test criterion when the null hypothesis of the parameter of interest holds, are some of the assumptions made by Neyman (1959).

Greenberg (1964) considered the following partially nonparametric one-way model:

$$X_{ij} = \mu + Y_i + \epsilon_{ij}, \ (i = 1, \ldots, \epsilon \text{ and } j = 1, \ldots, n),$$

where the Y_i's and ϵ_{ij}'s are assumed to be mutually independent, the Y_i's are normally distributed with mean zero and unknown variance σ_Y^2 and the ϵ_{ij}'s are i.i.d. according to an arbitrary continuous distribution F with density f and variance σ^2. The following practical situation might justify the above model.

Example 20.1.1. Consider a material such as steel which is manufactured in batches, a large number of bars being produced from each batch. The quality of the batches would be sufficiently controlled so that the assumption of normality of the batch effects is justified, while measurements on the individual bars produced might be expected to be subject to gross errors.

Thus, Greenberg (1964) develops a partially distribution-free test for H_0.

Because of the invariance property of the ranks under location shift of all the observations, we can set $\mu = 0$.

It is of interest to formulate the nonparametric version of the problem for one- and two-way experiments and derive locally most powerful (LMP) rank tests and study their distributions under H_0 and evaluate their asymptotic efficiencies relative to the relevant classical F tests.

20.2 LMP Tests for One-factor Models

Govindarajulu and Deshpande (1972) derived LMP rank tests for one-way random effects model. Let ϵ_{ij} be i.i.d. with distribution F_j and Y_j be independent with distribution G_j ($j = 1, \dots, c$). We will be interested in testing the null hypothesis

$$H_0 : G_j(y) = \begin{cases} 0 & \text{if } y < 0 \\ 1 & \text{if } y \geq 0,\ j = 1, \dots, c, \end{cases}$$

against the alternative hypothesis

$$H_1 : \text{at least one } G_j \text{ is nontrivial}.$$

In order to derive the LMP rank test, we consider the local alternative given by

$$H_\Delta : X_{jk} = \Delta Y_j + \epsilon_{jk} \text{ for small } \Delta,$$

where not all $E(Y_j)$ are the same.

Let $N = n_1 + \cdots + n_c$ and $W_1 < W_2 < \cdots < W_N$ denote the combined ordered sample and $Z = (Z_1, \dots, Z_N)$ denote the c-sample rank order, namely $Z_i = j$ if $W_i = X_{ik}$ for some $k = 1, \dots, n_j$. Let z be a possible realization of the $\prod_{i=1}^{c}(N_i!)$ possible rank orders. Then the following result was obtained by Govindarajulu and Deshpande (1972).

Result 20.2.1. *If (i) F has a density which is absolutely continuous on finite intervals, (ii) $f'(\cdot)$ is continuous at almost everywhere, (iii) G_j is such that $E|Y_j| < \infty$ for each j and the EY_i are not all equal, and (iv) $\int_{-\infty}^{\infty} |f'(t)|\, dt < \infty$, then the LMP rank test of H_0 versus H_Δ rejects H_0 when*

$$\sum_{j=1}^{c} E(Y_j) \sum_{k=1}^{n_j} E\left[\frac{-f'(W_{s_k^{(j)}})}{f(W_{s_k^{(j)}})}\right] \geq \tilde{k}(\alpha) \tag{20.2.1}$$

where α denotes the level of significance and $s_1^{(j)}, \ldots, s_{n_j}^{(j)}$ denote the ranks enjoyed by $(X_{j,1}, \ldots, X_{j,n_j})$.

The asymptotic normality of the test criterion in (20.2.1) when suitably standardized, follows from the Chernoff-Savage theorems or their generalizations.

Note that if all the random effects had the same expectations, none of our test statistics would be valid. This indicates that the preceding LMP rank test seems to be able to distinguish only among their locations. However, if

$$H_0 : X_{ik} = \epsilon_{jk}$$

and

$$H_\Delta : X_{ij} = \epsilon_{ik}(1 + \Delta Y_j),\ k = 1, \ldots, n_j \text{ and } j = 1, \ldots, c,$$

then the LMP rank test rejects H_0 when

$$\sum_{j=1}^{c} E(Y_j) \sum_{k=1}^{n_j} E\left[-W_{s_k^{(j)}} \frac{f'(W_{s_k^{(j)}})}{f(W_{s_k^{(j)}})}\right] > k(\alpha)\,. \tag{20.2.2}$$

When f is normal, the above test criterion is a linear combination of Capon (1961) type of test criterion for the problem of scales in the two-sample fixed-effects situation. Also note that each Y_j having a different distribution is tantamount to a Bayesian approach, whereas in the parametric case, the Y_j are assumed to be i.i.d. variables.

Govindarajulu (1975) derived LMP rank tests under the assumption that the Y_j are i.i.d. as G having zero mean. Then we wish to test

$$H_0 : G(y) = \left\{ \begin{array}{ll} 0 & \text{for } y < 0 \\ & \\ 1 & \text{for } y \geq 0\,, \end{array} \right\} \tag{20.2.3}$$

against the alternative

$$H_1 : G(y) \text{ is a nontrivial distribution}\,.$$

In order to derive an LMP rank test, we consider the local alternative hypothesis:

$$H_\Delta : X_{jk} = \Delta Y_{+}\epsilon_{jk} \text{ for small positive } \Delta\,.$$

Then we have the following result of Govindarajulu (1975).

Result 20.2.2. *If (i) the density f has a derivative that is absolutely continuous on finite intervals, (ii) $f''(x)$ is continuous almost everywhere, (iii) $EY_i^2 < \infty$ and (iv) $E\left|\frac{\partial^2}{\partial X^2}\ln f(X)\right| < \infty$, then the LMP rank test of H_0 against H_Δ is: Reject H_0 when*

$$T = \sum_{j=1}^{c}\left\{\sum_{i\neq k}^{N}\sum^{N} E_0\left[\frac{f'(W_i)f'(W_k)}{F(W_i)f(W_k)}\right]\delta_{j,z_i}\delta_{j,z_k} + \sum_{i=1}^{N} E_0\left[\frac{f''(W_i)}{F(W_i)}\right]\delta_{j,z_i}\right\} > K(\alpha) \tag{20.2.4}$$

where $K(\alpha)$ is determined by the level of significance α and $\delta_{j,k}$ denotes the Kronecker's delta (that is, $\delta_{j,k} = 1$ if $j = k$ and zero otherwise).

Special Cases

1. If f is the standard logistic density function, then

$$T = \sum_{j=1}^{c}\left[\sum_{i=1}^{N}\sum_{k=1}^{N} E\{(1-2U_i)(1-2U_k)\}\,\delta_{j,z_i}\delta_{j,z_k}\right] - 2\sum_{i=1}^{N} E\{U_i(1-U_i)\}\,, \tag{20.2.5}$$

   where U_i denotes the ith smallest standard uniform order statistic in a sample of size N.

2. If f is the standard normal density, then

$$\frac{T}{N}+1 = \sum_{j=1}^{c}\left[N^{-1/2}\sum_{i=1}^{N} E(V_i)\delta_{j,z_i}\right]^2 + \frac{1}{N}\sum_{j=1}^{c}\sum_{i=1}^{N}\sum_{k=1}^{N}\mathrm{cov}(V_i,V_k)\delta_{j,z_i}\delta_{j,z_k} \tag{20.2.6}$$

   where V_i denotes the ith smallest standard normal order statistic in a random sample of size N.

The preceding special cases can be embedded in a class of statistics of the form (except for the addition of a nonstochastic constant)

$$\frac{T}{N} = \frac{1}{N}\sum_{j=1}^{c}\sum_{i=1}^{N}\sum_{k=1}^{N} E_0\{g(W_i)g(W_k)\}\,\delta_{j,z_i}\delta_{j,z_k} \tag{20.2.7}$$

where

$$g(x) = \left\{ \begin{array}{ll} \text{sgn } x & \text{if } f \text{ is double exponential,} \\ 2F(x) - 1 & \text{if } f \text{ is logistic,} \\ x & \text{if } f \text{ is normal,} \\ \text{etc.} & \end{array} \right\}$$

and sgn $x = 1$, 0, or -1 according as $x > 0$, $x = 0$ or $x < 0$, respectively.

One can write

$$\frac{T}{N} = T^* + \tilde{T}$$

where

$$T^* = \sum_{j=1}^{c} \left[N^{-1/2} \sum_{i=1}^{N} E_0\left(g(W_i)\right) \delta_{j,z_i} \right]^2 \qquad (20.2.8)$$

$$\tilde{T} = \sum_{j=1}^{c} \tilde{T}_j,$$

$$\tilde{T}_j = N^{-1} \sum_{i=1}^{N} \sum_{k=1}^{N} \operatorname{cov}\left\{g(W_i), g(W_k)\right\} \delta_{j,z_i} \delta_{j,z_k} . \qquad (20.2.9)$$

Govindarajulu (1975, Lemma 3.3) has shown that

$$\tilde{T} - \sum_{j=1}^{c} \left\{n_j(n_j - 1)/N(N-1)\right\} \left[\operatorname{var}\left\{g(W)\right\}\right] \to 0 \qquad (20.2.10)$$

in probability as $N \to \infty$ in the three cases of interest to us. He has also obtained the following results.

Result 20.2.3. *Let* $c_{ik} = \operatorname{cov}\left[g(W_i), g(W_k)\right]$, $1 \le i, k \le N$. *If*

(i) $N^{-2} \sum^N \sum^N \sum^N{}_{i \ne k \ne l}\, c_{ik} c_{il} = o(1)$

and

(ii) $N^{-2} \sum^N \sum^N{}_{i \ne k}\, c_{ik}^2 = o(1)$,

then (20.2.10) holds.

Result 20.2.4. *If* $n_1 = \cdots = n_c$ *and* H_0 *is true, then* cT^*/I *is asymptotically distributed as noncentral chi-square with* $c-1$ *degrees of freedom, and noncentrality parameter* $\delta = (c-1)\eta^2$, $\eta = \int_0^1 J(u)du$ *where*

$$I = \int_0^1 J^2(u)du - \left[\int_0^1 J(u)du \right]^2, \; J(u) = gF^{-1}(u), \qquad (20.2.11)$$

and F^{-1} is the inverse of the distribution function of W. In particular, $\eta = 0$ if f is double exponential, normal or logistic, and

$$I = \begin{cases} 1 & \text{if } f \text{ is double exponential or normal,} \\ \frac{1}{3} & \text{if } f \text{ is logistic.} \end{cases}$$

Result 20.2.5. *Thus, in the three special cases of interest, the distribution of cT^*/I is asymptotically distributed as central chi-square with $c-1$ degrees of freedom.*

If the n_i are not all equal, if $\eta = \int_0^1 J(u)du = 0$ and if H_0 holds, then for sufficiently large N, $T^* \approx a\chi_d^2$ (in distribution) where

$$a = (1 - S_2)I/d\,,$$

$$d = (1 - S_2^2)/(S_2 + S_2^2 - 2S_3)$$

and

$$S_i = \sum_{j=1}^{c} \lambda_j^i,\ i = 1, 2, 3, \text{ and } \lambda_j = n_j/N\,.$$

Remark 20.2.1. *The above values for a and d are obtained by equating the mean and variance of T^* with those of $a\chi_d^2$.*

20.3 Asymptotic Distribution of Logistic Scores Test

The LMP rank test criterion for logistic f is given by (20.2.5) and can be rewritten as

$$\begin{aligned} \frac{T}{N} + \frac{1}{3} &= \sum_{j=1}^{c} \left[N^{-1/2} \sum_{i=1}^{N} E(1 - 2U_i)\delta_{j,z_i} \right]^2 \\ &+ \frac{4}{N} \sum_{j=1}^{c} \sum_{i=1}^{N} \sum_{k=1}^{N} \operatorname{cov}(U_i, U_k)\delta_{j,z_i}\delta_{j,z_k}\,. \end{aligned} \qquad (20.3.1)$$

Henceforth, assume that $n_1 = \cdots = n_c = n$. Also, let $R_k^{(j)}$ denote the rank of $X_{j,k}$ in the combined ranking of $X_{11}, X_{12}, \ldots, X_{nc}$. Since $\sum_{k=1}^{N} \delta_{j,z_k} =$

$n_j = n$ and $\sum_{k=1}^{N} k\delta_{j,z_k} = \sum_{k=1}^{n_i} R_k^{(j)}$ =sum of the ranks of the jth sample, we can write

$$\frac{1}{N}\sum_{j=1}^{c}\left[\sum_{i=1}^{N}\left(1-\frac{2i}{N+1}\right)\delta_{j,z_i}\right]^2 = \frac{4}{N(N+1)^2}\sum_{j=1}^{c}\left[\sum_{i=1}^{n} R_i^{(j)} - n\frac{(N+1)}{2}\right]^2 . \tag{20.3.2}$$

Next, consider

$$\begin{aligned}
&\frac{4}{N}\sum_{j=1}^{c}\sum_{i=1}^{N}\sum_{k=1}^{N}\operatorname{cov}(U_i,U_k)\delta_{j,z_i}\delta_{j,z_k} \\
&\quad = \frac{4}{N}\sum_{j=1}^{c}\sum_{i=1}^{N}(\operatorname{var} U_i)\delta_{j,z_i} \\
&\qquad + \frac{4}{N}\sum_{j=1}^{c}\left\{\sum\sum_{i<k} + \sum\sum_{i>k}\right\}\operatorname{cov}(U_i,U_k)\delta_{j,z_i}\delta_{j,z_k} \\
&\quad = T_1 + T_2 \text{ (say)},
\end{aligned}$$

where

$$\begin{aligned}
T_1 &= \frac{4}{N}\sum_{i=1}^{N}\frac{i(N+1-i)}{(N+1)^2(N+2)} \\
&= \frac{4}{N(N+1)(N+2)}\cdot\frac{N(N+1)}{2} - \frac{4}{N(N+1)^2(N+2)}\cdot\sum_{i=1}^{N} i^2 \\
&= \frac{2}{N+2} - \frac{4}{N(N+1)^2(N+2)}\sum_{1}^{N} i^2 .
\end{aligned} \tag{20.3.3}$$

Note that we deliberately left the second term on the right hand side of (20.3.2) without further simplification because it gets canceled with another term in T_2. Now we can write T_2 as

$$\begin{aligned}
T_2 &= \frac{4}{N(N+1)^2(N+2)} \sum_{j=1}^{c} \left[\sum\sum_{i<k} i(N+1-k) \right. \\
&\quad \left. + \sum\sum_{i>k} k(N+1-i) \right] \delta_{j,z_i}\delta_{j,z_k} \\
&= \frac{8}{N(N+1)^2(N+2)} \sum_{j=1}^{c} \sum\sum_{i<k} i(N-k+1)\delta_{j,z_i}\delta_{j,z_k} \\
&= \frac{8}{N(N+1)(N+2)} \sum_{j=1}^{c} \sum\sum_{i<k} i\delta_{j,z_i}\delta_{j,z_k} \\
&\quad - \frac{8}{N(N+1)^2(N+2)} \sum_{j=1}^{c} \sum\sum_{i<k} ki\delta_{j,z_i}\delta_{j,z_k} \\
&= S_1^* - S_2^* \text{ (say)} \qquad (20.3.4)
\end{aligned}$$

where

$$\begin{aligned}
S_1^* &= \frac{8}{N(N+1)(N+2)} \sum_{j=1}^{c} \sum_{i=1}^{N} i\delta_{j,z_i} \left(n - \sum_{k=1}^{i} \delta_{j,z_k} \right) \\
&= \frac{8n}{N(N+1)(N+2)} \left\{ \frac{N(N+1)}{2} - \sum_{i=1}^{c} \sum_{i=1}^{N} i\delta_{j,z_i} F_n^{(j)}(W_i) \right\}
\end{aligned}$$

since $\sum_{k=1}^{i} \delta_{j,z_k}$ = number of observations from the jth sample that are less than or equal to W_i.

Next consider

$$\begin{aligned}
S_2^* &= \frac{4}{N(N+1)^2(N+2)} \sum_{j=1}^{c} \sum\sum_{i \neq k} ik\delta_{j,z_i}\delta_{j,z_k} \\
&= \frac{4}{N(N+1)^2(N+2)} \left\{ \sum_{j=1}^{c} \left(\sum_{i=1}^{N} i\delta_{j,z_i} \right)^2 - \sum_{i=1}^{N} i^2 \sum_{j=1}^{c} \delta_{j,z_i} \right\} \\
&= \frac{4}{N(N+1)^2(N+2)} \sum_{j=1}^{c} \left\{ \sum_{i=1}^{N} i\delta_{j,z_i} - \frac{n(N+1)}{2} + \frac{n(N+1)}{2} \right\}^2 \\
&\quad - \frac{4}{N(N+1)^2(N+2)} \sum_{i=1}^{N} i^2 \\
&= \frac{4}{N(N+1)^2(N+2)} \sum_{j=1}^{c} \left\{ \sum_{i=1}^{N} \left(i - \frac{N+1}{2} \right) \delta_{j,z_i} \right\}^2 + \frac{n}{N+2} \\
&\quad - \frac{4}{N(N+1)^2(N+2)} \sum_{i=1}^{N} i^2 . \qquad (20.3.5)
\end{aligned}$$

Thus, using (20.3.2) – (20.3.5) in (20.3.1) and combining like terms, we have

$$\begin{aligned}
\frac{T}{N} + \frac{1}{3} - \frac{(3n+2)}{N+2} &= \frac{4}{N(N+1)(N+2)} \sum_{j=1}^{c} \left\{ \sum_{i=1}^{N} \left(i - \frac{N+1}{2} \right) \delta_{i,z_i} \right\}^2 \\
&\quad - \frac{8n}{N(N+1)(N+2)} \sum_{j=1}^{c} \sum_{i=1}^{N} iF_n^{(j)}(W_i)\delta_{j,z_i} .
\end{aligned} \qquad (20.3.6)$$

Further, the second term on the right-hand side can be written in the integral form as

$$\begin{aligned}
&\frac{n}{N(N+1)(N+2)} \sum_{j=1}^{c} \sum_{i=1}^{N} iF_n^{(j)}(W_i)\delta_{j,z_i} \\
&= \frac{n^2}{(N+1)(N+2)} \sum_{j=1}^{c} \int_{-\infty}^{\infty} F_n^{(j)}(x)H_N(x)dF_n^{(j)}(x)
\end{aligned}$$

where

$$\begin{aligned} H_N(x) &= c^{-1} \sum_{j=1}^{c} F_n^{(j)}(x) \\ &= \text{empirical d.f. of the combined sample of size } N. \end{aligned}$$

Now, assume a sequence of d.f.'s $G^{(n)}(x)$ for Y_j such that the following conditions are satisfied:

(a) $\text{var } Y_j = O(n^{-1})$,

(b) $E[Y_j/\sigma_Y]^4 = O(1)$,

(c) $\|f\| = \sup_x |f(x)| \le k_1 < \infty$ and

(d) $\|f'\| = \sup_x |f'(x)| \le k_2 < \infty$.

When (20.3.7) is satisfied, Shetty and Govindarajulu (1988) have shown that

$$\frac{n}{N(N+1)(N+2)} \sum_{j=1}^{c} \sum_{i=1}^{n} iF_n^{(j)}(W_i)\delta_{j,z_i} \to \frac{1}{3c}$$

in probability as $n \to \infty$.

Next, using empirical distribution functions, we can write

$$\begin{aligned} & 2\{N(N+1)(N+2)\}^{-1/2} \sum_{i=1}^{N} \left(i - \frac{N+1}{2}\right) \delta_{j,z_i} \\ &\approx 2N^{-1/2} \sum_{i=1}^{N} \left(\frac{i}{N+1} - \frac{1}{2}\right) \delta_{j,z_i} \\ &= 2nN^{-1/2} \int \left(H_N(x) - \frac{1}{2}\right) dF_n^{(j)}(x) \\ &= 2nN^{-1/2} \left[\int \left(H(x) - \frac{1}{2}\right) dF^{(j)}(x) \right. \\ &\quad \left. + \int \left(H(x) - \frac{1}{2}\right) d(F_n^{(j)}(x) - F^{(j)}(x))\right] \\ &\quad + \int (H_N(x) - H(x))\, dF^{(j)}(x) + R_{N,j} \qquad (20.3.7) \end{aligned}$$

where

$$H(x) = \frac{1}{c} \sum_{j=1}^{c} F^{(j)}(x), \quad H_N(x) = \frac{1}{c} \sum_{j=1}^{c} F_n^{(j)}(x)$$

and

$$R_{N,j} = 2nN^{-1/2} \int [H_N(x) - H(x)]\, d\left(F_n^{(j)}(x) - F^{(i)}(x)\right). \qquad (20.3.8)$$

Shetty and Govindarajulu (1988, pp. 179–181) have shown that $R_{N,j}$ tends to zero in probability as n tends to ∞. Furthermore,

$$\begin{aligned} \int \left\{H(x) - \frac{1}{2}\right\} dF^{(j)}(x) &= c^{-1} \sum_{k=1}^{c} \int \left\{F^{(k)}(x) - \frac{1}{2}\right\} dF^{(j)}(x) \\ &= c^{-1} \sum_{k=1}^{c} \int \left\{F^{(j)}(x) - \frac{1}{2}\right\} dF^{(j)}(x) = 0 \end{aligned}$$

since

$$F^{(1)}(x) = \cdots = F^{(c)}(x). \qquad (20.3.9)$$

Thus,

$$\begin{aligned} & 2\{N(N+1)(N+2)\}^{-1/2} \sum_{i=1}^{N} \left(i - \frac{N+1}{2}\right) \delta_{j,z_i} \\ \sim\ & 2nN^{-1/2} \left[\int \left(H(x) - \frac{1}{2}\right) d(F_n^{(j)} - F^{(j)}) + \int (H_N - H) dF^{(j)}(x)\right] \\ =\ & 2nN^{-1/2} \left[\int (H_N - H) dF^{(j)}(x) - \int (F_n^{(j)} - F^{(i)}) dH(x)\right] \\ =\ & \frac{2n}{cN^{1/2}} \sum_{\substack{k=1 \\ k \neq j}}^{c} \left[\int (F_n^{(k)} - F^{(k)}) dF^{(j)}(x) - \int (F_n^{(j)} - F^{(j)}) dF^{(k)}(x)\right]. \end{aligned} \qquad (20.3.10)$$

Writing

$$n \int F_n^{(k)}(x) dF^{(i)}(x) = \sum_{i=1}^{n} \left\{1 - F^{(j)}(X_{k,i})\right\},$$

$$n \int F_n^{(j)}(x) dF^{(k)}(x) = \sum_{i=1}^{n} \left\{1 - F^{(k)}(X_{j,i})\right\},$$

and since the marginal distributions $F^{(1)}, F^{(2)}, \ldots$ are identical, we can write

$$2\{N(N+1)(N+2)\}^{-1/2} \sum_{i=1}^{N} \left(i - \frac{N+1}{2}\right) \delta_{j,z_i} \approx c^{-1/2} B_{N,j} \text{ (say)}$$

where

$$c^{-1/2}B_{N,j} = \frac{2}{cN^{1/2}} \sum_{\substack{k=1 \\ k\neq j}}^{c} \sum_{i=1}^{n} \left\{ \left(F^{(k)}(X_{j,i}) - \frac{1}{2} \right) - \left(F^{(j)}(X_{k,i}) - \frac{1}{2} \right) \right\}, \tag{20.3.11}$$

which is a sum of i.i.d. random variables and consequently will have an asymptotic normal distribution. Then, under the assumptions in (20.3.7), Shetty and Govindarajulu (1988, p. 185) have shown that for sufficiently large n, $(c^{-1/2}B_{N,1}, \ldots, c^{-1/2}B_{N,c})$ has a multi-variate normal distribution with mean vector 0 and variance-covariance matrix $\sum$ where

$$\sum = \frac{1}{3}\{I - c^{-1}J\}\text{var}(Z^*),$$

I is a $c \times c$ identity matrix, J is a $c \times c$ matrix of 1's and

$$Z^* = n^{-1/2}(12)^{1/2} \sum_{i=1}^{n} \left(F^{(k)}(X_{ji}) - \frac{1}{2} \right) \text{ for } k \neq j. \tag{20.3.12}$$

Also note that the $B_{N,j}$ are subject to one linear constraint, namely that

$$\sum_{j=1}^{c} B_{N,j} = 0.$$

Now let

$$T^* = \sum_{j=1}^{c} \{c^{-1/2}B_{N,j}\}^2.$$

Then we have the main result of Shetty and Govindarajulu (1988).

Theorem 20.3.1. *Under the assumption (20.3.7), $3cT^*/var\, Z^*$ is asymptotically distributed as central chi-square with $c - 1$ degrees of freedom.*

PROOF: Use ANOVA transformation on the $B_{N,j}$ or Corollary 23.6.1.1.

Remark 20.3.1. *Since $F^{(1)} = F^{(2)} = \cdots = F^{(c)}$, Z^* can be construed as the average of standardized uniform $(0,1)$ variables for the following reason. Note that*

$$X_{j1} = \epsilon_{j1} + Y_j \text{ and } X_{k1} = \epsilon_{k1} + Y_k$$

are identically distributed. However, $F^{(k)}(X_{k1})$ is a uniform $(0,1)$ random variable. Hence, for $1 \leq j,k \leq c$,

$$E\left\{F^{(k)}(X_{ji})\right\} = \frac{1}{2}$$

and

$$var\left\{F^{(k)}(X_{ji}\right\} = \frac{1}{12} \text{ for } i = 1,\ldots,n. \tag{20.3.13}$$

Also we have the following result.

Result 20.3.1. *Under the assumption (20.3.7), we have*

$$\begin{aligned} var\, Z^* &= 12\left[\frac{1}{12} + (n-1)\text{cov}\left(F^{(k)}(X_{j1}), F^{(k)}(X_{j2})\right)\right] \\ &= 12\left[\frac{1}{12} + (n-1)\sigma_Y^2\left\{L^2 + O(n^{-1/2})\right\}\right]. \end{aligned}$$

PROOF:

$$\begin{aligned} &\text{cov}\left\{F^{(k)}(X_{j1}), F^{(k)}(X_{j2})\right\} \\ &= E\left[\left\{F^{(k)}(X_{j1}) - \frac{1}{2}\right\}\left\{F^{(k)}(X_{j2}) - \frac{1}{2}\right\}\right] \\ &= E\left[\int\left\{F^{(k)}(X_{j1} - t) - \frac{1}{2}\right\}dG(t)\right]\left[\int\left\{F^{(k)}(X_{j2} - s) - \frac{1}{2}\right\}dG(s)\right] \\ &= E\left[\int\int\left\{F(x + Y_j - t) - \frac{1}{2}\right\}dF(x)dG(t)\right] \\ &\quad \cdot\left[\int\left\{F^{(k)}(X_{j2} - s) - \frac{1}{2}\right\}dG(s)\right]. \end{aligned} \tag{20.3.14}$$

Now expand $F(x + Y_j - t)$ as follows:

$$F(x + Y_j - t) = F(x) + (Y_j - t)f(x) + \frac{1}{2}(Y_j - t)^2 f'(x^*) \tag{20.3.15}$$

where x^* lies between $x + Y_j - t$ and x. Then we obtain

$$\int\left\{F(x + Y_j - t) - \frac{1}{2}\right\}dF(x) = (Y_j - t)L + \frac{1}{2}\int(Y_j - t)^2 f'(x^*)dF(x)$$

and

$$\int \left\{ F(x + Y_j - t) - \frac{1}{2} \right\} dF(x)dG(t) = Y_j L + \frac{1}{2} \int\int (Y_j - t)^2 f'(x^*) dF(x)dG(t) .$$

Hence,

$$\left[\int\int \left\{ F(x + Y_j - t) - \frac{1}{2} \right\} dF(x)dG(t) \right]^2 = Y_j^2 L^2 + LY_j \int\int (Y_j - t)^2 f'(x^*) dF(x)dG(t) + \left[\frac{1}{2} \int\int (Y_j - t)^2 f'(x^*) dF(x)dG(t) \right]^2 \tag{20.3.16}$$

where $L = \int f^2(x)dx$ and $\sigma_Y^2 = \text{var } Y_j$. Thus,

$$\begin{aligned}
\text{cov}\left\{ F^{(k)}(X_{j1}), F^{(k)}(X_{j2}) \right\} - \sigma_Y^2 L^2 &\leq \|f'\| \, LE \int\int |Y_j|(Y_j - t)^2 dG(t) \\
&\quad + \frac{1}{4} \|f'\|^2 E \left[\int (Y_j - t)^2 dG(t) \right]^2 \\
&= \|f'\| LE \left\{ |Y_j|(Y_j^2 - 2Y_j Y_k + Y_k^2) \right\} \\
&\quad + \frac{1}{4} \|f'\|^2 E \left[Y_j^2 + \sigma_Y^2 \right]^2 \\
&= \|f'\| L \left\{ E|Y_j|^3 + \sigma_Y^2 E|Y_j| \right\} \\
&\quad + \frac{1}{4} \|f'\|^2 E \left[(Y_j^2 + \sigma_Y^2)^2 \right] \\
&\leq \|f'\| L \left\{ E|Y_j|^3 + \sigma_Y^3 \right\} \\
&\quad + \frac{1}{4} \|f'\|^2 \left[E(Y_j)^4 + 3\sigma_Y^4 \right] \\
&\leq K\sigma_Y^3 ,
\end{aligned}$$

since

$$E|Y_j|^3 \leq \left\{ E|Y_j|^4 \right\}^{3/4} = \sigma_Y^3 \left\{ \frac{\mu_4}{\sigma_Y^4} \right\}^{3/4} = O(\sigma_Y^3) .$$

Hence,

$$\sigma_Y^{-2}\text{cov}\left(F^{(k)}(X_{j1}), F^{(k)}(x)\right) - L^2 \leq O(1/\sigma_Y)\,. \tag{20.3.17}$$

Consequently,

$$\text{var } Z^* = \left[1 + 12(n-1)L^2\sigma_Y^2\left(1 + O(n^{-1/2})\right)\right]\,. \tag{20.3.18}$$

20.4 Asymptotic Distribution of F-test

As before, consider the one-way random effects model given by

$$X_{j,i} = \eta + Y_j + \epsilon_{ji},\ i = 1, \ldots, n \text{ and } j = 1, \ldots, c$$

where $\{Y_j\}$, $\{\epsilon_{ji}\}$ are completely independent with zero means, ϵ_{ji} are normal (O, σ^2) and Y_j are identically normally distributed with zero mean and variance σ_Y^2. The usual hypothesis that is tested in the present one-way set-up is H_0: $\sigma_Y^2 = 0$. The parametric test criterion is the well-known F-test based on

$$F = \{SS_Y/(c-1)\}\,/\,\{SS_e/c(n-1)\} \tag{20.4.1}$$

where

$$SS_Y = n\sum_{j=1}^{c}(\bar{X}_{j\cdot} - \bar{X}_{\cdot\cdot})^2,\quad SS_e = \sum_i\sum_j(X_{ji} - \bar{X}_{j\cdot})^2,$$

$$\bar{X}_{j\cdot} = n^{-1}\sum_{i=1}^{n}X_{ji},\ \bar{X}_{\cdot\cdot} = \sum_{j=1}^{c}\sum_{i=1}^{n}X_{jl}/nc\,. \tag{20.4.2}$$

Under the present model,

$$\bar{X}_{j\cdot} = \eta + Y_j + \bar{\epsilon}_{j\cdot},\quad \bar{X}_{\cdot\cdot} = \eta + \bar{Y}_{\cdot} + \bar{\epsilon}_{\cdot\cdot} \tag{20.4.3}$$

where $\bar{\epsilon}_{j\cdot}$ and $\bar{\epsilon}_{\cdot\cdot}$ are analogously defined on the ϵ_{ji}. Substituting for $\bar{X}_{j\cdot}$ and $\bar{X}_{\cdot\cdot}$ in the expressions for SS_Y and SS_e and writing $K_j = Y_j + \bar{\epsilon}_{j\cdot}$, we have

$$SS_Y = \sum_{j=1}^{c}(K_j - \bar{K}_{\cdot})^2$$

where the random variables $\{K_j\}$ are independent normal (O, σ_K^2) variables with $\sigma_K^2 = \sigma_Y^2 + n^{-1}\sigma^2$. Hence,

$$\sum_{j=1}^{c}(K_j - \bar{K}_{\cdot})^2/\sigma_K^2 \overset{d}{=} \chi_{c-1}^2$$

and, consequently,

$$SS_Y = n\sigma_K^2\chi_{c-1}^2 = (n\sigma_Y^2 + \sigma^2)\chi_{c-1}^2 \text{ (in distribution)}. \tag{20.4.4}$$

On the other hand,

$$SS_e/\sigma^2 = \sum_i \sum_j (\epsilon_{ji} - \bar{\epsilon}_{j.})^2/\sigma^2 \stackrel{d}{=} \chi_{(n-1)c}^2 \,. \tag{20.4.5}$$

Hence, $MS_e = SS_e/(n-1)c$ converges to its expected value σ^2 in probability since its variance $2\sigma^4/(n-1)c$ tends to zero as $n \to \infty$. Now, using Slutsky's theorem, we infer that F is asymptotically distributed as

$$\left\{(n\sigma_Y^2 + \sigma^2)/(c-1)\sigma^2\right\}\chi_{c-1}^2 = (c-1)^{-1}(1 + n\sigma_Y^2/\sigma^2)\chi_{c-1}^2 \,. \tag{20.4.6}$$

Under H_0, $F \to (c-1)^{-1}\chi_{c-1}^2$ and consequently we reject H_0 when $F > K_\alpha = (c-1)^{-1}\chi_{c-1,1-\alpha}^2$. Thus, the power at σ_A^2 is

$$\begin{aligned} \pi(\sigma_Y^2) &= P\left[\left\{\frac{n\sigma_Y^2 + \sigma^2}{(c-1)\sigma^2}\right\}\chi_{c-1}^2 > (c-1)^{-1}\chi_{c-1,1-\alpha}^2\right] \\ &= P\left[\chi_{c-1}^2 > \left\{\sigma^2/(n\sigma_Y^2 + \sigma^2)\right\}\chi_{c-1,1-\alpha}^2\right] \,. \end{aligned} \tag{20.4.7}$$

Asymptotic Efficiency of T^* relative to F

$ARE(T^*, F)$ is the limiting ratio of sample sizes required to achieve the same limiting power against the same Pitman alternatives when the limiting significance levels of the two test procedures are equal. Then we have the following theorem.

Theorem 20.4.1. *Under the assumption (20.3.7), we have*

$$ARE(T^*, F) = 12\sigma^2\left[\int f^2(x)dx\right]^2 .$$

PROOF: Note that because of Theorem 20.3.1, we reject H_0 when $T > T_\alpha$ where $T_\alpha = (3c)^{-1}\chi_{c-1,1-\alpha}^2$.

Asymptotic power of $T^* = P_{H_A}\left(T^* > (3c)^{-1}\chi_{c-1,1-\alpha}^2\right)$

$$= P\left(\chi_{c-1}^2 > \frac{\chi_{c-1,1-\alpha}^2}{\text{var } Z^*}\right) \tag{20.4.8}$$

where Z^* is defined by (20.3.13) and var Z^* is given by (20.3.19),

$$\text{var } Z^* = 12\left[\text{var } F^{(k)}(X_{j1}) + (n'-1)\text{cov}\left(F^{(k)}(X_{j1}), F^{(k)}(X_{j2})\right)\right] \quad (20.4.9)$$

where n' is the sample size on each treatment required by the test procedure based on T^*. Since var $F^{(k)}(X_{j1}) = 1/12$, from (20.4.7) and (20.4.8), equating the powers of the two test procedures, we have

$$n\sigma_Y^2/\sigma^2 = 12(n'-1)\text{cov}\left(F^{(k)}(X_{j1}), F^{(k)}(X_{j2})\right).$$

That is,

$$\frac{n}{n'-1} = 12\sigma^2\text{cov}\left\{F^{(k)}(X_{j1}), F^{(k)}(X_{j2})\right\}/\sigma_Y^2. \quad (20.4.10)$$

Then the proof of Theorem 20.4.1 will be complete if

$$\text{cov}\left\{F^{(k)}(X_{j1}), F^{(k)}(X_{j2})\right\} \approx \sigma_Y^2\left(\int f^2(x)dx\right)^2, \text{ as } n \to \infty, \quad (20.4.11)$$

which has been shown in (20.3.18).

Remark 20.4.1. *In Shetty and Govindarajulu (1988) it was shown that*

$$\text{cov}\left\{F^{(k)}(X_{j1}), F^{(k)}(X_{j2})\right\} - L^2\sigma_Y^2 \to 0 \text{ as } n \to \infty$$

assuming only that $\sigma_Y^2 = O(n^{-1})$. However, their computation of the covariance on p. 189 is slightly in error, and moreover, what is needed is

$$\sigma_Y^{-2}\text{cov}\left(F^{(k)}(X_{j1}), F^{(k)}(X_{j2})\right) - L^2 \to 0 \text{ as } n \to \infty.$$

Towards this, an additional assumption on $E(Y^4)$ is required and this is imposed in (20.3.7).

20.5 Null Distribution and Power Considerations

For computational purposes, one can rewrite T given by (20.3.1) as

$$\begin{aligned} T &= N + \sum_{j=1}^{c} n_j^2 - \frac{2N(2N+1)}{3(N+2)} + \frac{4}{(N+1)(N+2)}\sum_{j=1}^{c} S_j^2 \\ &\quad - \frac{4}{N-1}\sum_{j=1}^{c} n_j S_j + \frac{8}{(N+1)(N+2)}\sum_{j=1}^{c}\sum_{k=1}^{n_j}(n_j-k)\tilde{R}_{j,k} \end{aligned} \quad (20.5.1)$$

where $\tilde{R}_{j,k}$ denotes the rank in the combined sample of the kth smallest observation which received the jth treatment for $k = 1, \ldots, n_j$ and $j = 1, \ldots, c$. When $n_j \equiv n$, $\sum_{j=1}^{c} S_j = nc(nc+1)/2$ and consequently T takes the simpler form

$$T = -nc(n-1)\frac{2nc(2nc+1)}{3(nc+2)} + \frac{4}{(nc+1)(nc+2)} \sum_{j=1}^{c} \left\{ S_j^2 - 2\sum_{i=1}^{n} i\tilde{R}_{ji} \right\}. \tag{20.5.2}$$

Clemmens and Govindarajulu (1990) calculate the critical values of T for $\alpha = 0.01$, 0.05 and 0.10, $c = 2$, $N = 4(1)9$; $c = 3$, $N = 3(1)6$ and simulated critical values for $c = 2$, $n = 8(1)18$; $c = 3, 4, 5, 6$, $n = 2(1)15$. In the tables they retain only those sample sizes beyond which the asymptotic theory is meaningful. They also make the power comparison of the test for two populations with the classical F-test under a range of normal alternatives. If α is sufficiently large (say ≥ 0.10) the relative percentage difference of the power of T relative to F-test is positive when $\sigma_Y/\sigma \geq 3.5$. For $c > 2$, we expect the same or better behavior of T relative to F because the Student's t-test (when $c = 2$) has more optimal properties than the F-test (when $c > 2$).

20.6 LMP Tests in Two-way Layouts

The use of blocks in a design is an attempt to remove a source of variability in the observations and hence makes it possible to obtain a more accurate evaluation of the factor of interest. Then the model one would use is

$$X_{ijk} = \mu + \beta_i + Y_j + \epsilon_{ijk},\ k = 1, \ldots, n_{ij},\ j = 1, \ldots, c \text{ and } i = 1, \ldots, b,$$

where Y_j and ϵ_{ijk} are mutually independent. The null hypothesis we want to test is

$$H_0 : \text{var } Y = 0\,.$$

Assuming that the ϵ_{ijk} have a probability density function f, Clemmens and Govindarajulu (1994) derive an LMP rank test of H_0 under some regularity conditions on f and on the distribution of Y_j. The test criterion when specialized to the logistic scores (i.e., when f is standard logistic density) is given by

$$\begin{aligned}\Psi_L &= \sum_{j=1}^{c} n_{\cdot j}^2 - \sum_{i=1}^{b} \frac{2n_{i\cdot}(2n_{i\cdot}+1)}{3(n_{i\cdot}+2)} + 4\sum_{j=1}^{c}\left[\sum_{i=1}^{b} \frac{S_j^{(i)}}{n_{i\cdot}+1}\right]^2 \\ &\quad -4\sum_{i=1}^{b}\sum_{j=1}^{c} \frac{n_{\cdot j} S_j^{(i)}}{n_{i\cdot}+1} - 4\sum_{i=1}^{b}\sum_{j=1}^{c} \frac{S_j^{(i)2}}{(n_{i\cdot}+1)^2(n_{i\cdot}+2)} \\ &\quad +8\sum_{i=1}^{b}\sum_{j=1}^{c}\sum_{k=1}^{n_{ij}} \frac{(n_{ij}-k)R_{jk}^{(i)}}{(n_{i\cdot}+1)(n_{i\cdot}+2)} \end{aligned} \tag{20.6.1}$$

where $S_j^{(i)}$ is the sum of the ranks associated with the observations getting the jth treatment in the ith block, $R_{jk}^{(i)}$ is the kth smallest of the ranks assigned to $(i,j)^{\text{th}}$ cell when all observations in the ith block are ranked separately (for $i = 1, \ldots, b$) and

$$n_{i\cdot} = \sum_{j=1}^{c} n_{ij},\ n_{\cdot j} = \sum_{i=1}^{b} n_{ij}\,.$$

Further simplification results when $n_{ij} \equiv n$. Then

$$\begin{aligned}\Psi_L &= bnc - b^2n^2c - \frac{2bnc(2nc+1)}{3(nc+2)} + \frac{4bcn^2}{nc+2} \\ &\quad + \frac{4}{(nc+1)}\sum_{j=1}^{c}\left[\sum_{i=1}^{b} S_j^{(i)}\right]^2 - \frac{4}{(nc+1)^2(nc+2)}\sum_{j=1}^{c}\sum_{i=1}^{b} S_j^{(i)2} \\ &\quad - \frac{8}{(nc+1)(nc+2)}\sum_{j=1}^{c}\sum_{i=1}^{b}\sum_{k=1}^{n} kR_{jk}^{(i)}\,. \end{aligned} \tag{20.6.2}$$

The following properties have been established by Clemmens and Govindarajulu (1994).

Theorem 20.6.1.

i. *$E(\Psi_L|H_0) = 0$ and as $n \to \infty$, under H_0,*

ii. *$\frac{3c\Psi_L^*}{b} = \frac{3\Psi_L}{bn} - \frac{c(n-1)}{(nc-1)} + c$ is distributed as χ_{c-1}^2.*

PROOF: is analogous to the Proof of Result 20.2.4.

Tables of critical values of Ψ_L are provided for $b = 2,3,4$, $c = 2,3,4$ and $n = 2(1)10$ for $\alpha = 0.01$, 0.05 and 0.10. The most remarkable feature is the rapidity with which the critical values of $3c\Psi_L^*/b$ approach the corresponding points of the central chi-square distribution with $c-1$ degrees of freedom, especially when $\alpha = 0.10$ and $n \geq 10$. We conjecture that the power of the two-way test will be comparable to the power of the F-test for normal alternatives.

20.7 LMP Tests in Block Designs

Nugroho and Govindarajulu (1999) derive LMP rank tests for no treatment effect when there is only one observation in each cell of a randomized block design. Then the model is

$$X_{ij} = \mu + \beta_i + Y_j + \epsilon_{ij},\ i = 1,\ldots,b \text{ and } j = 1,\ldots,c, \tag{20.7.1}$$

with the usual assumptions on Y_j, ϵ_j and β_i. Then Nugroho and Govindarajulu (1999) derive LMP rank test criterion for H_0: var $Y = 0$ against local alternatives which, when specialized to the case of logistic density for the ϵ_{ij}, takes the form of

$$\Psi_L = 4(c+1)^{-2}\sum_{j=1}^{c}\left[\sum_{i=1}^{b} R_{ij}\right]^2 - b^2c - \frac{bc(c-1)}{3(c+1)} \tag{20.7.2}$$

where R_{ij} denotes the rank of X_{ij} among $X_{i1},\ldots,X_{ic}$ (for $i = 1,\ldots,b$). note that a linear version of this could have been gotten as a special case of (20.6.1) or (20.6.2) with $n_{ij} = n = 1$. The Friedman's (1937) test criterion for the null hypothesis of no difference among c treatments in the two-way fixed effects model by ranks is given by

$$F_R = \frac{12}{bc(c+1)}\sum_{j=1}^{c}\left[\sum_{i=1}^{b}\left(R_{ij} - \frac{c+1}{2}\right)\right]^2. \tag{20.7.3}$$

However, Ψ_L and F_R are linearly related as follows:

$$\Psi_L = \frac{bc}{3(c+1)}F_R - \frac{bc(c-1)}{3(c+1)}. \tag{20.7.4}$$

Hence, we conclude that Friedman's test criterion is LMP for H_0 against logistic alternatives.

Since $R_{1j}, R_{2j}, \ldots, R_{bj}$ are i.i.d. uniformly distributed under H_0, taking values $1, \ldots, c$. Thus,

$$E(R_{ij}|H_0) = (c+1)/2 \text{ and } \mathrm{var}(R_{ij}|H_0) = (c^2-1)/12\,.$$

Further, for large b,

$$\left(\frac{12}{b(c^2-1)}\right)^{1/2} \sum_{i=1}^{b} \left(R_{ij} - \frac{c+1}{2}\right)$$

is distributed as $N(0,1)$. Since

$$\sum_{j=1}^{c} \sum_{i=1}^{b} \left(R_{ij} - \frac{c+1}{2}\right) = 0\,, \tag{20.7.5}$$

it follows that

$$\sum_{j=1}^{c} \frac{12}{b(c^2-1)} \left\{\sum_{i=1}^{b} \left(R_{ij} - \frac{c+1}{2}\right)\right\}^2 \tag{20.7.6}$$

will be distributed under H_0 as chi-square with $c-1$ degrees of freedom. The authors carry out some simulations in order to study the power properties of Ψ_L in comparison with the parametric F test for selected values of $\theta = \sigma_Y/\sigma$, b and c and the efficiency of Ψ_L relative to F is 0.98 when $\alpha = 0.01$ or 0.05 and $b \geq 7$.

ARE of Friedman's Test

The usual parametric test procedure for the hypothesis of no treatment effect in a two-way experiment with one observation per cell or in a randomized complete block design is based on the statistic

$$F = \frac{MS_Y}{MS_E} = \frac{SS_Y/(c-1)}{SS_E/(bc-b-c+1)} \tag{20.7.7}$$

where

$$\begin{aligned} SS_Y &= b\sum_{j=1}^{c}(\bar{X}_{\cdot j} - \bar{X})^2 \\ SS_E &= \sum_{i=1}^{b}\sum_{j=1}^{c}(X_{ij} - \bar{X}_{i\cdot} - \bar{X}_{\cdot j} + \bar{X})^2\,, \end{aligned}$$

$\bar{X}_{i\cdot}$ denotes the ith row mean, $\bar{X}_{\cdot j}$ denotes the jth column mean and $\bar{X}$ denotes the grand mean of all the bc observations. It is well-known that MS_E converges to σ^2 which is the common variance of the errors ϵ_{ij} as $b \to \infty$ and MS_Y is equivalent in distribution to

$$(b\sigma_Y^2 + \sigma^2)\chi^2_{c-1}/(c-1) \text{ as } b \to \infty \, .$$

Then by Slutsky's theorem, $(c-1)F$ is asymptotically distributed as

$$\left(1 + \frac{b\sigma_Y^2}{\sigma^2}\right)\chi^2_{c-1} \, .$$

So, H_0 is rejected when $(c-1)F > \chi^2_{c-1,1-\alpha}$. Hence, the power of F at σ_Y^2 is

$$\pi(\sigma_Y^2) = P\left(\chi^2_{c-1} > \left(1 + \frac{b\sigma_Y^2}{\sigma^2}\right)^{-1} \chi^2_{c-1,1-\alpha}\right) , \tag{20.7.8}$$

where $\chi^2_{c-1,1-\alpha}$ denotes the upper α^{th} quantile of the chi-square distribution with $c-1$ degrees of freedom.

Next, we derive the distribution of

$$\frac{c(c+1)}{12} F_R = \sum_{j=1}^{c}\left[b^{-1/2}\sum_{i=1}^{b}\left(R_{ij} - \frac{c+1}{2}\right)\right]^2$$

given by (20.7.3). Towards this, let

$$T_j = b^{-1/2}\sum_{i=1}^{c}\left(R_{ij} - \frac{c+1}{2}\right) \tag{20.7.9}$$

where we can write

$$R_{ij} = \sum_{k=1}^{c} I(X_{ik} \le X_{ij}) = 1 + \sum_{k \ne j}^{c} I(X_{ik} \le X_{ij}) \tag{20.7.10}$$

and since $I(X_{ik} \le X_{ij}) = I(X_{ik} - \beta_i \le X_{ij} - \beta_i)$, we can, without loss of generality, set $\beta_i = 0$ $(i = 1, \ldots, b)$.

Let

$$T_j = b^{-1/2}\sum_{i=1}^{b}\left(R_{ij} - \frac{c+1}{2}\right), \; j = 1, \ldots, c \, .$$

Under H_0, R_{ij} has a discrete distribution on $1, \ldots, c$. Hence,

$$E(R_{ij}|H_0) = (c+1)/12 \text{ and } \operatorname{var}(R_{ij}|H_0) = (c^2-1)/12 \, .$$

Consequently,

$$E(T_j|H_0) = 0 \text{ and } \mathrm{var}(T_j|H_0) = (c^2-1)/12\,. \tag{20.7.11}$$

Next, consider, for $j \neq l$,

$$\begin{aligned}\mathrm{cov}(T_j, T_l) &= b^{-1}\mathrm{cov}\left(\sum_{i=1}^{b}\left(R_{ij}-\frac{c+1}{2}\right), \sum_{m=1}^{b}\left(R_{ml}-\frac{c+1}{2}\right)\right)\\ &= b^{-1}\left[\sum_{i=1}^{b}\mathrm{cov}(R_{ij}, R_{il}) + \sum_{i\neq m}^{b}\sum^{b}\mathrm{cov}(R_{ij}, R_{ml})\right].\end{aligned}$$

Note that, under H_0, R_{ij} and R_{ml} (for $i \neq m$) will be independent, and hence, the second sum in the preceding expression will be zero. So, consider, for $j \neq l$,

$$\mathrm{cov}(R_{ij}, R_{il}) = E(R_{ij}, R_{il}) - \left(\frac{c+1}{2}\right)^2$$

and

$$\begin{aligned}E(R_{ij}, R_{il}|H_0) &= \sum_{r\neq s}^{c}\sum^{c} rsP(R_{ij}=r_1, R_{il}=s|H_0)\\ &= \frac{1}{c(c-1)}\sum_{r\neq s}^{c}\sum^{c} rs\\ &= \frac{1}{c(c-1)}\left\{\sum_{1}^{c}\sum_{1}^{c} rs - \sum_{r=1}^{c} r^2\right\}\\ &= \frac{1}{c(c-1)}\left\{\frac{c^2(c+1)^2}{4} - \frac{c(c+1)(2c+1)}{6}\right\}\\ &= \frac{(c+1)}{12(c-1)}(3c^2-c-2)\,.\end{aligned}$$

Hence,

$$\mathrm{cov}(R_{ij}, R_{il}|H_0) = -(c+1)/12\,. \tag{20.7.12}$$

Now let

$$T_j^* = \left(\frac{12}{c(c+1)}\right)^{1/2} T_j\,, j = 1, \ldots, c\,.$$

Then,

$$E(T_j^*|H_0) = 0, \quad \text{var } T_j^* = 1 - \frac{1}{c} \text{ and } \text{cov}(T_j^*, T_l^*) = -1/2 .$$

Now applying Corollary 23.6.1.1, we infer that $\sum T_j^{*^2} \overset{d}{\approx} \chi^2_{c-1}$ under H_0 as b becomes large.

Because we can set $\beta_i = 0$ for $i = 1, \ldots, b$, we have

$$E\left\{R_{ij} - \frac{c+1}{2}\right\} = \sum_{\substack{k=1 \\ k\neq j}}^{c} \left\{\int_{-\infty}^{\infty} F^{(k)}(x)dF^{(j)}(x) - \frac{1}{2}\right\} = 0 \qquad (20.7.13)$$

for each $i = 1, \ldots, b$, since $F^{(1)} = \cdots = F^{(c)}$ where

$$F^{(j)}(x) = P(\epsilon_{ij} + Y_j \le x), \quad j = 1, \ldots, c .$$

Hence,

$$ET_j = 0$$

for each j under all hypotheses.

Further,

$$\text{var } T_j = b^{-1}\left\{\sum_{i=1}^{b} \text{var } R_{ij} + \sum_{i\neq l}^{b} \sum^{b} \text{cov}(R_{ij}, R_{il})\right\} \qquad (20.7.14)$$

where

$$\begin{aligned}
\text{var } R_{ij} &= E(R_{ij} - ER_{ij})^2 \\
&= E\left[\sum_{k\neq j}^{c} I^*(X_{ik} \le X_{ij})\right]^2, \quad I^*(\,) = I(\,) - \frac{1}{2} \\
&= E\left[\sum_{k\neq j}^{c} I^{*^2}(X_{ik} \le X_{ij}) + \sum_{k\neq j} \sum_{\substack{l\neq j \\ k\neq l}} I^*(X_{ik} \le X_{ij}) I^*(X_{il} \le X_{ij})\right] .
\end{aligned}$$

Since

$$I^{*^2}(X_{ik} \le X_{ij}) = I^2(X_{ik} \le X_{ij}) - I(X_{ik} \le X_{ij}) + \frac{1}{4},$$

we have

$$EI^{*^2}(X_{ik} \leq X_{ij}) = \frac{1}{4}. \tag{20.7.15}$$

Next, for $k \neq j$, $l \neq j$, $k \neq l$,

$$\begin{aligned} & EI^*(X_{ik} \leq X_{ij})I^*(X_{il} \leq X_{ij}) \\ = \;& E\left\{F^{(k)}(X_{ij}) - \frac{1}{2}\right\}\left\{F^{(l)}(X_{ij}) - \frac{1}{2}\right\} \\ = \;& \left[\int \left(F^{(j)}(x) - \frac{1}{2}\right)^2 dF^{(j)}(x)\right] \\ = \;& \frac{1}{12}, \text{ since } F^{(1)} = \cdots = F^{(c)}. \end{aligned} \tag{20.7.16}$$

Consequently, from (20.7.15) and (20.7.16), we have

$$\text{var } R_{ij} = \frac{c-1}{4} + \frac{(c-1)(c-2)}{12} = \frac{c^2-1}{12}. \tag{20.7.17}$$

Next consider, for $i \neq l$,

$$\text{cov}(R_{ij}, R_{lj}) = \sum_{k \neq j}^{c} \sum_{m \neq j}^{c} E\left\{I^*(X_{ik} \leq X_{ij})I^*(X_{lm} \leq X_{lj})\right\} \tag{20.7.18}$$

$$= E\sum_{k \neq j}^{c} I^*(X_{ik} \leq X_{ij})I^*(X_{lm} \leq X_{lj}) + E\sum_{k \neq j}^{c} \sum_{\substack{m \neq j \\ k \neq m}}^{2}. \tag{20.7.19}$$

Thus, consider

$$\begin{aligned}
&E\{I^*(X_{ik} \le X_{ij})I^*(X_{lk} \le X_{lj})\} \\
&= E\{I^*(\epsilon_{ik} \le \epsilon_{ij} + Y_j - Y_k)I^*(\epsilon_{lk} \le \epsilon_{lj} + Y_j - Y_k)\} \\
&= E\left[\int \left\{F(x) + Y_j - Y_k) - \frac{1}{2}\right\} dF(x)\right]^2 \\
&= E\left[\int \left\{F(x) - \frac{1}{2} + (Y_j - Y_k)f(x) + \frac{1}{2}(Y_j - Y_k)^2 f'(x^*)\right\} dF(x)\right]^2 \\
&= E\left\{L^2(Y_j - Y_k)^2 + L(Y_j - Y_k)^3 \int f'(x^*)dF(x) \right. \\
&\quad \left. + \frac{1}{4}(Y_j - Y_k)^4 \left[\int f'(x^*)dF(x)\right]^2\right\} \\
&\le 2L^2\sigma_Y^2 + L\|f'\|E|Y_j - Y_k|^3 + \frac{1}{4}\|f'\|^2 E(Y_j - Y_k)^4 . \qquad (20.7.20)
\end{aligned}$$

By the moment-inequality, we have

$$E|Y_j - Y_k|^3 \le \left[E\{|Y_j - Y_k|^4\}\right]^{3/4} = \sigma_Y^3 \left\{\frac{\mu_4}{\sigma_Y^4}\right\}^{3/4} = O(\sigma_Y^3),$$

because of assumption (b) in (20.3.7). Also,

$$E(Y_j - Y_k)^4 = 2\mu_4 + 6\sigma_Y^4 = O(\sigma_Y^4).$$

Hence, (20.7.18) becomes

$$E\{I^*(X_{ik} \le X_{ij})I^*(X_{lk} \le X_{lj})\} = 2L^2\sigma_Y^2 + O(\sigma_Y^3). \qquad (20.7.21)$$

Next consider, for $i \ne l$, $k \ne m$, $k, m \ne j$,

$$\begin{aligned}
&E\{I^*(X_{ik} \le X_{ij})I^*(X_{lm} \le X_{lj})\} \\
&= E\{I^*(\epsilon_{ik} + Y_k \le \epsilon_{ij} + Y_j)I^*(\epsilon_{lm} + Y_m \le \epsilon_{lj} + Y_j)\} \\
&= E\left[\int \left\{F(x + Y_j - Y_k) - \frac{1}{2}\right\} dF(x)\right] \\
&\quad \cdot \left[\int \left\{F(y + Y_j - Y_m) - \frac{1}{2}\right\} dF(y)\right].
\end{aligned}$$

Expanding each F about x or y and taking the expectation of the product of the two expansions, one can obtain

$$\begin{aligned} E\{I^*(X_{ik} \le X_{ij})I^*(X_{lm} \le X_{lj})\} &\le L^2\sigma_Y^2 + \frac{1}{2}L\|f'\|E|Y_j - Y_l|(Y_j - Y_m)^2 \\ &+ \frac{1}{2}L\|f'\|E|Y_j - Y_m|(Y_j - Y_k)^2 \\ &+ \frac{1}{4}\|f'\|^2 E|Y_j - Y_k|^2(Y_j - Y_m)^2 . \end{aligned}$$

Now, since

$$\begin{aligned} E|Y_j - Y_k|(Y_j - Y_m)^2 &\le \left\{E(Y_j - Y_k)^2 \cdot E(Y_j - Y_m)^4\right\}^{1/2} \\ &\le 2\sigma_Y^3\left\{3 + \left(\frac{\mu_4}{\sigma_Y^4}\right)\right\} = O(\sigma_Y^3) \end{aligned}$$

and

$$E(Y_j - Y_k)^2(Y_j - Y_m)^2 = \mu_4 + 3\sigma_Y^4 ,$$

we obtain

$$\text{L.H.S.} \le L^2\sigma_Y^2 + O(\sigma_Y^3) . \tag{20.7.22}$$

Hence,

$$\begin{aligned} \operatorname{var} T_j &= \frac{c^2-1}{12} + (b-1)\left[L^2\sigma_Y^2\{2(c-1) + (c-1)(c-2)\} + O(\sigma_Y^3)\right] \\ &= \frac{c^2-1}{12} + L^2\sigma_Y^2(b-1)c(c-1)\{1 + O(\sigma_Y)\} . \end{aligned} \tag{20.7.23}$$

Next, for $j \ne l$,

$$\begin{aligned} \operatorname{cov}(T_j, T_l) = E(T_jT_l) &= b^{-1}E\left[\sum_{r=1}^{b}\left(R_{rj} - \frac{c+1}{2}\right)\right]\left[\sum_{s=1}^{b}\left(R_{sl} - \frac{c+1}{2}\right)\right] \\ &= b^{-1}\sum_{r=1}^{b} E\left(R_{rj} - \frac{c+1}{2}\right)\left(R_{rl} - \frac{c+1}{2}\right) \\ &\quad + b^{-1}\sum_{r \ne s}^{b}\sum^{b} E()() . \end{aligned}$$

First, consider

$$
\begin{aligned}
&\operatorname{cov}(R_{rj}, R_{rl}) \\
&= E\left[\sum_{k \neq j, k \neq l}^{c} I^*(X_{rk} \leq X_{rj}) + I^*(X_{rl} \leq X_{rj})\right] \\
&\quad \cdot \left[\sum_{m \neq l, m \neq j}^{c} I^*(X_{rm} \leq X_{rl}) + I^*(X_{rj} \leq X_{rl})\right] \\
&= EI^*(X_{rl} \leq X_{rj}) I^*(X_{rj} \leq X_{rl}) + E(S_1^* + S_2^* + S_3^*) \text{ (say)}
\end{aligned}
$$

where

$$
\begin{aligned}
EI^*(X_{rl} \leq X_{rj}) I^*(X_{rj} \leq X_{rl}) &= EI^*(X_{rl} \leq X_{rj}) - EI^{*^2}(X_{rl} \leq X_{rj}) \\
&= 0 - E\left\{F^{(l)}(X_{rj}) - \frac{1}{2}\right\}^2 = -\frac{1}{4}.
\end{aligned}
$$

Next,

$$
\begin{aligned}
ES_1^* &= \sum_{\substack{k \neq j \\ k \neq l}}^{c} EI^*(X_{rj} \leq X_{rl}) I^*(X_{rk} \leq X_{rj}) \\
&= \sum_{\substack{k \neq j \\ k \neq l}}^{c} \int \left\{\frac{1}{2} - F^{(l)}(x)\right\}\left\{F^{(k)}(x) - \frac{1}{2}\right\} dF^{(i)}(x) \\
&= -(c-2)\int_0^1 \left(\frac{1}{2} - u\right)^2 du = -(c-2)/12\,.
\end{aligned}
$$

Analogously,

$$
ES_2^* = \sum_{\substack{m \neq j \\ m \neq l}}^{c} EI^*(X_{rl} \leq X_{rj}) I^*(X_{rm} \leq X_{rl}) = -(c-2)/12\,.
$$

Next one can write

$$\begin{aligned} ES_3^* &= E\left[\sum_{k\neq j,k\neq l}^{c} I^*(X_{rk} \leq X_{rj})\right]\left[\sum_{m\neq j,m\neq l}^{c} I^*(X_{rm} \leq X_{rl})\right] \\ &= ES_1 + ES_2 \end{aligned}$$

where

$$\begin{aligned} ES_1 &= \sum_{\substack{k\neq j\\ k\neq l}}^{c} EI^*(X_{rk} \leq X_{rj})I^*(X_{rk} \leq X_{rl}) \\ &= \sum_{\substack{k\neq j\\ k\neq l}}^{c} \int \left\{\frac{1}{2} - F^{(j)}(x)\right\}\left\{\frac{1}{2} - F^{(l)}(x)\right\} dF^{(k)}(x) \\ &= (c-2)\int_0^1 \left(\frac{1}{2} - u\right)^2 du = (c-2)/12\,. \end{aligned}$$

One can easily verify that

$$ES_2 = \sum_{k\neq j,l}\ \sum_{\substack{m\neq j,l\\ k\neq m}} EI^*(X_{rk} \leq X_{rj})EI^*(X_{rm} \leq X_{rl}) = 0\,.$$

Finally, consider the case $r \neq s$. Proceeding as in the computation of $\text{cov}(R_{rj}, R_{rl})$, we have

$$\text{cov}(R_{ij}, R_{sl}) = EI^*(X_{rl} \leq X_{rj})I^*(X_{sj} \leq X_{sl}) + E\tilde{S}_1 + E\tilde{S}_2 + E\tilde{S}_3\,,$$

where the first term on

$$\begin{aligned} \text{R.H.S.} &= E\left[\left\{\int F(x + Y_j - Y_k) - \frac{1}{2}\right\} dF(x)\right] \\ &\quad \cdot\left[\left\{\int \left\{F(y + Y_l - Y_j) - \frac{1}{2}\right\}\right\} dF(y)\right] \\ &= -L^2\sigma_Y^2 + O(\sigma_Y^3) \end{aligned}$$

upon expanding the F functions around x or y and using Holder's and moment inequalities. In a similar manner,

$$
\begin{aligned}
E\tilde{S}_1 &= \sum_{\substack{k\neq j\\ k\neq l}}^{c} EI^*(X_{rl} \leq X_{rj})I^*(X_{sj} \leq X_{sl}) \\
&= \sum_{\substack{k\neq j\\ k\neq l}}^{c} E\left[\int\left\{F(x+Y_j-Y_l)-\frac{1}{2}\right\}dF(x)\right] \\
&\qquad \cdot\left[\int\left\{F(x+Y_l-Y_j)-\frac{1}{2}\right\}dF(y)\right] \\
&= -2(c-2)L^2\sigma_Y^2 + O(\sigma_Y^3)\,.
\end{aligned}
$$

Also, $E\tilde{S}_2$ is obtained by interchanging m with k, s with r and l with j in $E\tilde{S}_1$. Thus,

$$
E\tilde{S}_2 = -2(c-2)L^2\sigma_Y^2 + O(\sigma_Y^3)\,.
$$

Next, one can write

$$
\begin{aligned}
E\tilde{S}_3 &= \sum_{k\neq j,l}^{c} EI^*(X_{rk} \leq X_{rj})I^*(X_{sk} \leq X_{sl}) + \sum_{k\neq j,l}^{c}\sum_{\substack{m\neq j,l\\ k\neq m}}^{c} \\
&= E\tilde{S}_{31} + E\tilde{S}_{32}\,.
\end{aligned}
$$

One can easily see that $E\tilde{S}_{32} = 0$ since it involves products of independent indicator functions, each having zero expectation. One can express $E\tilde{S}_{31}$ as

$$
\begin{aligned}
E\tilde{S}_{31} &= \sum_{k\neq j,l}^{c} E\left[\int\left\{F(x+Y_j-Y_k)-\frac{1}{2}\right\}dF(x)\right] \\
&\qquad \cdot\left[\int\left\{F\left(y+(Y_l-Y_k)\right)-\frac{1}{2}\right\}dF(y)\right] \\
&= (c-2)L^2\sigma_Y^2 + O(\sigma_Y^3)\,.
\end{aligned}
$$

Hence, putting all the computations together, we have for $j \neq l$,

$$\begin{aligned}
&\operatorname{cov}(T_j, T_l) \\
&= (b-1)\Bigg[-\frac{1}{4} - \frac{c-2}{12} - \frac{c-2}{12} + \frac{c-2}{12} \\
&\quad -L^2\sigma_Y^2\{1 + 2(c-2) + 2(c-2) - (c-2)\} + O(\sigma_Y^3)\Bigg] \\
&= -(b-1)\left[\frac{c+1}{12} + L^2\sigma_Y^2(1 + 3(c-2)) + O(\sigma_Y^3)\right]. \qquad (20.7.24)
\end{aligned}$$

Thus, the T_j are identically distributed, but are not independent. Also note that the R_{ij} are i.i.d. for each j, and hence,

$$T_j = b^{-1/2} \sum_{i=1}^{b} \left\{ R_{ij} - \frac{c+1}{2} \right\}$$

are asymptotically normal with zero mean and variance= $\operatorname{var}(R_{ij})$ for sufficiently large b. Thus, $\sum_{j=1}^{c} \gamma_j T_j$ is also asymptotically normal with mean 0 and certain variance. Since this is true for any fixed $(\gamma_1, \ldots, \gamma_c)$, we infer that $(T_1, \ldots, T_c)$ has asymptotically a c-variate normal distribution with mean zero and a certain variance-covariance matrix. However,

$$\sum_{1}^{c} T_j = 0 .$$

Hence, using an analysis of variance transformation, we can assert that

$$\sum_{j=1}^{c} T_j^2 / \operatorname{var} T_j$$

is asymptotically distributed as chi-square with $c-1$ degrees of freedom. Thus, we reject H_0 when

$$\sum_{j=1}^{c} \frac{12}{c^2-1} T_j^2 \geq \chi^2_{c-1,1-\alpha} \text{ (see (20.7.6))}$$

where α is the level of significance. Then the power at σ_Y^2 is

$$\begin{aligned}
&\pi^*(\sigma_Y^2) \\
&= P\left[\sum_{j=1}^{c} T_j^2/\text{var } T_j \geq \frac{(c^2-1)}{12\text{var } T_j}\chi^2_{c-1,1-\alpha}|\sigma_Y^2 > 0\right] \\
&= P\left[\chi^2_{c-1} \geq \left\{1 + 12(b'-1)\left(\frac{c}{c+1}\right)\sigma_Y^2 L^2 + O(\sigma_Y)\right\}^{-1} \chi^2_{c-1,1-\alpha}\right]
\end{aligned} \tag{20.7.25}$$

where b'denotes the number of blocks required by the Friedman's procedure. Now, equating the powers of π and π^* given by (20.7.8) and (20.7.23), we obtain

$$\frac{b}{\sigma^2} = 12(b'-1)\left(\frac{c}{c+1}\right)L^2 + O(\sigma_Y),$$

that is,

$$\frac{b}{b'-1} \to 12\sigma^2\left(\frac{c}{c+1}\right)L^2 \text{ as } b \text{ and } b' \to \infty \tag{20.7.26}$$

which coincides with (19.3.12), the asymptotic efficiency of Friedman's test relative to F test for the fixed effects case.

Remark 20.7.1. *One can proceed in an analogous manner in order to evaluate the asymptotic efficiency of the test procedure based on Ψ_L considered in Section 20.6.*

Remark 20.7.2. *Nugroho and Govindarajulu (2002) derive LMP rank tests for H_0 in balanced incomplete block designs and study the distributional properties of the test criterion for logistic alternatives. For details, the reader is referred to their Section 4.*

20.8 Problems

20.3.1 Consider the data in Problem 16.2.1 and view the plots as a random sample of treatments. We wish to test the null hypothesis given by (20.2.3) at $\alpha = 0.05$. Use the test criterion (20.3.1) and the asymptotic result given in Theorem 20.3.1 in order to carry out the test.

20.3.2 The following data pertains to ascorbic acid concentration (mg per 100g)in turnip greens. A leaf is taken from near the center of each of 5 plants and ascorbic acid concentration was determined for each leaf. This was repeated on each of 6 days with a new selection of plants obtained on each day. We have the following data[1].

	Plant				
Day	1	2	3	4	5
1	9.1	7.3	7.3	10.7	7.7
2	12.6	9.1	10.9	8.0	8.9
3	7.3	6.6	5.2	5.3	6.7
4	6.0	8.0	6.8	9.1	8.4
5	10.8	9.3	7.3	9.3	10.4
6	10.6	10.9	10.4	13.1	7.7

Treating the plants as random and the days as replications, carry out a test of H_0 given by (20.2.3) at $\alpha = 0.05$, using the test criterion (20.3.1) and the asymptotic result given in Theorem 20.3.1.

20.4.1 For the data in Problem 20.3.1, carry out an F test given by (20.4.1) with $\alpha = 0.05$.

20.4.2 For the data in Problem 20.3.2, carry out an F test given by (20.4.1) with $\alpha = 0.05$.

20.6.1 The following data pertains to ascorbic acid concentration (mg per 100g) in turnip greens. Two leaves were taken from near the center of each of 5 plants and ascorbic acid concentration was determined for each leaf. This experiment of plants obtained on each day, yielding the following data[2].

[1]This constitutes a part of the data in Problem 9.9 of Ostle. B. (1954). *Statistics in Research.* The Iowa State University Press, Ames, Iowa, p. 279.

[2]This data is taken from Problem 9.9 of Ostle. B. (1954). *Statistics in Research.* The Iowa State University Press, Ames, Iowa, p. 279.

Day	Leaf	Plant 1	2	3	4	5
1	A	9.1	7.3	7.3	10.7	7.7
	B	7.3	9.0	8.9	12.7	9.4
2	A	12.6	9.1	10.9	8.0	8.9
	B	14.5	10.8	12.8	9.8	10.7
3	A	7.3	6.6	5.2	5.3	6.7
	B	9.0	8.4	6.9	6.8	8.3
4	A	6.0	8.0	6.8	9.1	8.4
	B	7.4	9.7	8.6	11.2	10.3
5	A	10.8	9.3	7.3	9.3	10.4
	B	12.5	11.0	8.9	11.2	12.0
6	A	10.6	10.9	10.4	13.1	7.7
	B	12.3	12.8	12.1	14.6	9.4

Treating the plants as random, days as blocks and the leaves A and B as replicates, carry out a test of H_0 given by (20.2.3) using the test criterion given in (20.6.2) and its asymptotic distribution given in Theorem 20.6.1, with $\alpha = 0.05$.

20.7.1 For the data in Problem 20.3.2, treating plants as random and the days as blocks, test the null hypothesis H_0 with $\alpha = 0.05$ using test criterion given by (20.7.6).

20.7.2 For the data in Problem 19.4.2, treat the years as blocks and the four drugs as a random sample. Carry out a test of the null hypothesis that there is no difference in the various drugs used by high school seniors, with $\alpha = 0.05$ and the test criterion given by (20.7.6).

Chapter 21

[1]Estimation of Contrasts

21.1 Introduction and the Model

In the one- or two-way experiments, it is of much interest to estimate pairwise treatment effects. First, let us focus on one-way layout in which the model is

$$X_{ji} = \mu + \tau_j + \epsilon_{ji}, \quad i = 1, \ldots, n_i, \quad j = 1, \ldots, c, \tag{21.1.1}$$

where μ denotes the overall effect, τ_j is the jth treatment effect and $\sum_{j=1}^{c} \tau_j = 0$. Further, the ϵ_{ji} are the errors which are i.i.d. with an unknown distribution.

A contrast in the τ_j is defined as

$$\theta = \sum_{j=1}^{c} a_j \tau_j \tag{21.1.2}$$

where the a_j are some specified constants with $\sum_1^k a_j = 0$. Alternatively, one can express θ as

$$\theta = \sum_{h=1}^{c} \sum_{j=1}^{c} d_{hj} \Delta_{hj} \tag{21.1.3}$$

with $d_{hj} = \frac{a_h}{c}$, $j = 1, \ldots, c$ and $\Delta_{hj} = \tau_h - \tau_j$.

[1]Hollander and Wolfe (1973, pp. 133–136, 158, 159, 161, 178, 181) served as a source for parts of this chapter.

21.2 Estimation Procedure

In order to obtain an estimate of θ, proceed as follows.

Let

$$Z_{hj} = \text{median}\,\{X_{hr} - X_{js},\ r = 1, \ldots, n_h,\ s = 1, \ldots, n_j\}\,. \tag{21.2.1}$$

Notice that Z_{hj}, the unadjusted estimator of $\tau_h - \tau_j$, is the Hodges-Lehmann estimator of $\tau_h - \tau_j$. Further, since $Z_{hj} = -Z_{jh}$, only $c(c-1)/2$ estimators Z_{hj} with $h < j$ need to be calculated. Let

$$\bar{\Delta}_h = \frac{\sum_{j=1}^{c} n_j Z_{hj}}{\sum_{j=1}^{c} n_j}\,, \quad h = 1, \ldots, c \tag{21.2.2}$$

with $Z_{hh} = 0$ $(h = 1, \ldots, c)$. Then the weighted adjusted estimator of Δ_{hj} is defined by

$$W_{hj} = \bar{\Delta}_h - \bar{\Delta}_j\,. \tag{21.2.3}$$

Thus, the weighted adjusted estimator of θ is

$$\hat{\theta} = \sum_{j=1}^{c} a_j \bar{\Delta}_j\,. \tag{21.2.4}$$

We can also define $\hat{\theta}$ in terms of the d_{hj} as

$$\hat{\theta} = \sum_{h=1}^{c} \sum_{j=1}^{c} d_{hj} W_{hj} = \sum_{h=1}^{c} \sum_{j=1}^{c} d_{hj}(\bar{\Delta}_h - \bar{\Delta}_j)\,. \tag{21.2.5}$$

In the case of equal sample sizes (i.e., $n_1 = \cdots = n_c = n$), (21.2.2) takes the simpler form of

$$Z_{h\cdot} = \frac{1}{c} \sum_{1}^{c} Z_{hj} \tag{21.2.6}$$

and $\bar{\Delta}_h - \bar{\Delta}_j$ takes the form of

$$\bar{\Delta}_h - \bar{\Delta}_j = Z_{h\cdot} - Z_{j\cdot}\,. \tag{21.2.7}$$

21.3 Certain Remarks

The unadjusted estimators Z_{hj} given by (21.2.1) are somewhat ambiguous in contrast estimation because they do not satisfy the linear relations that are satisfied by the contrasts which they are designed to estimate. For instance, $\Delta_{13} = \Delta_{12} + \Delta_{23}$ whereas, in general, $Z_{13} \neq Z_{12} + Z_{23}$. Thus, the two estimators Z_{13} and $Z_{12} + Z_{23}$ of Δ_{13} may lead to different estimates. This was pointed out by Lehmann (1963a) and called the unadjusted estimators "incompatible". He removed the ambiguity by employing the estimators $Z_{h\cdot} - Z_{j\cdot}$ given by (21.2.6). These estimators were obtained by minimizing the sum of squares

$$\sum_{h \neq j}\sum [Z_{hj} - (\tau_h - \tau_j)]^2 .$$

Although the estimators given by (21.2.7) are compatible, additional difficulties arise as pointed out by Lehmann (1963a). The first difficulty is that the estimator $Z_{h\cdot} - Z_{j\cdot}$ of $\Delta_{hj} = \tau_h - \tau_j$ depends not only on the sample sizes n_h and n_j, but also on the other $c - 2$ sample sizes. Besides, for $c = 3$, the estimator $Z_{1\cdot} - Z_{2\cdot}$ (of $\tau_1 - \tau_2$) is not consistent when n_1 and n_2 tend to be large unless n_3 also tends to be large.

Lehmann (1963a, p. 960) has shown that the distribution of $\hat{\theta}$ given by (21.2.4) is symmetric about θ if either (i) the distributions F_j of X_{ij} are symmetric or (ii) all sample sizes, $n_1, n_2, \ldots, n_c$, are equal.

Let $N = \sum_{j=1}^{c} n_j$ and $\rho_j = n_j/N$ $(j = 1, \ldots, c)$. Then as n_j $(j = 1, \ldots, c)$ tend to infinity such that ρ_j are bounded away from zero and unity and $\tau_h - \tau_j = -a_h/N^{1/2}$, Lehmann (1963a) establishes the joint asymptotic normality of $V_h = (Z_{hc} - \Delta_{hc})N^{1/2}$, $h = 1, \ldots, c-1$, having

$$\begin{aligned} \operatorname{var}(V_h) &= \left(\frac{1}{12}\right)\left(\frac{1}{\rho_h} + \frac{1}{\rho_c}\right)\left(\int f^2(x)dx\right)^2 \\ \operatorname{cov}(V_h, V_j) &= \left(\frac{1}{12\rho_c}\right)\left(\int f^2(x)dx\right)^2 \end{aligned}$$

where f is the density of F, F denoting the distribution function of $X_{ij} - \mu$. Hence, he computes the asymptotic efficiency of the $\hat{\theta}$ based on Hodges-Lehmann estimating procedure relative to the procedure based on sample means as

$$e = 12\sigma^2 \left[\int f^2(x)dx\right]^2$$

where

$$\sigma^2 = \operatorname{var}(X_{ij}) ,$$

i.e., the asymptotic efficiency relative to the standard procedure is the same as that of the Wilcoxon test relative to Student's t-test. Lehmann (1963b) obtains confidence intervals for any contrast.

Bhuchongkul and Puri (1965) have extended the results of Hodges and Lehmann (1963) and Lehmann (1963a) by defining a class of estimates of all contrasts in terms of rank test statistics such as the Wilcoxon or normal scores statistic. They show that the asymptotic efficiency of these estimates relative to the standard least squares estimates is the same as the Pitman efficiency of the corresponding rank tests on which they are based relative to the t-test.

Spjøvoll (1968) removes the non-consistency difficulty by obtaining weighted adjusted estimators $\bar{\Delta}_h - \bar{\Delta}_j$ (21.2.3) which minimize the sum of squares

$$\sum_{h \neq j} \sum \rho_h \rho_j \left[Z_{hj} - (\tau_h - \tau_j)\right]^2 .$$

However, the above estimators still suffer from the disadvantage of having the estimator of $\tau_h - \tau_j$ depend on unrelated observations from other samples.

Spjøvoll (1968) also proposed weighted adjusted estimators that minimize

$$\sum_{h \neq j} \sum \left(\frac{1}{\rho_h} + \frac{1}{\rho_i}\right)^{-1} \left[Z_{hj} - (\tau_h - \tau_j)\right]^2 \qquad (21.3.1)$$

using the asymptotic variances of the Z_{hj}'s as weights in the sum of squares. Although the latter are more difficult to compute, Spjøvoll (1968) shows the asymptotic equivalence of the above estimators with those given in (21.2.3) provided $0 < \lambda_j = n_j/N < 1$ $(j = 1, \ldots, c)$ as the subsample sizes become large. However, when $n_1 = \cdots = n_c$ the estimators that minimize (21.3.1) coincide with Lehmann's (1963a) estimators (21.2.7).

Remark 21.3.1. *It is of much interest to find simultaneous confidence intervals for all possible contrasts since this enables us to assess any specific contrast the significance of which is suggested by the data. Procedures based on large sample sizes are given by Sen (1966), Marsascuilo (1966) and Crouse (1969).*

21.4 Contrasts in Two-way Layouts

Assume the model to be

$$X_{ji} = \mu + \beta_i + \tau_j + \varepsilon_{j,i}, \quad i = 1, \ldots, n \quad \text{and } j = 1, \ldots, c. \qquad (21.4.1)$$

As before, define the contrast to be

$$\theta = \sum_{j=1}^{c} a_j \tau_j \quad \left(\sum_{1}^{c} a_j = 0 \right) \tag{21.4.2}$$

where the a's are some given constants. We can also write

$$\theta = \sum_{h=1}^{c} \sum_{j=1}^{c} d_{hj} \Delta_{hj} \tag{21.4.3}$$

where

$$d_{hj} = \frac{a_h}{c}, \quad j, h = 1, \ldots, c$$

and

$$\Delta_{hj} = \tau_h - \tau_j, \quad j, h = 1, \ldots, c.$$

Doksum (1967) proposed the following procedure for estimating θ. Form the treatment u and v differences computed for each of the n blocks given by

$$D_{u,v}^{(i)} = X_{ui} - X_{vi}, \quad i = 1 \ldots, n. \tag{21.4.4}$$

Let

$$Z_{u,v} = \text{median}\left[D_{uv}^{(i)}, \quad i = 1, \ldots, n \right] \tag{21.4.5}$$

For example, Z_{23} is the median of the $X_{2i} - X_{3i}$ differences, which is an estimator of $\tau_2 - \tau_3$. $Z_{u,v}$ will be called the "unadjusted" estimator of $\Delta_{u,v} = \tau_u - \tau_v$. Note that we need to evaluate only $c(c-1)/2$ unadjusted estimators $Z_{u,v}$ with $u < v$ since $Z_{v,u} = -Z_{u,v}$. Now, let

$$\hat{\Delta}_{u,v} = Z_{u\cdot} - Z_{v\cdot} \tag{21.4.6}$$

where

$$Z_{u\cdot} = \sum_{j=1}^{c} Z_{uj}/c, \quad Z_{u,u} = 0, \quad u = 1, \ldots, c. \tag{21.4.7}$$

Then the adjusted estimator for θ is defined by

$$\tilde{\theta} = \sum_{h=1}^{c} \sum_{j=1}^{c} d_{hj} \hat{\Delta}_{hj}, \tag{21.4.8}$$

which is equivalent to

$$\tilde{\theta} = \sum_{j=1}^{c} a_j Z_{j\cdot}. \tag{21.4.9}$$

Remark 21.4.1. (i) *The unadjusted estimator $Z_{u,v}$ of $\Delta_{u,v}$ is the estimator associated with the sign test.*

(ii) *The unadjusted estimator $Z_{u,v}$ lacks the compatibility property discussed in Section 21.3.*

(iii) *The adjusted estimators $\hat{\Delta}_{u,v}$ of $\Delta_{u,v}$ are compatible and are as efficient or more efficient than the unadjusted ones.*

(iv) *The adjusted estimator of $\tau_u - \tau_v$ also depends on the observations from the other $c-2$ treatments.*

Asymptotic normality and efficiency of $\tilde{\theta}$ has been studied by Doksum (1967). Doksum (1967) has shown that $n^{1/2}\left[Z_{u,v} - (\tau_u - \tau_v) : u < v\right]$ is asymptotically normal with mean 0 and variance $\frac{1}{4}\left(\int f^2(x)dx\right)^2$. He also shows that the asymptotic efficiency of his procedure relative to the standard procedure is $12c\sigma^2\left(\int f^2(x)dx\right)^2/(c+1)$ where σ^2 denotes the variance of ε_{ji}.

Lehmann (1964) has investigated estimators of θ [see (21.4.8)] using estimators derived from the Wilcoxon signed rank test which will be discussed next.

21.5 Hodges-Lehmann Type of Estimator

Assume the model given by (21.4.1). Let $W_{u,v}$ be the median of the $n(n+1)/2$ Walsh averages defined by

$$(X_{ui} - X_{vi} + X_{uk} - X_{vk})/2, \quad i, k = 1, \ldots, n,$$

with $i \le j$. In other words,

$$W_{u,v} = \text{median}\left\{\frac{1}{2}(X_{ui} - X_{vi} + X_{uj} - X_{vj}),\ i \le j\right\}. \tag{21.5.1}$$

$W_{u,v}$ is the unadjusted estimator of $\tau_u - \tau_v$. Since $W_{u,v} = -W_{v,u}$, we need to calculate only $c(c-1)/2$, $W_{u,v}$ with $u < v$.

Next, let

$$\hat{\hat{\Delta}}_{u,v} = W_{u\cdot} - W_{v\cdot}\,, \tag{21.5.2}$$

where

$$W_{u\cdot} = \frac{1}{c}\sum_{j=1}^{c} W_{uj},\ \ W_{u,u} = 0,\ \ u = 1, \ldots, c. \tag{21.5.3}$$

Now, the adjusted estimator of θ is given by

$$\hat{\theta} = \sum_{h=1}^{c} \sum_{j=1}^{c} d_{hj} \; \hat{\hat{\Delta}}_{hj} \tag{21.5.4}$$

or

$$\hat{\theta} = \sum_{j=1}^{c} a_j W_{j.} \tag{21.5.5}$$

(i) The unadjusted estimator $W_{u,v}$ is incompatible.

(ii) The adjusted estimator $\hat{\hat{\Delta}}_{u,v}$ is compatible, but it also depends on the other $c-2$ treatment samples.

(iii) Doksum's estimator is relatively easy to compute than Lehmann's (1964) contrast estimator. However, Lehmann's estimator will be more efficient than Doksum's estimator.

The asymptotic standard deviation, normality and efficiency of the adjusted estimator have been studied by Lehmann (1964).

21.6 Problems

21.2.1 The following data constitute random samples of size 10 scores assigned by three different instructors in three sections of an elementary statistics course taught at Mars University. Maximum possible scores is 100.

Section 1	Section 2	Section 3
96	65	64
76	62	63
45	44	47
66	67	46
87	75	64
60	55	69
79	67	68
85	63	72
93	71	56
54	53	48

Assuming the model in (21.1.1) find an estimate of

$$\theta = \tau_1 - (\tau_2 + \tau_3)/2$$

of the form given by (21.2.4).

21.2.2 Analysis of samples (fictitious) of 5 different brands of diet/limitation margarine was made in order to determine the level of physiologically active polyunsaturated fatty acids (PAPFUA in percentages).

Imperial	13.9, 14.2, 14.5, 13.7
Blue Bonnet	13.4, 13.6, 14.1, 14.4
Chiffon	13.3, 12.8, 12.7, 13.8
Mazola	16.7, 17.1, 16.5, 17.4
Fleischman	18.0, 17.3, 18.6, 18.5

Note that Mazola and Fleischman's are corn-based, whereas others are soybean-based. Assuming the model (21.1.1), find an estimate of

$$\theta = \{(\tau_1 + \tau_2 + \tau_3)/3\} - \{(\tau_4 + \tau_5)/2\}$$

of the form given by (21.2.4).

21.4.1 Treat the data in Problem 21.2.1 as arising from a two-way layout with 3 treatments and 10 blocks. Assuming the model given in (21.4.1), obtain an estimate of

$$\theta = \tau_1 - (\tau_2 + \tau_3)/2$$

of the form given by (21.4.8) or (21.4.9).

21.4.2 Ten subjects were hypnotized and asked to regress to ages 15, 10, and 8. At each stage, an otis test of mental ability was administered. The following table gives the IQ scores, adjusted to the age suggested under hypnosis[2].

Subject	1	2	3	4	5	6	7	8	9	10
Waking	113	118	119	110	123	111	110	118	126	124
15 year	112	119	121	109	126	113	114	118	127	123
10 year	110	122	123	107	127	112	112	118	127	126
8 year	112	120	119	107	125	111	112	119	125	125

[2]Data from Kleine (1950). Hypnotic regression and intelligence. *J. Genet. Psychology.* 77, pp. 129–132.

Estimate the contrast θ where

$$\theta = \frac{1}{3}(\tau_2 + \tau_3 + \tau_4) - \tau_1.$$

21.4.3 The following fictitious data gives gains in weight (in pounds) of 40 steers fed different rations. (Data coded for easy calculations.)

Rations

Blocks	A	B	C	D
1	2	5	8	6
2	3	4	7	12
3	3	5	10	5
4	5	5	9	4
5	1	3	8	5
6	2	5	8	6
7	1	6	7	7
8	2	7	8	8
9	4	4	6	4
10	3	4	5	7

Chapter 22

Regression Procedures

22.1 Introduction

Regression analysis is extensively used by research workers. In this chapter we provide nonparametric methods for simple linear regression problems. First, we consider the case of linear regression and then briefly consider the multiple regression case.

22.2 Brown-Mood Method

Suppose we have n pairs of observations that are denoted by (X_i, Y_i), $i = 1, \ldots, n$. We wish to fit to the data a regression line of the form

$$Y = a + bx\,.$$

Brown and Mood (1951) and Mood (1950) provide a method for determining the values of a and b. First we divide the Y values into two groups.

(i) Those values of Y with x-values that are less than or equal to the median of x, and

(ii) those values of Y associated with x-values that exceed the median of x. Then the desired values of a and b are those that yield a line for which the median of the deviations about the line is zero in each of the two groups.

Graphical Procedure

(a) Prepare a scatter diagram of the sample data.

(b) Draw a line parallel to the Y-axis through the median of the x-values. Shift this line to the left or right if one or more points fall on this line so that the number of points on either side of the median is roughly equal.

(c) Determine the median of x and the median of Y in each of the two groups of observations formed in (b), i.e., compute the four medians.

(d) In the first group of points, plot a point denoting the median of x and the median of Y. Plot a similar point for the second group of observations.

(e) Draw a line through the two points plotted in step (d).

If the median of the vertical deviations of the points from this line is not zero in both groups, move the line to a new position until the deviations in each group have a median of zero. If greater accuracy is needed, follow the iterative procedure described by Mood (1950).

22.3 Case of a Single Regression Line

Let us observe a random variable Y at each of n distinct values of the x-variable, namely $x_1, \ldots, x_n$ where we assume that $x_1 < \cdots < x_n$.

Let us consider the model

$$Y_i = \alpha + \beta x_i + e_i, \quad i = 1, \ldots, n \tag{22.3.1}$$

where the x's are assumed to be known constants, α, β are unknown parameters and the errors e_i are mutually independent having an unknown continuous distribution.

Suppose we are interested in testing

$$H_0 : \beta = \beta_0 \text{ (when } \beta_0 \text{ is specified)}$$

against

$$H_a : \beta > \beta_0 \,.$$

Then Theil (1950a) proposed the following test procedure.

Calculate the n differences

$$D_i = Y_i - \beta_0 x_i, \quad i = 1, \ldots, n \,. \tag{22.3.2}$$

Let

$$C = \sum_{i<j}\sum c(D_j - D_i) \tag{22.3.3}$$

where

$$c(t) = \begin{cases} 1 & \text{if } t > 0 \\ 0 & \text{if } t = 0 \\ -1 & \text{if } t < 0. \end{cases} \tag{22.3.4}$$

Then for a specified level of significance α,

$$\begin{array}{c} \text{reject } H_0 \text{ when } C \geq k(\alpha, n) \\ \text{accept } H_0 \text{ when } C < k(\alpha, n)\,, \end{array} \tag{22.3.5}$$

where the constant $k(\alpha, n)$ is such that

$$P_0\,(C > k(\alpha, n)) = \alpha\,.$$

An analogous test holds for $H_a : \beta < \beta_0$. For $H_a : \beta \neq \beta_0$, we reject H_0 when $C \geq k(\alpha_2, n)$ or $C \leq -k(\alpha_1, n)$ and accept H_0 when $-k(\alpha_1, n) < C < k(\alpha_2, n)$ where $\alpha = \alpha_1 + \alpha_2$.

22.4 Large Sample Approximation

When H_0 holds, the D_i are i.i.d. random variables. Hence,

$$P_0\,(c(D_j - D_i) = 1) = P_0(D_j > D_i) = \frac{1}{2}$$

and

$$P_0\,(c(D_j - D_i) = -1) = P_0(D_j < D_i) = \frac{1}{2}\,.$$

Consequently,

$$\begin{aligned} E_0(C) &= \sum_{i=1}^{n} Ec(D_j - D_i) = 0 \\ \text{var}(C|H_0) &= E_0(C^2)\,. \end{aligned}$$

Note that under H_0, the D_i are i.i.d. random variables. Also, Theil's (1950a) statistic is analogous to Kendall's tau.

Next, we will obtain a closed form expression for var C under H_0. Consider

$$E(C^2) = E\left[\sum_{i<j}^{n}\sum^{n} c(D_j - D_i)\right]^2$$
$$= \sum_{i<j}^{n}\sum^{n}\sum_{k<l}\sum Ec(D_j - D_i)c(D_l - D_k)\,.$$

We consider the following cases.

Case 1: $k = i$, $l = j$: RHS becomes $n(n-1)/2$ since $c^2 \equiv 1$.

Case 2: $i = l$, $k = j$: This cannot happen.

Case 3: $k = i$, $l \neq j$, $i < j$:

$$Ec(D_j - D_i)c(D_l - D_i) = \int_0^1 (1-F)^2 dF + \int F^2 dF - 2\int F(1-F)dF$$
$$= 1/3$$

and the number of terms are

$$\sum_{i=1}^{n-1}\sum_{j=i+1}^{n}\sum_{l=i+1,l\neq 1}^{n} 1 = \sum_{i=1}^{n-1}(n-i)(n-i-1) = 2\binom{n}{3}$$

which can also be obtained using a combinatorial argument.

Case 4: $l = j$, $i < j$, $k < j$. the number of terms as in case 3 is $2\binom{n}{3}$ and

$$Ec(D_j - D_i)c(D_j - D_k) = 1/3\,.$$

Case 5: $j = k$, $i < j = k < l$. The number of terms are $\binom{n}{3}$ and each expectation is equal to $-1/3$.

Case 6: $l = i$, $k < l = i < j$. The number of terms is $\binom{n}{3}$ and each expectation is equal to $-1/3$.

Case 7: All subscripts are distinct and each expectation is equal to zero. The number of terms is $6\binom{n}{4}$. Note that the total number of terms in C^2 is $\binom{n}{2}^2$. Thus, putting all the cases together, we have

$$\begin{aligned} E(C^2|H_0) &= \frac{n(n-1)}{2} + 2\binom{n}{3}\left[\frac{1}{3}+\frac{1}{3}\right] - \binom{n}{3}\left[\frac{1}{3}+\frac{1}{3}\right] \\ &= \binom{n}{2} + \frac{2}{3}\binom{n}{3} \\ &= \frac{n(n-1)}{18}\left[9+2(n-2)\right] = \frac{1}{18}n(n-1)(2n+5)\,. \end{aligned}$$

Asymptotic Normality of C under H_0

One can establish the asymptotic normality of C as in the case of Kendall's tau. Thus,

$$C^* = C/\left[\operatorname{var}(C|H_0)\right]^{1/2}$$

is distributed as normal$(0,1)$ when H_0 is true.

Note that C is large when $D_j > D_i$ for several pairs (i,j). Also,

$$\begin{aligned} D_j - D_i &= \left[Y_j - \beta_0 x_j - (Y_i - \beta_0 x_i)\right] \\ &= \left[Y_j - Y_i + \beta_0(x_i - x_j)\right]\,. \end{aligned}$$

Further, since

$$Y_j - Y_i = \beta(x_j - x_i) + (e_j - e_i)\,,$$

the median of $Y_j - Y_i$ is $\beta(x_j - x_i)$. Hence, the median of $D_j - D_i$ is $(\beta - \beta_0)(x_j - x_i)$. Consequently, $D_j - D_i$ will tend to be positive when $\beta > \beta_0$ and thus the value of C tends to be large when $\beta > \beta_0$. This provides a motivation for rejecting H_0 against $H_a : \beta > \beta_0$ for large values of C.

Please note that the statistic C is Kendall's tau computed between the x and $Y - \beta_0 x$ values. In particular, a test of $\beta_0 = 0$ can be viewed as a test of zero correlation between the Y and x sequence. Thus, for sufficiently large n, H_0 is rejected when

$$C^* > z_\alpha\,.$$

22.5 Theil's Estimator for Slope

In order to estimate the slope β of the regression model, compute $N = n(n-1)/2$ individual slope values

$$S_{ij} = (Y_j - Y_i)/(x_j - x_i),\ 1 \le i < j \le n\,.$$

Then Theil's (1950c) estimator of β is given by

$$\hat{\beta} = \text{median}\{S_{ij},\ 1 \le i < j \le n\}\,. \tag{22.5.1}$$

Let $S^{(1)} < \cdots < S^{(N)}$ denote the ordered values of the sample slopes S_{ij}. Then if N is odd, say, $N = 2k+1$,

$$\hat{\beta} = S^{(k+1)}\,. \tag{22.5.2}$$

If N is even, then

$$\hat{\beta} = \{S^{(k)} + S^{(k+1)}\}/2\,. \tag{22.5.3}$$

When the x's are not distinct, let n' denote the number of positive $x_j - x_i$ differences for $1 \le i < j \le n$. Sen (1968) proposes the estimator of β to be the median of the n' sample slope values that can be computed from the data. In the special case when $x_1 = \cdots = x_m = 0$ and $x_{m+1} = \cdots = x_{m+q} = 1$ (with $n = m+q$ and $m < n$), then Sen's (1968) estimator of β reduces to the median of the mq differences $(Y_j - Y_i)$ where $i = 1, \ldots, m$ and $j = m+1, \ldots, n$. That is, the estimator becomes the Hodges-Lehmann two-sample estimator applied to the two samples $Y_1, \ldots, Y_m$ and $Y_{m+1}, \ldots, Y_n$. It should also be noted that the estimator $\hat{\beta}$ is more robust to gross errors in the observations than is the classical least squares estimator given by

$$\overline{\beta} = \frac{\sum_{i=1}^{n}(Y_i - \bar{Y})x_i - \bar{x})}{\sum_{j=1}^{n}(x_j - \bar{x})^2}$$

where $\bar{x} = \sum_1^n x_i/n$ and $\overline{Y} = \sum_{j=1}^{n} Y_j/n$.

22.6 Tests for Regression Parameters

Let $Y = (Y_{n1}, \ldots, Y_{nn})$ be a sequence of random vectors, where Y_{nj} ($j = 1, \ldots, n$) are independent with distributions

$$P_{\alpha,\beta}(Y_{nj} \le y) = P(y - \alpha - \beta x_{nj}) \tag{22.6.1}$$

where $P_{\alpha,\beta}(\cdot)$ denotes the probability of the event $(\cdot)$ computed for the parameter values α and β. The x_{nj} are known constants which depend on n. We suppress this dependence for the sake of simplicity. The problem is to construct rank score tests for the hypothesis $H : \alpha = \beta = 0$ when the form of F is unknown. Daniels (1954) proposed a distribution-free test of H which is related to the Hodge's bivariate sign test [see Hill (1960)]. Adichie (1967a) proposed a class of nonparametric tests for H and studied the asymptotic efficiency of the class. In the following we present his results.

Let us assume that $F \in \mathcal{F}$ where

$$\begin{aligned} &\mathcal{F} = \text{absolutely continuous } F \text{ with:} \\ &\text{(i)} \quad F'(y) = f(y) \text{ is absolutely continuous,} \\ &\text{(ii)} \quad \int_{-\infty}^{\infty} \left(f'(y)/f(y)\right)^2 f(y)dy \quad \text{is finite,} \\ &\text{(iii)} \quad f(-y) = f(y). \end{aligned} \tag{22.6.2}$$

Also, assume that the constants x_j, $j = 1, \ldots, n$ satisfy:

$$\lim_{n\to\infty} \left[\left\{ \max_j (x_j - \bar{x}_n)^2 \right\} / \left\{ \sum_{j=1}^{n} (x_j - \bar{x}_n)^2 \right\} \right] = 0\,, \tag{22.6.3}$$

$$\lim \left\{ n^{-1} \sum_{j=1}^{n} (x_j - \bar{x}_n)^2 \right\} < \infty, \quad (\lim\ \bar{x}_n) < \infty\,, \tag{22.6.4}$$

$$\lim\ n^{-1} \sum_{j=1}^{n} (x_j - \bar{x}_n)^2 > 0 \tag{22.6.5}$$

where

$$\bar{x}_n = \frac{1}{n} \sum_{j=1}^{n} x_j\,.$$

Also, define a class of functions

$$\psi(u) = -\left[g'\left(G^{-1}\left(\frac{u}{2} + \frac{1}{2}\right)\right) / g\left(G^{-1}\left(\frac{u}{2} + \frac{1}{2}\right)\right)\right]\,, \quad 0 < u < 1\,, \tag{22.6.6}$$

where G^{-1} is the inverse of G and G is any known distribution function belonging to the class $\mathcal{F}$. The (22.6.6) function corresponding to F is denoted by $\phi(u)$ and is given by

$$\phi(u) = -\left[f'\left(F^{-1}\left(\frac{u}{2} + \frac{1}{2}\right)\right) / f\left(F^{-1}\left(\frac{u}{2} + \frac{1}{2}\right)\right)\right]\,. \tag{22.6.7}$$

Note that unlike $\psi(u)$, $\phi(u)$ is not known since F is unknown.

A Class of Test Statistics

Let R_j denote the rank of $|Y_j|$ in the sequence of absolute values $|Y_1|, \ldots, |Y_n|$. Then define the pair of statistics T_1 and T_2 by

$$T_1(Y) = n^{-1} \sum_{j=1}^{n} \psi_n\left(\frac{R_j}{n+1}\right) \text{ Sign } Y_j \tag{22.6.8}$$

and

$$T_2(Y) = n^{-1} \sum_{j=1}^{n} x_j \psi_n\left(\frac{R_j}{n+1}\right) \text{ Sign } Y_j \tag{22.6.9}$$

where

$$\psi_n(u) = \psi\left(\frac{j}{n+1}\right), \quad \frac{j-1}{n} \le u \le \frac{j}{n}. \tag{22.6.10}$$

Then

$$\lim_{n\to\infty} \int_0^1 [\psi_n(u) - \psi(u)]^2 \, du = 0 \tag{22.6.11}$$

by a result of Hájek (1961). Define the 2×2 matrix $((\gamma_{kl}))$ by

$$\begin{aligned} \gamma_{11} &= \int_0^1 \psi^2(u) du, \quad \gamma_{22} = n^{-1} \sum_{j-1}^{n} x_j^2 \gamma_{11} \\ \gamma_{21} &= \gamma_{12} = n^{-1} \sum_{j=1}^{n} x_j \gamma_{11} . \end{aligned} \tag{22.6.12}$$

Also, let

$$M(\psi) = n(T_1, T_2)((\gamma_{kl}))^{-1}(T_1, T_2)' . \tag{22.6.13}$$

Adichie (1967a) proposes $M(\psi)$ as a test criterion for H. Note that M is well defined since both $((\gamma_{kl}))^{-1}$ and its limit as $n \to \infty$ exist because of (22.6.5).

So, $M(\psi)$ is a class of test statistic generated by $G \in \mathcal{F}$. For instance, if $G(y) = \{1 + \exp(-y)\}^{-1}$ (namely, the logistic distribution), $\psi(u) = u$ and, consequently,

$$T_1(Y) = n^{-1} \sum_{j=1}^{n} \left\{\frac{R_j}{(n+1)}\right\} \text{Sign } Y_j$$

and

$$T_2(Y) = n^{-1} \sum_{j=1}^{n} \left\{ \frac{x_j R_j}{(n+1)} \right\} \text{Sign } Y_j$$

in which case M is said to be of the Wilcoxon type. If G is the double exponential distribution, then $\psi(u) \equiv 1$,

$$T_1(Y) = n^{-1} \sum_{j=1} \text{Sign } Y_j, \qquad T_2(Y) = n^{-1} \sum_{j=1} x_j \text{ Sign } Y_j\,,$$

and M is said to be of the sign type; G being the standard normal distribution leads to M of the Van der Waerden type. Also observe that T_1 is equivalent to the usual rank score statistic for the one-sample problem [see Govindarajulu (1960)], while T_2 is analogous to Hájek's (1962) statistic for the test of symmetry.

Adichie (1967a) establishes that under H_0 a linear combination of T_1 and T_2 is asymptotically normal and thus, (T_1, T_2) has an asymptotic bivariate normal distribution with mean 0 and variance-covariance matrix $((\gamma_{kl}))$. Thus, for large n, H is rejected when $M > \chi^2_{2,1-\alpha}$.

Next, in order to evaluate the asymptotic efficiency of M, Adichie (1967a, Section 4) employs contiguity arguments for establishing the asymptotic bivariate normality of $n^{1/2}(T_1, T_2)$ under a sequence of alternatives tending to H at the rate of $n^{-1/2}$. Thus, it is shown that the Pitman efficiency of M relative to the classical F test is equivalent to the efficiency of the corresponding rank score tests relative to Student's t-test in the two-sample problem. For details, see Adichie (1967a, Section 5).

22.7 Estimates of Regression Coefficients

Hodges and Lehmann (1963) proposed a general method of obtaining nonparametric (i.e., robust) estimates for the location parameter based on statistics used for testing the hypothesis that the parameter has a specified value. To illustrate this, in the following lemma, we will show the relation between Hodges-Lehmann estimator of the location of symmetry of a distribution and the Wilcoxon signed rank test statistic. Without loss of generality, let us assume that we wish to test the null hypothesis that the point of symmetry of the distribution is zero.

Let $Y_1, \ldots, Y_n$ be a random sample from $F(x)$. Let

$$T = \sum_{j=1}^{n} R_j \text{ Sign } (Y_j) \tag{22.7.1}$$

where R_j denotes the rank of $|Y_j|$ among the ranked absolute Y-values. Then we have the following lemma.

Lemma 22.7.1 *With the above notation, we can write*

$$T = 2N^+ - n(n+1)/2 \tag{22.7.2}$$

where N^+ *=number of pairs* (i, j), $1 \le i \le j \le n$ *such that* $Y_i + Y_j$ *is positive.*

PROOF: One can write

$$T = \sum_{j=1} R_j^+ - \sum_{j=1} R_j^-$$

where R_j^+ denotes the rank of positive Y_j. However,

$$\sum R_j^+ + \sum R_j^- = \sum_{j=1}^{n} R_j = n(n+1)/2\,.$$

Consequently,

$$T = 2\sum R_j^+ - n(n+1)/2\,.$$

Thus, it suffices to show that $\sum R_j^+ = N^+$. Towards this, consider

$$\begin{aligned} \sum R_j^+ &= \sum_{j=1}^{n}\left[\sum_{i=1}^{n} I\left(|Y_i| < |Y_j|\right)\right] I(Y_j > 0) \\ &= \sum_{i=1}^{n}\left[\sum_{j=1}^{n} I\left(|Y_i| < |Y_j|\right) I(Y_j > 0)\right] = \sum_{i=1}^{n} N_i^+ = N^+ \end{aligned}$$

where N_i^+ denotes the number of Y_j's which are positive and which exceed Y_i in absolute value.

Adichie (1967b) proposes Hodges-Lehmann type of estimators for the regression coefficients in a linear regression setup and studies their asymptotic properties. In particular, he shows that the Brown-Mood median procedures considered in Section 22.2 are inefficient relative to Hodges-Lehmann type of estimators. In the following, we will provide Adichie's results suppressing some of the technical details. As in Section 22.6, let $Y_1, \ldots, Y_n$ be independent random variables having distributions

$$P_{\alpha,\beta}(Y_j \le y) = F(y - \alpha - \beta x_j) \tag{22.7.3}$$

where the x_j are known regression constants that are not all equal which satisfy the limiting conditions (22.6.3)–(22.6.5). As before, we assume that the underlying distribution function F belongs to a class $\mathcal{F}$ satisfying (22.6.2).

Let $T_1(Y_1, \ldots, Y_n)$ and $T_2(Y_1, \ldots, Y_n)$ (to be respectively denoted by $T_1(Y)$ and $T_2(Y)$ for the sake of simplicity) be two statistics for testing hypotheses about α and β in (22.7.3). Throughout we assume that T_1 and T_2 satisfy:

(A) For fixed b, $T_1(y + a + bx)$ is nondecreasing in a; and for every a, $T_2(y + a + bx)$ is nondecreasing in b, for every y and x, where

$$y + a + bx = (y_1 + a + bx_1, \ldots, y_n + a + bx_n)\,.$$

(B) When $\alpha = \beta = 0$, the distributions of $T_1(Y)$ and $T_2(Y)$ are symmetric about fixed points μ and ν, independent of $F \in \mathcal{F}$.

Let

$$\begin{aligned}
\beta^* &= \sup\{b : T_2(y - a - bx) > \nu, \text{ for all } a\}\ ;\\
\beta^{**} &= \inf\{b : T_2(y - a - bx) < \nu, \text{ for all } a\}\ ;\\
\hat{\beta} &= (\beta^* + \beta^{**})/2;\\
\alpha^* &= \sup\left\{a : T_1(y - a - \hat{\beta}x) > \mu\right\}\ ;\\
\alpha^{**} &= \inf\left\{a : T_1(y - a - \hat{\beta}x) < \mu\right\}\ ;
\end{aligned} \tag{22.7.4}$$

and

$$\hat{\alpha} = (\alpha^* + \alpha^{**})/2\,. \tag{22.7.5}$$

For suitable functions T_1 and T_2, $\hat{\alpha}$ and $\hat{\beta}$ are proposed as estimates of α and β. It should be noted that several existing estimates of α and β belong to the class of (22.7.4) and (22.7.5) estimates. In particular, the least squares estimates $\tilde{\alpha}$ and $\tilde{\beta}$ constitute special cases of $\hat{\alpha}$ and $\hat{\beta}$. Towards this, let $T_1(Y) = \sum_1^n Y_j$ and $T_2(Y) = \sum_1^n (x_j - \bar{x}_n)(Y_j - \bar{Y}_n)$ where as before, $\bar{x}_n$ and $\bar{Y}_n$ denote the averages of the x's and Y's. For this choice, it is easy to see that conditions (A) ad (B) are satisfied with $\mu = \nu = 0$. Furthermore,

$$\begin{aligned}
\sup\{b\colon T_2(y - a - bx) > 0, \text{ for all } a\} &= \inf\{b\colon T_2(y - a - bx) < 0, \text{ for all } a\}\\
&= \frac{\{\sum_1^n (x_j - \bar{x}_n)(y_j - \bar{y}_n)\}}{\{\sum_1^n x_i - \bar{x}_n)^2\}}\\
&= \bar{\beta}\,.
\end{aligned} \tag{22.7.6}$$

Also,

$$\sup\left\{a : T_1(y - a - \tilde{\beta}x) > 0\right\} = \inf\left\{a : T_1(y - a - \tilde{\beta}x) < 0\right\} \\ = (\bar{y}_n - \tilde{\beta}\bar{x}_n) = \tilde{\alpha}\,. \tag{22.7.7}$$

Estimates Based on Rank Tests

Since robustness is the criterion, Adichie (1967b) considers $\hat{\alpha}$ and $\hat{\beta}$ based on rank (or mixed rank) statistics. Let $\psi(u)$ be defined by (22.6.6), $\phi(u)$ by (22.6.7) and

$$\psi_0(u) = -\left[g'\left(G^{-1}(u)\right)/g\left(G^{-1}(u)\right)\right], \quad 0 < u < 1 \tag{22.7.8}$$

and

$$\varphi_0(u) = -\left[f'\left(F^{-1}(u)\right)/f\left(F^{-1}(u)\right)\right], \quad 0 < u < 1\,. \tag{22.7.9}$$

Let

$$T_1(Y) = n^{-1}\sum_1^n \psi_n\left(\frac{\mathcal{R}_i}{n+1}\right) \text{ Sign } Y_j \tag{22.7.10}$$

and

$$T_2(Y) = n^{-1}\sum_1^n (x_j - \bar{x}_n)\psi_{0n}\left(\frac{R_j}{n+1}\right) \tag{22.7.11}$$

where $\mathcal{R}_j$ is the rank of $|Y_j|$ among $|Y_1|, \ldots, |Y_n|$ while R_j denotes the rank of Y_j in the ordered sample $V_1 < \cdots < V_n$. That is,

$$Y_j = V_{R_j}, \quad (j = 1, \ldots, n),$$

and

$$\psi_n(u) = \psi\left(\frac{j}{n+1}\right), \ \psi_{0n}(u) = \psi_0\left(\frac{j}{n+1}\right), \quad \text{for } \frac{j-1}{n} < u \le \frac{j}{n}\,. \tag{22.7.12}$$

Notice that T_1 given by (22.7.10) is the statistic studied in Section 22.6, while T_2 given by (22.7.11) is studied by Hájek (1962).

Using arguments similar to those in Hodges and Lehmann (1963), one can show that T_1 and T_2 given by (22.7.10) and (22.7.11) satisfy conditions (A) and (B). If we choose G to be the logistic distribution in (22.6.6) and (22.7.8), then $\psi(u) = u$ and hence T_1 and T_2 coincide with the Wilcoxon one-sample (signed rank) and two-sample statistics, respectively. Denoting the resulting estimates of α and β by $\hat{\alpha}_W$ and $\hat{\beta}_W$, we have

$$
\begin{aligned}
T_1(y - a - \hat{\beta}_W x) &= [n(n+1)]^{-1} \sum_{j=1}^{n} \hat{\mathcal{R}}_j \text{ Sign } (y_j - a - \hat{\beta}_W x_j) \\
&= [n(n+1)]^{-1} \left[2N^+ - n(n+1)/2\right] \text{ (see Lemma 22.7.1)}
\end{aligned}
$$

where N^+ is the number of pairs (i, j) with $1 \le i \le j \le n$ such that $y_i + y_j - \hat{\beta}_W(x_i + x_j) - 2a$ is positive and $\hat{\mathcal{R}}_j$ is the rank of $|y_j - a - \hat{\beta}_W x_j|$ in the sequence of absolute values $|y_1 - a - \hat{\beta}_W x_1|, \ldots, |y_n - a - \hat{\beta}_W x_n|$. The estimate $\hat{\alpha}_W$ is then given by

$$
\hat{\alpha}_W = \text{med}_{i \le j} \frac{1}{2} \left\{ Y_i + Y_j - \hat{\beta}_W(x_i + x_j) \right\} \tag{22.7.13}
$$

where $\hat{\beta}_W$ is obtained from (22.7.4) with

$$
T_2(Y) = n^{-1} \sum (x_j - \bar{x}_n) \left(\frac{R_j}{n+1} \right) . \tag{22.7.14}
$$

But, it is not easy to obtain an explicit expression for $\hat{\beta}_W$ in terms of Y's without some further assumptions on the regression constants x_j. For instance, if $x_j = c_1$ for $1 \le j \le k$ and $x_j = c_2$ for $k+1 \le j \le n$, the resulting estimate of $\hat{\beta}_W$ would coincide with the Hodges-Lehmann estimate for shift in the two-sample problem, i.e., $\hat{\beta}_W = \text{med}(Y - Z)$ where $Y = (Y_1, \ldots, Y_k)$ and $Z = (Y_{k+1}, \ldots, Y_n)$. However, for arbitrary constants c_j, one can utilize an iterative method for computing $\hat{\beta}_W$ for moderate values of n [see Adichie (1967b, pp. 897–898)]. The estimates $\hat{\alpha}$ and $\hat{\beta}$ defined by (22.7.4) and (22.7.5) have some invariance and symmetry properties given by the following lemma.

Lemma 22.7.2 *For any real a and b, the (22.7.4) and (22.7.5) estimates are such that*

$$
\hat{\beta}(y + a + bx) = \hat{\beta}(y) + b \tag{22.7.15}
$$

and

$$
\hat{\alpha}(y + a + bx) = \hat{\alpha}(y) + a . \tag{22.7.16}
$$

Furthermore, let T_1 and T_2 be given by (22.7.10) and (22.7.11) with ψ non-decreasing and $\hat{\alpha}$ and $\hat{\beta}$ be the (22.7.5) and (22.7.4) estimates. If $F \in \mathcal{F}$, then the distributions of $\hat{\alpha}$ and $\hat{\beta}$ are symmetric about α and β respectively, and hence, $\hat{\alpha}$ and $\hat{\beta}$ are unbiased.

Adichie (1976b, Sections 5 and 6) establishes the joint asymptotic normality of $\hat{\alpha}$ and $\hat{\beta}$ when $\bar{x}_n \to \bar{x}$ as $n \to \infty$, $\alpha = an^{-1/2}$ and $\beta = bn^{-1/2}$ where a and b are some real constants. He shows that the asymptotic efficiency of the Brown-Mood estimate relative to estimates $(\hat{\alpha}_s, \hat{\beta}_s)$ based on the sign statistics

$$\begin{aligned} T_1(Y) &= n^{-1}\sum \text{Sign } Y_j \text{ and} \\ T_2(Y) &= n^{-1}\sum(x_j - \bar{x}) \text{ Sign } Y_j \end{aligned} \tag{22.7.17}$$

is $(1/2)3^{1/2} < 1$ which implies that there is some loss in the efficiency of the median estimates. This loss can perhaps be attributed to the fact that some information is lost in the process of ordering observations in two separate groups.

Jurecková (1971) has extended these ideas to constructing an estimate of regression parameter vector in multiple regression problems based on suitable rank statistics. Asymptotic linearity of these rank statistics associated with the multiple regression setup has also been established. It is also shown that the asymptotic distribution of the regression vector estimates is the same as that of maximum likelihood estimates. Koul (1969) constructs a class of confidence regions based on rank statistics for the multiple regression vector. He also studies the asymptotic behavior of the center of gravity of a region corresponding to the Wilcoxon type rank statistic.

22.8 Estimates Based on Residuals

An appealing approach to estimating regression coefficients in a linear model is to obtain those values of the coefficients which make the residuals as small as possible. Such an approach has been proposed by Jaeckel (1972) who defines the estimates as those values of the parameters which minimize the dispersion of the residuals.

Let $Y_1, \ldots, Y_n$ be independent random variables with continuous distribution functions

$$F(y - \alpha^0 - \beta^0 c^i), \quad i = 1, \ldots, n, \tag{22.8.1}$$

where the c^i are known K-vectors and α^0 and the K-vector β^0 are unknown. Only estimation of β^0 and not that of α^0 is possible by the following method.

Let $D(z)$ be a translation-invariant measure of the dispersion of $z = (z_1, \ldots, z_n)$; that is $D(z + b) = D(z)$ where $b = (b, \ldots, b)$. The estimate of β^0 is any β that minimizes $D(Y - \beta C)$ where $Y = (Y_1, \ldots, Y_n)$ and $C = ((c_{ji}))$ is the $K \times n$ matrix having c^i as its columns. If $D(z)$ is the variance of the z_i, then our estimates are the usual least squares estimates of β_i^0. Since least squares estimates are somewhat sensitive to large deviations, it is of interest to consider "more robust" estimates. The particular class of dispersion measures considered by Jaeckel (1972) is certain linear combinations of the ordered z_i.

Let $a_n(k)$, $k = 1, \ldots, n$ be a nondecreasing set of scores, not all equal, such that

$$\sum_{k=1}^{n} a_n(k) = 0\,. \tag{22.8.2}$$

Such scores may be generated by a function $\phi(u)$, $0 < u < 1$ as follows:

$$a_n(i) = E\phi(U_n^{(i)}) \text{ or } a_n(i) = \phi\left(i/(n+1)\right)\,. \tag{22.8.3}$$

For any $z = (z_1, \ldots, z_n)$, let $z_{(1)} \leq \cdots \leq z_{(n)}$ be the ordered z_i. Define

$$D(z) = \sum_{k=1}^{n} a_n(k) z_{(k)}\,. \tag{22.8.4}$$

Since $D(z)$ is small when the z_i are close to each other, we will use it as a measure of dispersion.

For fixed $Y_1, \ldots, Y_n$ and for any β, the residuals are

$$z_i = Y_i - \beta c^i\,, \quad i = 1, \ldots, n\,. \tag{22.8.5}$$

The ordered residuals are

$$z_{(k)} = Y_{i(k)} - \beta c^{i(k)}\,, \quad k = 1, \ldots, n\,,$$

where $i(k)$ denotes the index of the observation giving rise to the kth ordered residual. (When two residuals are equal, there is an ambiguity in $i(k)$, but not in $z_{(k)}$.) The dispersion of the residuals, as a function of β, is

$$D(Y - \beta C) = \sum_{k=1}^{n} \theta_n(k) \left[Y_{i(k)} - \beta c^{i(k)}\right]\,. \tag{22.8.6}$$

In the following we list the properties of the dispersion and for proofs, see Theorems 1 and 2 of Jaeckel (1972).

Theorem 22.8.1

(i) *For fixed Y, $D(Y - \beta C)$ is nonnegative, continuous and a convex function of β.*

(ii) *Let E be the $n \times n$ matrix all of whose entries are $1/n$. Let $\bar{C} = CE$. If $C - \bar{C}$ has rank K, then for any D_0, the set $\{\beta : D(Y - \beta C) \leq D_0\}$ is bounded.*

Then the proposed estimate of β^0 is β_D which minimizes $D(Y - \beta C)$. Note that β_D may not be unique. Jaeckel (1972, p. 1452) claims that since the diameter of the set of possible values of $n^{1/2}\beta_D$ tends to zero as n gets large, it is immaterial how β_D is chosen when n is large. The estimate is invariant in the sense that if β' is any K-vector, then $\beta_D(Y + \beta' C) = \beta_D(Y) + \beta'$.

Computation of B_D involves minimizing a convex function of β, which is relatively easy. One possible method is the method of steepest descent which employs first derivatives. Minimization procedures based on second derivatives cannot be used since the second derivatives of D are identically zero whenever they exist.

Jaeckel (1972, Section 2) shows that β_D is asymptotically equivalent to Jureckovás (1971) estimate. Hence, $n^{1/2}\beta_D$ is asymptotically normal.

For the special case of Wilcoxon scores, namely,

$$a_n(k) = \frac{k}{k+1} - \frac{1}{2},$$

when $K = 1$ (single parameter case), β_0 is a median of the set of pairwise slopes

$$\beta_{ij} = (Y_j - Y_i)/(c^j - c^i) \text{ for } c^j > c^i,$$

which coincides with Theil's (1950c) estimate of β. Bickel (1973) examines the large sample behavior of some estimates of the regression parameters which in general depend on a preliminary estimate and the ordered residuals based on it.

So far, we have assumed that the regression function belongs to the parametric family of linear functions in $R^{(K)}$, but no parametric form is assumed for the distribution of the residuals. In fact, one can consider the case of estimating $E(Y|X)$ where no functional form for the conditional expectation is assumed. Stone (1975, 1977, 1982) considers the rate of convergence of such nonparametric estimates. Also, there is a lot of literature on the use of method of splines for estimating such conditional expectation. For a survey of these results, the interested reader is referred to Stone, Hansen, Kooperberg and Troung (1977) and Wahba and Wold (1975).

22.9 Problems

22.3.1 The following data represents the heights (x) and the weights (Y) of several men. The heights are fixed in advance and then observe the weights of a random group of men having the selected heights.

x (in inches)	60	61	63	62	64	65	68	70	69	72
Y (in pounds)	120	132	140	120	150	152	160	175	180	186

Assuming that a linear regression model of the form (22.3.1) fits the data, test

$$H_0 : \beta = 1 \quad \text{versus} \quad H_a : \beta > 1$$

using a level of significance $\alpha = 0.05$.

22.3.2 The following data represents $x=$ the number of weeks a certain person was in a Weight-Watcher's Program and $Y=$ weight of that person in pounds.

x	1	2	3	4	5	6	7	8
Y	100	96	94	94	90	88	86	85

Assuming a linear regression model for x and Y, test

$$H_0 : \beta = -1 \quad \text{versus} \quad H_a : \beta < -1$$

using a level of significance $\alpha = 0.05$.

(**Hint:** In Problems 22.3.1 and 22.3.2 use the asymptotic normality of the test criterion discussed in Section 22.4.)

22.5.1 The following data deals with trips made by a random sample of 10 faculty members of a certain college to professional meetings giving the number of days stayed and the total cost of the trip.

Faculty Member	Length of the trip (in days)	Total Living Expenses (in dollars)
1	4	250
2	3	195
3	5	305
4	2	126
5	2.5	150
6	3.5	210
7	4.5	260
8	1.5	85
9	4.75	285
10	3.75	245

Let x denote the length of the trip and Y denote the cost of the trip. Find Theil's esimator of the slope of the regression of Y on x.

22.5.2 The following gives Y = the tensile strength of a certain steel product and x = the concentration of a compound thought to be related to the tensile strength.[1]

Test	1	2	3	4	5	6	7	8	9	10
x	21	8	22	18	15	10	19	9	20	16
Y	184	70	94	120	117	26	110	47	110	88

Find Theil's esimator of the slope of the regression of Y on x.

22.6.1 The following fictitious data gives years in service and rating for a random sample of employees in a certain company. We wish to know whether there is a tendency to rate old employees higher than more recent employees.

Employee	1	2	3	4	5	6	7	8	9	10	11	12
Years in Service (X)	1	8	9	3	4	2	5	6	7	6	5	3
Rating (Y)	4	8	9	5	6	6	4	6	5	9	5	7

a) Plot the scatter diagram.

[1]Wallis, W.A. and Roberts, H.V. (1956). *Statistics: A New Approach.* The Free Press, New York, p. 557.

b) We wish to test the hypothesis $H_0 : \beta = 0$ at level of significance $\alpha = 0.05$. Carry out a test of H_0 using the statistic M given by (22.6.13) with T_1 and T_2 that are based on the logistic distribution.

22.6.2 Suppose you are preparing a report for the chief executive officer (CEO) of a certain manufacturing company, who is interested in the average cost per unit of production. Units of production can easily be measured but average cost requires lengthy and difficult computations. An estimation procedure can be employed, provided some relationship between these two quantities can be determined empirically. From past records the following data are available[2]:

Units Produced: x (in thousands)	9	13	5	17	18	8	14	12	7	15	10	11	4	23
Average Cost: Y (in dollars)	1.1	1.9	3.5	5.9	7.4	1.4	2.6	1.4	1.9	3.5	1.0	1.1	4.6	17.9

We are interested in testing $H : \alpha = \beta = 0$ at level of significance $\alpha = 0.05$. Carry out a test of H based on the statistic M given by (22.6.13) with T_1 and T_2 that are based on the logistic distribution.

22.6.3 The following data is on Y = average blood lead level of white children age 6 months to 5 years and x = amount of lead used in gasoline production (in 1000 tons) for ten 6-month periods[3]:

x	48	59	79	80	95	95	97	102	102	107
Y	9.3	11.0	12.8	14.1	13.6	13.8	14.6	14.6	16.0	18.2

Test the hypothesis $H : \alpha = \beta = 0$ at the level of significance $\alpha = 0.05$ using the statistic M given by (22.6.13) with T_1 and T_2 that are based on the logistic distribution.

[2]Ostle, B. (1954). *Statistics in Research,* The Iowa State Univ. Press, Ames, Iowa, Problem 6.15 on p. 170.

[3]Chronological trend in blood lead levels. *New England Journal of Medicine.* 1983, p. 1373.
See also, Devore, J.L. (2000). *Probability and Statistics for Engineering and the Sciences.* Duxbury, Pacific Grove, CA. Problem No. 65 on p. 538.

Chapter 23

Useful Asymptotic Results

23.1 Introduction

In the proceeding chapters we provide the asymptotic results and certain inequalities that are useful in nonparametric inference and for a complete understanding of the material covered in the preceding chapters. The proceeding results, although of interest in themselves, are not required to follow and implement the testing or estimation procedures presented in earlier chapter.

23.2 Probability Inequalities

Markov inequality. If X is a positive random variable having a finite mean μ, then for any $a > 0$

$$P(X > a) \leq \frac{\mu}{a}.$$

Chebyshev's inequality. If X is a random variable having finite mean μ and variance σ^2, then for any $a > 0$

$$P\Big[|X - \mu| > a\Big] \leq \frac{\sigma^2}{a^2}.$$

Kolmogorov's inequality. If $\{X_i\}$ is an independent sequence having zero means and variances $\sigma_i^2 (i = 1, 2, \ldots)$, then

$$P(\max_{k \leq n} |S_k| \geq \varepsilon) \leq \varepsilon^{-2} \sum_{i=1}^{n} \sigma_i^2$$

where $S_k = X_1 + \ldots + X_k$.

Jensen's inequality. If $g(x)$ is convex and the relevant expectations exist, then for any random variable X,

$$Eg(X) \geq g(EX).$$

Hölder's inequality. Let $p > 0$ and $q > 0$ such that $\frac{1}{p} + \frac{1}{q} = 1$. Then for any random variables X and Y

$$E|XY| \leq \{E|X|^p\}^{\frac{1}{p}}\{E|Y|^q\}^{\frac{1}{q}}.$$

Expansion for the Tail Probability of a Normal Variable. Without loss of generality we can assume the normal variable to have mean 0 and variance 1. Let $\phi(t) = (2\pi)^{-\frac{1}{2}}\exp(-\frac{t^2}{2})$, for $-\infty < t < \infty$. So, consider, $x > 0$. Then we have

$$\int_x^\infty \phi(t)dt = \phi(x)(\frac{1}{x} - \frac{1}{x^3} + \frac{3}{x^5} - \frac{3\cdot 5}{x^7} + \ldots).$$

Proof.

$$\begin{aligned}
\int_x^\infty \phi(t)dt = \int \frac{1}{t}d(-\phi) &= \frac{\phi(x)}{x} - \int_x^\infty t^{-2}\phi(t)dt \\
&= \frac{\phi(x)}{x} + [\int_x^\infty t^{-3}d(\phi)] \\
&= \frac{\phi(x)}{x} - \frac{\phi(x)}{x^3} + 3\int_x^\infty t^{-4}\phi(t)dt.
\end{aligned}$$

So, by repeated integration by parts, we can generate the rest of the series. In particular, we obtain the following bounds:

$$\phi(x)(\frac{1}{x} - \frac{1}{x^3}) \leq \int_x^\infty \phi(t)dt \leq \phi(x)(\frac{1}{x} - \frac{1}{x^3} + \frac{3}{x^5}).$$

$R(x) = \frac{\{1-\Phi(x)\}}{\phi(x)}$ is called Mill's ratio. An alternate lower bound for $R(x)$ is given by Gnedenko (1962, p.134, Problem #16) and Pitman(2003, p.26) using Hermite functions, provides a system of recurring relations from which one can obtain lower and upper bounds. These will be provided in the following. The Hermite function of index $-2q$ ($q > 0$) is defined by the integral (with $h_0(x) = 1$)

$$\begin{aligned}
h_{-2q}(x) &= \frac{1}{\Gamma(2q)}\int_0^\infty t^{2q-1}e^{-\frac{1}{2}t^2 - xt}dt \\
&= \frac{2^{q-1}}{\Gamma(2q)}\int_0^\infty v^{q-1}e^{-v-x\sqrt{2v}}dv.
\end{aligned}$$

Then $h_{-1}(x)$ equals the Mill's ratio $R(x)$. By performing integration by parts we obtain

$$\begin{aligned} h_{-2}(x) = \int_0^\infty te^{-\frac{1}{2}t^2 - xt}dt &= \int_0^\infty (t + x - x)e^{-\frac{1}{2}t^2 - xt}dt \\ &= 1 - xh_{-1}(x). \end{aligned}$$

Similarly

$$\begin{aligned} 2!h_{-3}(x) = \int_0^\infty t^2 e^{-\frac{1}{2}t^2 - xt}dt &= \int_0^\infty t(t + x)e^{-\frac{1}{2}t^2 - xt}dt - xh_{-2}(x) \\ &= -te^{-\frac{1}{2}t^2 - xt}\Big|_0^t + \int_0^\infty e^{-\frac{1}{2}t^2 - xt}dt - xh_{-2}(x) \\ &= 0 + h_{-1}(x) - xh_{-2}(x) \\ &= -x + (1 + x^2)h_{-1}(x). \end{aligned}$$

Proceeding analogously and using the preceding recurrence relation, one can obtain

$$3!h_{-4}(x) = 2 + x^2 - (3x + x^3)h_{-1}(x), (\text{see Pitman(2003, Eq. (109))}).$$

Now since all the h functions are nonnegative, we obtain the series of inequalities

$$\begin{aligned} h_{-1} &\le \frac{1}{x} \\ h_{-1}(x) &\ge \frac{x}{1 + x^2} \quad \text{(Gnedenko (1962))} \\ \text{and} \qquad h_{-1}(x) &\le \frac{(2 + x^2)}{(3x + x^3)}. \end{aligned}$$

The first upper bound coincides with the earlier one. Gnedenko's lower bound will be sharper than $\frac{(x^2-1)}{x^3}$ provided

$$\frac{x}{1 + x^2} \ge \frac{x^2 - 1}{x^3}.$$

Or when $x^4 \ge x^4 - 1$ which is true. Next the Pitman's upper bound to $R(x)$ will be sharper than the earlier bound, namely, $x^{-1} - x^{-3} + 3x^{-5}$ provided

$$\begin{aligned} \frac{2 + x^2}{3x + x^3} &< \frac{x^4 - x^2 + 3}{x^5}. \\ \text{or when} \quad x^4(2 + x^2) &< (3 + x^2)(x^4 - x^2 + 3). \\ \text{or when} \quad 2x^4 &< 2x^4 + 9 \end{aligned}$$

which is always true.

23.3 Laws of Large Numbers

23.3.1 Weak Law of Large Numbers

Theorem 23.3.1. *Let $\{X_i\}$ be i.i.d random variables with distribution function F possessing a finite mean μ. Then $n^{-1}\sum_{i=1}^{n} X_i$ converges in probability to μ.*

Remark 23.3.1. *If the variance of X is finite, then the weak law can be proved using the Chebyshev's inequality in a straight forward fashion.*

23.3.2 Strong Law of Large Numbers

Theorem 23.3.2 (Kolmogorov). *If the $\{X_i\}$ are i.i.d with finite mean μ, then $n^{-1}\sum_{i=1}^{n} X_i \to \mu$ with probability one (wp1.).*

23.3.3 Convergence of a Function of Variables

In nonparametric statistics one is often faced with the problem of asserting convergence in probability of a continuous function of a sequence of random variables converging to a random variable in probability. In the following we shall provide an elementary proof of the same.

Theorem 23.3.3. *Let $\{X_N\}$ be a sequence of random variables converging to X in probability. Let g be a continuous function that maps X_N onto the real line. Then $g(X_N)$ converges to $g(X)$ in probability.*

PROOF: Let $\varepsilon > 0$ and choose another constant r_1 such that $P(|X| > r_1) < \varepsilon$. Now, for some r, $\delta > 0$ consider the event

$$\begin{aligned}\{|X_N - X| < \delta, |X| < r\} &\subseteq \{|X_N - X| < \delta, |X_N| < r + \delta, |X| < r\} \\ &\subseteq \{|X_N - X| < \delta, |X| < r_1;\ |X_N| < r_1\} \\ &\subseteq \{|g(X_N) - g(X)| < \varepsilon\}\end{aligned}$$

where $r_1 = r + \delta$ and because g is uniformly continuous in a compact interval. Taking complements of the events on both sides we obtain,

$$\{|g(X_N) - g(X)| > \varepsilon\} \subseteq \{|X_N - X| > \delta\} \cup \{|X| > r_1\}.$$

Hence

$$P(|g(X_N) - g(X)| > \varepsilon) \subseteq P(|X_N - X| > \delta) + P(|X| > r_1) \leq 2\varepsilon.$$

It should be noted that X_N and X could be finite dimensional vectors in Theorem 23.3.3.

23.4 Central Limit Theorems

Theorem 23.4.1 (Lindeberg-Levy). *If* $\{X_i\}$ *is an i.i.d. sequence having finite* μ *and variance* σ^2, *then, for large* n, $n^{\frac{1}{2}}\frac{(n^{-1}\sum_1^n X_i-\mu)}{\sigma}$ *is distributed as normal* $(0,1)$. *The multi-variate extension of Theorem 23.4.1 is*

Theorem 23.4.2. *Let* $\{\mathbf{X}_i\}$ *be i.i.d. random vectors having mean* $\mu = (\mu_1,\ldots,\mu_p)'$ *(where p denotes the dimension of* $\mathbf{X}_i$*) and covariance matrix* $\mathbf{\Sigma}$. *Then, for large* n, $n^{\frac{1}{2}}(n^{-1}\sum_{i=1}^n \mathbf{X_i} - \mu)$ *is distributed as normal* $(0,\mathbf{\Sigma})$.

Theorem 23.4.3 (Lindeberg-Feller). *Let* X_i *be an independent random variable with mean* μ_i, *variance* σ_i^2 *and distribution* $F_i(i = 1,\ldots)$. *Also let* $B_n^2 = \sum_1^n \sigma_i^2$ *such that* $B_n \to \infty$ *and* $\frac{\sigma_n^2}{B_n^2} \to 0$ *as* $n \to \infty$. *Then* $B_n^{-1}\sum_1^n(X_i - \mu_i)$ *is asymptotically standard normal if and only if*

$$B_n^{-2}\sum_{i=1}^n \int_{|x-\mu_i|>\varepsilon B_n} (x-\mu_i)^2 dF_i(x) \to 0$$

as $n \to \infty$ *for any* $\varepsilon > 0$.

Theorem 23.4.4 (Liapunov). *Let* X_i *be independent random variables having mean* μ_i, *variance* σ_i^2 *and* $\rho_i^3 = E|X_i - \mu_i|^3 (i = 1,\ldots)$. *Let* $B_n^2 = \sum_1^n \sigma_i^2$ *and* $C_n^3 = \sum_1^n \rho_i^3$. *Then* $B_n^{-1}\sum_1^n(X_i - \mu_i)$ *is asymptotically standard normal provided* $\frac{C_n}{B_n} \to 0$ *as* $n \to \infty$.

A useful multivariate extension of Theorem 23.4.3 given by Rao (1973, p. 147) is as follows.

Theorem 23.4.5. *Let* $\mathbf{X}_i$ *be independent random vectors with mean* μ_i, *covariance matrix* $\mathbf{\Sigma}_i$ *and distribution function* F_i $(i = 1,2,\ldots)$. *Assume that* $n^{-1}(\mathbf{\Sigma}_1 + \ldots + \mathbf{\Sigma}_n) \to \mathbf{\Sigma}$ *as* $n \to \infty$, *and*

$$n^{-1}\sum_{i=1}^n \int_{\|x-\mu_i\|>\varepsilon\sqrt{n}} \|\mathbf{x} - \boldsymbol{\mu}_i\|^2 dF_i(\mathbf{x}) \to 0$$

as $n \to \infty$ *for any* $\varepsilon > 0$.

Then $n^{\frac{1}{2}}(\mathbf{X}_i - \boldsymbol{\mu}_i)$ *is asymptotically multivariate normal with mean* $\mathbf{0}$ *and covariance matrix* $\mathbf{\Sigma}$.

The following theorem is quite useful in asserting the limiting distribution of a random variable that is slightly modified.

Theorem 23.4.6 (Slutsky(1925)). *Let* X_n *tend in distribution to* X *and* Y_n *tend in probability to a constant* $c < \infty$. *Then* *(i)* $X_n + Y_n \xrightarrow{D} X + c$;

(ii) $X_n Y_n \xrightarrow{D} cX$; *and*

(iii) $\frac{X_n}{Y_n} \xrightarrow{D} \frac{X}{c}$ *provided* $c \neq 0$, *where tending in distribution is abbreviated by* $\xrightarrow{D}$.

Remark 23.4.1. *Note that there are no restrictions on the possible dependence of the variables* X_n *and* Y_n.

Remark 23.4.2. *Also note that convergence in probability to a random variable implies convergence in distribution. Further, convergence in probability to a constant is equivalent to convergence in distribution to the specified constant.*

23.5 Dependent Random Variables

When a set of random variables are not independent one often tries to characterize the nature of dependence among the random variables. One of the simple kind of dependence is known as m-dependence which is defined below.

Definition 23.5.1. *Let* $\{X_i\}$ *be a sequence of random variables and* m *be a fixed positive integer. Then the sequence* $\{X_i\}$ *is said to be* m*-dependent if* $X_1, X_2, \ldots, X_r$ *is independent of* $X_s, X_{s+1}, \ldots$ *provided* $s - r > m$. *A similar definition applies to finite sequences of random variables.*

Several contributions have been made to the asymptotic normality of a sequence of m-dependent random variables Hoeffding and Robbins(1948) were the first to formulate the problem and provide sufficient conditions for the asymptotic normality of a suitably standardized sum of m-dependent random variables. Their basic assumption being the boundednes of the third absolute moment of the summands. Diananda (1953-55) has extended and generalized the results of Hoeffding and Robbins(1948) in several directions. Orey (1958) has strengthened the results of Diananda (1955) for the non-identical triangular array of random variables. In the following we will present the results for a single sequence of random variables with the least restrictive conditions.

One simple way m-dependent sequence of random variables can arise is as follows. Let $Z_1, Z_2, \ldots$ be a sequence of independent random variables and

$g(Z_1, \ldots, Z_{m+1})$ be any Borel-measurable function of $m+1$ real variables. Define

$$X_i = g(Z_1, \ldots, Z_{m+i}), i = 1, 2, \ldots.$$

Then, the sequence $\{X_i\}$ is a m-dependent sequence of variables. In particular if $g(Z_1, Z_2) = Z_1, Z_2$ or $Z_2 - Z_1$ then the resultant $\{X_i\}$ is a 1-dependent sequence.

We also need the following definition.

Definition 23.5.2. *The sequence $\{X_i\}$ is said to be stationary if for any $i \geq 1$ and $r \geq 0$ the joint distribution of $X_i, \ldots, X_{i+r}$ does not depend on i.*

Then we have the following result of Diananda (1953) which is an extension of Lindeberg-Levy form of the central limit theorem to the case of stationary m-dependent variables, which is an improvement of the result of Hoeffding and Robbins (1948).

Theorem 23.5.1. *Let $\{X_i\}(i = 1, 2, \ldots)$ be a sequence of stationary m-dependent variables with means zero and $E(X_iX_j) = C_{i-j}$. Then the distribution of $\frac{(X_1+\ldots+X_n)}{\sqrt{n}}$ tends to the normal distribution having mean zero and variance $\sum_{-m}^{m} C_p$ as $n \to \infty$.*

Next, we state the result of Orey (1958) for m-dependent random variables which is a corollary to his general result for the triangular array of variables.

Theorem 23.5.2. *Let $\{X_i\}, i = 1, 2, \ldots$ be a sequence of m-dependent random variables with zero mean and finite variance. Let B_n^2 be the variance of $X_1 + \ldots + X_n$. Then the distribution of $\frac{(X_1+\ldots+X_n)}{B_n}$ converges to normal$(0, 1)$ provided*

(a) $B_n^{-2} \sum_{k=1}^{n} \int_{|x|>\varepsilon B_n} x^2 dF_k(x) \to 0$, *for every* $\varepsilon > 0$

(b) $B_n^{-2} \sum_{k=1}^{n} \int_{-\infty}^{\infty} x^2 dF_k(x) = 0(1)$, *where $F_k(x)$ denotes the distribution of $X_k (k = 1, \ldots, n)$.*

Remark 23.5.1. *Note that Theorem 23.5.2 is a generalization of a result of Diananda (1955) who assumes the natural generalization of the Lindeberg condition, uniform boundedness of the variances and a certain growth rate for the variances of the partial sums.*

Vector Variable Case.
Theorems 23.5.1 and 23.5.2 can be extended to the vector variable case since if $\mathbf{X_i} = (X_{1i}, \ldots, X_{pi}), i = 1, 2, \ldots$ is an m-depedents sequence of vector random variables, then $\{\sum_{j=1}^{p} a_j X_{ji}\}, i = 1, 2, \ldots$ is an m-dependent sequence of scalar variables to which Theorems 23.5.1 and 23.5.2 are applicable. Then the asymptotic normality of $n^{-\frac{1}{2}} \sum_{i=1}^{n} \sum_{j=1}^{p} a_j X_{ji}$ implies the asymptotic p-variate normality of $n^{-\frac{1}{2}} \sum_{i=1}^{n} \mathbf{X}_i$.

23.6 Chi-Square for Correlated Variables

In this section we shall provide the distribution of a quadratic form in correlated normal variables that is useful in testing homogeneity of k-populations. Let $Z = (Z_1, \ldots, Z_k)$ be normally distributed having mean $\boldsymbol{\mu} = (\mu_1, \ldots, \mu_k)$ and variance–covariance matrix $\boldsymbol{\Sigma}$.

Theorem 23.6.1. *Let A be a real symmetric matrix. Then a necessary and sufficient condition for $Z'AZ$ to have a chi-square distribution is that $A \sum A = A$.*

PROOF: See C.R. Rao (1965 p. 152).

Corollary 23.6.1.1. *Let $\sum = \mathbf{I} - \mathbf{aa}'$ where $\mathbf{a}' = (a_1, \ldots, a_k)$ $i = 1, \ldots, k$ (with $a_i \geq 0$) is such that $\mathbf{a}'\mathbf{a} = 1$. Then $\sum_1^k Z_i^2 - (\sum_1^k a_i Z_i)^2$ is distributed as chi-square with $k-1$ degrees of freedom and noncentrality parameter $\sum_1^k \mu_i^2 - (\sum_1^k a_i \mu_i)^2$.*

PROOF: One can write $\sum Z_i^2 - (\sum_1^k a_i Z_i)^2 = \mathbf{Z}'A\mathbf{Z}$ where $A = I - \mathbf{aa}' = \boldsymbol{\Sigma}$. Now $\boldsymbol{\Sigma\Sigma} = \mathbf{I} - 2\mathbf{aa}' + \mathbf{aa}'\mathbf{aa}' = \mathbf{I} - \mathbf{aa}' = \boldsymbol{\Sigma}$. That is $\boldsymbol{\Sigma}$ is idempotent. Hence, by the above theorem $Z'AZ$ is chi-square with noncentrality parameter $\boldsymbol{\mu}'A\boldsymbol{\mu}$ and the degrees of freedom is equal to the rank of $A = \boldsymbol{\Sigma}$. Since $\mathbf{a}'\boldsymbol{\Sigma} = \mathbf{a}'\mathbf{I} - \mathbf{a}'\mathbf{aa}' = \mathbf{a}' - \mathbf{a}'; = 0$, the rank of $\boldsymbol{\Sigma}$ is at most $k - 1$. Further, since the rank of $\sum Z_i^2$ is k and the rank of $(\sum a_i Z_i)^2$ is one, the rank of $\mathbf{Z}'(\mathbf{I} - \mathbf{aa}')\mathbf{Z}$ is at least $k - 1$. Hence, the rank of $\mathbf{Z}'(I - \mathbf{aa}')\mathbf{Z}$ is $k - 1$. Alternatively, let B be an orthogonal transformation such that its first row is $\mathbf{a}'$. Then letting $\mathbf{Y} = B\mathbf{Z}$ we find $\sum Z_i^2 - (\sum_1^k a_i Z_i)^2 = \sum_1^k Y_i^2 - Y_1^2 = Y_2^2 + \ldots + Y_k^2$ which has rank $k - 1$. Thus the degrees of freedom associated with the chi-square variable is $k - 1$.

Remark 23.6.1. *Hájek and Sidák (1967, p. 31) try to prove Corollary 23.6.1.1. However, the statement of their theorem and the proof thereof are slightly in error.*

Theorem 23.6.2. *Let A_1 and A_2 be real symmetric matrices. Then the quadratic forms $\mathbf{Z}'A_1\mathbf{Z}$ and $\mathbf{Z}'A_2\mathbf{Z}$ are independent if and only if $A_1\mathbf{\Sigma}A_2 = 0$.*

PROOF: See Rao (1965, pp. 151 and 152).

23.7 Projection Approximations

Hájek (1968) has provided a method of approximating nonlinear statistics by linear statistics, thereby enabling one to apply the asymptotic theory available for the linear statistics to the nonlinear statistics.

Theorem 23.7.1 (Hájek (1968)). *Let $X_1, \ldots, X_N$ be independent random variables and $T = t(X_1, \ldots, X_N)$ be a statistic such that $ET^2 < \infty$. Let*

$$\hat{T} = \sum_{i=1}^{N} E(T|X_i) - (N-1)ET. \tag{23.7.1}$$

Then

$$E\hat{T} = ET \tag{23.7.2}$$

and

$$E(T-\hat{T})^2 = varT - var\hat{T}. \tag{23.7.3}$$

Furthermore, if $L = \sum_{i=1}^{N} l_i(X_i)$ with $El_i^2(X_i) < \infty$ $(i = 1, \ldots, N)$ then

$$E(T-L)^2 = E(T-\hat{T})^2 + E(\hat{T}-L)^2. \tag{23.7.4}$$

PROOF: We will provide the proof since it is so elegant and short. (23.7.2) trivially follows by taking expectations on both sides of (23.7.1) and (23.7.3) follows from (23.7.4) by setting $L = ET = E\hat{T}$. Thus, it is sufficient to prove (23.7.4). Also, without loss of generality, we can assume that $ET = E\hat{T} = 0$. Then, we obtain

$$E[(T-\hat{T})(\hat{T}-L)] = \sum_{i=1}^{N} E[\{E(T-\hat{T})|X_i\}\{E(T|X_i) - l_i(X_i)\}]\,. \tag{23.7.5}$$

Now since $X_1, \ldots, X_n$ are independent

$$\begin{aligned} E[E(T|X_j)|X_i] &= ET = 0 \text{ if } j \neq i \\ &= E(T|X_i) \text{ if } j = i. \end{aligned}$$

Hence

$$E(\hat{T}|X_i) = \sum_{j=1}^{N} E[E(T|X_j)|X_i] = E(T|X_i).$$

That is

$$E(T - \hat{T}|X_i) = 0 \text{ for } 1 \leq i \leq N. \tag{23.7.6}$$

Using (23.7.6) in (23.7.5) we obtain $E[(T - \hat{T})(\hat{T} - L)] = 0$ which implies (23.7.4). That is, in the class of linear statistics, $\hat{T}$ minimizes the mean square error between T and the linear statistic.

Remark 23.7.1. *If $E(T - \hat{T})^2 \to 0$ (when T is suitably normalized) as $N \to \infty$, then $\hat{T}$ is equivalent to T in probability and hence in distribution.*

As a special case, Hájek and Sidák (1967) specialized Theorem 23.7.1 to rank statistics. Let $X_1, \ldots, X_N$ be random variables having a continuous joint distribution which is symmetric in its arguments. Let $(R_1, \ldots, R_N)$ denote the vector of ranks of $(X_1, \ldots, X_N)$ and let $T = t(R_1, \ldots, R_N)$ be a rank statistic. Let L be the class of linear rank statistics of the form

$$S = \sum_{i=1}^{N} a(i, R_i). \tag{23.7.7}$$

Also, let

$$\hat{T} = \frac{N-1}{N} \sum_{i=1}^{N} \hat{a}(i, R_i) - (N-2)ET \tag{23.7.8}$$

where

$$\hat{a}(i,j) = E(T|R_i = j),\ 1 \leq i,\ j \leq N. \tag{23.7.9}$$

Then we have the following corollary.

Corollary 23.7.1.1. *With the above notation*

$$E\hat{T} = ET, \tag{23.7.10}$$

and

$$E(T - \hat{T})^2 = varT - var\hat{T}. \tag{23.7.11}$$

Also, $E(T - S)^2$ is minimized when $S = \hat{T}$.

PROOF: As before, we can, without loss of generality, assume that $ET = 0$. For any linear rank statistic of the form (23.7.7).

$$E(T-S)^2 = E(T-\hat{T})^2 + E(\hat{T}-S)^2 + 2E(T-\hat{T})(\hat{T}-S). \quad (23.7.12)$$

Now (23.7.11) follows from (23.7.12) by setting $S = ET$, provided we can show that $E(T-\hat{T}|R_i = j) = 0$ for $1 \le i,\ j \le N$. Or equivalently it suffices to show that $E(\hat{T}|R_i = j) = \hat{a}(i,j)$, $1 \le i,\ j \le N$. Thus consider

$$\begin{aligned} & E(\hat{T}|R_i = j) \quad (23.7.13) \\ = & \frac{(N-1)}{N}\sum_{k=1}^{N} E\{\hat{a}(k,R_k)|R_i = j\} \\ = & \frac{(N-1)}{N}\hat{a}(i,j) + \frac{(N-1)}{N}\sum_{k=1,k\neq i}^{N}(N-1)^{-1}\sum_{l\neq j}\hat{a}(k,l). \end{aligned}$$

Note that the events $\{R_k = j\}$, $1 \le j \le N$ denote a partition of the sample space for any k. Also the events $\{R_k = j\}$, $1 \le k \le N$ denote a partition of the sample space, for every j. Now we have

$$\begin{aligned} \hat{a}(k,l) &= E(T|R_k = l) \\ &= \sum_{s_1}\cdots\sum_{s_n} t(s_1,\ldots,l,\ldots,s_n)P(R_i = s_i, i\neq k|R_k = l) \\ &= \sum_{s_1}\cdots\sum_{s_n} t(s_1,\ldots,l,\ldots,s_n)\frac{P(R_i = s_i, i\neq k, R_k = l)}{P(R_k = l)}. \end{aligned}$$

Hence $\sum_{l=1}^{N}\hat{a}(k,l) = N\sum_{l=1}^{N}\sum_{s_1}\cdots\sum_{s_n} t(s_1,\ldots l,\ldots,s_n)P(R_i = s_i, i\neq k, R_k = l)$, since $P(R_k = l) = \frac{1}{N}$, becausc of the symmetry of the joint distribution. Thus

$$\sum_{l=1}^{N}\hat{a}(k,l) = NET = 0. \quad (23.7.14)$$

Similarly

$$\sum_{k=1}^{N}\hat{a}(k,j) = \sum_{k=1}^{N} E(T|R_k = j) = NET = 0\,. \quad (23.7.15)$$

Now, using (23.7.14) and (23.7.15) in (23.7.13) we obtain

$$\begin{aligned} E(\hat{T}|R_i = j) &= \frac{(N-1)}{N}\hat{a}(i,j) - \frac{(N-1)}{N}\sum_{k=1,k\neq i}^{N}(N-1)^{-1}\hat{a}(k,j) \\ &= \frac{(N-1)}{N}\hat{a}(i,j) + \frac{1}{N}\hat{a}(i,j) = \hat{a}(i,j) \end{aligned}$$

which completes the proof of the assertion.

Remark 23.7.2. *If T is suitably standardized with respect to N and if $E_{H_0}(T-\hat{T})^2 \to 0$ as $N \to \infty$, then $\hat{T}$ is equivalent to T in probability and consequently in distribution when the null hypothesis, H_0, is true.*

Hájek and Sidák (1967), applying Theorem 23.7.2, show that Kendall's tau for testing the null hypothesis of independence of two variables is equivalent to Spearman's rank correlation coefficient when the null hypothesis holds. We will elaborate this in the following.

Let (X_i, Y_i), $i = 1, \ldots, N$ be i.i.d pairs of random variables having an unknown continuous distribution function $H(x,y)$. Let the null hypothesis be denoted by $H_0 : H(x,y) = F(x)G(y)$ for all x and y where F and G denote the marginal distributions. The test statistic proposed by Kendall (1962) is given by τ where

$$\tau = \{N(N-1)\}^{-1} \sum_{i \neq k}\sum \operatorname{sgn}(i-k)\operatorname{sgn}(R_i - R_k)$$

where $(R_1, \ldots, R_N)$ denote the ranks of the associated Y's when the X's are ordered in the increasing order. Let $\hat{\tau}$ denote the projection of τ to the space of linear rank statistics. Then $E(\tau|H_0) = 0$. Now consider $E\{\operatorname{sgn}(R_i - R_k)|R_l = j\}$.

Case 1. $l \neq i$ and $l \neq k$.
Then the expectation is zero.

Case 2. $l = i, l \neq k$. Then

$$\begin{aligned} E\{\operatorname{sgn}(R_i - R_k)|R_i = j\} &= \sum_{l=1, l\neq j}^{N} \operatorname{sgn}(j-l) \cdot \frac{1}{N-1} \\ &= (N-1)^{-1}[\sum_{l=1}^{j-1} - \sum_{l=j+1}^{N}] \\ &= \frac{(2j-N-1)}{(N-1)}, \ 1 \leq j \leq N. \end{aligned}$$

Case 3. $l = k, l \neq i$
As in case 2, one can show that

$$E\{\operatorname{sgn}(R_i - R_k)|R_k = j\} = \frac{(N+1-2j)}{(N-1)}, \ 1 \leq j \leq N.$$

Hence

$$
\begin{aligned}
E(\tau|R_k = j) &= \{N(N-1)\}^{-1}\Big[\sum_{h=1,h\neq k}^{N} \operatorname{sgn}(k-h)\frac{(2j-N-1)}{N-1} \\
&\quad + \sum_{i=1,i\neq k}^{N} \operatorname{sgn}(i-k)\frac{(N+1-2j)}{N-1}\Big] \\
&= \frac{(2j-N-1)}{n(N-1)^2}\Big[\sum_{h=1,h\neq k}^{N} \operatorname{sgn}(k-h) - \sum_{i=1,i\neq k}^{N} \operatorname{sgn}(i-k)\Big] \\
&= \frac{2(2j-N-1)}{N(N-1)^2}\sum_{h=1,h\neq k}^{N} \operatorname{sgn}(k-h) \\
&= \frac{2(2j-N-1)}{N(N-1)^2}(2k-N-1) \\
&= \frac{8}{N(N-1)^2}(j-\frac{N+1}{2})(k-\frac{N+1}{2}) = \hat{a}(k,j).
\end{aligned}
$$

From (23.7.9) we have, for the projection of τ,

$$
\hat{\tau} = \frac{8}{N(N-1)^2}\sum_{i=1}^{N}(i-\frac{N+1}{2})(R_i-\frac{N+1}{2})
$$

which is Spearman's rank correlation coefficient except for a constant multiplier. Also since

$$
\frac{\text{var}\hat{\tau}}{\text{var}\tau} = \frac{\frac{4}{9}(\frac{N+1}{N})^2(N-1)^{-1}}{\frac{2(2N+5)}{9N(N-1)}} = \frac{2(N+1)^2}{N(2N+5)} \to 1
$$

as $N \to \infty$, τ and $\hat{\tau}$ are asymptotically equivalent under H_0.

23.8 U-Statistics

A statistic could be used either for estimating a real valued parameter or for testing a statistical hypothesis about the real-valued parameter. For the time being let us treat a U-statistic as an estimator.

Definition 23.8.1. *A real-valued parameter $g(\theta)$ is* **estimable** *if it has an unbiased estimator; that is, if there exists a* **statistic** *$f(x_1, \ldots, x_n)$ such that*

$$
E_\theta\{f(X_1, \ldots, X_n)\} = g(\theta), \text{ for all } \theta \in \Omega .
$$

An estimable parameter is sometimes called a **regular parameter**. *Notice that θ as well as $g(\theta)$ could be vectors, and hence f also will be a vector.*

Definition 23.8.2. The degree m of an estimable parameter is **defined to be the smallest** sample size for which the parameter has an unbiased estimator.

Definition 23.8.2. *A kernel is an unbiased estimator of a parameter based on the minimum sample size m.*

For every kernel there is a symmetric kernel. For, if $f(x_1, \ldots, x_n)$ is a kernel, then the symmetric kernel is given by

$$f_s(x_1, \ldots, x_n) = \frac{1}{m!} \sum_P f(x_{i_1}, \ldots, x_{i_m})$$

where the summation is over all the permutations $(i_1, \ldots, i_m)$ of $(1, \ldots, m)$. Since the symmetric kernel is an average of $m!$ unbiased estimates, it is an unbiased estimate of the parameter.

We also have the following properties of estimable parameters. If $g_1(\theta), g_2(\theta)$ are estimable parameters of degrees m_1, m_2 then the sum $g_1(\theta) + g_2(\theta)$ and the product $g_1(\theta)g_2(\theta)$ are also estimable parameters and have degrees, respectively, less than or equal to $m = \max(m_1, m_2)$ and $m_1 + m_2$. (For proof, see Fraser (1957) pages 136–137.) It also follows that any polynomial in estimable parameters is also an estimable parameter. If the parameters are vectors, the addition and multiplication are interpreted as the addition and multiplication of corresponding coordinates.

Definition 23.8.3. *Corresponding to any estimator $f(x_1, \ldots, x_m)$ of an estimable parameter $g(\theta)$, we define a U-statistic for a sample of n $(n > m)$ by*

$$U(x_1, \ldots, x_n) = \binom{n}{m}^{-1} \sum_C f_s(x_{i_1}, \ldots, x_{i_m}) \tag{23.8.1}$$

where the summation C is overall $\binom{n}{m}$ combinations $(i_1, \ldots, i_m)$ of m integers chosen from $(1, \ldots, n)$ and f_s is the symmetrized statistic corresponding to $f(x_1, \ldots, x_m)$. Also, one could write

$$U(x_1, \ldots, x_n) = \frac{(n-m)!}{n!} \sum_P f(x_{i_1}, \ldots, x_{i_m})$$

where P denotes all permutations of m integers $(i_1, \ldots, i_m)$ chosen from $(1, \ldots, n)$. Thus U is the symmetrized form of $f(x_1, \ldots, x_m)$ considered as a function of $(x_1, \ldots, x_n)$. Clearly $EU = g(\theta)$ since $Ef(X_1, \ldots, X_m) = g(\theta)$. The sample mean with $h(x) = x$, the sample variance with kernel $h(x_1, x_2) = \frac{1}{2}(x_1 - x_2)^2$ and the empirical distribution function with kernel $h(x) = I(x \le t)$ are examples of U-statistics. Other examples of U-statistics are given by Serfling (1980, pp. 173–174). Hoeffding (1948) was the first to propose one-sample U-statistics and establish their asymptotic normality. k-sample U-statistics were considered by Lehmann (1951) and Dwass (1956). Also, as pointed out by Serfling (1981, p. 176) a U-statistic can be represented as the conditional expectation of the kernel given the order statistic. That is

$$U_n = E\{h(x_1, \ldots, x_m) | \mathbf{X}_n\}$$

where $\mathbf{X}_n$ denotes the order statistic $(X_{1n}, \ldots, X_{nn})$. Toward its asymptotic normality, Hoeffding (1948) has established the following result.

Theorem 23.8.1. *If $E_F h^2 < \infty$ and $\xi_1 > 0$, then $n^{\frac{1}{2}}(U_n - \theta)$ is asymptotically normal with mean 0 and variance $m^2 \xi_1$, where*

$$\xi_1 = var_F\{h_1(X_1)\},\ h_1(x_1) = E_F\{h(x_1, X_2, \ldots, X_m)\}.$$

Furthermore one can show, for instance, see Serfling (1981, p. 183) that

$$var_F(U_m) = \frac{m\xi_1^2}{n} + O(n^{-2}) \text{ as } n \to \infty. \tag{23.8.2}$$

For other properties of U-statistics, the reader is referred to Serfling (1980).

23.9 Problems

23.2.1 Lex $X = \pm 1$ with probability 1/2. Show that the Chebyshev's inequality is sharp in the sense that the inequality becomes equality when $a = 1$.

23.2.2 Using Jensen's inequality show that the variance of a random variable is always nonnegative.

23.2.3 Using Jensen's inequality show that for any random variable X such that $EX < \infty$.

$$E(\frac{1}{X}) \ge \frac{1}{EX}.$$

23.2.4 If (X, Y) has a bivariate normal distribution with correlation ρ show that $|\rho| \leq 1$.
(HINT: It suffices to assume that $EX = EY = 0$ and $varX = varY = 1$. Then use Hölder's inequality with $p = q = 2$.)

23.3.1 Let $X_1, X_2, \cdots$, be a sequence of independent random variables such that $var(X_i) < \infty$ for $i = 1, 2, \cdots$ and $n^{-2} \sum_{i=1}^{n} varX_i \to 0$ as $n \to \infty$. Show that the weak law of large numbers holds.

23.3.2 Check whether the weak law of large numbers holds for the following sequences of independent-random variables.

(i) $P(X_n = \pm 2^n) = 2^{-(2n+1)}$ and $P(X_n = 0) = 1 - 2^{-2n}$, $n = 1, 2, \cdots$

(ii) $P(X_n = \pm \frac{1}{n}) = \frac{1}{2}$, $n = 1, 2, \cdots$.

23.4.1 Check whether the Lindeberg-Feller condition is satisfied for the following sequences of independent random variables.

(i) $P(X_n = \pm 2^{-n}) = \frac{1}{2}$, $n = 1, 2, \cdots$

(ii) $P(X_n = \pm 2^{-(n+1)}) = 2^{-(n+3)}$, $P(X_n = 0) = 1 - 2^{-(n+2)}$, $n = 1, 2, \ldots$.

23.4.2 Let $X_1, X_2, \cdots$ be a sequence of independent Bernoulli random variables such that

$$P(X_i = 1) = p_i \text{ and } P(X_i = 0) = 1 - p_i, \; i = 1, 2, \ldots.$$

Find a sufficient condition on the p_i such that $\sum_1^n X_i$ is approximately standard normal for sufficiently large n when suitably standardized.

23.4.3 Let X_n be an i.i.d. sequence of random variables with mean μ and variance σ^2. Let $Y_1, \ldots, Y_n$ be an i.i.d. sequence of random variables with mean a. Find the limiting distribution of

$$Z_n = \frac{\sqrt{n}(\overline{X}_n - \mu)}{\overline{Y}_n}$$

where $\overline{X}_n$ and $\overline{Y}_n$ denote the sample mean of size n drawn from the X and Y sequences respectively.

23.4.4 Let $X_1, X_2, \ldots$ be a sequence of independent random variables having finite mean μ and variance σ^2. Let s_n^2 denote the sample variance given by

$$(n-1)^{-1}\sum_1^n (X_i - \overline{X})^2.$$

Show that $\sqrt{n}(\overline{X}_n - \mu)/s_n$ has asymptotically, a standard normal distribution.

23.5.1 Let $Z_1, Z_2, \ldots$ be a sequence of i.i.d. random variables with mean 0 and variance σ^2. Let

$$X_i = Z_{i+1} - Z_i, \qquad i = 1, 2, \ldots$$

Then, find the asymptotic distribution of

$$n^{-\frac{1}{2}}(X_1 + \cdots + X_n).$$

23.7.1 $X_1, \ldots, X_n$ be a random sample from F. Let

$$T = \frac{2}{n(n-1)}\sum\sum\nolimits_{i<j} X_i X_j\,.$$

Obtain T' the projection of T onto the space of statistics which are linear in observations.

23.8.1 Let $X_1, \ldots, X_n$ be i.i.d. random variables having finite second moments. If $h(x_1, x_2) = |x_2 - x_1|$, then the statistic

$$U = \frac{2}{n(n-1)}\sum^n\sum\nolimits^n_{i<j} |X_j - X_i|$$

is known as Gini's mean difference. Using Theorem 23.8.1, establish the asymptotic normality of U.

23.8.2 Let $X_1, \ldots, X_n$ be i.i.d. random variables distributed as F. If $h(x_1, x_2) = x_1 x_2$, then the corresponding U-statistic is

$$U = \frac{2}{n(n-1)}\sum\sum\nolimits_{i<j} X_i X_j$$

which is supposed to estimate $\theta(F) = \mu^2 = (EX)^2$. Find the asymptotic distribution of U.

23.8.3 Under the same hypothesis as in Problem 23.8.2, if $h(x_1, x_2) = \frac{1}{2}(x_1 - x_2)^2$, then the corresponding U-statistic is

$$\begin{aligned} U &= \frac{2}{n(n-1)} \sum\sum\nolimits_{i<j} (X_i - X_j)^2 \\ &= \frac{1}{n-1}\left(\sum X_i^2 - n\overline{X}^2\right) \\ &= s^2 \end{aligned}$$

where s^2 denotes the sample variance. Here $\theta(F) = varX$. Find the asymptotic distribution of U assuming the existence of suitable moments.

Chapter 24

Asymptotic Theory of CS-class of Statistics

24.1 Introduction

Let $X_1, \ldots, X_m$ and $Y_1, \ldots, Y_n$ be random samples from continuous distributions functions $F(x)$ and $G(y)$ respectively. Let $N = m + n$ and $Z_{N,i} = 1$ when the ith smallest of N observations is an X and $Z_{N,i} = 0$. Otherwise, then the class of linear rank statistics (of which many nonparametric test statistics are members) is given by:

$$mT_N = \sum_{i=1}^{N} E_{N,i} Z_{N,i} \tag{24.1.1}$$

where $E_{N,i}$ are some known constants. The results of Wald and Wolfowitz (1944), Noether (1949), Hoeffding (1948), Lehmann (1951), Madow (1948), Dwass (1956) and Hájek (1961) provide sufficient conditions for the asymptote normality of T_N. Chernoff and Savage (1958) have significantly extended the above results to cover more sitatuations when $F \neq G$. In particular, they establish the asymptotic normality of the normal scores test statistic (known also as Fisher-Yates-Terry Hoeffding statistic) and show that it is at least as efficient as the two-sample t-test. The results of Chernoff and Savage (1958) have been generalized by three sets of authors. Govindarajalu, LeCam and Raghavachari (to be abbreviated as GLR) (1967), Hájek (1968), Pyke and Shorack (1968). Here we will be focusing on the results of GLR (1967) and briefly mention the other generalizations.

24.2 Formulation of Problem

Let $F_m(x)$ and $G_n(y)$ respectively denote the empirical distribution functions based on the X and Y samples. Let

$$H_N(x) = \lambda_N F_m(x) + (1 - \lambda_N) G_N(x) \text{ where } \lambda_N = m/N \tag{24.2.1}$$

be the combined empirical d.f. Furthermore, let

$$H(x) = \lambda_N F(x) + (1 - \lambda_N) G(x) \tag{24.2.2}$$

If we define a random variable W as

$$W = \begin{cases} X & \text{with prob. } \lambda_N \\ Y & \text{with prob. } 1 - \lambda_N \end{cases}$$

then H denotes the d.f of W. Let $W_{IN} \leq \ldots \leq W_{NN}$ denote the combined ordered sample. Chernoff and Savage provide the following integral representation of T_N:

$$T_N = \int_{-\infty}^{\infty} (\frac{N}{N+1} H_N(x))\, dF_m(x)$$

since

$$T_N = \frac{1}{m} \sum_{i=1}^{N} J_N(\frac{R_i}{N+1}) = \frac{1}{m} \sum_{i=1}^{N} J_N(\frac{i}{N+1}) Z_i \tag{24.2.3}$$

where $(R_1, \ldots, R_m)$ denote the ranks of $(X_1, \ldots, X_m)$ in the combined ordered sample which is denoted by $(W_{IN} \leq \ldots \leq W_{NN})$.

Examples of the weight function J_N are as follows:

$$\begin{aligned} a_{N,i} &= J_N(\frac{i}{N+1}) \\ &= E(W_{iN}) \text{ in the case of normal scores test} \\ &\quad \text{where } W_{iN} \text{ are standard normal order statistics} \\ &= \frac{i}{N+1} \text{for the wilcoxon test} \\ &= \begin{cases} i/(N+1) & \text{for } i < (N+1)/2 \\ 1 - (i/((N+1)) & \text{for } i > (N+1)/2 \end{cases} \\ &\quad \text{for the Ansani-Bradley test.} \end{aligned}$$

Hereafter, whenever there is no ambiguity, the subscript N on J will be suppressed.

24.3 Regularity Assumptions

Chernoff and Savage (1958) assume that F and G are continuous. Then we can avoid ties among X-observations, among Y observations and among the X and Y observations. This assumption is somewhat restrictive. If F (and/or G) have jump points, then it can be made continuous by a continuization method described by GLR (1967, p. 61). All that is needed is that F and G distributions do not have mutual discontinuities. If the X and Y distributions have mutual discontinuities, then ties among X and Y observations will occur with positive probability and hence the statistic T_N is not well defined. It is further assumed that there exists a $\lambda_0 \leq \frac{1}{2}$ such that

$$0 < \lambda_0 < \lambda_N < 1 - \lambda_0 < 1.$$

Let us define a class of functions for which the asymptotic normality of T_N can be established.

Definition 24.3.1. *A function f, $f \geq 1$ defined on the interval (0,1) is said to belong to the class $\mathcal{U}$ if there exists some α $(0 < \alpha < 1)$ such that f is monotonically decreasing in $(0, \alpha)$ and monotonically increasing in $[\alpha, 1]$.*

Definition 24.3.2. *A function $f \in \mathcal{U}$ is said to belong to the class $\mathcal{U}_1(\mathcal{U}_2)$ if it is integrable (square integrable).*

Let b denote a finite positive constant. Let f_0, f and g be nonnegative functions when $f \in u_1$, $g \in u_2$ and f_0 is integrable, and J defined by integrals of the type

$$J(u) = \int_{\frac{1}{2}}^{u} J'(\zeta) d\zeta.$$

One could also introduce functions J which differ from the above integrals by a constant; however, this will not affect the differences that would be considered below. Notice that, in particular, f could be of the form $K[u(1-u)]^{-1+\delta}$ and g could be of the form $K[u(1-u)]^{-\frac{1}{2}+\delta}$ for some $\delta(0 < \delta < \frac{1}{2})$.

Definition 24.3.3. *J is said to belong to the class $\mathcal{S}_1$ if J is absolutely continuous with $J' = J_1' + J_2'$, $|J_2'(u)| \leq f(u)g(u)$ and $\int |J_1'(u)| du \leq b$.*

Definition 24.3.4. *J is said to belong to the class $\mathcal{S}_0$ if J is absolutely continuous with $|J'| \leq fg$ and J is said to belong to the class $\mathcal{S}$ if $|J'| \leq f_0 + fg$.*

Note that $\mathcal{S}_0 \subset \mathcal{S}_1 \subset \mathcal{S}$. If $J \in \mathcal{S}_1$, then one can take $|J_2'| = fg$ and $|J_1'| = f_0$. That is, $J \in \mathcal{S}$. Hence $\mathcal{S}_1 \subset \mathcal{S}$.

Then the following lemma is satisfied by the function $g(u)$.

Lemma 24.3.1. *Let J belong to $\mathcal{S}$, then*

$$\lim_{u\to 0} u|J(u)| = \lim_{u\to 1}(1-u)|J(u)| = 0.$$

PROOF: Since $J(u) = \int_u^{1/2} J'(v)dv$ for small u

$$|J(u)| \le \int_u^{1/2} |J'(v)|dv \le \int_u^{1/2} f(v)g(v)dv + \int_u^{1/2} f_0(v)dv\,.$$

Since f_0 is integrable, the last term on right-hand side can be neglected. Hence,

$$u|J(u)| \le ug(u)\int_u^{1/2} f(v)dv = cug(u).$$

Thus it suffices to show that $ug(u)$ tends to zero as u goes to zero. Now since g is square integrable $\int_0^u g^2(v)dv$ goes to zero as u tends to zero. However, $\int_0^u g^2(v)dv \ge ug^2(u) \ge ug(u)$. This completes the proof that $ug(u) \to 0$.

Analogously, one can show that $(1-u)|J(u)|$ tends to zero as $u \to 1$.

This completes the proof of the lemma.

Also J is uniformly square integrable for every $J \in \mathcal{S}$ (see lemma 2 of GLR (1967, p. 614).

24.4 Partition of the Statistic

Let

$$T_N^* = N^{1/2}\{\int J(\tfrac{N}{N+1}H_N)dF_m - \int J(H)dF\} \tag{24.4.1}$$

where J_N is assumed to be absolutely continuous on (0,1).

One can expand T_N^* as

$$T_N^* = B_N + C_{1N} + C_{2N} \tag{24.4.2}$$

where B_N, the first order random term is given by

$$B_N = N^{1/2} \int J(H)d(F_m - F) + N^{1/2} \int J'(H)(H_N - H)dF \qquad (24.4.3)$$

and the higher order random terms are given by

$$C_{1N} = N^{1/2} \int \{J(\tfrac{N}{N+1} H_N) - J(H)\} d(F_m - F) \qquad (24.4.4)$$

and

$$C_{2N} = N^{1/2} \int \{J(\tfrac{N}{N+1} H_N) - J(H) - (H_N - H)J'(H)\} dF. \qquad (24.4.5)$$

By performing integration by parts once, we obtain

$$\begin{aligned} &N^{1/2} \int J(H)d(F_m - F) \\ &\quad = N^{1/2} J(H)(F_m - F)|_{F=0}^{1} - N^{1/2} \textstyle\int (F_m - F)J'(H)dH\,. \end{aligned}$$

However,

$$\begin{aligned} |N^{1/2}(F_m - F)J(H)| &\leq KN^{1/2}|(H_N - H)J(H)| \\ &\sim KN^{1/2}H|J(H)| \to 0 \text{ as } H \to 0 \\ &\sim KN^{1/2}(1 - H)|J(H)| \to 0 \text{ as } H \to 1 \end{aligned}$$

after applying Lemma 24.3.1. Thus

$$\begin{aligned} B_N &= -N^{1/2} \int (F_m - F)J'(H)dH + N^{1/2} \int (H_N - H)J'(H)dF \\ &= (1 - \lambda_N)N^{1/2} \left\{ \int (G_n - G)J'(H)dF - \int (F_m - F)J'(H)dG \right\}. \end{aligned} \qquad (24.4.6)$$

Thus B_N is an independent sum of i.i.d. random variables which is asymptotically normally distributed with mean zero and variance σ_N^2 where

$$\begin{aligned} \sigma_N^2 = \\ &2(1 - \lambda_N)^2 \{(1 - \lambda_N)^{-1} \iint_{x<y} G(x)[1 - G(y)]J'(H(x))J'(H(y))dF(x)dF(y) \\ &+ \lambda_N^{-1} \iint_{x<y} F(x)[1 - F(y)]J'[H(x)]J'[H(y)]dG(x)dG(y)\}\,. \end{aligned} \qquad (24.4.7)$$

Also

$$(\sigma_N^2 | F = G) = 2\{(1-\lambda_N)/\lambda_N\} \iint_{u<v} u(1-v)J'(u)J'(v)dudv \,. \quad (24.4.8)$$

So, in order to assert the asymptotic normality of $T_N^\star$, it suffices to show that C_{1N} and C_{2N} will tend to zero in probability as $N \to \infty$

The following lemma is not only of independent interest but also has been used in showing the negligibility of C_{2N}.

Lemma 24.4.1. *$N^{1/2}g(H)(H_N - H)$ is bounded in probability.*

PROOF: See Lemma 7 of GLR (1967, p. 618).

The negligibility of C_{1N} and C_{2N} in probability is accomplished in two stages. In stage one, it is shown that the tails of the statistic T_N go to zero. That is, let $A = \{(0,\tau] \bigcup [1-\tau, 1)\}$. Then

$$\begin{aligned} \Delta_N(J,\tau) &= N^{1/2} \int_A J(\frac{N}{N+1}H_N)dF_m - \int_A J(H)dF \\ &= N^{1/2} \int_A \left\{ J(\frac{N}{N+1}H_N) - J(H) \right\} dF_m \\ &\quad + N^{1/2} \int_A J(H)d(F_m - F) = \Delta_1 + \Delta_2 \ (say) \,. \end{aligned}$$

Notice that Δ_2 is a normalized sum with zero expectation and its variance can be made arbitrarily small by choosing τ small. Δ_1 is shown to be negligible in probability in Section 5 of GLR (1967). The following lemma is not only of independent interest but also plays a crucial role in establishing that C_{1N} goes to zero in probability for every $J \in \mathcal{S}_1$, as $N \to \infty$ (see Proposition 3 of GLR (1967, p. 622)).

Lemma 24.4.2. *Let $H(x)$ be uniform on $(0,1)$.*

$$K_N(u) = \inf\{x : H_N(x) \geq u\} \quad (24.4.9)$$

and

$$Z_N(u) = N^{1/2}[K_N(u) - u]. \quad (24.4.10)$$

Then the distribution of the vector $\{Z_N(u_j),\ j = (1,...,r)\}$ for any finite set $0 = u_o < u_1 < ... < u_{r-1} < u_r = 1$ is asymptotically normal with mean $\underline{0}$ and covariance function

$$C_N(u,v) = \lambda_N F(u)[1-F(v)] + (1-\lambda_N)G(u)[1-G(v)] \ , \quad u \leq v. \quad (24.4.11)$$

PROOF: It follows from the equivalence

$$\{K_N(u) > x\} \Leftrightarrow \{H_N(x) \le u\}$$

and the asymptote normality of the process $N^{1/2}(H_N - H)$. The asymptotic negligibility of C_{2N} is shown in pages 629–630 of GLR (1967) for every $J \in \mathcal{S}$.

24.5 Alternative Form of the First Order Terms

Without loss of generality we can set H to be uniform on $(0,1)$ and also assume that $J(\frac{1}{2}) = 0$. Let

$$B_N = B_{1N} + B_{2N}$$

where

$$B_{1N} = N^{1/2} \int J(H) d(F_m - F)$$

and

$$B_{2N} = N^{1/2} \int (H_N - H) J'(H) dF\,.$$

Let

$$L(x) = \int_{1/2}^{x} J'(\xi)\, dF(\xi) = \int_{1/2}^{x} J'(\xi)\phi(\xi)\, d\xi \tag{24.5.1}$$

and

$$M(x) = \int_{1/2}^{x} J'(\xi)\, dG(\xi) = \int_{1/2}^{x} J'(\xi)\psi(\xi)\, d\xi \tag{24.5.2}$$

where $\phi(x) = dF/dx$.

By performing integration by parts once, we obtain $B_{2N} = N^{1/2} \int (H_N - x)\, dL(x) = N^{1/2}(H_n - x)L(x)|_0^1 - N^{1/2} \int L(x)\, d(H_N - x)$. Since $|H_N - x|\,|J(x)| \to 0$ as $x \to 0$ or 1 (by Lemma 24.3.1), we can write

$$B_{2N} = -N^{1/2} \int L(x)\, d(H_N - x). \tag{24.5.3}$$

Now, since

$$\lambda_N \phi + (1-\lambda_N)\psi = 1, \tag{24.5.4}$$

we have

$$J(x) = \lambda_N L(x) + (1-\lambda_N) M(x). \tag{24.5.5}$$

Hence

$$\begin{aligned} B_N &= B_{1N} + B_{2N} \\ &= N^{1/2}(1-\lambda_N)\left[\int M(x)\, d(F_m - F) - \int L(x)\, d(G_m - G)\right]. \end{aligned} \tag{24.5.6}$$

Thus

$$\begin{aligned} &E(B_N) = 0 \quad \text{and} \\ &\text{var}(B_N) = (1-\lambda_N)^2 \lambda_N^{-1}\, \text{var} M(X) + (1-\lambda_N)\, \text{var} L(Y) \end{aligned} \tag{24.5.7}$$

when $F = G$, $\varphi = \psi = 1$ and hence $M(x) = L(x) = J(x)$. Thus

$$(\sigma_N^2 |\, F = G) = \left(\frac{1-\lambda_N}{\lambda_N}\right) \text{var} J(U),\ U \stackrel{d}{=} \text{unif}\ (0,1). \tag{24.5.8}$$

That is,

$$\text{var}\ J(U) = \int_0^1 J^2(u)\, du - \left\{\int_0^1 J(u)\, du\right\}^2. \tag{24.5.9}$$

In order to show the equivalence of (24.4.8) and (24.5.8) we employ the ingenious trick of Chernoff and Savage (1958, p. 978).

One can write

$$\begin{aligned}
\iint_{x<y} x(1-y)J'(x)J'(y)\,dxdy &= \iiiint_{0<u<x<y<v<1} J'(x)J'(y)\,dudvdxdy \\
&= \iiint_{0<u<y<v<1} [J(y)-J(u)]\,J'(y)\,dydudv \\
&= \frac{1}{2}\iint_{0<u<v<1} \left\{[J(y)-J(u)]^2\,|_u^v\right\}\,dudv \\
&= \frac{1}{2}\iint_{0<u<v<1} [J(v)-J(u)]^2\,dudv \\
&= \frac{1}{4}\int_0^1\int_0^1 [J(v)-J(u)]^2\,dudv \\
&= \frac{1}{2}\left[\int_0^1 J^2(u)\,du - \left\{\int_0^1 J(u)\,du\right\}^2\right].
\end{aligned}$$

Thus the main result of GLR (1967, Theorem 1) can be stated as follows.

Theorem 24.5.1. *Let J be a member of $\mathcal{S}_1$ and T_N^* be defined as* (24.4.1). *Let σ_N^2 given by* (24.4.7) *be bounded away from zero. Then*

$$\sup_x \left|P\left(T_N^*/\sigma_N \le x\right) - \Phi(x)\right| \to 0 \text{ as } N \to \infty$$

for every $J \in$ S (where S is a relatively compact subset of $\mathcal{S}_1$) and every triple $\{(F,G),\ \lambda_N\}$, or for every $J \in \mathcal{S}_1$, and every triple $\{(F,G),\ \lambda_N\} \in \mathcal{F}$ where $\mathcal{F}$ is a relatively compact[1] *subset of triples $\{(F,G),\ \lambda_N\}$.*

The following remarks are in order.

Remark 24.5.1. *GLR point out that the convergence of the distribution of T_N to the normal distribution is not uniform on the set of system $\{(F,G),\lambda_N,J\}$ (see Remark 1 and the counter example of GLR (1967, p. 630)).*

Remark 24.5.2. *The assumption on J' is very basic to Theorem 24.5.1. The assumption on J' puts a limit on the growth of J. Chernoff and Savage (1958) in their main result assume that the second derivative of J satisfies a smoothness condition:*

[1]Let $\mathcal{X}$ be a topological space. A subset S is said to be relatively compact if every sequence contained in S admits a convergent subsequence.

$$|J''(u)| \le k[u(1-u)]^{-(\frac{5}{2})+\delta} \text{ for some } \delta \ (0 < \delta < \frac{1}{2})$$

which implies that the available family $\{J'\}$ is relatively compact for uniform convergence on the compact intervals of (0,1). GLR's (1967).

Theorem 1 asserts uniformity of the convergence on that class. Chernoff and Savage (1958, p. 975) conjecture that their basic result is true without the smoothness condition on J''.

Remark 24.5.3. *Test statistics such as proposed by Ansari and Bradley (1960) and Siegel and Tukey (1960) do not satisfy the regularity condition imposed on the J function by Chernoff and Savage (1958).*

Remark 24.5.4. *The uniformity asserted by Theorem 23.5.1 is of interest in the following particular case. Suppose that the distribution of X is $F = \Psi$ which admits a density and the distribution of Y is $G(x) = \Psi((x - \theta_N)/\beta_N)$. If $\theta_N \to 0$ and $\beta_N \to 1$, then the uniformity in the convergence of the distribution of T_N^*/σ_N to the standard normal d.f. is uniform for any set $S \in \mathcal{S}$ such that $J \in S$ implies that var $J(U)$ given by (24.5.9) is bounded away from zero.*

24.6 Scores: Expectations of Order Statistics

In several cases, the functions J_N are obtained using expectations of suitable order statistics and these are not included in the class of U-Statistics. In this regard, the following lemma and theorem will be of interest. The role of $f(x)g(x)$ is taken over by the role of $K[x(1-x)]^{-(\frac{3}{2})+\delta}$ of Chernoff and Savage (1958).

Let $\xi_{k,N}$ denote the kth smallest order statistic in a random sample of size N from a standard uniform distribution. For any function J_N let $\overline{J}_N$ be the function defined on (0,1) such that if $y = k/(N+1), \ \ k = 1, ..., N$

$$\overline{J}_N(y) = J_N(\xi_{k,N}) = \int_0^1 J_N(x)\beta_N(x,k)dx \tag{24.6.1}$$

where

$$\beta_N(x,k) = \frac{N!}{(k-1)!(N-k)!}x^{k-1}(1-x)^{N-k}. \tag{24.6.2}$$

The definition of $\overline{J}_N(y)$ is completed by interpolating linearly between $\{k(N+1)^{-1}, (k+1)(N+1)^{-1}\}$. Then we have the following results of GLR (1967, p. 631).

Theorem 24.6.1. *With the above notation, let*

$$\left|\overline{J}'_N(x)\right| \le K\left[x(1-x)\right]^{-3/2+\delta}$$

and J'_N converge in Lebesgue measure to a limit J'. Let $\widetilde{T_N}$ be defined by

$$\widetilde{T_N} = N^{1/2}\left\{\int \overline{J_N}\left(\frac{N}{N+1}H_N\right)\, dF_m - \int J_N(H)\, dF\right\}. \qquad (24.6.3)$$

Then as $N \to \infty$

$$\sup_x \left|P\left(\tilde{T_N}/\sigma_N \le x\right) - \Phi(x)\right| \to 0$$

uniformly in $[(F,G),\ \lambda_N]$.

PROOF: Let

$$T_N^* = N^{1/2}\left\{\int \overline{J}_N\left(\frac{N}{N+1}H_N\right)\, dF_m - \int \overline{J}_N(H)\, dF\right\}. \qquad (24.6.4)$$

Then according to Lemma 1 of GLR (1967, p. 612) the function $\overline{J}'_N$ satisfies the conditions of Theorem 24.5.1, with $f(x) = k\left[x(1-x)\right]^{-1+(\delta/2)}$ and $g^2(x) = k\left[x(1-x)\right]^{-1+\delta}$. Besides J'_N converges in measure to a limit J'; consequently, $\overline{J}'_N$ converges to J' according to Lemma 1 of GLR (1967). Thus it follows that Theorem 24.6.1 would be proved if $\widetilde{T_N}$ was replaced by T_N^*. Hence it suffices to show that the difference $\widetilde{T_N} - T_N^* = N^{1/2}\int\left[\overline{J}_N(H) - J_N(H)\right]\, dF$ will go to zero. However, this is smaller than

$$\lambda_N^{-1} N^{1/2}\int \left|\overline{J}_N(x) - J_N(x)\right|\, dx$$

which was shown to go to zero as $N \to \infty$ by GLR (1967, pp. 631–632).

Remark 24.6.1. *Theorem 24.6.1 corresponds to Theorem 2 of Chernoff and Savage (1958). As a particular application, let us provide the following corollary.*

Corollary 24.6.1.1. *Let k be a fixed integer and let a_j, $j = 1, 2, \ldots, k$ be bounded constants. Let*

$$J_N\left(\frac{i}{N+1}\right) = \sum_{j=1}^{k} a_j E\left(\xi_{i,N}^j\right)$$

where $\xi_{i,N}$ denote the ith smallest order statistic in a random sample of size N from a population whose cumulative distribution function is the inverse of L. If

$$\left|\frac{dL^j(x)}{dx}\right| \leq K\,[x(1-x)]^{-(3/2)+\delta}\,, \quad \textit{for } j=1,\ldots,k,$$

then the functions J_N satisfy the conditions of Theorem 24.5.1.

PROOF: This follows from the linearity of the transformation $J \hookrightarrow \overline{J}_N$ used to define the functions occurring in Theorem 24.6.1.

For the case where $a_1 = 0,\ a_2 = 1,\ h = 2,$ and $L = \Phi^{-1}$, the resultant test statistic is the one considered by Capon (1961) and Klotz (1962).

24.7 Extension to c-sample Case

Without additional assumptions, Puri (1964) had extended the Chernoff-Savage results to c-sample cases. GLR (1967) have extended theorems 24.5.1 and 24.6.1 to the c-sample cases and these will be considered below.

Let $X_{j,k},\ k = 1,\ldots,n_j$ be a random sample from population having a continuous distribution $F^{(j)}$. Assume that the c-samples obtained for $j = 1,\ldots,c$ are mutually independent. Let $N = \sum_{j=1}^{c} n_j$ and $\lambda_j = n_j/N$. Further assume that there is a $\lambda_0 < 1/c$ such that $0 < \lambda_0 < \lambda_j < 1-\lambda_0 < 1$ for $j = 1,\ldots,c$ and every N. Let $H = \sum \lambda_j F^{(j)}$ and $H_N = \sum \lambda_j F^{(j)}_{n_j}$ be respectively the combined cumulative distribution and empirical distribution based on the samples $\{X_{j,k}\}$.

Let $T_{N,j} = n_j^{-1} \sum_{i=1}^{N} E_{N,i,j}\, Z_{N,i,j};\ \ j = 1,\ldots,c$ where $Z_{N,i,j} = 1$ if the ith smallest observation in the combined sample of size N comes from the jth sample, and it is equal to zero otherwise. The $E_{N,i,j}$ are specified constants. One can represent $T_{N,j}$ in an integral form given by

$$T_{N,j} = \int_{-\infty}^{\infty} J_{N,j}\left[\frac{N}{N+1} H_N(x)\right] dF^{(j)}_{n_j} \qquad (24.7.1)$$

For the sake of simplicity, we assume that the functions $F^{(j)}$ are continuous and state a result analogous to Theorem 24.5.1 in a form similar to the form of Theorem 1 of Chernoff and Savage.

Theorem 24.7.1. *Assume that for all $j = 1, \ldots, c$ the following conditions are satisfied:*

(1) $J_j(H) = \lim J_{N,j}(H)$ *exists for* $0 < H < 1$ *and this limit is not a constant and each* $J_j(H)$ *is absolutely continuous on* $(0,1)$.

(2)
$$\int_{0<H_N\leq 1}\left\{J_{N,j}\left(\frac{N}{N+1}H_N\right) - J_j\left(\frac{N}{N+1}H_N\right)\right\}\, dF_{n_j}^{(j)} = o_p\left(N^{-1/2}\right)$$

(3) $\left|J_j'(u)\right| \leq K\left[u(1-u)\right]^{-(3/2)+\delta}$ *with* $0 < 2\delta < 1$.

Let

$$\mu_{N,j} = \int_{-\infty}^{\infty} J_j\left(H(x)\right)\, dF^{(j)}(x) \tag{24.7.2}$$

and

$$\begin{aligned}\sigma_{N,j}^2 = 2 &\sum_{i=1, i\neq j}^{c} \lambda_i \iint_{x<y} F^{(i)}(x)\left[1 - F^{(i)}(y)\right] J_j'(H(x))J_j'(H(y)) \\ &\cdot dF^{(j)}(x) dF^{(j)}(y) \\ &+ \frac{2}{\lambda_j}\iint_{x<y} F^{(j)}(x)\left[1 - F^{(j)}(y)\right] J_j'((H(x))J_j'(H(y))\cdot \\ &d\left[H(x) - \lambda_j F^{(j)}(x)\right] d\left[H(y) - \lambda_j F^{(j)}(y)\right].\end{aligned} \tag{24.7.3}$$

Then if $\liminf_{N\to\infty} \sigma_{N,j}^2 > 0$, *one has*

$$\lim_{N\to\infty} P\left\{N^{1/2}\left[T_{N,j} - \mu_{N,j}\right] \leq t\sigma_{N,j}\right\} = \Phi(t).$$

PROOF: Conditions (1) and (2) of the theorem imply that one may consider instead of $(T_{N,j} - \mu_{N,j})\, N^{1/2}$, the statistic

$$T_{N,j}^{\star} = N^{1/2}\left\{\int J_j\left[\frac{N}{N+1}H_N\right]\, dF_{n_j}^{(j)} - \int J_j(H)\, dF^{(j)}\right\} \tag{24.7.4}$$

which is similar to the statistic considered in Theorem 24.5.1. One can proceed exactly as in the Proof of Theorem 24.5.1.

However, for testing null hypothesis of the type $H_0: \quad F_1 = F_2 = \ldots = F_c$ for all x, one needs the joint asymptotic distribution of statistics $T_{N,j}$, $j = 1, \ldots, c$.

Theorem 24.7.2. *Let*

$$T_N = N^{1/2}\left\{(T_{N,1} - \mu_{N,1}), \ \cdots, (T_{N,c} - \mu_{N,c})\right\}. \tag{24.7.5}$$

If assumptions (1), (2), and (3) of Theorem 24.7.1 are satisfied, then for sufficiently large N, T_N has a c-variate normal distribution with mean $\underline{0}$ and variance-covariance matix Γ_N, the diagonal elements of which are given by (24.7.3) and the covariances are given by Eq. (7.11) of GLR. However, when $F_1 = F_2 = \ldots = F_c$, Γ_N takes a much simpler form, the diagonal elements of which are

$$\frac{(1-\lambda_j)}{\lambda_j} I^2, \quad j = 1, \ldots, c \tag{24.7.6}$$

and each off-diagonal element-of Γ_N is $-I^2$, where

$$I^2 = \int_0^1 J^2(u)\, du - \left\{\int_0^1 J(u)\, du\right\}^2. \tag{24.7.7}$$

Remark 24.7.1. *The conclusion of theorems 24.7.1 and 24.7.2 still hold if the assumption (3) of Theorem 24.7.1 is replaced by the condition that the function $J'_j = J'_{j,1} + J'_{j,2}$ with $\int \left|J'_{j,1}(x)\right| \, dx < b$ and $\left|J'_{j,2}\right| \le fg$ with $f \in U_1$ and $g \in U_2$ for $j = 1, \ldots, c$. The uniformity statements of Theorem 24.5.1 extend to this case.*

Remark 24.7.2. *Remark 24.5.4 concerning location and scale families is still applicable here and that Theorem 24.6.1 provides a class of functions J_N which satisfy assumptions (1) and (2) of Theorem 24.7.1. The latter statement is justified by arguments which do not depend on the observations but only on properties of order statistics in a sample from the uniform distribution.*

Remark 24.7.3. *Certain typographical errors in GLR (1967) have been corrected in this presentation.*

24.8 Dependent Samples Case

In the earlier sections, we have focussed in the direction of relaxing the assumptions on the second derivative and its growth of the score function. The other direction in which some contributions have been made in the literature to the Chernoff-Savage class of Statistics, was the assumption of a type of dependence (like ϕ-maxing) among the observations in each sample; however, maintaining the independence between the two samples. (For instance see Fears and Mehta (1974) and Ahmed and Lin (1980).) It is of much interest to test the equality of univariate or bivariate marginal distributions when the samples arise from a continuous bivariate or multivariate population, the joint distribution of which is unknown. For instance, consider the following practical application given by Podgor and Gastwirth (1996). This pertains to the study of the change in visual aquity scores at follow up visit which was about 20 months after initial treatment by comparing treated and untreated eyes. That is, let (X_i, Y_i), $(i = 1, \ldots, n)$ denote the bivariate data in which

X_i = change in visual aquity score in the treated eye of the ith subject, and

Y_i = change in visual aquity score in the untreated eye of the ith subject.

We will be interested in testing whether the distributions of X and Y are the same under the null hypothesis of no treatment effect. The reference for the data is Ederer, Podgor and the Diabetic Retinopathy Study Group (1984). Podger and Gastwirth (1996), ignoring the correlation between the eyes, apply linear rank test procedures that are based on the assumption that X and Y are independent. One can apply the test procedure proposed by the present author (1997, Section 7) which takes into account the dependence between X and Y.

Also, heuristic nonparametric tests for bivariate symmetry have been proposed in the literature. That is, to test

$$H_0 : L(x, y) = L(y, x) \quad \text{for all } x \text{ and } y.$$

In this regard one can easily show by letting $x \to 0$ that bivariate symmetry implies equality of marginal distributions. However, equality of marginals does not imply bivariate symmetry. For example, consder the following family of bivariate distributions:

$$L(x,y) = F(x)G(y) + \rho\{F(x) - F^2(x)\}\{G(y) - G^2(y)\}^2$$

for $-\infty < x, y < \infty$ and $0 < \rho < 3\sqrt{3}$. (Here the bounds for ρ are chosen so that the bivariate density is nonnegative for all x and y.) One can easily see that when $F = G$, $L(x,y) \neq L(y,x)$ for all x and y. Thus tests for bivariate symmetry are not appropriate to test for the equality of the marginal distributions. Thus, we are interested in testing for the equality of univariate or bivariate marginals in the case of multivariate set up. The present-author (1997) has extended the results of GLR (1967) so as to take care of the bivariate or multivariate dependence. He then compares these results with those of Bhuchongkul (1964), Puri and Tran (1981) and Ruymgaart and van Zuijlen (1978), pertaining to the asymptotic normality of a class of rank-based test statistics designed to test for independence of the variables. The assumptions made by Bhuchongkul (1964) and Puri and Tran (1981) are too stringent. They assume the existence and smoothness conditions on the second derivatives of both the score functions. The latter assumption excludes some possible score functions such as Ansari-Bradley or Siegel-Tukey score functions. Also when the scores $E_{in} = E\{J(U_{i,2n})\}$ where $U_{i,2n}$ denote the standard uniform order statistics in a sample of size $2n$, the results of Ruymgaart and van Zuijlen (1978) are applicable provided the existence of J'' and a growth condition on it is further assumed. For further details and applications, the reader is referred to Sections 5, 6, and 7 of the author (1997). Shirahata (1973) has shown that in several multivariate models, LMP rank tests are based on linear combinations of the Chernoff-Savage statistics which are equivalent in distrubution to the same linear combination of the $B_{n,j}$ (see Eq. (6.7) of the author (1997)).

24.9 Results of Hájek, Pyke and Shorack

Hájek (1968) has provided a systematic way of approximating a linear rank statistic by the projection method and also provided an alternate set of sufficient conditions under which the linear rank statistic has an asymptotic normal distribution when it is suitably standardized. This will be considered in the following.

Let $X_1, \ldots, X_N$ be independent random variables with continuous distribution functions $F_1, \ldots, F_N$ and let $R_1, \ldots, R_N$ denote the corresponding ranks. Let

$$u(x) = \begin{cases} 1 & \text{for } x \geq 0 \\ 0 & \text{for } x < 0 . \end{cases} \tag{24.9.1}$$

Then we can express R_i as

$$R_i = \sum_{j=1}^{N} u\left(X_i - X_j\right) \quad 1 \leq i \leq N .$$

Let us consider a class of simple linear rank statistics given by

$$S = \sum_{i=1}^{N} c_i a_N \ (R_i) \tag{24.9.2}$$

where $c_1, \ldots, c_N$ are arbitrary "regression constants" and $a_N(1), \ldots, a_N(N)$ are scores generated by a function $\psi(t),\ 0 < t < 1$, in either of the following two-ways:

$$a_N(i) = \psi\left(i/\left(N+1\right)\right), \qquad 1 \leq i \leq N \tag{24.9.3}$$

$$a_N(i) = E\ \psi\left(U_{i,N}\right), \qquad 1 \leq i \leq N \tag{24.9.4}$$

where $U_{1N} \leq \ldots \leq U_{N,N}$ denote standard uniform order statistics in a sample of size N. Scores given by (24.9.4) are generated by locally most powerful rank test statistics. We assume that the scores are fixed whereas $N,\ (c_1, \ldots, c_N)$, and $(F_1, \ldots, F_N)$ do change.

Let

$$\overline{c} = N^{-1} \sum_{i=1}^{N} c_i,\ \overline{\psi} = \int_0^1 \psi(t)\, dt\,,\ H(x) = N^{-1} \sum_{i=1}^{N} F_i(x)\,. \tag{24.9.5}$$

Then Hájek's (1968) best result can be summarized as follows.

Theorem 24.9.1. *Let $\psi(t) = \psi_1(t) - \psi_2(t),\ \ 0 < t < 1$, where the $\psi_i(t)$ both are nondecreasing, square integrable, and absolutely continuous inside $(0,1)$. Then*

$$\max_{-\infty < x < \infty} \left| P\left(S - ES\right) < x(\text{var } S)^{1/2} - \Phi(x) \right| \to 0 \text{ as } N \to \infty$$

provided

$$\text{var} S > \eta N \max_{1 \leq i \leq N} \ (c_i - \overline{c})^2\ . \tag{24.9.6}$$

Note that var$S$ *can be replaced by*

$$\sigma^2 = \sum_{i=1}^{N} \text{var}\,[\ell_i X_i] \tag{24.9.7}$$

where

$$\ell_i(x) = N^{-1} \sum_{j=1}^{N} (c_j - c_i) \int [u(y-x) - F_i(y)]\; \varphi'(H(y))\, dF_j(y)\,. \tag{24.9.8}$$

Remarks

(1) If $c_1 = \ldots = c_m = 1,\;\; c_{m+1} = \ldots = c_N = 0$, and $F_1 = \ldots = F_m = F$ and $F_{m+1} = \ldots = F_N = G$ then the statistic S given by (24.9.2) coincides with the Chernoff-Savage statistic.

(2) Similarly, the c-sample statistic can be generated by setting $F_1 = \ldots = F_{n_1} = F_1, \ldots, F_{N-n_c+1} = \ldots = F_N = F_c$. Let $s_1, \ldots, s_c$ be a decomposition of the set $\{1, \ldots, N\}$. Let η_j denote the size of the set s_j $(1 \le j \le c)$ and consider the statistics

$$S_j = \sum_{i \in s_j} a_N\,(R_i) \tag{24.9.9}$$

where the scores are given either by (24.9.3) or (24.9.4). Then any linear combination of the statistics S_j, for instance

$$S = \sum_{j=1}^{c} \lambda_j S_j \tag{24.9.10}$$

is of the form (24.9.2) with $c_j = \lambda_j$ if $i \in s_j$ $(1 \le j \le c)$. However, when $\lambda_j = 1/c$ $(1 \le j \le c)$ S reduces to a non-stochastic constant.

(3) The assumption of Hájek on the score function φ is slightly weaker than that of GLR.

(4) Let

$$\sigma^2 = \sum_{i=1}^{N} \text{var}\;\ell_i\,(X_i)\,.$$

Then var S in Therem 24.9.1 can be replaced by σ^2.

(5) It is not always easy, especially when the alternative hypothesis is true, to find a nice closed-form expression for ES. GLR and Chernoff-Savage(1958) show that S is asymptotically normal (μ, σ^2) with $\mu = \sum_{i=1}^{N} c_i E\left[\varphi\left(H(X_i)\right)\right]$. However, Hájek (1968) did not succeed in showing that this is true under the conditions of Theorem 24.9.1. We may need additional assumptions in order to replace ES by μ in Theorem 24.9.1.

(6) **GLR (1967, Theorem 1)** provides a certain uniformity with respect to the class of J functions which is missing in the results of Hájek (1968).

The Results of Pyke and Shorack (1968)

Pyke and Shorack (1968) define an empirical stochastic process for two sample problems and study its weak convergence. They establish an identity (see their Lemma 3.1 on p. 762) which relates the two-sample empirical process to the usual one-sample empirical process. Based on this identity, Pyke and Shorack obtain a version of Chernoff-Savage Theorem (see their Theorem 5.1 on p. 767). They seem to impose differentiablity conditions on the d.f.'s F and G.

Example 24.9.1. In the case of the normal scores test, let

$$J(u) = \Phi^{-1}(u), \quad 0 < u < 1.$$

Then

(i) $J(u)\left[-2\ln u(1-u)\right]^{-1/2} \to 1$
(ii) $J'(u)\left[-2\ln u(1-u)\right]^{1/2}\{u(1-u)\} \to 1$
(iii) $J''(u)\left[u(1-u)\right]^{-2}\left[-2\ln u(1-u)\right]^{1/2} \to 1$ as $u \to 0$ or 1.

In order to establish (i) consider

$$1 - u = \frac{1}{\sqrt{2\pi}} \int_{J(u)}^{\infty} e^{-t^2/2}\, dt \sim \frac{\varphi(J)}{J}.$$

Thus

$$-\ln(1-u) = -\text{const} + \tfrac{1}{2}J^2 + \ln J$$

so $J \sim [-2\ln(1-u)]^{1/2}$ as $u \to 1$.

Analogously one can show that $J \sim [-2\ln(1-u)]^{1/2}$ as $u \to 0$. Consequently

$$J(u) \sim [-2\ln u(1-u)]^{1/2} \quad \text{as } u \to 0 \text{ or } 1.$$

Also we have

$$J'(u) = (2\pi)^{1/2} \exp\left(J^2/2\right).$$

Hence

$$1 - u \sim \frac{1}{JJ'}$$

or

$$J' \sim \frac{(1-u)^{-1}}{J} \approx (1-u)^{-1}\left[-2\ln(1-u)\right]^{-1/2}.$$

By symmetry, one can get the bound when u is near zero. Thus

$$J' \sim \{u(1-u)\}^{-1}\left[-2\ln(1-u)\right]^{-1/2}.$$

Also, differentiating the exact expression for $\ln J'$, we obtain

$$\begin{aligned} J''(u) =& J(u)\left[J'(u)\right]^2 \\ &\sim \left[u(1-u)\right]^{-2}\left[-2\ln u(1-u)\right]^{-1/2}. \end{aligned}$$

Notice that $-\ln u(1-u) \sim [u(1-u)]^{-\delta}$ for any $\delta > 0$. Gavin (1977, Eq. (3.5)) obtained the asymptote expression for J and J'.

24.10 Asymptotic Equivalence of Procedures

Capon's (1961) two-sample test for scale is based on

$$T_N = N^{-1/2}\sum_{i=1}^{N} E(W_{iN}^2) Z_{N,i}$$

where W_{iN}, $(i = 1, \ldots, N)$ are standard normal order statistics in a random sample of size N. Klotz's (1962) test for the same hypothesis is based on

$$T_N^* = N^{-1/2}\sum_{i=1}^{N}\left[\Phi^{-1}\left(\frac{i}{N+1}\right)\right]^2 Z_{N,i}.$$

For Capon's test $J(u) = \left[\Phi^{-1}(u)\right]^2$ and

$$\begin{aligned} \left|J'(u)\right| &= 2\left|\Phi^{-1}(u)\right| / \phi\left(\Phi^{-1}(u)\right) \\ &\leq K\left[u(1-u)\right]^{-\delta}\left[u(1-u)\right]^{-1} = K\left[u(1-u)\right]^{-1-\delta} \end{aligned}$$

thus satisfying the conditions of Theorem 24.7.1. Hence the two tests are asymptotically equivalent in probability.

Gibbons and Chakrabarti (1992, p. 273) has conjectured that the test

$$T_N^{**} = N^{-1/2} \sum_{i=1}^{N} \{E\left(W_{iN}\right)\}^2 \ Z_{N,i}$$

is equivalent in distribution to T_N as $N \to \infty$.

Consider

$$T_N - T_N^{**} = N^{-1/2} \sum_{i=1}^{N} \left(\text{var } W_{iN}\right) \ Z_{N,i} \, .$$

Hence it suffices to show that $N^{-1/2} \sum_{i=1}^{N} \text{var } W_{iN} \to 0$ as $N \to \infty$. Towards this we need the following lemma of Mason (1977).

Lemma 24.10.1. *Let $J(u) = F^{-1}(u)$ and let $|J'(u)| \leq K\left[u(1-u)\right]^{-1}$, $0 < u < 1$. Then, there exists a constant $k > 0$ independent of N $(N \geq 1)$ and i $(1 \leq i \leq N)$ such that*

$$\text{var } W_{iN} \leq KN^{-1}\left[\frac{i}{N+1}\left(1 - \frac{i}{N+1}\right)\right]^{-1} .$$

PROOF: Let $U_{1N} \leq \ldots \leq U_{NN}$ denote the standard uniform order statistics in a sample of size N and let $g(u) = F^{-1}(u)$ for $0 < u < 1$. Then

$$\begin{aligned}\text{var } W_{iN} &= \text{var}\left[\int_{i/(N+1}^{U_{iN}} g'(u)\,du\right]\\ &\leq E\left(\int_{i/(N+1}^{U_{iN}} g'(u)\,du\right)^2\\ &\leq KE\left[\int_{i/(N+1}^{U_{iN}} (u(1-u))^{-1}\,du\right]^2\\ &\leq KE\left\{\int_{i/(N+1}^{U_{iN}} \left[u^{-1}+(1-u)^{-1}\right]du\right\}^2\\ &= KE\left\{\left[\ln U_{iN} - \ln\frac{i}{N+1}\right] + \left[\ln(1-U_{iN}) - \ln\left(1-\frac{i}{N+1}\right)\right]\right\}^2.\end{aligned}$$

Now, applying Minkowski's inequality, we obtain

$$\begin{aligned}\text{var } W_{iN} \quad \leq \quad & K\left\{\left[E(\ln U_{iN}) - \ln(\frac{i}{N+1})^2\right]^{1/2}\right.\\ & \left. + \left[E(\ln(1-U_{iN})) - \ln(1-(\frac{i}{N+1}))^2\right]^{1/2}\right\}^2.\end{aligned}$$

Note that $-\ln U_{iN}$ is equivalent to the $(N-i+1)^{\text{th}}$ smallest standard negative exponential order statistic in a sample of size N. Hence

$$E\left[-\ln U_{iN}\right] = \textstyle\sum_{j=1}^{N} j^{-1} \qquad \text{and} \qquad \text{var }(-\ln U_{iN}) = \textstyle\sum_{j=1}^{N} j^{-2}.$$

Consequently,

$$\begin{aligned}& E\left(-\ln U_{iN} + \ln(\frac{i}{N+1})\right)^2\\ &\quad = \text{var }(-\ln U_{iN}) + \left\{E(-\ln U_{iN}) + \ln(\frac{i}{N+1})\right\}^2\\ &\quad = \sum_{j=i}^{N} j^{-2} + \left\{\sum_{j=i}^{N} j^{-1} + \ln(\frac{i}{N+1})\right\}^2.\end{aligned}$$

Next

$$\sum_{j=i}^{N} j^{-2} \leq \int_{i}^{N+1} u^{-2}\,du = \frac{1}{i} - \frac{1}{N+1} \leq \frac{1}{i}$$

$$\begin{aligned}
\sum_{j=i}^{N} j^{-1} &\le \int_{i-\frac{1}{2}}^{N+1} j^{-1} = \int_{i-\frac{1}{2}}^{N+1} u^{-1}\,du \\
&= -\ln\left(\frac{i-\frac{1}{2}}{N+1}\right) \\
&= -\ln\left(\frac{i}{N+1}\right) - \ln\left(1-\frac{1}{2i}\right); \\
&\le -\ln\left(\frac{i}{N+1}\right) + \frac{(2i)^{-1}}{1-(2i)^{-1}} \\
&= -\ln\left(\frac{i}{N+1}\right) + \frac{1}{2i-1}.
\end{aligned}$$

Thus

$$E\left[-\ln U_{iN} + \ln\left(\frac{i}{N+1}\right)\right]^2 \le \frac{1}{i} - \frac{1}{2i-1} \le \frac{2}{i}$$

since $1-U_{iN}$ is distributed as $U_{N-i+1,N}$, replacing i by $N-i+1$, in the above expression, we obtain

$$E\left[-\ln(1-U_{iN}) + \ln\left(\frac{N-i+1}{N+1}\right)\right]^2 \le 2/(N-i+1).$$

Hence

$$\begin{aligned}
\text{var } W_{iN} &\le K\left[(2/i)^{1/2} + (2/(N-i+1))^{1/2}\right]^2 \\
&\le 4K\left(\frac{1}{i} + \frac{1}{N+1-i}\right) = \frac{4K(N+1)}{i(N-i+1)} \\
&\le \frac{K}{N}\left[\frac{i}{N+1}\left(1-\frac{i}{N+1}\right)\right]^{-1},
\end{aligned}$$

after using the fact $(a+b)^2 \le 2\,(a^2+b^2)$. Thus

$$\begin{aligned}
\sum_{i=1}^{N} \text{var } W_{iN} &\le K\sum_{i=1}^{N}\left(\frac{1}{i} + \frac{1}{N+1-i}\right) \\
&= K\sum_{i=1}^{N}\frac{1}{i} = K \quad (\ln N + \text{Euler's constant});
\end{aligned}$$

consequently

$$N^{-1/2}\sum_{i=1}^{N} \text{var } W_{iN} \to 0 \text{ as } N \to \infty.$$

Remark 24.10.1. *One can readily obtain upper bounds for covariances because*

$$\text{cov}\ (W_{iN},\ W_{jN}) \leq [(\text{ var } W_{iN})\ (\text{var } W_{jN})]^{1/2}$$

Remark 24.10.2. *If* $g'(u) \leq K\,[u(1-u)]^{-(3/2)+\delta}$ *for some* $0 < \delta < 1/2$, *Mason (1977, Proposition 3.6) has shown that*

$$\text{var } W_{iN} \leq \frac{c}{N}\left[\frac{i}{N+1}\left(1-\frac{i}{N+1}\right)\right]^{-1+\delta} \quad \textit{for } i = 1, \ldots, N.$$

24.11 Problems

24.4.1 Evaluate the normalizing constants when $F = G$ for the following members of the Chernoff-Savage class of statistics.

(i) Wilcoxon rank sum test statistic

(ii) The normal scores test statistic.

24.5.1 Verify the regularity conditions of Theorem 24.5.1 for the Ansari-Bradley or Siegel-Tukey statistic generated by the following score function.

$$J_N\left(\frac{i}{N+1}\right) = \left|\frac{i}{N+1} - \frac{1}{2}\right|, \qquad i = 1, \ldots, N.$$

24.5.2 Verify the regularity conditions of Theorem 24.5.1 for the score function

$$J_N(x) = x^k \qquad (k > 1),\ 0 < x < 1.$$

24.7.1 Let $X_{i,j} (j = 1, \ldots, n_i)$ be a random sample from the distribution $F_i(x) (i = 1, \ldots, c)$. We wish to test

$$H_0 :\ F_1(x) = \cdots = F_c(x) \quad \text{for all } x$$

against the alternative

$$H_1 :\ F_i(x) \neq F_j(x) \qquad \text{for some } x \text{ and some } i \neq j.$$

Devise a test procedure for H_0 versus H_1 that is based on the test statistic T_N given by (24.7.5) with $J_j(u) \equiv u$.
(HINT: Let

$$\tilde{T}_{N,j} = \sqrt{\lambda_j}(T_{N,j} - \mu_{N,j}), \qquad j = 1, \cdots, c.$$

Then by Theorem 24.7.2, the $\tilde{T}_{N,j}$ will have a joint asymptotic normal distribution with variances

$\sigma_j^2 = (1-\lambda_j)I^2$ and covariances $\sigma_{ij} = -\sqrt{\lambda_i\lambda_j}I^2 (i \neq j)$.

Also note that

$$\begin{aligned}\sum_{j=1}^{c} \sqrt{\lambda_j}\tilde{T}_{N,j} &= \sum_{j=1}^{c} \lambda_j (T_{N,j} - \mu_{N,j}) \\ &= N^{\frac{1}{2}} \left[\int J_N(\frac{N}{N+1}H_N)dH_N - \int J(H)dH \right] \\ &= N^{\frac{1}{2}} \left[\frac{1}{N}\sum_{i=1}^{N}(\frac{i}{N+1}) - \frac{1}{2} \right] \\ &= 0 .\end{aligned}$$

Now applying Corollary 24.6.1.1 with $a_j = \lambda_j^{\frac{1}{2}}$ we can assert that $\sum_{j=1}^{c} \tilde{T}_{N,j}^2 \overset{d}{\approx} \chi_{c-1}^2$ (since $\sum_1^c \lambda_j^{\frac{1}{2}} \tilde{T}_{N,j} = 0$). Thus, the rule is: reject H_0 when $\sum_{j=1}^{c} \tilde{T}_{N,j}^2 > \chi_{c-1,1-\alpha}^2$ and accept H_0 otherwise.)

24.7.2 Consider the following artificial data obtained on three teaching methods in an undergraduate course on statistics. Method 1 consists of a professor tcaching in person a medium sized class. Method 2 consists of the same professor teaching a large section of the same course with recitation classes handled by a teaching assistant. Method 3 consists of teaching a course from a distance using televideo. The response variable is the score out of 50 a student obtains on a comprehensive final examination. The following gives the scores of random samples of students taken from each class.

Method 1	Method 2	Method 3
35	36	39
30	31	43
44	37	34
46	41	39
49	42	43
40	42	32
40	38	29
48	38	28
46	31	27
30		32
44		

Assuming the samples are large, carry out a test of the null hypothesis that there is no difference among the methods against the alternative that different teaching methods have different effects on the students' performance. Use $\alpha = 0.05$ and employ the chi-square test procedure proposed in Problem 24.7.1.

24.8.1 The following data gives the weights in pounds of a random sample of size 16 women before and after they have completed a Weight Watchers program.

before	120	131	152	116	136	101	145	138	132	128	110	125	127	130	133	140
after	100	101	134	95	142	94	126	106	107	105	120	130	123	145	147	131

Test the hypothesis of equality of marginal distributions against the alternative that the distribution function for "before" is smaller than the distribution function for "after".

24.8.2 Musculoskeletal neck-and-shoulder disorders are common among office staff who perform repetitive tasks using visual display units. The paper "Upper-arm elevation during office work" (published in *Ergonomics*, 1966: pp. 1221–1230) reported a study for determining whether more varied work conditions will have any impact on arm movement and the following data was obtained.[2] Each observation is the amount of time expressed as a proportion of total time observed during which arm elevation was below 30°. The time duration for the two measurements from each subject was 18 months, during which the work conditions were changed, and subjects were allowed to engage in a wider variety of work tasks. Does the data suggest that the true average time during which elevation is below 30° differs after the change from what it was before?

Subject	1	2	3	4	5	6	7	8
Before	81	87	86	82	90	86	96	73
After	78	91	78	78	84	67	92	70
Subject	9	10	11	12	13	14	15	16
Before	74	75	72	80	66	72	56	82
After	58	62	70	58	66	60	65	73

[2]See also: Devore, J.L. (2000). *Probability and Statistics for Engineering and the Sciences.* Duxbury Thompson Learning, Pacific Grove, CA. p. 376 (Example 9.9).

Chapter 25

CS Class for One Sample Case

25.1 Introduction

Let $X_{j,1}, \ldots, X_{j,N_j}$ denote the observations belonging to the jth category $(j = 1, \ldots, c)$ such that $N_1 + \ldots + N_c = N$ which is non-random. In particular, the N_j $(j = 1, \ldots, c)$ can have a multinomial or multi-variate hypergeometric distribution. Let $F^{(j)}(x)$ denote the distribution function of $h(X_{j,1}), \ldots, h(X_{j,N_j})$ for some specified Borel measurable h $(j = 1, \ldots, c)$. Suppose we wish to test $H_0 : F^{(1)}(x) = \ldots = F^{(c)}(x)$ for all x against $H_a : F^{(i)}(x) \neq F^{(j)}(x)$ for some x and for some $i \neq j$. Also, let

$$\begin{aligned} Z_{N,j,i} &= 1, \text{if the ith smallest of } N \text{ ordered } h(X_{j,i}),\ 1 \le i \le N \text{ belongs} \\ &\qquad \text{to } j^{\text{th}} \text{ category} \\ &= 0, \text{otherwise, for } i = 1, \ldots, N. \end{aligned}$$

For some fixed constants $E_{N,j,i}$, consider the nonparametric test statistics of the form

$$T_{N,j} = \sum_{i=1}^{N} E_{N,j,i} Z_{N,j,i}, \quad j = 1, \ldots, c. \tag{25.1.1}$$

The class of statistics given by (25.1.1) when, in particular, $c = 2$ includes the class of nonparametric test statistics for the problem of location with symmetry. Significant contributions have been made to the problem of location with symmetry by Govindarajulu (1960), Puri and Sen (1969), Huskova (1970), Vorlickova (1972). Also, Albers, Bickel and Van Zwet (1976), Pyke and Shorack (1968) obtain very interesting weak convergence results for Chernoff-Savage type of statistics based on random sample sizes. The sign test, the Wilcoxon signed rank test and the absolute normal scores test will be special cases of the class of two category nonparametric test statistics. Since the exact distribution of $T_{N,1}$ will be quite complicated except for very small sample sizes, asymptote distribution of $T_{N,1}$ will be quite useful either in carrying out the test procedure and/or the computation of power of the test under the alternative hypothesis. Towards this goal, first we develop the asymptotic distribution of $T_{N,1}$ for the case $c = 2$ and these results will be extended to the c-category problem. We simplify the notation in the case of $c = 2$ as follows.

25.2 Regularity Assumptions

Let $X_1, \ldots, X_{\mathbf{m}}$ $[Y_1, \ldots Y_{\mathbf{n}}]$ denote a random sample of size $\mathbf{m}\,[\mathbf{n}]$ drawn from a continuous distribution $F(x)$ $[G(x)]$ where $\mathbf{m}$ and $\mathbf{n}$ are random but $N = \mathbf{m}+\mathbf{n}$ is non-random. Continuity of F and G is not necessary provided F and G can be made continuous by the Continuization Procedure given in GLR (1967). As a special case, $V_1, \ldots, V_N$ is a random sample drawn from a distribution $L(x)$, $X_1, \ldots X_{\mathbf{m}}$ $[Y_1, \ldots, Y_{\mathbf{n}}]$ are those V's or Borel measurable functions of V's which belong to category I [category II]. In a further special case, the X's can represent the absolute values of the negative V's and the Y's denote the positive V's. (Generally, values other than zero could have been used as the breaking point.)

Let $F_{\mathbf{m}}(x)$ $[G_{\mathbf{n}}(y)]$ denote the empirical distribution function (edf) based on $X_1, \ldots, X_{\mathbf{m}}$ $[Y_1, \ldots, Y_{\mathbf{n}}]$. Further let

$$H_N(x) = \underline{\lambda}_N F_{\mathbf{m}}(x) + (1 - \underline{\lambda}_N) G_{\mathbf{n}}(x), \quad \underline{\lambda}_N = \mathbf{m}/N. \tag{25.2.1}$$

(i.e. $H_N(x)$ is the edf of the combined sample of size N).

$$H(x) = \underline{\lambda}_N F(x) + (1 - \underline{\lambda}_N) G(x), \tag{25.2.2}$$

and

$$H^*(x) = p_N F(x) + (1 - p_N) G(x), \quad p_N = E(\underline{\lambda}_N), \tag{25.2.3}$$

where $L(x)$ and consequently F and G may depend on N and we suppress this for the sake of simplicity. m and n denote some specified values assumed by the random variables $\mathbf{m}$ and $\mathbf{n}$ respectively. Furthermore,

$$\lambda_N = m/N, \quad \mu_{2,N} = NE\left\{(\underline{\lambda}_N - p_N)^2\right\} \tag{25.2.4}$$

and

$$s = N^{-1/2}(m - Np_N)/\mu_{2,N}^{1/2}. \tag{25.2.5}$$

In all the definitions, we implicitly assume that the relevant quantities exist. We also assume that there exists a p_0 $(p_0 \leq 1/2)$ such that $0 < p_0 \leq p_N \leq 1 - p_0 < 1$. We will be concerned with test statistics of the form

$$T_N = \sum_{i=1}^{N} E_{N,i} Z_{N,i} \tag{25.2.6}$$

where

$$\begin{aligned} Z_{N,i} &= 1, \text{if the ith smallest observation in the combined sample is an } X \\ &= 0, \text{otherwise, for } i = 1, \ldots, N. \end{aligned}$$

and $E_N = (E_{N,1}, \ldots, E_{N,N})$ is a given vector of constants for each N. In the case of the location problem with symmetry,

$$T_N = \begin{cases} \text{Sign statistic when } E_{N,i} \equiv 1 \\ \text{Wilcoxon signed rank statistic when } E_{N,i} = i/(N+1) \\ \text{Absolute normal scores test when } E_{N,i} \text{ is the} \\ \text{expected value of the ith smallest order statistic} \\ \text{in a sample of } N \text{ from the chi-population with 1 d.f.} \end{cases}$$

We use the following integral representation of T_N:

$$T_N = \mathbf{m} \int J_N\left(\frac{N}{N+1} H_N(x)\right) \, dF_{\mathbf{m}}(x) \tag{25.2.7}$$

where (25.2.1) and (25.2.7) are equivalent when

$$E_{N,i} = J_N\left(\frac{i}{N+1}\right), \quad i = 1, \ldots, N.$$

We will be interested in studying the asymptotic behavior of the differences of the form

$$T_N^* = N^{1/2}\left[\underline{\lambda}_N \int J_N\left(\frac{N}{N+1} H_N(x)\right) dF_{\mathbf{m}}(x) - p_N \int J_N\left(H^*(x)\right) dF(x)\right]. \tag{25.2.8}$$

The function $J_N(u)$ may depend on N, although the subscript N will be suppressed whenever there is no ambiguity. As in Chapter 24, although it suffices to define $J(u)$ at $1/(N+1), \ldots, N/(N+1)$ we extend its domain of definition to $(0, 1)$ by letting $J(u)$ be constant over

$(1/(N+1), \ldots, N/(N+1))$ $(i = 0, \ldots, N)$. Also without loss of generality, we can assume that $H^*(x)$ is uniform on $(0,1)$. Hereafter, whenever there is no ambiguity, the subscript N in $J_N, \underline{\lambda}_N$ and p_N will be suppressed. Throughout, K will be used as a generic finite constant which will not depend on F, G and N.

Let f and g be U-shaped functions (≥ 1) defined on the interval $(0,1)$ such that f is integrable and g is square integrable. Consider functions J defined by $J(x) = \int_{1/2}^{x} J'(v)dv$. Define the following classes of functions. We say that $J \in \mathcal{S}_0$ if $|J'| \leq fg$, $J \in \mathcal{S}$ if $J' = J_1' + J_2'$ with $|J_1'| \leq fg$ and $\int |J_2'(u)|\, du \leq b < \infty$, and $J \in \mathcal{S}_1$ if $|J'(u)| \leq K\,[u(1-u)]^{-(3/2)+\delta}$ for some δ $(0 < \delta < 1/2)$. Notice that $\mathcal{S}_1 \subseteq \mathcal{S}_0 \subseteq \mathcal{S}$.

Let $U_{k,N}$ denote the kth smallest uniform order statistic on $(0,1)$ in a random sample of size N. For any function J, define $\overline{J}_N(y)$ as follows: if $y = k/(N+1)$ $(k = 1, \ldots, N)$, let

$$\overline{J}_N(y) = E\,|J(U_{k,N})| = \int_0^1 J(x)\beta_N(x,k)\, dx \tag{25.2.9}$$

where $\beta_N(x,k)$ denotes the density function of $U_{k,N}$. Interpolate linearly for $y \in \{k/(N+1), (k+1)/(N+1)\}$ and leave $\overline{J}_N(y)$ constant below $1/(N+1)$ and above $N/(N+1)$. Then the main results pertaining to the asymptotic normality of T_N^* will be given in Section 25.3.

25.3 Main Theorems

With the notation and assumptions of Section 25.2, let T_N^* be given by (25.2.8). Then we have the following Theorem of Govindarajulu (1985).

Theorem 25.3.1. *Let $N^{1/2}(\underline{\lambda}_N - p_N)$ be asymptotically normal with $f(m) = P(\mathbf{m} = m) = P(N^{1/2}(\underline{\lambda} - p)/\sqrt{\mu_2} = s) = (N\mu_2)^{-1/2}\phi(s) + o(N^{-1/2})$ where $s = (m - Np)/\sqrt{N\mu_2}$ and ϕ denotes the standard normal density function. Let J belong to $\mathcal{S}$, P_N be the distribution of T_N^* and $\Phi(x)$ be the standard normal distribution function. If $\sigma_N^2\,[(F,G), J, p]$ stays bounded away from zero, then*

$$\sup_x |P_N(x) - \Phi(x/\sigma_N)| \to 0 \text{ as } N \to \infty$$

for every $J \in \mathcal{S}$ and every triple $((F,G),p)$ where

$$\sigma_N^2 = p(1-p)^2 I_1 + p^2(1-p)I_2 + \mu_2 I_3^2, \tag{25.3.1}$$

$$I_1 = 2 \iint_{x<y} F(x)[1 - F(y)] J'(H^*(x)) J'(H^*(y))\, dG(x) dG(y), \tag{25.3.2}$$

$$I_2 = 2 \iint_{x<y} G(x)[1 - G(y)] J'(H^*(x)) J'(H^*(y))\, dF(x) dF(y), \tag{25.3.3}$$

and

$$I_3 = \int \left[J(H^*(x)) + p(F - G) J'(H^*) \right] dF. \tag{25.3.4}$$

PROOF: We will give a rough sketch of the proof. One can write

$$\begin{aligned} T_N^* &= N^{1/2} \underline{\lambda} \left[\int J\left(\frac{N}{N+1} H_N \right) dF_{\mathbf{m}} - \int J(H^*) dF \right] \\ &\quad + N^{1/2} (\underline{\lambda} - p) \int J(H^*) dF \\ &= B_N + R_{N,1} + R_{N,2} \end{aligned}$$

where

$$B_N = N^{1/2} \underline{\lambda} \left[\int J(H^*) d(F_{\mathbf{m}} - F) \right] + \int (H_N - H^*) J'(H^*) dF + N^{1/2} (\underline{\lambda} - p) \int J(H^*) dF, \tag{25.3.5}$$

$$R_{N,1} = N^{1/2} \underline{\lambda} \int \left[J\left(\frac{N}{N+1} H_N \right) - J(H^*) \right] d(F_{\mathbf{m}} - F) \tag{25.3.6}$$

and

$$R_{N,2} = N^{1/2} \underline{\lambda} \int \left[J\left(\frac{N}{N+1} H_N \right) - J(H^*) - (H_N - H^*) J'(H^*) \right] dF. \tag{25.3.7}$$

After performing integration by parts once in the first term in B_N, one can rewrite B_N as

$$\begin{aligned} N^{-1/2} B_N &= -\underline{\lambda} \int (F_{\mathbf{m}} - F) J'(H^*) dH^* + \underline{\lambda} \int (H_N - H^*) J'(H^*) dF \\ &\quad + (\underline{\lambda} - p) \int J(H^*) dF. \end{aligned}$$

Now writing

$$\underline{\lambda}(H_N - H^*) = p(H_N - H) + p(H - H^*) + (\underline{\lambda} - p)(H_N - H^*)$$

and

$$\begin{aligned} -\underline{\lambda} \int (F_{\mathbf{m}} - F)J'(H^*)dH^* &+ p\int (H_N - H)J'(H^*)dF \\ &= -\underline{\lambda}(1-p)\int (F_{\mathbf{m}} - F)J'(H^*)dG + p(1-\underline{\lambda})\int (G_{\mathbf{n}} - G)J'(H^*)dF \end{aligned}$$

we have

$$B_N = B_N^* + R_N^*$$

where

$$\begin{aligned} N^{-1/2}B_N^* = -\underline{\lambda}(1-p)\int (F_{\mathbf{m}}-F)J'(H^*)dG + p(1-\underline{\lambda})\int (G_{\mathbf{n}}-G)J'(H^*)dF \\ + (\underline{\lambda} - p)\int \left[J(H^*) + p(F-G)J'(H^*)\right] dF \quad (25.3.8) \end{aligned}$$

and

$$R_N^* = N^{1/2}(\underline{\lambda} - p)\int (H_N - H^*)J'(H^*)dF. \quad (25.3.9)$$

Now,

$$E\left(B_N^* | \mathbf{m} = m\right) = N^{1/2}(\lambda - p)I_3$$

and

$$\text{var}\left(B_N^* | \mathbf{m} = m\right) = \lambda(1-p)^2 I_1 + p^2(1-\lambda)I_2 = \sigma^{*^2}\,(say).$$

Consider

$$\begin{aligned} P\left(B_N^*/\sigma \le x\right) &= \sum_{m=0}^{N} P\left(B_N^*/\sigma \le x,\ \mathbf{m} = m\right) \\ &= \sum_{m:|s|\le\varepsilon} + \sum_{m:|s|>\varepsilon} \end{aligned}$$

where $s = N^{1/2}(\lambda - p)/\mu_2^{1/2}$. By Chebyshev's inequality, the second sum is bounded above by ε^{-2}. Also, the first sum can be written as (for sufficiently large N)

$$\sum_{m_i|s|\le\varepsilon} P\left(B_N^* \le x\sigma|\mathbf{m}=m\right) f(m)$$
$$= \sum \Phi\left(\frac{\sigma x - \sqrt{\mu_2} I_3 s}{\sigma^*}\right)\left\{(N\mu_2)^{-1/2}\phi(s) + o(N^{-1/2})\right\}$$
$$= \int_{-\varepsilon}^{\varepsilon} \Phi\left(\frac{\sigma x - \sqrt{\mu_2} I_3 s}{\sigma^*}\right)\phi(s)ds + o(1)$$
$$= \int_{-\varepsilon}^{\varepsilon} \Phi\left\{\frac{\sigma x - \sqrt{\mu_2} I_3 s}{(\sigma^2 - \mu_3 I_3^2 + o(1))^{1/2}}\right\}\phi(s)ds + o(1)\,.$$

Now, appealing to the continuity of Φ and letting ε and N tend to infinity we have

$$\lim_{N\to\infty} P\left(B_N^*/\sigma \le x\right)$$
$$= \int_{-\infty}^{\infty} \Phi\left(\frac{x - as}{(1-a^2)^{1/2}}\right)\phi(s)ds, \quad a = \sqrt{\mu_2} I_3/\sigma$$
$$= (2\pi)^{-1}\int_{s=-\infty}^{\infty}\int_{y=-\infty}^{(x-as)(1-a^2)^{-1/2}} \exp\left\{-\frac{1}{2}(y^2+s^2)\right\} dyds.$$

Letting $u = y(1-a^2)^{1/2} + as$, we have

LHS
$$= (2\pi)^{-1/2}\int_{u=-\infty}^{x} e^{-u^2/2}\left[\int_{-\infty}^{\infty}(2\pi)^{-1/2}(1-a^2)^{-1/2}e^{-(s-au)^2/2(1-a^2)}ds\right]du$$
$$= \Phi(x).$$

Remark 25.3.1. *In particular* $\mathbf{m}$ *can have a binomial or hypergeometric distribution because the hypergeometric probability function tends to the binomial probability which has the desired asymptotic representation.*

Towards the completion of the proof of Theorem 25.3.1, it suffices to show that the higher order terms, namely, $R_{N.1}, R_{N,2}$ *and* R_N^* *go to zero in probability as* N *tends to infinity. By Lemma 7 of GLR (1967),* $N^{1/2}(H_N - H^* g(H^*)$ *is bounded in probability. Also* $N^{1/2}(\underline{\lambda} - p)$ *is bounded in probability. Thus* R_N^* *is bounded by* $(KN^{-1/2})\int f(H^*)dH^* = K/N^{1/2}$ *for all* $J \in \mathcal{S}$*. The asymptotic negligibility of* $R_{N,1}$ *and* $R_{N,2}$ *will be considered in the next section.*

25.4 Bounds for Tails and Higher Order Terms

Before we consider the higher order terms, let us bound the tails of the statistic. Let $J \in \mathcal{S}$ and τ be a number such that $0 < 2\tau < 1$. Let $W_{1N} \leq \ldots \leq W_{NN}$ denote the combined ordered X's and Y's and let $R_1, \ldots, R_{\mathbf{m}}$ denote the ranks of X's in the combined ordered sample. Furthermore, let $U_{1N} \leq \ldots \leq U_{NN}$ denote the standard uniform order statistics in a random sample of size N. Then consider the statistic $\widetilde{T}_N$ given by

$$\widetilde{T}_N = N^{-1/2} \sum_{i=1}^{N} \left\{ J\left(\frac{N}{N+1} H_N(W_{iN}) \right) - J(H^*(W_{iN})) \right\} Z_{N,i}.$$

Now without loss of generality, set $H^*(x)$ to be the uniform $(0, 1)$ distribution function. Let

$$\Delta_N^*(J, \tau) = \Delta_{N,1}^*(J, \xi) + \Delta_{N,2}^*(J, \xi)$$

where

$$\Delta_{N,1}^*(J, \xi) = \sum_{i:U_{iN} \leq \xi} \left\{ J\left(\frac{i}{N+1} \right) - J(U_{iN}) \right\} Z_{N,i}$$

and an analogous expression for $\Delta_{N,2}^*(J, \xi)$ which denotes the upper tail of the statistic. If $U_1, \ldots, U_N$ denote the unordered $U_{1N}, \ldots, U_{NN}$, since $U_i = U_{R_i,N}$ $(i = 1, \ldots, N)$ we have

$$\left|\Delta_{N,1}^*(J, \xi)\right| \leq N^{-1/2} \sum_{i:U_i \leq \xi} \left| J\left(\frac{R_i}{N+1} \right) - J(U_i) \right|. \tag{25.4.1}$$

Now, proceed as in GLR (1967) in order to assert that, for every $\varepsilon > 0$ there exists a ξ_0 such that

$$P\left[\sup\left\{ \left|\Delta_{N,1}^*(J, \xi)\right|, \; 0 < \xi < \xi_0, \; J \in \mathcal{S} \right\} > \varepsilon \right] < \varepsilon \tag{25.4.2}$$

for every N and every pair (F, G).

By symmetry one can analogously bound $\Delta_{N,2}^*(J, \xi)$. The tail of the statistic $T_{N,i}^*$ s not quite equal to $\Delta_N^*(J, \xi)$ and is given by

$$\Delta_N(J, \xi) = \Delta_N^*(J, \xi) + N^{1/2} \lambda_N \int_A J(H^*) \, d(F_{\mathbf{m}} - F) + N^{1/2}(\lambda_N - p) \int_A J(H^*) \, dF, \tag{25.4.3}$$

where $A = (0, \xi) \cup (1 - \xi, 1)$.

Now $N^{-1/2} \int_A J(H^*) d(F_{\mathbf{m}} - F)$ is a normalized sum with zero expectation and variance bounded by expressions of the form $p \int_A J^2(H^*)\, dF \leq \int_A J^2(u)\, du$ which, because $J^2(u)$ is uniformly integrable (since $J \in \mathcal{S}$), can be made arbitrarily small . Furthermore, $N^{1/2}(\underline{\lambda}_N - p)$ is bounded in probability . Hence it follows that for every $\varepsilon > 0$ there is a ξ_0 such that

$$P\left\{\left|\Delta_N - \Delta_N^*\right| > \varepsilon\right\} < \varepsilon$$

for every $J \in \mathcal{S}$ and all $\xi < \xi_0$.

Consequently, for every $\varepsilon > 0$ there exists a number $\xi_0 > 0$ such that for $\xi \leq \xi_0$ and $J \in \mathcal{S}$, we have

$$P\left(|\Delta_N(J, \xi)| > \varepsilon\right) < \varepsilon \text{ for every } N \text{ and every pair } (F, G). \tag{25.4.4}$$

Next, since the tails of the statistic T_N^* are bounded, it suffices to establish the negligibility of the higher order terms $R_{N,1}$ and $R_{N,2}$ for $H^* \in (\xi, 1-\xi)$. Now for $J \in \mathcal{S}$, we can assume that $J(x) = \mu(0, x)$ where μ is a signed measure. So, consider

$$\begin{aligned} P\left(|R_{N,1}| > \varepsilon\right) &= \sum_{m=0}^{N} P\left(|R_{N,1}| > \varepsilon | \mathbf{m} = m\right) P(\mathbf{m} = m) \\ &= \sum_{\{m:|s| \leq \nu\}} + \sum_{\{m:|s| > \nu\}} \\ &= \sum_{\{m:|s| \leq \nu\}} P\left(|R_{N,1}| > \varepsilon | \mathbf{m} = m\right) P(\mathbf{m} = m) + \nu^{-2} \end{aligned}$$

where $s = N^{-1/2}(m - Np_N)/\mu_2^{1/2}$ and $\mu_2 = NE\left\{(\underline{\lambda}_N - p_N)^2\right\}$.

Now, apply Proposition 3 of GLR (1967) and assert that $P(|R_{N,1}| > \varepsilon | \mathbf{m} = m) < \varepsilon$ uniformly in m and n and set $\nu = \varepsilon^{-1/2}$. Thus $R_{N,1}$ is asymptotically negligible in probability.

Towards $R_{N,2}$ since $H^* \in (\xi, 1-\xi)$ it suffices to assume that J' is bounded by an integrable function. Noting that for given $\varepsilon > 0$ there is a c such that

$$P\left(N^{1/2} \sup_x \left|H_N(x) - H^*(x)\right| \geq c\right) < \varepsilon$$

and by repeating the arguments of GLR (1967, pp. 629–630) assert that

$$P\left(|R_{N,2}| > \varepsilon \| s| \leq \nu\right) < \varepsilon \tag{25.4.5}$$

for every $J \in \mathcal{S}$ and every pair (F, G). Hence, it follows that $R_{N,2}$ is negligible in probability.

Remark 25.4.1. *In the following situation, the uniformity asserted in Theorem 25.3.1 is of much interest. Let*

$$F(x) = P\left(|V| \le x | V < 0\right) = \left\{\Psi(-\theta) - \Psi(-x-\theta)\right\}/\Psi(-\theta),$$

and

$$G(x) = P\left(|V| \le x | V > 0\right) = \left\{\Psi(x-\theta) - \Psi(-\theta)\right\}/\Psi(\theta)$$

where Ψ *is symmetric about zero and* θ *is in a bounded interval which depends on* N*. Furthermore, if* $\theta = \theta_N$ *where* $\theta_N \to 0$ *as* $N \to \infty$*, then the Kolmogorov distance* $|P_N - \Phi(x/\sigma_N)|$ *also tends to zero uniformly for any* $\mathcal{S}^* \subset \mathcal{S}$ *such that* $J \in \mathcal{S}^*$ *implies*

$$\int J^2(u)\,du \ge a > 0.$$

Note that as $\theta_N \to 0$, $\lim F(x) = \lim G(x) = 2\Psi(x) - 1$ for $x \ge 0$ and

$$\sigma_N^2\left[F, G, J, p\right] \to \sigma_N^2\left[(F,F), J, 1/2\right] = (1/4)\int_0^1 J^2(u)\,du. \qquad (25.4.6)$$

25.5 Absolute Normal Scores Test Statistic

In certain cases, the functions J are obtained using expectations of suitable order statistics. For each integer N let J_N' be a nonnegative function such that $0 < J_N' \le K[x(1-x)]^{-(3/2)+\delta}$ for some fixed constant K and some δ ($0 < 2\delta < 1$). Let J_N be an integral of J_N' and $\overline{J}_N\left(i/(N+1)\right) = E\{J_N(U_{i,N})\}$. Complete the definition of $\overline{J}_N$ as suggested in Section 25.2. Then we have the following theorem

Theorem 25.5.1. *Suppose that* $\left|J_N'(u)\right| \le K[u(1-u)]^{-(3/2)+\delta}$ *for some* $0 < \delta < 1/2$ *and* J_N' *converges in Lebesgue measure to a limit* J'*. Furthermore, assume that* $P(\mathbf{m} = m)$ *satisfies the condition stated in Theorem 25.3.1. Let* τ_N *be the difference*

$$\tau_N = N^{1/2}\left[\underline{\lambda}\int \overline{J}_N\left(\frac{N}{N+1}H_N(x)\right)dF_{\mathbf{m}} - p\int J(H^*)dF\right]. \qquad (25.5.1)$$

Let $\widetilde{P}_N$ *denote the distribution of* τ_N*. Then*

$$\sup_x\left|\widetilde{P}_N(x) - \Phi(x/\sigma_N)\right|$$

tends to zero uniformly in (F, G, p) *as* $N \to \infty$ *provided* σ_N^2 *is bounded away from zero.*

PROOF: Let

$$\tau_N^* = N^{1/2}\left[\underline{\lambda}\int \overline{J}_N\left(\frac{N}{N+1}H_N(x)\right)dF_{\mathbf{m}} - p\int \overline{J}(H^*)dF\right]. \tag{25.5.2}$$

From Lemma 6.1 of Govindarajulu (1985) or Lemma 1 of GLR(1967) we have that $\overline{J}_N'$ satisfies the regularity conditions of Theorem 25.3.1 with $f(u) = K[u(1-u)]^{-1+(\delta/2)}$ and $g^2(u) = [u(1-u)]^{-1+\delta}$. Besides J_N' converges in measure to a limit J'. Hence, from the above mentioned lemma, it follows that $\overline{J}_N'$ converges to J' in measure. Thus the proof of Theorem 25.5.1 would be complete if τ_N is replaced by τ_N^*. That is, it suffices to show that the difference

$$\tau_N - \tau_N^* = N^{1/2}p\int\left\{\overline{J}_N(H^*) - J_N(H^*)\right\}dF$$

tends to zero as $N \to \infty$. However, this difference is smaller than

$$N^{1/2}\int_0^1 \left|\overline{J}_N(u) - J_N(u)\right|du$$

which GLR (1967, pp. 631–632) have bounded by a negligible quantity.

Corollary 25.5.1.1. *Let k be a fixed integer and a_j, $j = 1,\ldots,k$ be bounded constants. Let*

$$J_N(i/(N+1)) = \sum_{j=1}^{k} a_j E\left[(\xi_{i,N})^j\right] \tag{25.5.3}$$

where $\xi_{i,N}$ denotes the ith smallest order statistic in a sample of size N drawn from a population having for its distribution function, the inverse of a function $S(u)$. If

$$\left|dS^j(u)/du\right| \le K[u(1-u)]^{-(3/2)+\delta}, \quad j = 1,\ldots,k \tag{25.5.4}$$

then the functions J_N satisfy the conditions of Theorem 25.3.1.

PROOF: This follows from the linearity of the transformation $J \hookrightarrow \overline{J}_N$ employed to define the functions occurring in Theorem 25.5.1. For the special case when $k = 1, a_1 = 1$ and $S(u) = \chi_1^{-1}(u)$ where $\chi_1(x) = 2\Phi(x) - 1$ for $x > 0$ and zero for $x < 0$, the resultant statistic is Fraser's (1957a) or absolute normal scores test statistic.

25.6 Relative Efficiency of Tests for Symmetry

In this section we will obtain an explicit expression for the asymptotic relative efficiency (ARE) of the class of test procedures for symmetry when compared with student's t-test. Let $V_1, \ldots, V_N$ be a random sample from the distribution $\Psi(x-\theta)$, the X's denoting the absolute values of the negative V's and the Y's denoting the positive V's. Also, let σ^2 denote the variance of V. We wish to test $H_0 : \theta = 0$ versus $H_1 : \theta > 0$ assuming that $\Psi(x)$ is symmetric about zero.

It is well known that the Pitman efficacy of Student's t-test of H_0 is σ^{-2}. Also, the efficacy of T_N^* defined by (25.2.8) is

$$\lim_{N\to\infty} \left\{ \frac{\left[N^{1/2} p \int_0^\infty J(H^*)\, dF - N^{1/2} 2^{-1} \int_0^1 J(u)\, du\right]^2}{\xi^2 \mathrm{var}(T_N^* | H_0)} \right\} \tag{25.6.1}$$

where

$$\begin{aligned} \theta &= \xi N^{-1/2} \text{ for some fixed } \xi \\ p &= P(V<0) = \Psi(-\theta) \text{ as } N \to \infty \text{ and} \\ \mathrm{var}(T_N^*|H_0) &= (1/4) \int_0^1 J^2(u)\, du. \end{aligned}$$

Also

$$\begin{aligned} pF(x) = P(|V| \le x, V<0) &= P(V > -x, V < 0) \\ &= \Psi(-\theta) - \Psi(-x-\theta), \end{aligned}$$

and

$$H^* = P\left[|V| \le x\right] = \Psi(x-\theta) - \Psi(-x-\theta) = \Psi(x+\theta) + \Psi(x-\theta) - 1.$$

Also, we can write

$$\begin{aligned} B(\theta) &= N^{1/2} p \int J(H^*)\, dF - N^{\frac{1}{2}} \cdot \frac{1}{2} \int J(u)\, du \\ &= N^{1/2} \int_0^\infty \left[J\left\{\Psi(y) + \Psi(y-2\theta) - 1\right\} - J\left\{2\Psi(y) - 1\right\}\right] d\Psi(y) \\ &\quad - N^{1/2} \int_0^\theta J\left\{\Psi(y) + \Psi(y-2\theta) - 1\right\} d\Psi(y) \\ &= B_1(\theta) + B_2(\theta) \;\; (say) \end{aligned} \tag{25.6.2}$$

where noting $\theta = N^{-1/2}\xi$ and applying the mean value theorem, we have

$$\lim_{N\to\infty} B_2(\theta) = \xi\psi(0)J(0), \qquad \psi(x) = d\Psi(x)/dx. \tag{25.6.3}$$

Let $J \in \mathcal{S}_1$ and

$$N^{1/2}\left|\Psi(x) - \Psi(x-2\theta)\right| \le K\left[\Psi(x)|1-\Psi(x)\right]^{\frac{1}{2}-\delta'} \tag{25.6.4}$$

for some $0 \le \delta' \le \delta$ where $\theta = \xi N^{-1/2}$. Then using Theorem 3.1 of Govindarajulu (1980) for interchange of limit and integration, we obtain

$$\lim_{N\to\infty} B_1(\theta) = -2\int_0^\infty J'\left[2\Psi(y)-1\right]\psi^2(y)\,dy. \tag{25.6.5}$$

Thus, from (25.6.3) and (25.6.5) we are led to the following result.

Result 25.6.1. *The asymptotic efficiency of T_N^* relative to Student's t is*

$$e(T^*,t) = 4\sigma^2\left[J(0)\psi(0) + 2\int_0^\infty J'\left[2\Psi(y)-1\right]\psi^2(y)\,dy\right]^2 \Big/ \left[\int_0^1 J^2(u)\,du\right], \tag{25.6.6}$$

provided (25.6.4) holds. In particular, for the sign test

$$J(u) = 1 \quad \text{and} \quad e\left(T^*,t\right) = 4\sigma^2\psi^2(0), \tag{25.6.7}$$

for the signed rank test

$$J(u) = u \quad \text{and} \quad e\left(T^*,t\right) = 48\sigma^2\left[\int_0^\infty \psi^2(y)\,dy\right]^2. \tag{25.6.8}$$

25.7 Absolute Normed Scores Test

For the problem of location with symmetry, an invariant locally most powerful test against normal alternatives is equivalent to (see Fraser (1957a))

$$\tau = \sum_{i=1}^N E_{N,i}Z_{N,i} \tag{25.7.1}$$

where $E_{N,i}$ is the expected value of the ith smallest order statistic in a random sample of size N drawn from the absolute normal or chi-distrubution with one degree of freedom, the density of which is given by

$$2\phi(y) = (2/\pi)^{1/2}\exp(-y^2/2) \quad \text{for } y > 0 \text{ and zero elsewhere.}$$

The asymptotic normality of τ under H_0 follows from Liapounov's form of the central limit theorem. Its asymptotic normality under all the alternatives follows from Theorem 25.5.1. We shall verify the regularity assumptions. Let

$$\chi(x) = 2\Phi(x) - 1. \text{ Then } J(u) = \chi^{-1}(u), 0 \le u \le 1. \tag{25.7.2}$$

Since $J(u)$ is bounded for u near $u = 0$, Theorem 25.5.1 is applicable provided

$$|J(u)| \le K(1-u)^{-(3/2)+\delta} \quad \text{for some } 0 < \delta < 1/2. \tag{25.7.3}$$

By letting $J(u) = x$, we have $J'(u) = dx/du = \{2\phi(x)\}^{-1}$, since $u = \chi(x) = 2\Phi(x) - 1$. Further, it is well known that

$$1 - u = 2\int_x^\infty \phi(y)\,dy \le 2\phi(x)/x \le Kx^{-j} \text{ for any } j > 1. \tag{25.7.4}$$

That is $(1-u)^{1/j} \le Kx^{-1}$. Hence, for large x, say > 1

$$(1-u)(1-u)^{1/j} \le \frac{1}{J'(u)x}\cdot\frac{K}{x} = \frac{K}{J'(u)\cdot x^2} \le \frac{K}{J'(u)}. \tag{25.7.5}$$

Now setting $1/j = 1/2 - \delta$ we obtain (25.7.3). Also (25.6.4) trivially holds since $\phi(x)$ is bounded.

Pitman efficiency of absolute normal scores test relative to Student's t test is obtained from (25.6.6) by setting

$$J(0) = 0 \text{ and } \int_0^1 J^2(u)\,du = 1 \text{ as}$$

$$16\sigma^2\left[\int_0^\infty J'(2\Psi(y) - 1)\psi^2(y)\,dy\right]^2. \tag{25.7.6}$$

When $\Psi(y) = \Phi(y/\sigma)$, (25.7.6) reduces to unity. However, Govindarajulu (1960, 1985) obtains the following stronger result.

Theorem 25.7.1. *Let $\Psi(x-\theta)$ be a distribution function symmetric about θ, having density $\psi(x-\theta)$ and a finite second moment (which, without loss of generality be set equal to unity). If $J \in \mathcal{S}_1$, $\theta = \xi/N^{1/2}$ and (25.6.4) holds, that is*

$$N^{1/2}\left|\Psi(x)-\Psi(x-\theta)\right| \le K\left[\Psi(x)|1-\Psi(x)\right]^{\frac{1}{2}-\delta'}, \quad 0<\delta'<\delta,$$

then $E_{T,t} \ge 1$ for all Ψ and $E_{T,t}=1$ only if Ψ is normal.

PROOF: Since it has been verified that $J \in \mathcal{S}_1$ it follows from Result 25.6.1 that the ARE of Fraser's test is $I^2(2\Psi-1)$ where

$$I\left(2\Psi-1\right) = \int_0^\infty J'\left(2\Psi(y)-1\right)\psi^2(y)\,dy \tag{25.7.7}$$

and J is the inverse of the chi-distribution function with one degree of freedom. When $\Psi=\Phi$ it is easy to see that $I\left(2\Psi-1\right)$ is unity. Thus it suffices to show that $I\left(2\Psi-1\right)$ is greater than unity when Ψ is nonnormal and symmetric about zero with

$$\int_{-\infty}^{\infty} x^2\psi(x)\,dx = 1.$$

Let J_* be the inverse of Φ and let

$$\begin{aligned} I_1^*(\Psi) &= \int_{-\infty}^{\infty} J_*'\left(\Psi(x)\right)\psi^2(x)\,dx \\ &= \int_0^\infty \left\{J_*'\left(\Psi(x)\right)+J_*'(1-\Psi(x))\right\}\psi^2(x)\,dx. \end{aligned} \tag{25.7.8}$$

However, since $\Psi(x)$ is symmetric about zero,

$$J_*(\Psi) = -J_*(1-\Psi) \text{ and hence } J_*'(\Psi) = J_*'(1-\Psi).$$

Consequently

$$I_1^*(\Psi) = 2\int_0^\infty J_*'\left(\Psi(x)\right)\psi^2(x)\,dx. \tag{25.7.9}$$

Also, from the definition of J_* we have

$$(1+u)/2 = \int_{-\infty}^{J_*((1+u)/2)} \phi(x)dx$$

or

$$u = 2\int_0^{J_*((1+u)/2)} \phi(x)\, dx.$$

Also, from the definitions of J we have

$$u = 2\int_0^{J(u)} \phi(t)\, dt.$$

Hence, we infer that

$$\begin{aligned} J_*\left((1+u)/2\right) &= J(u) \text{ or} \\ J_*'\left((1+u)/2\right) &= 2J'(u). \end{aligned} \tag{25.7.10}$$

Using (25.7.10) in (25.7.9) we have (with $\Psi = (1+u)/2$)

$$I_1^*(\Psi) = \int_0^\infty J'(2\Psi - 1)\psi^2(x)\, dx = I(2\Psi - 1). \tag{25.7.11}$$

Chernoff and Savage (1958) have considered the problem of minimizing $I_1^*(\Psi)$ given by (25.7.8) over all Ψ. Without the assumption of symmetry about zero, they have established in their Theorem 3 (using a variational argument) that $I_1^*(\Psi) \geq 1$ for any arbitrary $\Psi(x)$ having a density and finite second moment and equality holds only when $\Psi(x)$ is standard normal. If I_1^* exceeds unity for any arbitrary $\Psi(x)$, it also exceeds unity for any Ψ that is symmetric about zero. Thus $I(2\Psi - 1) \geq 1$ for any arbitrary Ψ which is symmetric about zero and has a finite second moment. Also, for an elementary proof of Chernoff-Savage assertion, see Gastwirth and Wolfe (1968).

Remark 25.7.1. *The reader should be aware that there are some typos in Govindarajulu (1985), especially regarding confusion about Ψ and ψ. Hopefully, these are corrected in this chapter.*

25.8 Application to Halperin's Statistic

Halperin (1960) proposes a Wilcoxon test procedure for censored samples and establishes its asymptotic normality under the null hypothesis. In the following we describe the statistic, its asymptotic normality under all hypotheses and its Pitman efficacy.

Let $X_1, \ldots, X_m^*$ $\left[Y_1, \ldots, Y_n^*\right]$ denote a random sample from a continuous distribution function $F(x)[G(y)]$. Assume that the X's and Y's are mutually independent. Let t_0 be a fixed point and we truncate the X's and Y's at the same point. Let the uncensored samples be denoted by $X_1, \ldots, X_{\mathbf{m}}$ and $Y_1, \ldots, Y_{\mathbf{n}}$, where $m^* - \mathbf{m}$ X's and $n^* - \mathbf{n}$ Y's are censored. Halperin (1960) further assumes that $m^* + n^* - (\mathbf{m} + \mathbf{n})$ is fixed. Note that this can happen in life testing where one stops after observing a combined fixed number of failures of items from two populations. Let

$$\begin{aligned} &p_1 = P(X \le t_0) = F(t_0), q_1 = 1 - p_1, \\ &p_2 = P(Y \le t_0) = G(t_0), q_2 = 1 - p_2, \ N = \mathbf{m} + \mathbf{n} \text{ and } N^* = m^* + n^* . \end{aligned} \tag{25.8.1}$$

Note that N and N^* are nonrandom.

Haperin's test statistic is an extension of the Mann-Whitney-Wilcoxon statistic and is given by

$$\sum_{i=1}^{\mathbf{m}} R_i - [\mathbf{m}(\mathbf{m}+1)/2) + \mathbf{n}(m^* - m) \tag{25.8.2}$$

where $R_1, \ldots, R_{\mathbf{m}}$ denote the ranks of the ordered uncensored X's in the combined uncensored sample. After dividing by $N^{3/2}$ and considering only the random component, Halperin's (1960) statistic can be written as (with $p_N = E(\underset{\sim}{\lambda}_N = E(\mathbf{m}/N))$

$$N^{-3/2} \sum_{i=1}^{\mathbf{m}} R_i - (N + \mathbf{m} + 2^{-1} - Np)(\underset{\sim}{\lambda} - p)N^{-1/2} + N^{1/2}(\underset{\sim}{\lambda} - p)^2/2 \tag{25.8.3}$$

where the first term belongs to the class of statistics T_N defined by (25.2.6). Also one can easily see that $N^{1/2}(\underset{\sim}{\lambda} - p)^2$ converges to zero in probability since $N^{1/2}(\underset{\sim}{\lambda} - p)$ is bounded in probability.

Now, combining the $(\underset{\sim}{\lambda} - p)$ term with the appropriate term in B_N^* given by (25.3.8), the asymptotic normality of Halperin's statistic under all the hypotheses follows from Theorem 25.3.1. The asymptote mean and the variance are explicitly given by Govindarajulu (1985, pp. 167–168).

It should be noted under H_0 (i.e. $F = G$), $\mathbf{m}$ has the hypergeometric distribution given by

$$P(\mathbf{m} = m | \ \mathbf{m} + \mathbf{n} = N) = \binom{m^*}{m}\binom{n^*}{n} \Big/ \binom{N^*}{N}, \quad m = 0, \ldots, m^* \tag{25.8.4}$$

which tends to be asymptotically normal if its variance tends to infinity (see Govindarajulu (1966, Result 4.4.1)).

When $p_2 = p_1$ (i.e. H_0 holds) the asymptotic conditional variance of $\mathbf{m}$ for given $\mathbf{m} + \mathbf{n} = N$ is $m^* n^* p_1 q_1 / N^*$ (see Lemma 5.3 of the author (1985)). However, substituting $p_1 = N/N^*$ and $q_1 = 1 - p_1$ since N/N^* converges to p_1 almost surely under H_0, this variance is equivalent to the conditional variance of $\mathbf{m}$ computed from (25.8.4). Thus the test statistic is distribution-free under H_0.

Asymptotic Efficiency of Halperin's Test Procedure

The asymptotic mean of the test statistic is

$$N^{1/2} p \int_{-\infty}^{t_0} \widetilde{H}^* d\widetilde{F}(x), \tag{25.8.5}$$

where

$$\begin{aligned} \widetilde{F}(x) &= F(x)/[1 - F(t_0)], \quad -\infty < x < t_0, \\ \widetilde{G}(x) &= G(x)/[1 - G(t_0)], \quad -\infty < x < t_0, \\ \widetilde{H}(x) &= p\widetilde{F} + (1-p)\widetilde{G} \ \text{ and } p = E(\mathbf{m}/N) = m^*/N^* . \end{aligned} \tag{25.8.6}$$

Its variance under H_0 simplifies to

$$\sigma_N^2(H_0) = (1/12)p(1-p)\left\{1 + 3(1-\eta)^{-1}\left[1 + \eta(2p-1)^2\right]^2\right\} \tag{25.8.7}$$

where $p = m^*/N^*$ and $\eta = 1 - (N/N^*)$.

Then the Pitman efficacy of Halperin's test for location [scale] changes is

$$p^2 \left(\int_{-\infty}^{t_0} \widetilde{f}^2(x)\, dx\right)^2 \Big/ \sigma_N^2(H_0) \left[p^2 \left(\int_{-\infty}^{t_0} x\widetilde{f}^2(x)\, dx\right)^2 / \sigma_N^2(H_0)\right] \tag{25.8.8}$$

from which one can compute its efficiency relative to Student's t-test [F-test], the efficacy of the latter being $\sigma^{-2}\left[(2+\gamma_2)^{-1}\right]$.

Remark 25.8.1. *Regarding other contributions to the problem of location with symmetry, the sufficient conditions imposed by Puri and Sen (1969) and Albers et al. (1976) especially on the J function are somewhat stronger than those given here. For the same problem, Huskova (1970) and Vorlickova (1972) took the approach of Hájek (1968). Pyke and Shorack (1968) assume that the weight function J is of bounded variation on $(\varepsilon, 1-\varepsilon)$ for every $\varepsilon > 0$ and the underlying distributions are absolutely continuous.*

25.9 c-Sample Case with Random Allocation

The author (1960) has formulated a generalization of the Chernoff-Savage problem to the c-sample or c-category case where the sub-sample sizes are non-random or random respectively. Also, the asymptotic normality in the c-sample case with non-random sample sizes has been considered by Puri (1964) and GLR (1967). The c-category case when the sub-sample sizes are random which is described in Section 25.2, is considered by the author (1985, Section 9). Assuming that the joint probability function of the c-subsample sizes can be approximated by the corresponding joint normal density function (which holds when the sub-sample sizes have either a multinomial or multi-variate hypergeometric distribution) the joint asymptotic normality of the $T_{N,j}$, $(j = 1, \dots, c)$ has been established, when suitably standardized. Next, when we wish to test the null hypothesis of equality of c-category distributions against translation or scale change alternatives, the author proposes to reject the null hypothesis when

$$S_N = \sum_{j=1}^{c} \underset{\sim}{\lambda}_j \left(\underset{\sim}{\lambda}_j^{-1} T_{N,j} - \overline{T}_N \right)^2 \Big/ N\widetilde{I} > \chi^2_{c-1,1-\alpha} \tag{25.9.1}$$

where

$$\begin{aligned} \overline{T}_N &= N^{-1/2} \sum_{j=1}^{c} T_{N,j} = N^{-1/2} \sum_{i=1}^{N} E_{N,i} \\ &= \text{a nonstochastic constant when} \\ &\quad \text{we set } E_{N,j,i} = E_{N,i} \text{ (i.e. } J_{N,j} \equiv J_N). \end{aligned}$$

$\widetilde{I} = \int_0^1 J^2(u)\, du$ and $\chi^2_{c-1,1-\alpha}$ denotes the $(1-\alpha)^{\text{th}}$ quantile of the central chi-square distribution with $c-1$ degrees of freedom. When compared with the usual F-test the Pitman relative efficiency is given by

$$\sigma^2 \left(\int J'(F) f^2(x)\, dx \right)^2 \Big/ \widetilde{I} \tag{25.9.2}$$

and the Pitman efficiency of S_N relative to its parametric analogue is given by

$$(2+\gamma_2) \left(\int x J'(F) f^2(x)\, dx \right)^2 \Big/ 4\widetilde{I} \tag{25.9.3}$$

where σ^2 and γ_2 denote the variance and kurtosis of F, the common distribution. Note that (25.10.2) and (25.10.3) coincide with the well known corresponding expressions of Pitman efficiency for the c-sample location and scale problems respectively when the sub-sample sizes are nonrandom. For further details the reader is referred to the author (1985, Section 9).

25.10 Problems

25.3.1 For the following random sample of weight loss in pounds (weight before - weight after) due to a weight-watcher's program, test the hypothesis that the location of symmetry of the distribution is zero versus it is greater than zero under the alternative hypothesis.

10, 30, 26, 21, -6, 7, 19, 22, 25, 23, -10, -5, 4, -15, 14, 9

Hint: Use the Wilcoxon signed rank test procedure.

25.3.2 For the data in Problem 24.8.2, we obtain the following difference between before and after.

Subject	1	2	3	4	5	6	7	8
Difference	3	-4	8	4	6	19	4	3

Subject	9	10	11	12	13	14	15	16
Difference	16	13	2	22	0	12	-9	9

Assuming that the data comes from a distribution that in symmetric about 0, test the hypothesis $H_0 : \theta = 0$ versus the alternative hypothesis $H_1 : \theta > 0$ using the Wilcoxon-signed rank test.

25.3.3 Mascular weakness due to adrenal or pituitary dysfunction is called Cushing's disease. Effective treatment can be provided if childhood Cushing's disease is detected as early as possible. Age at onset of symptoms and age of diagnosis for 15 children suffering from the disease were given in the paper entitled "Treatment of Cushing's disease in childhood and adolescence by transphenoidal microadenomectomy" published in New England Journal of Medicine (1984) p. 889. The following are the values of the differences between age at onset of symptoms and age of diagnosis:

-24, -12, -55, -15, -30, -60, -14, -21, -48, -12, -25, -53, -61, -69, -80.

Using Wilcoxon-signed rank test procedure, test the hypothesis H_0 : no difference in the two ages, against H_1 : there is a difference in the ages.

(See also J. Devore (2000), p. 383.)
Using a normal probability plot, show that the distribution of the differences does not conform to normality.

25.3.4 The following data represent artificially generated differences in weight of before and after a lapse of time of 20 men. Gain below is defined as (after-before) in lbs.

Subject	1	2	3	4	5	6	7	8	9	10
Gain	4	5	-2	7	1	3	-3	2	6	5
Subject	11	12	13	14	15	16	17	18	19	20
Gain	-3	-1	6	-1	2	0	-2	8	5	4

Using the Wilcoxon signed rank test procedure, test

H_0: the difference is zero
versus H_1: the difference is positive. Use $\alpha = 0.05$.

25.8.1 The following data constitutes failure times of GE bulbs (denoted by Xs) and the Westinghouse bulbs (denoted by Ys) in hundreds of hours. Let $m = 15$ and $n = 15$ (hence $N = 30$) $\mathbf{m} = 10$, $\mathbf{n} = 12$ (hence $N = 22$). Also assume that t_0 (the common truncation point) is 750 hours.

X: 4.5, 4.7, 5.1, 5.3, 5.7, 6.1, 6.2, 6.3, 7.1, 7.3
Y: 3.8, 4.2, 4.4, 5.2, 5.4, 5.6, 5.8, 6.0, 6.4, 6.6, 6.8, 7.0

Using Halperin's test criterion, test the null hypothesis that the failure time distributions of X and Y are the same, against the alternative hypothesis that X is stochastically larger than Y. (Use $\alpha = 0.05$).

25.9.1 Suppose we wish to test the null hypothesis that the c-category distributions are the same versus the alternative hypothesis that they are different. For the data in Problem 24.7.2, assume that $c = 3$ and the sample sizes arose from a multinomial population with $N = 30$, $p_1 = p_2 = p_3 = \frac{1}{3}$. Using the test criterion S_n given by (25.9.1) with $J(u) \equiv u$, test the above hypothesis for $\alpha = 0.05$.

Chapter 26

A Class of Statistics

26.1 Introduction

In the class of statistics considered by Chernoff and Savage (1958) the ranks of one ordered sample within the combined ordered sample have been weighted with functions of the combined empirical distribution function. Furthermore, even with generalizations of the results of Chernoff and Savage (1958) by GLR (1967) it is assumed that there are no mutual discontinuities among the underlying distribution functions. Also, it was assumed that the ratios of the sample sizes to the total sample size are bounded away from zero and unity. The above two conditions can be weakened if we consider perhaps a smaller class of statistics. However, one can weight the ranks of one ordered sample with functions of the other empirical distribution function, which might be easier from the computational point of view. A class of similar statistics called weighted rank sum tests for dispersion has been considered by Sen (1963). Also, Sen and Govindarajulu (1966) proposed a class of c-sample ($c \geq 2$) nonparametric tests for the homogeneity of location or scale parameters, which may be regarded as the c-sample extension of the two-sample tests considered by Sen (1963). Govindarajulu (1976) considered a class of two and c-sample tests which, in particular, include those proposed by Sen (1963). In the following we will consider in detail the tests of Govindarajulu (1976) and its extension to the bi-variate and multi-variate cases.

26.2 Regularity Assumptions

Let $X_1, \ldots, X_m (Y_1, \ldots, Y_n)$ denote the random sample of size $m(n)$ drawn from a population having continuous $F(x)\,[G(x)]$ for its distribution functions (df). Also, let $F_m(G_n)$ denote the empirical df based on the X's (Y's). Define

$$\lambda_m = n/m \tag{26.2.1}$$

where $0 < \lambda_0 < \lambda_m$ for some λ_0. F and G may depend on m and we suppress this for the sake of simplicity. Define a class of statistics by

$${}_mT_n = \sum_{j=1}^{m} E(j,m) G_n(X_i) \tag{26.2.2}$$

where $E(j,m)$ are some specified constants. The case of $E(j,m) = J(j/(m+1))$ can easily be handled by the following methods. When $E(j,m) \equiv 1$ the statistic reduces to the Mann-Whitney test statistic. When $E(j,m) = \{b(j-1,m) - b(j,m)\}\, m,\ \ 2 \le j \le [m/2],\ \ E(1,m) = mb(0,m)$, $E\{[m/2]+1, m\} = mb([m/2], m)$ and $E(j,m) = 0$ for $j > [m/2]+1$, the class of test statistics given by (26.2.2) reduces to the class considered by Sen (1963).

The following representation of T_m will be used.

$$T_m = \int_{-\infty}^{\infty} G_n(x) J_m\left(\frac{m}{m+1} F_m\right) dF_m(x). \tag{26.2.3}$$

Representations (26.2.2) and (26.2.3) are equivalent when $E(j,m) = J_m(j/(m+1))$. Also, throughout K may be used as a generic constant which is free of F, G, m and n.

Although it suffices to define $J_m(u)$ at $u = \frac{1}{m+1}, \ldots, \frac{m}{m+1}$, we can extend its domain of definition to $(0,1)$ by letting $J_m(u)$ be constant on $(i/(m+1), (i+1)/(m+1))\quad (i = 0, 1, \ldots, m)$. Although continuity of F and G is assumed it is not necessary to do so. If they have a finite or denumerable number of discontinuities, they can be made continuous by the Continuization Process described by GLR (1967), which preserves the probability distributions of the order relations among the X's and among the Y's. Furthermore, GLR (1967) require that F and G have no common discontinuities which does not seem to be necessary for the present investigation, since the statistic T_m is well defined. Although without loss of generality, we can set $F(x)$ to be the uniform distribution on $[0,1]$. Hereafter, wherever there is no ambiguity, the subscript m on J will be suppressed, as done in chapters 24 and 25. Let $f \geq 1$ $[g \geq 1]$ be U-shaped and be integrable [square integrable] for the Lebesgue measure. Also, let b be a finite constant. Consider functions J defined by

$$J(x) = \int_{1/2}^{x} J'(u)du. \tag{26.2.4}$$

We say that $J \in \mathcal{S}_0$ if $|J'| \leq fg$, J belongs to $\mathcal{S}_1$ if $J' = J_1' + J_2'$ with $|J_2'| \leq fg$ and $\int |J_1'(u)|du \leq b$ and J belongs to $\mathcal{S}$ if $|J'| \leq f_0 + fg$ with f_0 integrable. $J \in \mathcal{S}_0^*$ if $|J'(u)| \leq K[u(1-u)]^{(-3/2)+\delta}$ for some $0 < \delta < 1/2$ and $K > 0$. Note that $\mathcal{S}_0^* \subseteq \mathcal{S}_0 \subseteq \mathcal{S}_1 \subseteq \mathcal{S}$.

Let $\xi_{m,k}$ be the kth smallest standard uniform order statistic in a random sample of size m. For any function J_m let $\overline{J}_m(y)$ be the function defined on $(0,1)$ as follows: If $y = k/(m+1), k = 1, \ldots, m$ let

$$\overline{J}_m(y) = EJ_m(\xi_{m,k}) = \int_0^1 J_m(x)\beta_m(x,k)dx \tag{26.2.5}$$

where $\beta_m(x,k)$ denotes the density of $\xi_{m,k}$.

Complete the definition of $\overline{J}_m(y)$ by linearly interpolating between successive values $\{k/(m+1), (k+1)/(m+1)\}$ and leaving $\overline{J}_m$ constant below $1/(m+1)$ and above $m/(m+1)$.

26.3 Statement of Main Results

With the preceding notation, the main results of Govindarajulu (1976) can be stated as follows: Let

$$T_m^* = m^{1/2}\left\{\int G_n J\left(\frac{m}{m+1}F_m\right)dF_m - \int GJ(F)dF\right\}. \tag{26.3.1}$$

Then one can expand T_m^* as

$$\begin{aligned} m^{-\frac{1}{2}}T_m^* = \int (G_n - G)J(F)dF + \int GJ(F)d(F_m - F) \\ + \int G\left[J\left(\frac{m}{m+1}F_m\right) - J(F)\right]dF_m \\ + m^{-\frac{1}{2}}c_{1m} + m^{-\frac{1}{2}}c_{3m} \end{aligned} \tag{26.3.2}$$

where

$$m^{-1/2}c_{1m} = \int (G_n - G)J(F)d(F_m - F) \quad \text{and}$$
$$m^{-1/2}c_{3m} = \int (G_n - G)\left[J\left(\tfrac{m}{m+1}F_m\right) - J(F)\right] dF_m.$$

Now, integrating by parts once, we obtain

$$\int GJ(F)d(F_m - F) = -\int (F_m - F)J(F)dG - \int (F_m - F)GJ'(F)dF.$$

Hence, we can rewrite

$$T_m^* = m^{1/2}B_m + c_{1m} + c_{2m} + c_{3m} + c_{4m} \tag{26.3.3}$$

where

$$\begin{aligned} B_m &= \int (G_n - G)J(F)dF - \int (F_m - F)J(F)dG, \\ c_{2m} &= m^{1/2}\int G\left[J\left(\frac{m}{m+1}F_m\right) - J(F)\right] d(F_m - F) \quad \text{and} \\ c_{4m} &= m^{1/2}\int G\left[J\left(\frac{m}{m+1}F_m\right) - J(F) - (F_m - F)J'(F)\right] dF. \end{aligned}$$

$m^{1/2}B_m$ is a sum of two independent sums of i.i.d. random variables centered at zero means. Hence by the Central Limit Theorem $m^{1/2}B_m$ is asymptotically normal with mean zero and variance given by

$$\begin{aligned} \sigma_m^2\left[F, G, J, \lambda_m\right] = (2/\lambda_m) &\iint_{x<y} G(x)[1 - G(y)]J(F(x))J(F(y))dF(x)dF(y) \\ &+ 2\iint_{x<y} F(x)[1 - F(y)]J(F(x))J(F(y))dG(x)dG(y)\,. \end{aligned} \tag{26.3.4}$$

Govindarajulu (1976, Proposition 3.1) shows that σ_m^2 is uniformly bounded for all F, G, λ_m $(0 < \lambda_0 < \lambda_m)$ and every $J \in \mathcal{S}$.

Then the following results have been established by Govindarajulu (1976).

Proposition 26.3.1. *Let P_m be the distribution function of $m^{1/2}B_m/\sigma_m$. Then there exists an $m(\varepsilon, a)$ such that $m \geq m(\varepsilon, a)$ and $\sigma_m^2(F, G, J, \lambda_m) > a$ implies*

$$\sup_x |P_m(-\infty, x) - \Phi(x)| < \varepsilon$$

for every $J \in \mathcal{S}$ and all triples $[(F, G), \lambda_m]$.

Corollary 26.3.1.1. *Assume that $J \not\equiv 0$. Let $\{J_k\}$ be a sequence such that $J_k \in \mathcal{S}$ and $J_k \to J \in \mathcal{S}$ in Lebesgue measure. Let $\{(F_k, G_k)\}$ be a sequence of mutually absolutely continuous pairs converging to a pair (F_0, G_0) at all points of continuity of the pair (F_0, G_0). Then if $F_0 = G_0$, then, uniformly in m*

$$\begin{aligned}
&\lim_{k\to\infty} \frac{n}{m+n}\sigma_m^2[F_k, G_k, J_k, \lambda_m) \\
&\quad = 2 \iint\limits_{0<u<v<1} u(1-v)J(u)J(v)dudv \\
&\quad = \int_0^1 M^2(u)du - \left[\int_0^1 M(u)du\right]^2 \qquad (26.3.5)
\end{aligned}$$

where $M(u) = \int_0^u J(t)dt$.

The main result of Govindarajulu (1976) is:

Theorem 26.3.1. *Let J be an element of $\mathcal{S}_1$ and let*

$$T_m^* = m^{1/2}\left\{\int G_n J\left(\frac{m}{m+1}F_m\right)dF_m - \int GJ(F)dF\right\}. \qquad (26.3.6)$$

Let P_m be defined as in Proposition 26.3.1. Let σ_m^2 be bounded away from zero. Then there is an $m(\varepsilon)$ such that $m \geq m(\varepsilon)$ implies

$$\sup_x |P_m(-\infty, x) - \Phi(x)| < \varepsilon$$

for every $J \in S$ and every triple $\{(F,G), \lambda_m\}$ where S is a relatively compact subset of $\mathcal{S}_1$. Furthermore, if $\mathcal{F}$ is a relatively compact subset of triples $\{(F,G), \lambda_m\}$, then $m \geq m(\varepsilon)$ implies that

$$\sup_x |P_m(-\infty, x) - \Phi(x)| < \varepsilon$$

for every $J \in \mathcal{S}_1$ and every triple $\{(F,G), \lambda_m\} \in \mathcal{F}$.

Brief Sketch of the Proof. For some τ such that $0 < 2\tau < 1$, let $A = \{(0,\tau] \cup [1-\tau, 1)\}$. Then the tails of the statistic T_m^* is

$$\begin{aligned}
\Delta_m(J,\tau) &= m^{1/2}\left\{\int_A G_n J\left(\frac{m}{m+1}F_m\right)dF_m - \int_A GJ(F)dF\right\} \\
&= \Delta_m^*(J,\tau) + m^{1/2}\int_A GJ(F)d(F_m - F) \\
&\quad + m^{1/2}\int_A (G_n - G)J(F)dF_m \qquad (26.3.7)
\end{aligned}$$

where $\Delta_m^*(J,\tau) = m^{1/2}\int_A G_n\left\{J\left(\frac{m}{m+1}F_m\right) - J(F)\right\}dF_m$.

First proceeding as in GLR (1967) it is shown that $\Delta_m^*(J,\tau)$ is negligible. Further the other two terms in the expansion of $\Delta_m(J,\tau)$ are normalized sums with expectation zero and variance bounded by $\int_A J^2(u)du$ and $n^{-1}\int_A J^2(u)du$ respectively. It follows that the tails of the statistic can be neglected. Then c_{1m} is shown to go to zero in probability in Proposition 4.1 for every $J \in \mathcal{S}$, c_{2m} is shown to be negligible in Proposition 4.2 for every $J \in \mathcal{S}_1$, c_{3m} is shown to be negligible in Proposition 4.3 for every $J \in \mathcal{S}$, and c_{4m} is shown to be negligible in Proposition 4.4 for every $J \in \mathcal{S}_1$. For doing so the techniques developed by GLR (1967) are heavily used. Please note that Theorem 26.3.1 is valid for $J \in \mathcal{S}_1$ and not $J \in \mathcal{S}$ as claimed in the paper.

Remark 26.3.1. *The uniformity asserted in Theorem 24.5.1 is of interest in the case considered in Remark 24.5.4 provided* $\text{var}M(U)$ *is bounded away from zero where* U *has a standard uniform distribution.*

Also, results analogous to Theorem 24.6.1 and Corollary 24.6.1.1 are obtained by Govindarajulu (1976, pp. 552–553). They can be reproduced by replacing N by m in Theorem 24.6.1 and Corollary 24.6.1.1.

26.4 An Application

Let $X_1,\ldots,X_m$ $(Y_1,\ldots,Y_m)$ be a random sample from a binomial population with parameters k and p_1 (k and p_2). We wish to test $H_0 : p_1 = p_2$ against the alternative $H_1 : p_1 < p_2$. Suppose we use the Mann-Whitney test criterion. Note that even the GLR's (1967) version of Chernoff-Savage theorem does not cover the asymptotic normality of the test statistic, since the distributions F and G have common discontinuities. However, its asymptotic normality follows from Theorem 26.3.1. Moreover, this approximating normal distribution is free of F and G when H_0 holds. Chanda (1963) has applied the Mann-Whitney test procedure for discriminating between two purely discrete populations. However, this test is not distribution-free since he used the score function

$$C(u) = \begin{cases} 1 & \text{if } u > 0 \\ 1/2 & \text{if } u = 0 \\ 0 & \text{if } u < 0 \end{cases}$$

whereas we use $C(u) = 1$ for $u \geq 0$ and zero elsewhere. That is, we do not randomize at the ties between X's and Y's.

26.5 Case of Random Sample Size

Tests based on random sample sizes do naturally arise in inferential problems. For instance, the random samples are truncated from a certain point or the sample sizes are determined by a random process. So, assume that there exists a nonrandom sequence $\{N^*\}$. However, it is assumed that the dfs F, G and H do not involve m and n; they may depend on N^*. Then we obtain the following result by an application of Slutsky's Theorem.

Theorem 26.5.1. *If there exists a nonrandom sequence $\{N^*\}$ of integers such that m/N^* and n/N^* converge in probability to p_1 and p_2 respectively, where $0 < p_0 \le p_1, p_2$ and the hypothesis of Theorem 26.3.1 holds, then*

$$\left[\int G_n J\left(\frac{m}{m+1}F_m\right) dF_m - \int GJ(F)dF\right] \Big/ \sigma_{N^*}$$

converges in law to a standard normal variable as $N^* \to \infty$ where

$$\begin{aligned} N^*\sigma_{N^*}^2 &= 2p_1^{-1} \iint\limits_{x<y} G(x)[1-G(y)]J(F(x))J(F(y))dF(x)dF(y) \\ &\quad + 2p_2^{-1} \iint\limits_{x<y} F(x)[1-F(y)]J(F(x))J(F(y))dG(x)dG(y), \end{aligned} \tag{26.5.1}$$

provided σ_{N^*} is bounded away from zero.

26.6 c-Sample Case

Let $X_{j,k}$, $k = 1, \ldots, n_j$ be a random sample from a population having $F^{(j)}(x)$ for its distribution function ($j = 1, \ldots, c$). Assume that the c-samples are mutually independent. Define $\lambda_j = n_j/n_1$ and assume $0 < \lambda_0 \le \lambda_j$ ($j = 1, \ldots, c$) for some fixed λ_0.

Further, let $N = n_1 + \ldots + n_c$ and $c_j = n_j/N$ ($j = 1, \ldots, c$) and $F_{n_j}^{(j)}(x)$ denote the empirical df based on $X_{j,1}, \ldots, X_{j,n_j}$. Now define the statistics

$$\begin{aligned} T_{n_1,j} &= n_1^{-1} \sum_{i=1}^{n_1} E(i, n_1) F_{n_j}^{(j)}(X_{1,i}) \\ &= \int F_{n_j}^{(j)}(x) J\left(\frac{n_1}{n_1+1} F_{n_1}^{(1)}(x)\right) dF_{n_1}^{(1)}(x), \quad j = 1, \ldots, c. \end{aligned} \tag{26.6.1}$$

Let

$$\mu_{n_1,j} = \int F^{(j)} J(F^{(1)}) dF^{(1)}, \quad j = 1, \ldots, c. \tag{26.6.2}$$

Using Theorem 26.3.1, one can readily establish that $(T_{n_1,j} - \mu_{n_1,j})$, $j = 1, \ldots, c$ have a degenerate c-variate normal distribution with zero for the mean and $\boldsymbol{\Sigma}$ = variance-covariance matrix which is too long to be presented here. (In order to compute the var-cov matrix, just look at the first order random components of the statistics.)

Suppose we wish to test $H_0 : F^{(1)}(x) = \ldots = F^{(c)}(x)$ for all x versus the alternative $H_1 : F^{(j)}(x) \neq F^{(k)}(x)$ for some x and some pair (j, k). Let

$$\overline{T}_N = \sum_{j=1}^{c} c_j T_{n_1,j}, \quad \overline{\mu}_n = \sum_{1}^{c} c_j \mu_{n_1,j}\,. \tag{26.6.3}$$

Consider rejecting H_0 for large values of S_N where

$$S_N = \sum_{j=1}^{c} n_j \left(T_{n_1,j} - \overline{T}_N\right)^2 \Big/ I \tag{26.6.4}$$

where $I = 2 \iint\limits_{u<v} u(1-v) J(u) J(v) du dv$.

Computations yield

$$\begin{aligned} \sigma_{jj} &= (1 - c_j) I \quad \text{and} \\ \sigma_{j,k} &= -\sqrt{c_j c_k} I \quad \text{for } 1 \le j \neq k \le c. \end{aligned} \tag{26.6.5}$$

Now applying Theorem 23.6.1, one asserts that S_N under H_0 is distributed as chi-square with $c-1$ degrees of freedom. Also one can show that if $F^{(j)}(x) = F(x - \theta_j N^{-1/2})$ $\left[F(x(1 + \theta_j N^{-1/2}))\right]$ for $j = 1, \ldots, c$, then S_N is approximately distributed as noncentral chi-square with $c - 1$ degrees of freedom and noncentrality parameter equal to

$$\begin{gathered} \sum_{j=1}^{c} c_j \left(\theta_j - \sum_{k=1}^{c} c_k \theta_k\right)^2 \left\{\int f^2(x) J(F) dx\right\}^2 \Big/ I \\ \left[\sum_{j=1}^{c} \left(\theta_j - \sum_{j=1}^{c} c_k \theta_k\right)^2 \left\{\int x f^2(x) J(F) dx\right\}^2 \Big/ I\right]. \end{gathered} \tag{26.6.6}$$

Straightforward computations yield the following Pitman efficiency of the two-sample test based on T_m^* (see (26.3.6)) for location [scale] alternatives relative to Student's t [F-test] is given by

$$\sigma^2 \left\{ \int_{-\infty}^{\infty} f^2(x) J(F) dx \right\}^2 \Big/ I$$

$$\left[\left\{ \int_{-\infty}^{\infty} x f^2(x) J(F) dx \right\}^2 (2 + \gamma_2) \Big/ 4I \right]$$

where σ denotes the common scale parameter of F and G and $\gamma_2$denotes the kurtosis of $F(x)$. Note that the Pitman efficiency of the c-sample test will be the same as in the two-sample case.

Remark 26.6.1. *It should be noted that the class of c-sample test statistics proposed here include as a special case the class of test statistics considered by Sen and Govindarajulu (1966).*

26.7 Case of Dependent Samples

Let (X_i, Y_i), $i = 1, \ldots, n$ be a random sample from a continuous bivariate distribution $H(x, y)$ having marginals $F(x)$ and $G(y)$ where H, F and G are unknown. Then consider the class of test statistics proposed by Govindarajulu (1976, Section 7)

$$\begin{aligned} T_n &= n^{-1} \sum_{j=1}^{n} E(j, n) G_n(X_j) \\ &= \int_{-\infty}^{\infty} G_n(x) J_n \left(\frac{n}{n+1} F_n(x) \right) dF_n(x) \end{aligned} \tag{26.7.1}$$

where $F_n(x)$ and $G_n(x)$ denote the empirical distribution functions based on $(X_1, \ldots, X_n)$ and $(Y_1, \ldots, Y_n)$ respectively. Let

$$T_n^* = n^{1/2} \left\{ \int G_n J \left(\frac{n}{n+1} F_m \right) dF_m - \int G J(F) dF \right\}. \tag{26.7.2}$$

Then one can expand T_n^* as in (26.3.3) where m is replaced by n. The first order random terms given by

$$\begin{aligned} B_n &= \left[\int (G_n - G) J(F) dF - \int (F_n - F) J(F) dG \right] \\ &= n^{-1} \sum_{i=1}^{n} \{ B^*(Y_i) - B(X_i) \} \end{aligned}$$

where

$$B(X) = \int (F_1 - F)J(F)dG \quad \text{and}$$
$$B^*(Y) = \int (G_1 - G)J(F)dF$$

where F_1 and G_1 are empirical d.f.'s based on a single X and a single Y observation respectively. Since B_n constitutes a sum of i.i.d. random variables, by the classical central limit theorem, $n^{1/2}B_n$ is asymptotically normal with mean 0 and variance σ_n^2 where

$$\begin{aligned}\sigma_n^2 &= \text{var}B(X) + \text{var}B^*(Y) - 2\text{cov}(B(X), B^*(Y)) \\ &= 2\Bigg[\iint_{x<y} F(x)[1 - F(y)]J(F(x))J(F(y))dG(x)dG(y) \\ &\quad + \iint_{x<y} G(x)(1 - G(y))J(F(x))J(F(y))dF(x)F(y) \\ &\quad - \int_{-\infty}^{\infty}\int_{-\infty}^{\infty} \{H(x,y) - F(x)G(y)\}\, J(F(x))J(F(y))dG(x)dF(y)\Bigg].\end{aligned} \tag{26.7.3}$$

Thus, in order to assert the asymptotic normality of T_n^*, it suffices to show that the terms c_{in} $(i = 1, \ldots, 4)$ tend to zero in probability. The arguments of Govindarajulu (1976) can be repeated verbatim in order to show that $c_{2,n}$ and $c_{4,n}$ tend to zero in probability. The term c_{1n} was shown to go to zero in probability by Govindarajulu (1995, pp. 379–382).

Also, Govindarajulu (1976) states that the tails of the statistic T_n^* can be bounded even when the variables are dependent. In the following we provide the details for bounding both the c_{1n} and the tails of the statistic.

One can express c_{1n} as

$$c_{1n} = n^{-1/2}\sum_{i=1}^{n}\left[\{G_n(X_i) - G(X_i)\}\, J(F(X_i)) - \int (G_n - G)J(F)dF\right]$$

where

$$G_n(x) = n^{-1}\sum_{j=1}^{n}\chi_j(x)$$

and

$$\chi_j(x) = \begin{cases} 1 & \text{if } Y_j \le x \\ 0 & \text{if } Y_j > x. \end{cases}$$

One can show that

$$\begin{aligned} Ec_{1n} &= n^{-1/2}\left[E\chi_1(X_1)J(F(X_1)) - \int GJ(F)dF\right] \\ &= n^{-1/2}\left[\int F_{21}(x|x)J(F(x))dF(x) - \int GJ(F)dF\right] \end{aligned} \quad (26.7.4)$$

where

$$F_{21}(y|x) = P(Y \le y|X = x).$$

Since both the integrals are finite, Ec_{1n} tends to zero as $n \to \infty$. Next consider

$$\begin{aligned} Ec_{1n}^2 &= n^{-1}E\left\{\sum_{i=1}^{n}\left[\{G_n(X_i) - G(X_i)\}J(F(X_i)) - \int (G_n - G)J(F)dF\right]\right\}^2 \\ &= n^{-1}\sum_{i=1}^{n} E\left[\{G_n(X_i) - G(X_i)\}J(F(X_i)) - \int (G_n - G)J(F)dF\right]^2 \\ &\quad + n^{-1}\sum_{i\neq j}^{n}\sum^{n} E\left[\{G_n(X_i) - G(X_i)\}J(F(X_i)) - \int (G_n - G)J(F)dF\right] \\ &\quad \cdot\left[\{G_n(X_j) - G(X_j)\}J(F(X_j)) - \int (G_n - G)J(F)dF\right] \\ &= D_1 + D_2 \ (say) \end{aligned}$$

(26.7.5)

where

$$\begin{aligned} D_1 &= E\left[\{G_n(X_1) - G(X_1)\}J(F(X_1)) - \int (G_n - G)J(F)dF\right]^2 \\ &\le 2E\{[G_n(X_1) - G(X_1)]J(F(X_1))\}^2 + 2E\left[\int (G_n - G)J(F)dF\right]^2. \end{aligned}$$

Write

$$G_n(x) - G(x) = \frac{1}{n}\sum_{j=1}^{n} \chi_j^*(x) \tag{26.7.6}$$

where

$$\chi_j^*(x) = \begin{cases} 1 - G(x) & \text{if } Y_j \le x \\ -G(x) & \text{if } Y_j > x. \end{cases}$$

Then one can show that

$$\begin{aligned} &E\left[\{G_n(X_1) - G(X_1)\}\, J(F(X_1))\right]^2 \\ &= n^{-2}\Big[E\chi_1^{*2}(X_1)J^2(F(X_1)) + (n-1)EG(X_1)\{1 - G(X_1)\}\, J^2(F(X_1)) \\ &\qquad + \sum^{n}\sum^{n}{}_{j\ne k} E\chi_j^*(X_1)\chi_k^*(X_1)J^2(F(X_1))\Big] \end{aligned} \tag{26.7.7}$$

where

$$\begin{aligned} &E\chi_1^{*2}(X_1)J^2(F(X_1)) \\ &\quad = \int F_{21}(x|x)J^2(F)dF - 2\int G(x)F_{21}(x|x)J^2(F)dF + \int G^2(x)J^2(F)dF \end{aligned}$$

and all the product terms will be zero upon taking conditional expectation for given X_1. Thus L.H.S. of (26.7.7) tends to zero as n becomes large since J is square integrable. Also, one can easily show that

$$\begin{aligned} &E\left[\int (G_n - G)J(F)dF\right]^2 \\ &\qquad = \frac{2}{n}\iint_{x<y} G(x)[1 - G(y)]J(F(x))J(F(y))dF(x)dF(y) \to 0. \end{aligned} \tag{26.7.8}$$

So, $D_1 = O\left(\frac{1}{n}\right)$.

Since the X's are i.i.d., one can write D_2 as

$$D_2 = \frac{n(n-1)}{n} E\left[\{G_n(X_1) - G(X_1)\} J(F(X_1)) - \int (G_n - G)J(F)dF\right]$$

$$\cdot \left[\{G_n(X_2) - G(X_2)\} J(F(X_2)) - \int (G_n - G)J(F)dF\right]$$

$$= \frac{n-1}{n^2} E\left[\sum_{j=1}^{n}\left\{\chi_j^*(X_1)J(F(X_1)) - \int \chi_j^*(x)J(F(x))dF(x)\right\}\right]$$

$$\cdot \left[\sum_{k=1}^{n}\left\{\chi_k^*(X_2)J(F(X_2)) - \int \chi_k^*(y)J(F(y))dF(y)\right\}\right]$$

$$= \frac{n-1}{n^2}\sum_{j=1}^{n} E\left\{\chi_j^*(X_1)J(F(X_1)) - \int \chi_j^*(x)J(F(x))dF(x)\right\}$$

$$\cdot \left\{\chi_j^*(X_2)J(F(X_2)) - \int \chi_j^*(y)J(F(y))dF(y)\right\}$$

$$+ \frac{n-1}{n^2}\sum^{n}\sum^{n}_{j\neq k} E\left\{\chi_j^*(X_1)J(F(X_1)) - \int \chi_j^*(x)J(F(x))dF(x)\right\}$$

$$\cdot \left\{\chi_k^*(X_2)J(F(X_2)) - \int \chi_k^*(y)J(F(y))dF(y)\right\}. \tag{26.7.9}$$

Since Y_j is independent of (X_1, Y_1) and (X_2, Y_2), in the first summation all the terms with $j \geq 3$ will be zero. Straightforward computations yield the terms corresponding to $j = 1$ and 2 will also vanish. Thus the first summation is identically equal to zero for all F, G and $H(x, y)$. Next consider the summation on j and k with $j \neq k$. Here also all the terms when either j or k exceed 2 will yield zero. Thus it suffices to consider the terms with $(j, k) = (1, 2)$ and $(j, k) = (2, 1)$. Let $j = 1$ and $k = 2$. Then the corresponding term is

$$E\left[\chi_1^*(X_1)J(F(X_1)) - \int \chi_1^*(x)J(F)dF\right]$$

$$\cdot \left[\chi_2^*(X_2)J(F(X_2)) - \int \chi_2^*(x)J(F(x))dF(x)\right]$$

$$= E\left[\chi_1^*(X_1)\chi_2^*(X_2)J(F(X_1))J(F(X_2))\right] = \{E[\chi_1^*(X_1)J(F(X_1))\}^2$$

$$= \left\{\int [F_{21}(x|x) - G(x)]J(F(x))dF(x)\right\}^2 < \infty$$

since all other terms in the cross product will have zero expectation.

The term with $(j, k) = (2, 1)$ simplifies to

$$E\left[\chi_2^*(X_1)J(F(X_1)) - \int \chi_2^*(x)J(F)dF\right]$$
$$\cdot\left[\chi_1^*(X_2)J(F(X_2)) - \int \chi_1^*(x)J(F(x))dF(x)\right]$$
$$= E\left\{\chi_2^*(X_1)\chi_1^*(X_2)J(F(X_1))J(F(X_2))\right\}$$
$$= \int_x\int_y [F_{21}(x|y)-G(x)][F_{21}(y|x)-G(y)]J(F(x))J(F(y))dF(x)dF(y) < \infty\,.$$

Hence $D_2 = O(1/n)$.

Bounding the Tails of the Statistic

Let $A = \{(0, \tau] \cup [1-\tau, 1)\}$ for some $0 < 2\tau < 1$. Then consider

$$\Delta_n(J,\tau) = n^{1/2}\left[\int_A G_n J\left(\frac{n}{n+1}F_n\right)dF_n - \int_A GJ(F)dF\right]$$
$$= \Delta_n^*(J,\tau) + n^{1/2}\int_A GJ(F)d(F_n - F) + n^{1/2}\int (G_n - G)J(F)dF_n\,, \tag{26.7.10}$$

where

$$\Delta_n^*(J,\tau) = n^{1/2}\int G_n\left\{J\left(\frac{n}{n+1}F_n\right) - J(F)\right\}dF_n$$

which can be handled as in the case of independent samples. Further, the second term in the expansion of $\Delta_n(J,\tau)$ is a normalized sum with zero expectation and variance bounded by $\int_A J^2(u)du$. All that remains is to show that the last term is negligible in probability. Towards this, let

$$D_3 = n^{1/2}\int (G_n - G)J(F)dF_n$$
$$= n^{-1/2}\sum_{i=1}^{n}\left\{G_n(X_i) - G(X_i)\right\}J(F(X_i)).$$

Write $G_n(x) = n^{-1}\sum_{j=1}^{n}\chi_j(x)$, where $\chi_j(x) = 1$ if $Y_j \le x$ and zero otherwise. Thus

$$
\begin{aligned}
E|D_3| &\leq n^{1/2} E\left|G_n(X_1) - G(X_1)\right| \left|J(F(X_1))\right| \\
&\leq n^{1/2} \left[E\left\{G_n(X_1)) - G(X_1)\right\}^2 J^2(F(X_1))\right]^{1/2} .
\end{aligned}
$$

Next, consider

$$
\begin{aligned}
E\left[\{G_n(X_1) - G(X_1)\}^2 |X_1\right] &= n^{-2}\left[E\left\{\sum_{j=1}^{n}(\chi_j(X_1) - G(X_1))\right\}^2 |X_1\right] \\
&= n^{-2}\left[E\left\{\sum_{j=1}^{n}(\chi_j(X_1) - G(X_1))^2 |X_1\right\}\right. \\
&\quad + \sum\sum_{j\neq k} E\{\{\chi_j(X_1) - G(X_1)\} \\
&\quad \left. \cdot \{\chi_k(X_1) - G(X_1)\} |X_1\}\}\right].
\end{aligned}
$$

Straightforward computations yield

$$
\begin{aligned}
E\left[\{\chi_1(X_1) - G(X_1)\}^2 |X_1\right] &= \{1 - G(X_1)\}^2 F_{21}(X_1|X_1) \\
&\quad + G^2(X_1)\{1 - F_{21}(X_1|X_1)\} \\
&= F_{21}(X_1|X_1) - 2G(X_1)F_{21}(X_1|X_1) + G^2(X_1),
\end{aligned}
$$

and

$$
E\left[\{\chi_j(X_1) - G(X_1)\}^2 |X_1\right] = G(X_1)(1 - G(X_1)) \quad \text{for } j \geq 2.
$$

Also one can easily verify that all the cross product terms will be equal to zero. Thus

$$
\begin{aligned}
E|D_3| &\leq n^{-3/2} E\left\{F_{21}(X_1|X_1) - 2G(X_1)F_{21}(X_1|X_1) + G^2(X_1)\right. \\
&\quad \left. + (n-1)G(X_1)(1 - G(X_1))\right\} \\
&= n^{-1/2} K \int J^2(u)du \to 0 \text{ as } n \to \infty.
\end{aligned}
$$

Thus by Markov inequality D_3 goes to zero in probability. This completes the proof of the assertion that T_n^* is asymptotically normal.

Remark 26.7.1. *If* $J_n\left(\frac{i}{n+1}\right)$ *equals the expectation of the* i*th smallest order statistic in a sample of size* n *drawn from a specified continuous distribution, then one can appeal to Lemma 1 of GLR (1967) and assert the asymptotic normality of* T_n^*.

Remark 26.7.2. *Fubini's theorem has been used when interchanging the order of integration, especially in computing* Ec_{1n} *and* Ec_{1n}^2.

Remark 26.7.3. *One can estimate* σ_n^2 *by* $\hat{\sigma}_n^2$ *which is obtained by replacing* F, G *and* H *by their empirical values, namely* F_n, G_n, *and* H_n *respectively. Note that the empirical distributions converge in probability to their true values (see for instance, Gaenssler and Stute (1979, Section 2.1)). In testing situations, one can use the fact that* $T_n^*/\hat{\sigma}_n$ *will be approximately standard normal for sufficiently large* n.

A Simple Form for σ_n^2 when $G = F$ and its consistent estimate

When $G = F$, σ_n^2 given by (26.7.3) takes the form of

$$\begin{aligned}(\sigma_n^2 | F = G) &= 4 \iint\limits_{u<v} u(1-v)J(u)J(v)dudv \\ &\quad - 2\int_0^1\int_0^1 [H(F^{-1}(u), F^{-1}(v)) - uv)]J(u)J(v)dudv \\ &= 2\int_0^1\int_0^1 \min(u,v)J(u)J(v)dudv \\ &\quad - 2\int_0^1\int_0^1 H(F^{-1}(u), F^{-1}(v))J(u)J(v)dudv\,,\end{aligned} \tag{26.7.11}$$

when once J is specified, one can evaluate the first integral except $I(F, F)$ where

$$\begin{aligned}I(F,G) &= \int_0^1\int_0^1 H(F^{-1}(u), G^{-1}(v))J(u)J(v)dudv \\ &= \int_{-\infty}^{\infty}\int_{-\infty}^{\infty} H(x,y)dM(F(x))dM(G(y)), \quad \text{with } M(u) = \textstyle\int_0^u J(t)dt. \\ &= \int_{-\infty}^{\infty} dM(F(x))\left[H(x,y)M(G(y))|_{y=-\infty}^{\infty} - \int M(G(y))\partial_y H(x,y)\right]\end{aligned}$$

where $\partial_y H(x,y)$ denotes the partial differential of $H(x,y)$ w.r.t y. Thus

$$
\begin{aligned}
I(F,G) &= M(1)\int_{-\infty}^{\infty} F(x)dM(F(x)) \\
&\quad - \int_{y=-\infty}^{\infty} M(G(y))\int_{x=-\infty}^{\infty} \partial_y H(x,y)dM(F(x)) \\
&= M(1)\int_0^1 udM(u) - \int_{-\infty}^{\infty} M(G(y))[M(F(x))\partial_y H(x,y)|_{x=-\infty}^{\infty} \\
&\quad - \int M(F(x))dH(x,y)] \\
&= M(1)\left[\int_0^1 udM(u) - \int_0^1 M(u)du\right] \\
&\quad + \iint M(F(x))M(G(y))dH(x,y) \\
&= M(1)\left[2\int_0^1 uJ(u)du - M(1)\right] + \iint M(F(x))M(G(y))dH(x,y).
\end{aligned} \tag{26.7.12}
$$

A consistent estimate of the last integral on the right is

$$
n^{-1}\sum_{i=1}^{n} M(F_n(X_i)M(G_n(Y_i)). \tag{26.7.13}
$$

26.8 Applications

In the following we will consider some applications of the above results.

26.8.1. Testing for equality of the Marginals

Suppose we are interested in testing $H_0 : F(x) = G(x)$ against the alternative $H_1 : G(x) = F(x-\theta)$ for some $\theta \neq 0$ or $H_2 : G(x) = F(\theta x)$, for some $\theta > 0$. One can use the Mann-Whitney-Wilcoxon test procedure, the asymptotic normality of which has been established and its robustness has been studied by Govindarajulu (1975, 1991). Also one can use T_n defined by (26.3.3). However, because of the dependence in the variables one cannot hope to get a distribution-free test. However, using a consistent estimator $\hat{\sigma}_n^2$ in the place of σ_n^2 one can construct an asymptotically distribution-free test. For instance, for testing H_0 versus H_1 with $\theta > 0$, we reject H_0 when

$$
\left(T_n - \int_0^1 uJ(u)du\right)\sqrt{n} > \hat{\sigma}_n \Phi^{-1}(1-\alpha). \tag{26.8.1}
$$

where, using (26.7.12)

$$\hat{\sigma}_n^2 = 2\Bigg[\int_0^1\int_0^1 \min(u,v)J(u)J(v)dudv - 2M(1)\int_0^1 uJ(u)du + M^2(1)$$
$$-n^{-1}\sum_{i=1}^{n} M(F_n(X_i))M(G_n(Y_i))\Bigg]. \qquad (26.8.2)$$

For testing H_0 versus H_2 with $\theta < 1$, (i.e. $H_2 : F(x) \geq G(x)$), we reject H_0 when $(T_n - \int_0^1 uJ(u)du)\sqrt{n} > \hat{\sigma}_n \Phi^{-1}(1-\alpha)$.

26.8.2 Testing for Independence of the Variables

Suppose we wish to test the null hypothesis that the variables X and Y are independent i.e.

$$H_0 : H(x,y) = F(x)G(y) \quad \text{for all } x,y$$

vs.

$$H_1 : H(x,y) > F(x)G(y) \quad \text{for some } x,y \text{ (i.e. positive association).}$$

Cifarelli and Regazzini (1990) and Cifarelli, Conti and Ragazzini (1996) have defined a general index of cograduation given by

$$\gamma_g(H) = \frac{\int_{-\infty}^{\infty}\int_{-\infty}^{\infty} \{g(|G(y)+F(x)-1|) - g(|G(y)-F(x)|)\}\, dH(x,y)}{\int_0^1 g(x)dx} \qquad (26.8.3)$$

where $g(x) = 0$ for $x < 0$ and $g(x) > 0$ for $x > 0$. Since $\int_0^1 g(x)dx$ is a non-stochastic constant depending only on g, we can say that $\gamma_g(H)$ is proportional to the numerator in (26.8.2). When $g(x) = x^2$,

$$\gamma_g(H) \propto \int_{-\infty}^{\infty}\int_{-\infty}^{\infty} F(x)G(y)dH(x,y). \qquad (26.8.4)$$

When $g(x) = x$,

$$\begin{aligned}\gamma_g(H) &\propto \int_{-\infty}^{\infty}\int_{-\infty}^{\infty} [|F(x)-G(y)-1| - |F(x)-G(y)|]\, dH(x,y)\\ &= \int_0^1\int_0^1 [|u+v-1| - |u-v|]\, dH(F^{-1}(u), G^{-1}(v))\\ &= J_1 - J_2 \ (say)\end{aligned}$$

where

$$J_1 = \int_0^1 \int_0^1 |u+v-1| dH$$
$$= \iint_{u+v>1} (u+v-1)dH + \iint_{u+v<1} (1-u-v)dH = J_{11} + J_{12} \ (say).$$

Now

$$J_{11} = \iint_{u+v>1} u\, dH - \iint_{u+v>1} (1-v)dH$$
$$= \int_{u=0}^1 u \left[\int_{v=1-u}^1 dH \right] - \int_0^1 (1-v) \left[\int_{1-v}^1 dH \right]$$
$$= \int_{u=0}^1 u \left[du - dH(F^{-1}(u), G^{-1}(1-u)) \right]$$
$$- \int_0^1 (1-v) \left[dv - dH(F^{-1}(1-v), G^{-1}(v)) \right]$$
$$= - \int_0^1 udH(F^{-1}(u), G^{-1}(1-u)) - \int_0^1 udH(F^{-1}(u), G^{-1}(1-u))$$
$$= -2 \int_0^1 udH(F^{-1}(u), G^{-1}(1-u)).$$

Next, consider

$$J_{12} = \int_0^1 (1-u) \int_{v=0}^{1-u} dH - \int_0^1 v \int_{u=0}^{1-v} dH$$
$$= \int_0^1 (1-u)d(H(F^{-1}(u), G^{-1}(1-u))) - \int_0^1 vdH(F^{-1}(1-v), G^{-1}(v))$$
$$= -2 \int_0^1 vdH(F^{-1}(1-v), G^{-1}(v)).$$

Similarly,

$$J_2 = \iint_{u>v} (u-v)dH + \iint_{u<v} (v-u)dH = J_{21} + J_{22} \ (say)$$

where, proceeding in an analogous manner, we obtain

$$J_{21} = 2 \int_0^1 udH(F^{-1}(u), G^{-1}(u)) - \frac{1}{2} \quad \text{and}$$
$$J_{22} = 2 \int_0^1 vdH(F^{-1}(v), G^{-1}(v)) - \frac{1}{2}.$$

Thus, putting them all together, we have

$$\gamma_g(H) = 1-2\left[2\int_0^1 vdH(F^{-1}(v),G^{-1}(v)) + \int_0^1 vdH(F^{-1}(1-v),G^{-1}(v)) + \int_0^1 vdH(F^{-1}(v),G^{-1}(1-v))\right].$$

Now, performing integration by parts once in each of the integrals, we have

$$\begin{aligned}\gamma_g(H) &= -3+2\left[\int_0^1 H(F^{-1}(1-v),G^{-1}(v))dv + \int_0^1 H(F^{-1}(v),G^{-1}(1-v))dv + 2\int_0^1 H(F^{-1}(v),G^{-1}(v))dv\right]\\ &= -3+4\left[\int_0^1 H(F^{-1}(v),G^{-1}(1-v))dv + \int_0^1 H(F^{-1}(v),G^{-1}(v))dv\right].\end{aligned}$$

Thus

$$\gamma_g(H) \propto \int_0^1 \left\{H(F^{-1}(x),G^{-1}(x)) + H(F^{-1}(x),G^{-1}(1-x))\right\}dx. \quad (26.8.5)$$

Notice that (26.8.4) and (26.8.5) are proportional to the continuous versions of the Spearman rank correlation index and Gini's (1914) cograduation index. If $R_i[S_i]$ is the rank of X_i among $X_1,\ldots,X_n$ [of $Y_1,\ldots,Y_n$], then a sample version that is proportional to $\gamma_g(H)$ is

$$\begin{aligned}\gamma_{g,n} &= \frac{1}{n}\sum_{i=1}^n\left\{g\left(\left|\frac{n+1-R_i-S_i}{n}\right|\right) - g\left(\left|\frac{R_i-S_i}{n}\right|\right)\right\}\\ &= \frac{1}{n}\sum_{i=1}^n\left\{g\left(\left|1-\frac{R_i+S_i}{n}\right|\right) - g\left(\left|\frac{R_i-S_i}{n}\right|\right)\right\} + O\left(\frac{1}{n}\right) \quad (26.8.6)\end{aligned}$$

for a sufficiently smooth function g. If $g(x)=x^2$, then

$$\gamma_{g,n} \propto \gamma_n^{(1)} = \frac{1}{n}\sum_{i=1}^n R_iS_i \quad (26.8.7)$$

which is equivalent to Spearman's rank correlation coefficient. Also $g(x)=x$ yields

$$\gamma_{g.n} \propto \gamma_n^{(2)} = n^{-2}\sum_{i=1}^n\left\{|n-R_i-S_i| - |R_i-S_i|\right\} \quad (26.8.8)$$

which is equivalent to Gini's (1914) cograduation index. Also, without loss of generality, we can assume that the X_i's are ordered and hence take $R_i = i$ and let S_i^* be the rank of Y_i $(i = 1, \ldots, n)$. The asymptotic normality of $\gamma_n^{(1)}$ follows because it can be written as T_n (with $J(u) = u$). So, using $\gamma_n^{(1)}$ we can set up a distribution-free test of H_0 against H_1.

Since $|u| - |v| = \frac{u^2-v^2}{|u|+|v|}$ we can assert the asymptotic normality of $\gamma_n^{(2)}$ from that of $\gamma_n^{(1)}$ using Slutsky's theorem and replacing $|u| + |v|$ by the corresponding nonstochastic constant. Further, if g is an arbitrary function, we can expand

$$g(|u|) - g(|v|) = (|u| - |v|)h(|u|, |v|)$$

and via Slutsky's theorem, one can replace $h(|u|, |v|)$ by its nonstochastic counterpart, and thereby assert the asymptotic normality of $\gamma_{g,n}$ when suitably standardized.

26.9 Multivariate Case

Let $\tilde{X} = (X_1, \ldots, X_c)$ have a continuous c-variate distribution $H(x_1, \ldots, x_c)$ with marginals $F^{(j)}(x)$, $(j = 1, \ldots, c)$. Let $\tilde{X}_1, \tilde{X}_2, \ldots, \tilde{X}_n$ constitute a random sample from this distribution. Let $F_n^{(\ell)}$ denote the empirical distribution based on $(X_{\ell 1}, \ldots, X_{\ell n})$ for $\ell = 1, \ldots, c$. Now define the set of statistics by

$$\begin{aligned} T_{n,\ell} &= n^{-1} \sum_{i=1}^{n} E^{(\ell)}(i,n) F_n^{(\ell)}(X_{\ell,i}) \\ &= \int_{-\infty}^{\infty} F_n^{(1)}(x) J_n\left(\frac{n}{n+1} F_n^{(\ell)}(x)\right) dF_n^{(\ell)}(x), \quad \ell = 2, \ldots, c. \end{aligned} \tag{26.9.1}$$

Let

$$\mu_{n,\ell} = \int_{-\infty}^{\infty} F^{(1)}(x) J_n(F^{(\ell)}) dF^{(\ell)}, \;\; \ell = 2, \ldots, c. \tag{26.9.2}$$

Notice that $T_{n,1}$ is nonstochastic. In order to assert the joint normality of $n^{1/2}(T_{n,\ell} - \mu_{n\ell})$ it suffices to look at the first order random term $n^{1/2} B_{n,\ell}^{(1)}$ where

$$B_{n,\ell}^{(1)} = \int (F_n^{(\ell)} - F^{(\ell)}) J(F^{(\ell)}) dF^{(1)} - \int (F_n^{(1)} - F^{(1)}) J(F^{(\ell)}) dF^{(\ell)}, \tag{26.9.3}$$

for $\ell = 2, \ldots, c$ where the subscript n on the J function has been deleted since we assume that the J function does not depend on n. For some fixed constants $(a_2, \ldots, a_c)$ consider

$$\begin{aligned} L &= \sum_{\ell=2}^{c} a_\ell B_{n,\ell}^{(1)} \\ &= \frac{1}{n}\sum_{i=1}^{n}\int\sum_{\ell=2}^{c} a_\ell \chi_{\ell i}^{*} J(F^{(\ell)})dF^{(1)} - \frac{1}{n}\sum_{i=1}^{n}\int \chi_{1,i}^{*}\sum_{\ell=2}^{c} a_\ell J(F^{(\ell)})dF^{(\ell)} \\ &= \frac{1}{n}\sum_{i=1}^{n} W_i \end{aligned} \tag{26.9.4}$$

where

$$\begin{aligned} W_i &= \sum_{\ell=2}^{c} a_\ell \int \chi_{\ell i}^{*}(J(F^{(\ell)}))dF^{(1)} - \int \chi_{1,i}^{*} d\left(\sum_{\ell=2}^{c} a_\ell \int_{v}^{F^{(\ell)}} J'(u)du\right), \\ \chi_{\ell i}^{*} &= \left(F_i^{(\ell)}(x) - F^{(\ell)}(x)\right), \quad \ell = 1, \ldots, c \end{aligned} \tag{26.9.5}$$

and the W_i are i.i.d. random variables. Hence by the central limit theorem L has an asymptotic normal distribution when suitably standardized. Since this is true for every set of constants $(a_2, \ldots, a_c)$, the $B_{n,\ell}^{(1)}$ $(\ell = 2, \ldots, c)$ will have an asymptotic multivariate normal distribution with mean $\underline{0}$ and a certain variance-covariance matrix.

Standard computations yield

$$\begin{aligned} nE(B_{n,\ell}^{(1)})^2 &= nE\left[-\int (F_n^{(1)} - F^{(1)})J(F^{(\ell)})dF^{(\ell)} + \int (F_n^{(\ell)} - F^{(\ell)})J(F^{(\ell)})dF^{(1)}\right]^2 \\ &= \iint \left\{F^{(1)}(\min(x,y)) - F^{(\ell)}(x)F^{(\ell)}(y)\right\} J(F^{(\ell)}(x))J(F^{(\ell)}(y)) \\ &\qquad \cdot dF^{(\ell)}(x)dF^{(\ell)}(y) \\ &+ \iint \left\{F^{(1)}(\min(x,y) - F^{(1)}(x)F^{(1)}(y))J(F^{(\ell)}(x))\right\} J(F^{(\ell)}(y)) \\ &\qquad \cdot dF^{(\ell)}(x)dF^{(\ell)}(y) \\ &- 2\iint \left[H_{1\ell}(x,y) - F^{(1)}(x)F^{(\ell)}(y)\right] J(F^{(1)}(x))J(F^{(\ell)}(x))dF^{(1)}(y) \end{aligned}$$

and for $\ell \neq k \neq 1$

$$\begin{aligned} nE(B_{n,\ell}^{(1)} B_{n,k}^{(1)}) = & \iint \left[F^{(1)}(\min(x,y) - F^{(1)}(x)F^{(1)}(y)\right] J(F^{(\ell)}(x))J(F^{(k)}(y) \\ & \quad \cdot d(F^{(\ell)}(x))d(F^{(k)}(y)) \\ & + \iint \left[H_{\ell k}(x,y) - F^{(\ell)}(x)F^{(k)}(y)\right] J(F^{(\ell)}(x))J(F^{(k)}(y)) \\ & \quad \cdot dF^{(1)}(x)dF^{(1)}(y) \\ & - \iint \left[H_{1k}(x,y) - F^{(1)}(x)F^{(k)}(y)\right] J(F^{(\ell)}(x))J(F^{(k)}(y)) \\ & \quad \cdot dF^{(\ell)}(x)dF^{(1)}(y) \\ & - \iint \left[H_{1\ell}(x,y) - F^{(1)}(x)F^{(\ell)}(y)\right] J(F^{(k)}(x))J(F^{(\ell)}(y)) \\ & \quad \cdot dF^{(k)}(x)dF^{(1)}(y). \end{aligned}$$

Suppose $F_1 = \ldots = F_c = F$, then the preceding expressions become:

$$\begin{aligned} n(B_{n,\ell}^{(1)})^2 = & \, 2\iint [\min(u,v) - uv]\, J(u)J(v)dudv \\ & - 2\iint \left[H_{1\ell}(F^{-1}(u), F^{-1}(v)) - uv\right] J(u)J(v)dudv \\ = & \, 2\iint \left[\min(u,v) - H_{1\ell}(F^{-1}(u), F^{-1}(v))\right] J(u)J(v)dudv\,. \end{aligned}$$

For $\ell \neq k$

$$\begin{aligned} n(B_{n,\ell}^{(1)}, B_{n,k}^{(1)}) = & \iint \left[\min(u,v) + H_{\ell k}(F^{-1}(u), F^{-1}(v)) - \right. \\ & \left. H_{1k}(F^{-1}(u), F^{-1}(v)) - H_{1\ell}(F^{-1}(u), F^{-1}(v))\right] J(u)J(v)dudv. \end{aligned}$$

Let $\mathbf{\Sigma}$ denote the variance-covariance matrix of $(B_{n,2}, \ldots, B_{n,c})$. Then we reject H_0 when

$$(B_{n,2}, \ldots, B_{n,c})\ \mathbf{\Sigma}^{-1} \begin{pmatrix} B_{n,2} \\ \vdots \\ B_{n,c} \end{pmatrix} \geq \chi^2_{c-1,1-\alpha}.$$

However, since $\mathbf{\Sigma}$ is unknown we can replace it by $\mathbf{\Sigma}_n$ by estimating the variances and covariances as follows. For instance, a consistent estimate of

$$\iint H_{\ell k}(x,y)J(F^{(\ell)}(x))J(F^{(k)}(y))dF^{(\ell)}(x)dF^{(k)}(y)$$

is

$$M(1)\left[2\int_0^1 uJ(u)du - M(1)\right] + n^{-1}\sum_{i=1}^{n} M(F_n^{(\ell)}(X_{\ell,i}))M(F_n^{(k)}(X_{k,i}))$$

where $(X_{\ell,1}, X_{k,1}), \ldots, (X_{\ell,n}, X_{k,n})$ denote the $(l,k)^{\text{th}}$ bivariate sample.

Remark 26.9.1. *Govindarajulu (1995, p. 389) considers the problem of testing $H_0 : H_{ij}(x,y) = H(x,y)$ for all $1 \le i \ne j \le c$ and all (x,y) against $H_1 : H_{ij}(x,y) \ne H_{k\ell}(x,y)$ for some $i \ne k$, $j \ne \ell$ and some (x,y) and obtains a chi-squared test for sufficiently large n. For details, see Govindarajulu (1995, pp. 389-391). The case of random sample sizes is also considered by him in Section 7.*

26.10 Problems

26.2.1 When $J(u) \equiv 1$, show that T_m^* given by (26.2.3) reduces to the Mann-Whitney statistic.

26.3.1 When $F = G$, find the asymptotic mean and variance of T_m^* given by (26.3.1) with $J(u) = u$.

26.3.2 Verify the regularity conditions of Theorem 26.3.1 when the score function is given by

$$J_N(u) = u^2, \quad 0 < u < 1.$$

26.5.1 Let X denote the failure time of a light bulb made by manufacturer 1 (in hundreds of hours) and Y denote the failure time of a light bulb made by manufacturer 2. Suppose we originally set to test $m^* = n^* = 20$ bulbs of each brand and we obtain the following data truncated at 500 hours.

X: 3.1, 2.7, 2.5, 2.6, 3.2, 1.8, 2.3, 3.4, 3.6, 3.1
Y: 4.1, 4.2, 3.7, 3.3, 4.3, 4.4, 3.8, 4.5, 4.4, 4.6

We wish to test H_0: The distributions of X and Y are the same versus H_1: Y is stochastically larger than X. Use Theorem 26.5.1 with $J(u) \equiv 1$ in order to test H_0 against H_1.

26.6.1 For the data in Problem 24.7.2, test the hypothesis that there is no difference among the three methods against the alternative that the different teaching methods have different effects on the students' performance, using $\alpha = 0.05$.
(Hint: use test criterion given by (26.6.4).)

26.8.1 For the data in Problem 24.8.1, test the hypothesis of equality of marginal distributions against the alternative that the distribution function for "before" is smaller than the distribution function for "after". (Hint: Use Eqs. (26.8.1) and (26.8.2) with $\alpha = 0.05$.)

26.8.2 For the data in Problem 24.8.2, test the same hypotheses as in Problem 26.8.1.

26.8.3 For the data in Problem 24.8.1, test

$$H_0\colon\ H(x,y) = F(x)G(y) \text{ for all } x \text{ and } y, \text{ versus}$$
$$H_1\colon\ H(x,y) > F(x)G(y) \text{ for some } x, y$$

using test criterion given by Eq. (26.8.7) with $\alpha = 0.05$.

26.8.4 For the data in Problem 24.8.2, test the same hypothesis as in Problem 26.8.3.

Chapter 27

Systematic Statistics

27.1 Introduction

Linear combinations of order statistics (sometimes called systematic statistics) naturally arise in best linear unbiased estimation of location and scale parameters of a distribution, based on order statistics (see Blom (1958)): Jung (1955) considered linear combinations of order statistics with continuous weight functions for estimating location and scale parameters and studied their asymptotic efficiencies. Trimmed and Winsorized means of samples are proposed in robust estimation of parameters. Hence, it is of interest and importance to study the large-sample properties of such statistics. Asymptotic normality of sample quantities has been studied by several authors (for instance, see Hájek (1964)). If the distribution function has a continuous density function which is strictly positive on the range of variation, using Renyi's (1953) representation of order statistics in random samples drawn from continuous populations, one can obtain simple sufficient conditions for the asymptotic normality of linear combinations of order statistics (LCOS's) based on trimmed samples, for example, sample quantile, α-trimmed mean, Winsorized mean, etc. (for definitions of these see Tukey (1962)). However, inclusion of the tails of the ordered sample in the linear combination makes the study of its asymptotic normality nontrivial. Imposing a smoothness condition on the tails of the distribution, Chernoff, Gastwirth and Johns (1967) (to be abbreviated as CGJ (1967)) studied the asymptotic normality of linear combinations of functions of order statistics for the one-sample case. Also, the conditions of Hájek (1964) are somewhat too restrictive. Hence it is of much interest to make a systematic study of the asymptotic normality of LCFOS's in one and c-samples without the smoothness condition on the tails of the distribution. The following results constitute strengthening of the author's (1965) earlier results. In the end we briefly review the recent contributions to this problem. First we consider the one-sample case.

27.2 Regularity Assumptions

Let $X_1, \ldots, X_N$ be a random sample of size N drawn from a population having the continuous distribution function (df) $F(x)$. Also, let $F_N(x)$ denote the empirical distribution function (edf) based on the random sample $X_1, \ldots, X_N$. $F(x)$ may depend on N, but this will not be stated explicitly. Define the class of statistics by

$$N\overline{T}_N = \sum_{i=1}^{N} E_{N,i}\, g^*(X_{i,N}) \tag{27.2.1}$$

where $g^*(x)$ is an arbitrary continuous function, $E_{N,i}$ are some given constants and the $X_{i,N}$ are the ordered values of the X's. When $g^*(x) = x$, $\overline{T}_N$ generates the systematic statistics. Alternatively, one can write

$$\overline{T}_N = \int_{-\infty}^{\infty} g(F(x)) J_N\left(\frac{N}{N+1} F_N(x)\right) dF_N(x) \tag{27.2.2}$$

where $J_N(i/(N+1)) = E_{N,i}$ and $g(F(x)) = g^*(x)$.

As suggested in chapters 24–26, one can extend the domain of the definition of $J_N(\cdot)$ to $(0,1)$ by adopting the convention that J_N is constant on $\left(\frac{i}{N+1}, \frac{i+1}{N+1}\right)$ $(i = 0, \ldots, N)$. $J_N(u)$ might depend upon N, but the subscript N in $J_N(u)$ will be suppressed whenever convenient.

We impose no restrictions on F. If it is discontinuous having a denumerable numbering of jump points, then F can be made continuous as suggested in GLR (1967). If F has a jump of size α at a point t, remove the point t from the real line and insert in its place a closed interval of length α. Distribute the probability mass α uniformly over this interval. The new cumulative distribution F^* so obtained is continuous. The relative order relations among X's have the same probability distribution as if the sample was drawn from F.

As a further simplification, without loss of generality, one can assume that F is uniform on $[0, 1]$. To see this, note that if after removal of discontinuities, the function F^* is constant over certain intervals, no observations will occur in these intervals which can be deleted from the real line without affecting the order of the observations. Then we are left with a continuous strictly increasing distribution function which can be transformed to the uniform distribution on $[0, 1]$ by a strictly increasing continuous transformation.

We need to define the classes of functions for which the asymptotic results hold.

Definition 27.2.1. *A function h, $h \geq 1$ defined on $(0, 1)$ which is U-shaped, is said to belong to the class $\mathcal{U}_1(\mathcal{U}_2)$ if it is integrable (square integrable) for the Lebesgue measure. Note that in particular, $h_1(u) = K[u(1-u)]^{-1+\delta}$ belongs to $\mathcal{U}_1$ and $h_2(u) = K[u(1-u)]^{-\frac{1}{2}+\delta}$ belongs to $\mathcal{U}_2$ for some $0 < \delta < 1/2$ and $0 < K < \infty$. Let J be defined by an integral of the form $J(u) = \int_{1/2}^{u} J'(v)dv$.*

Definition 27.2.2. *(g, J) is said to belong to $\mathcal{S}$ if g is continuous with $|g| \leq h^*$ where h^* is U-shaped, J is absolutely continuous with $|J'| \leq h$ (h is U-shaped) and $hh^* \leq h_1 h_2$.*

We also say that $(g, J) \in \mathcal{S}_0$ if $|g(u)| \leq K[u(1-u)]^{-\alpha+\delta}$ and $|J'(u)| \leq K[u(1-u)]^{-\frac{3}{2}+\alpha+\delta}$ for some $0 < \delta < 1/4$ and $\alpha \leq 3/2$ where K is a finite and positive generic constant. Notice that $\mathcal{S}_0 \subset \mathcal{S}$ since, in particular, we can have $h^*(u) = K[u(1-u)]^{-\alpha+\delta}$ and $h(u) = K[u(1-u)]^{-\frac{3}{2}+\alpha+\delta}$.

27.3 Main Results

In the following we state the main results of the author (1980) and briefly sketch the proofs.

Theorem 27.3.1. *Let $(g, J) \in \mathcal{S}$ and let*

$$T_N = N^{1/2} \left\{ \int g(F) J\left(\frac{N}{N+1} F_N\right) dF_N - \int g(F) J(F) dF \right\}. \quad (27.3.1)$$

Let $P_N(\cdot)$ denote the distribution of the T_N and Φ denote the standard normal d.f. Then, for every $\varepsilon > 0$ there is an $N(\varepsilon)$ such that $N \geq N(\varepsilon)$ implies that

$$|P_N(y) - \Phi\left(y/\sigma_N(F, J, g)\right)| < \varepsilon$$

for every y, every $(g, J) \in \mathcal{S}$ and every F. On sets where σ_N^2 is bounded away from zero, we have

$$\sup_y |P_n(y) - \Phi(y/\sigma_N)| < \varepsilon$$

where $\sigma_N^2(F, J, g)$ denotes the asymptotic variance of T_N.

PROOF: One can expand T_N as

$$T_N = B_N + C_{1N} + C_{2N}$$

where

$$B_N = N^{1/2} \int g(F)J(F)d(F_N - F) + N^{1/2} \int g(F)J'(F)(F_N - F)dF \quad (27.3.2)$$

$$C_{1N} = N^{1/2} \int g(F)\left[J\left(\frac{N}{N+1}F_N\right) - J(F)\right] d(F_N - F) \quad (27.3.3)$$

and

$$C_{2N} = N^{1/2} \int g(F)\left[J\left(\frac{N}{N+1}F_N\right) - J(F) - (F_N - F)J'(F)\right] dF, \quad (27.3.4)$$

where $A = (0, \tau] \cup [1 - \tau, 1)$.

The tails of the statistic T_N for some τ $(0 < \tau < 1)$ is given by

$$\begin{aligned}\Delta_N(J, g, \tau) &= N^{1/2}\left\{\int_A g(F)J\left(\frac{N}{N+1}F_N\right) dF_N - \int_A g(F)J(F)dF\right\} \\ &= \Delta_N{}^* + N^{1/2} \int_A g(F)J(F)d(F_N - F)\end{aligned} \quad (27.3.5)$$

where

$$\Delta_N{}^* = N^{1/2} \int_A g(F)\left[J\left(\frac{N}{N+1}F_N\right) - J(F)\right] dF_N. \quad (27.3.6)$$

The author (1980, Proposition 6.1) establishes that there exists a number τ_0 such that

$$P\left(\sup\left[|\Delta_N{}^*(J, g, \tau)| : 0 < \tau \le \tau_0, (J, g) \in \mathcal{S}\right] > \varepsilon\right) < \varepsilon \quad (27.3.7)$$

for every F and N. Next the difference $\Delta_N - \Delta_N{}^*$ is a normalized sum having zero expectation and a variance bounded by $[c^*(\tau)]^4$ where

$$c^{*2}(\alpha) = \max\left[\int_0^\alpha h_2{}^2(u)du, \left\{\int_0^\alpha h_1(u)du\right\}^2\right]. \quad (27.3.8)$$

Thus, for every $\varepsilon > 0$ there exists a τ_0 such that

$$P(|\Delta_N - \Delta_N{}^*| > \varepsilon) < \varepsilon \text{ for every } (g, J) \in \mathcal{S} \text{ and } \tau \leq \tau_0 . \tag{27.3.9}$$

Hence, it follows that for every $\varepsilon > 0$ there is a number $\tau_0 > 0$ such that $\tau \leq \tau_0$ and $(g, J) \in \mathcal{S}$ imply that

$$P(|\Delta_N(J, g, \tau)| > \varepsilon) < \varepsilon \tag{27.3.9}$$

for every N and every F.

One can write B_N as $N^{-1/2} \sum_{i=1}^{N} B(X_i)$ where

$$\begin{aligned} B(X) &= \int g(F)J(F)d(F_1 - F) + \int g(F)J'(F)(F_1 - F)dF \\ &= B_1 + B_2 \ \ (say)\,, \end{aligned} \tag{27.3.10}$$

and F_1 denotes the empirical distribution function based on a single observation. One can compute the variance of B as

$$\text{var } B = \text{var } B_1 + \text{var } B_2 + 2 \text{ cov}(B_1, B_2) \tag{27.3.11}$$

and obtain explicit expressions for the quantities on the right hand side of (27.3.11). (See (3.7) and (3.8) of the author (1980).)

However, by performing integration by parts once in the first term of B_N, we obtain

$$B_N = -N^{1/2} \int J(F)(F_N - F)dg(F). \tag{27.3.12}$$

Now the variance of B_N given by (27.3.12) is

$$\text{var } B_N = 2 \iint\limits_{0<u<v<1} u(1 - v)J(u)J(v)dg(u)dg(v). \tag{27.3.13}$$

Note that when $g(F(x)) = x$, the form of the variance of B_N as given by (27.3.13) agrees with Jung's (1955) expression for the asymptotic variance of B_N.

Further, if $g(u)$ is differentiable, (27.3.5) takes the form of

$$\text{var } B_N = \int_0^1 M^2(u)du - \left[\int_0^1 M(u)du\right]^2 \tag{27.3.14}$$

where

$$M(u) = \int_{1/2}^{u} J(x)g'(x)dx$$

since

$$\text{var } B_N = 2 \iint_{x<y} \left[\iint_{x<u<v<y} J(u)J(v)g'(u)g'(v)dudv \right] dxdy$$
$$= \iint_{0<x<y<1} \left[\int_x^y J(u)g'(u)du \right]^2 dxdy. \qquad (27.3.15)$$

It is of interest to note that when $|J(u)g'(u)| \leq h_1 h_2$ with $h_1 \in \mathcal{U}_1$ and $h_2 \in \mathcal{U}_2$, var B_N as given in (27.3.15) is finite. To see this, first integrate on x and y in (27.3.15) for fixed u and v and then note that for $u < v$, $u(1-v) \leq \{uv(1-u)(1-v)\}^{1/2}$. Hence

$$\text{var } B_n \leq \iint_{u<v} \{u(1-u)v(1-v)\}^{1/2} h_1(u)h_1(v)h_2(u)h_2(v)dudv$$
$$= \left[\int_0^1 \{u(1-u)\}^{1/2} h_1(u)h_2(u)du \right]^2. \qquad (27.3.16)$$

However, since

$$\alpha {h_2}^2(\alpha) \leq \int_0^\alpha {h_2}^2(y)dy,$$

$$\int_0^\alpha \sqrt{u} h_1(u)h_2(u)du \leq \int_0^\alpha h_1(u) \left[\int_0^\alpha {h_2}^2(v)dv \right]^{1/2} du \leq {c^*}^2(\alpha).$$

Thus, the first part of Theorem 27.3.1 follows by applying the central limit theorem to B_N and the second statement follows from the first by considering B_N/σ_N instead of B_N provided the higher order terms, namely C_{1N} and C_{2N} are negligible in probability. In propositions 5.1 and 5.2 of the author (1980, Section 5), these higher order terms are shown to be negligible. This completes the proof of Theorem 27.3.1.

Remark 27.3.1. *In the following situation, the uniformity asserted in Theorem 27.3.1 may be of interest.*

Suppose that for each N, $F = \Psi((x-\theta)/\beta)$ where Ψ admits a density, with $\beta > 0$. If $\theta_v \to \theta_0$ and $\beta_v \to \beta_0 \neq 0$, the density of $\Psi((x-\theta_v)/\beta_v)$ converges in measure to that of $\Psi((x-\theta_0)/\beta_0)$. The convergence of $|P_N - \Psi(x/\sigma_N|$ to zero asserted in Theorem 27.3.1 is therefore uniform for $(g, J) \in \mathcal{S}$ and every bounded set of values $[\theta, \ln \beta]$. Also, if for every integer N, the value of (θ, β) is (θ_N, β_N) where $\theta_N \to 0$ and $\beta_N \to 1$, then the Kolmogorov *distance* $\sup |P_N(y) - \Phi(y/\sigma_N)|$ *also tends to zero uniformly for any set* $S \in \mathcal{S}$ *such that* $(g, J) \in S$ *implies that* $\lim_{N\to\infty} \sigma_N^2$ *is finite and bounded away from zero.*

In some instances, the weights J_N can be expectations of suitable order statistics. Towards this we introduce the additional notation and state some theorems of interest. Let $U_{k,N}$ denote the kth smallest order statistic in a random sample of size N from the uniform distribution on $[0, 1]$. For any function J_N, let $\overline{J}_N\left(\frac{i}{N+1}\right) = E[J_N(U_{i,N})], \quad i = 1, \ldots, N$. The definition of $\overline{J}_N$ is completed by interpolating linearly between successive values $\left\{\frac{k}{N+1}, \frac{k+1}{N+1}\right\}$ and leaving $\overline{J}_N$ constant below $1/(N+1)$ and above $N/(N+1)$. Then the author (1980) obtains the following theorem which we state here and the reader is referred for its proof to the author (1980, pp. 334–335).

Theorem 27.3.2. *Let*

$$|g(u)| \leq K[u(1-u)]^{-\alpha+\delta} \quad \text{and} \quad |J_N'(u)| \leq K[u(1-u)]^{-\frac{3}{2}+\alpha+\delta} \tag{27.3.17}$$

for some $0 < \delta < 1/4$, $\alpha \leq 3/2$ *and fixed* $K < \infty$. *Let* J_N' *converge in Lebesgue measure to a limit* J'. *Also, let* T_N *be the expression*

$$T_N = N^{1/2}\left\{\int g(F)\overline{J}_N\left(\frac{N}{N+1}F_N\right)dF_N - \int g(F)J_N(F)dF\right\}. \tag{27.3.18}$$

Let P_N *be the distribution of* T_N. *Then the conclusion of Theorem 27.3.1 holds uniformly in* F *as* $N \to \infty$.

Corollary 27.3.2.1. *Let* k *be a fixed integer and let* a_j $(j = 1, \ldots, k)$ *be bounded constants. Let*

$$J_N\left(\frac{i}{N+1}\right) = \sum_{j=1}^{k} a_j E(W_{i,N}^j) \tag{27.3.19}$$

where $W_{i,N}$ *is the* i^{th} *smallest order statistic in a random sample of size* N *drawn from a population having the inverse of a function* V *for its distribution function. If* (g, V^j) *satisfy (27.3.15) for* $j = 1, \ldots, k$, *then the functions* J_N *given by (27.3.17) satisfy the conditions of Theorem 27.3.2.*

PROOF: It follows from the linearity of the transformation $J \hookrightarrow \overline{J}_N$ which is used for defining the functions occurring in Theorem 27.3.2.

27.4 Random Sample Size

One is faced with random sample sizes in dealing with truncated samples, in sequential estimation or when the sample size is determined by an independent random process. Hence it is of interest to study the asymptotic normality of T_N defined by (27.3.1) when N is random. Let N^* be a nonstochastic integer tending to infinity. Assume that $N = N(N^*)$ and N/N^* tends to unity in probability as $N^* \to \infty$.

First, the author (1980, Lemma 8.1) shows that $N^{1/2}h_2(u)[F_N(F^{-1}(u) - u)]$ is bounded in probability as $N^* \to \infty$ whenever $h_2 \in \mathcal{U}_2$. Then he extends Khintchine's weak law of large numbers to the random sample size case (see his Lemma 8.2). Using these two lemmas, he obtains the following theorem which we still state without proof.

Theorem 27.4.1. *With the above notation and assumptions we have*

$$\lim_{N^* \to \infty} |P(T_N \le x) - \Phi(x/\sigma)| = 0$$

for every x, for every F and every $(g, J) \in \mathcal{S}$ where σ^2 is given by (27.3.2) and the d.f. $F(x)$ and the function J do not depend on N but may depend on N^*.

Remark 27.4.1. *The analogues of Theorem 27.3.2 and Corollary 27.3.2.1 also hold when N is random. Remark 27.3.1 is applicable to Theorem 27.4.1 and the analogues of Theorem 27.3.2 and Corollary 27.3.2.1.*

27.5 c-Sample Case

Let a rank sum statistic be a linear combination of expected values of order statistics in a random sample, the coefficients being random. If in the rank sum statistic, the expected values of order statistics are replaced by the corresponding order statistics, the resultant statistic, called the randomized statistic, under mild regularity assumptions will be asymptotically equivalent to the rank sum statistic (see, for instance, Bell and Doksum (1965, Theorem 2.5)). Hence, it is of interest to extend the results of sections 27.3 and 27.4 to c-sample situations. We need the following additional notation.

Let $X_{j,k}$ $(k = 1, 2, \ldots, n_j)$ denote the random sample drawn from a population having for its d.f. $F^{(j)}(x)$ $(j = 1, \ldots, c)$. We assume that $F^{(1)}, \ldots, F^{(c)}$

have no common points of discontinuity. Also, let $F_{n_j}^{(j)}(x)$ denote the empirical distribution function (e.d.f.) based on the random sample $X_{j,1},\ldots,X_{j,n_j}$ $(j=1,\ldots,c)$. Further, let

$$N = n_1 + \ldots + n_c \tag{27.5.1}$$

$$H(x) = \sum_{j=1}^{c} \lambda_j F^{(j)}(x), \tag{27.5.2}$$

and

$$H_N(x) = \sum_{j=1}^{c} \lambda_j F_{n_j}^{(j)}(x) \tag{27.5.3}$$

where

$$\lambda_j = n_j/N \quad \text{and} \quad 0 < \lambda_0 \le \lambda_j \le 1 - \lambda_0 < 1 \quad (j=1,\ldots,c) \tag{27.5.4}$$

for some fixed $\lambda_0 \le 1/c$. Thus $H_N(x)$ denotes the combined e.d.f. and $H(x)$ denotes the combined population d.f. Although $F^{(j)}(x)$ may depend on N, this will be suppressed for the sake of notational case. Now define the class of statistics of interest by

$$n_j T_{N,j} = \sum_{i=1}^{N} E_{N,i,j} g_j^*(W_{i,N}) Z_{N,i,j} \quad (j=1,\ldots,c) \tag{27.5.5}$$

where $E_{N,i,j}$ are some given constants and

$$\begin{aligned} Z_{N,i,j} &= 1 \text{ if the } i^{\text{th}} \text{ smallest observation in the combined ordered sample} \\ &\qquad \text{belongs to } j^{\text{th}} \text{ sub-sample,} \\ &= 0 \text{ otherwise,} \end{aligned}$$

$W_{1N} \le \ldots \le W_{N,N}$ denotes the combined ordered sample and the $g_j^*(x)$ are some continuous functions. Alternatively, one can write (27.5.5) as

$$T_{N,j} = \int_{-\infty}^{\infty} g_j(H) J_{N,j}\left(\frac{N}{N+1} H_N\right) dF_{n_j}^{(j)}(x) \quad (j=1,\ldots,c) \tag{27.5.6}$$

when $J_{N,j}\left(\frac{i}{N+1}\right) = E_{N,i,j}$ and $g_j(H) = g_j^*(x)$.

Also, for the sake of simplicity, the subscript N is $J_{N,j}$ will be dropped. The author (1980) has established the joint asymptotic normality of

$$N^{1/2}\left\{T_{N,j} - \int g_j(H) J_j(H) dF^{(j)}(x)\right\}, \qquad j=1,\ldots,c.$$

The variance-covariance matrix, namely $((\sigma_{N,i,j}))$ is too complicated to be presented here. However, its elements assume simpler forms when $g_1 = \ldots = g_c$ and $F^{(1)} = \ldots = F^{(c)}$. c-sample analogues of Theorems 27.3.2 and 27.4.1 are also obtained.

27.6 Applications

In this section, we will give some statistics the large-sample properties of which follow from the results developed in Sections 27.3, 27.4 and 27.5. Jung's (1955) asymptotically optimal estimate of the scale parameter σ of a normal distribution is given by

$$\hat{\sigma} = N^{-1} \sum_{i=1}^{N} \Phi^{-1}\left(\frac{i}{N+1}\right) X_{i,N}. \tag{27.6.1}$$

In our notation for this statistic $g^*(x) = x$, $g(u) = \Phi^{-1}(u)$ and $J(u) = \Phi^{-1}(u)$, where Φ denotes the standard normal distribution function. The asymptotic normality of $\hat{\sigma}$ follows from Theorem 27.3.1 since

$$g(u) \sim \{-2\ln u(1-u)\}^{1/2} \quad , \quad J'(u) = \left\{u(1-u)[-2\ln u(1-u)]^{1/2}\right\}^{-1}$$

and hence

$$g(u)J'(u) \sim 1/u(1-u) \quad \text{as } u \to 0 \text{ or } 1.$$

As a modification for small sample sizes, let us modify $\hat{\sigma}$ to $\hat{\hat{\sigma}}$ where

$$\hat{\hat{\sigma}} = N^{-1} \sum_{i=1}^{N} \mu_{i,N} X_{i,N} \tag{27.6.2}$$

where the $\mu_{i,N}$ are expected values of standard normal order statistics in a sample of size N. The asymptotic normality of $\hat{\hat{\sigma}}$ follows from Theorem 27.3.2. The statistics (27.6.1) and (27.6.2) can be extended to c-sample situations and their large-sample properties studied. Also, the asymptotic normality of the Bell-Doksum (1965) statistics arising from two or more samples can be established.

For another application set $F^{(1)} = \ldots = F^{(c)} = F$, $n_1 = \ldots = n_c = n$ and $c = 2$. Then the bivariate asymptotic normality of $T^*_{N,1}$ and $T^*_{N,2}$ follows from the results of Section 27.5. Such pairs of statistics and the ratio of $T^*_{N,1}$ to $T^*_{N,2}$ have been considered by Sarkadi (1981) in goodness of fit problems.

Concluding Remarks. Bickel (1967), Govindarajulu (1968), Moore (1968), Stiegler (1969, 1974), Shorack (1972), Ruymgaart and Van Zuijlen (1977) and Mason (1979) have also made significant contributions to the asymptotic normality of linear combinations (or of functions) of order statistics. Some of them do not seem to require the absolute continuity of the J-function. However, they work with smooth score functions and impose some regularity conditions. Stiegler (1974), Ruymgaart and Van Zuijlen (1979) consider the case when the X's are independent but are not necessarily identically distributed. Wellner (1977) and Van Zwet (1980) have obtained strong laws for linear functions of order statistics.

27.7 Problems

27.3.1 Using Theorem 27.3.1 find the normalizing constants of Jung's statistic and its modified version given in Section 27.6.

27.3.2 α-trimmed mean of a sample is defined by

$$T = \sum_{i=[\alpha N]+1}^{N-[N\alpha]} X_{i,N}/N(1-2\alpha), \quad 0 < \alpha < 1/2.$$

Establish T as a member of the class given by (27.3.1) and establish its asymptotic normality.

27.3.3 Specialize the J function such that the statistic T_N coincides with $\overline{X}$, the sample mean and find the normalizing constants for the same.

27.3.4 Establish the asymptotic normality of the statistic

$$T_N = \sum_{i=1}^{N} iX_{i,N}$$

and evaluate the normalizing constants.
(Hint: let $J_N\left(\frac{i}{N+1}\right) = \frac{i}{N+1}$ in (27.2.3).)

27.3.5 As a special case of Theorem 27.3.1, establish the asymptotic normality of αth sample quantile given by $X_{[N\alpha]+1,N}$ and evaluate the normalizing constants.

Hint: Asymptotic normality of the sample quantile follows from Theorem 27.3.1 upon setting $g(F(u)) = F^{-1}(u)$ and

$$\begin{aligned} E_{N,i} &= 1, \quad \text{if } i = [N\alpha] + 1 \\ &= 0, \quad \text{otherwise} \end{aligned}$$

or equivalently, setting $J(u) = 1$ when $u = \alpha$ and zero otherwise. In order to obtain the explicit expressions for the normalizing constants, set

$$\widetilde{M}(u) = \int_0^u J(v)\,dv; \quad \text{i.e. } d\widetilde{M}(u) = J(u)du.$$

Then the asymptotic mean of the sample quantile is

$$\int_0^1 F^{-1}(u)J(u)\,du = \int_0^1 F^{-1}(u)\,d\widetilde{M}(u) = F^{-1}(\alpha).$$

The asymptotic variance given by (27.3.14) is

$$\int_0^1 M^2(u)\,du - \left[\int_0^1 M(u)\,du\right]^2$$

$$\begin{aligned} M(u) = \textstyle\int_{1/2}^u g'(u)J(u)\,du &= \textstyle\int_{1/2}^u g'(u)\,d\widetilde{M}(u) \\ &= g'(\alpha) \qquad \text{if } u > \alpha \\ &= 0 \qquad \text{if } u < \alpha. \end{aligned}$$

Hence the asymptotic variance of the sample quantile is

$$\int_\alpha^1 [g'(\alpha)]^2\,du - [g'(\alpha)(1-\alpha)]^2 = \alpha(1-\alpha)[g'(\alpha)]^2$$

with $g'(\alpha) = \frac{d}{d\alpha}[F^{-1}(\alpha)] = 1/f(F^{-1}(\alpha))$.
Thus the normalizing constants for the sample quantile coincide with the traditional (well known) normalizing constants.

Appendices

Appendix I: Best Estimate of Normal Standard Deviation

The coefficients of the best linear estimates of the standard deviation using ordered samples of sizes up to and including 15 drawn from the normal population with mean θ and variance σ^2. The estimate of $\hat{\theta}$ is the arithmetic mean of the order observations in the uncensored sample. These values are taken from A.E. Sarhan and B.G. Greenberg (1956). "Estimation of location and scale parameters by ordered statistics from singly and doubly censored samples". Part I, *Ann. Math. Statist.* **27**, 427–451. Part II, *Ann. Math. Statist.* **29**, 79–105.

Sample Size N	$X_{1,N}$	$X_{2,N}$	$X_{3,N}$	$X_{4,N}$	$X_{5,N}$	$X_{6,N}$	$X_{7,N}$	var$(\hat{\sigma}/\sigma)$
2	−.6662							.5706
3	−.5908							.2755
4	−.4539	−.1102						.1800
5	−.3724	−.1352						.1333
6	−.3175	−.1386	−.0432					.1057
7	−.2776	−.1351	−.0625					.0875
8	−.2476	−.1294	−.0713	−.0230				.0746
9	−.2237	−.1232	−.0751	−.0360	−.0142			.0650
10	−.2044	−.1172	−.0762	−.0436	−.0234			.0576
11	−.1883	−.1115	−.0760	−.0481	−.0294			.0517
12	−.1748	−.1061	−.0749	−.0506	−.0335	−.0097		.0469
13	−.1632	−.1013	−.0735	−.0520	−.0335	−.0164		.0429
14	−.1532	−.0968	−.0717	−.0526	−.0362	−.0212	−.0070	.0395
15	−.1444	−.0927	−.0699	−.0526	−.0379	−.0247	−.0122	.0366

Other coefficients can be obtained by symmetry and multiplying by a minus sign. Coefficient of $X_{i,N} = -(\text{coefficient of } X_{N-i+1,N})$, $i = 1, 2, \ldots, [N/2]$. Also, coefficient of $X_{[(N+1)/2],N} = 0$ and covariance between $\hat{\theta}$ and $\hat{\sigma}$ is zero.

Appendix II: Confidence Intervals for Median

Table of confidence interval for the median in samples from any continuous population, showing the values of k such that

$$P(X_{k,N} < \xi_{0.5} < X_{N-k+1,N}) = 1 - 2I_{0.5}(N-k+1,k) = \gamma\,.$$

Sample	$\gamma \geq 0.95$		$\gamma \geq 0.99$	
Size N	k	$I_{0.5}(N-k+1,k)$	k	$I_{0.5}(N-k+1,k)$
6	1	.0156	...	...
7	1	.0078	...	...
8	1	.0039	1	.0039
9	2	.0195	1	.0020
10	2	.0107	1	.0010
11	2	.0059	1	.0005
12	3	.0193	2	.0032
13	3	.0112	2	.0017
14	3	.0065	2	.0009
15	4	.0176	3	.0037
16	4	.0106	3	.0021
17	5	.0245	3	.0012
18	5	.0154	4	.0038
19	5	.0096	4	.0022
20	6	.0207	4	.0013
21	6	.0133	5	.0036
22	6	.0085	5	.0022
23	7	.0173	5	.0013
24	7	.0113	6	.0033
25	8	.0216	6	.0020
26	8	.0145	7	.0047
27	8	.0096	7	.0030
28	9	.0178	7	.0019
29	9	.0121	8	.0041
30	10	.0214	8	.0026
35	12	.0205	10	.0030
40	14	.0192	12	.0032
45	16	.0178	14	.0033
50	18	.0164	16	.0033
55	20	.0150	18	.0032
60	22	.0137	20	.0031
65	25	.0232	22	.0030
70	27	.0207	24	.0028
75	29	.0185	26	.0026
80	31	.0165	...	...

This table is taken from K.R. Nair (1940): Table of confidence interval for the median in samples from any continuous population. *Sankhyā*, **4**, 551–558.

Appendix III: Sample Size for Tolerance Limits

Smallest sample size N required in order to assert with confidence at least γ that 100β percent of the population is contained within the sample range, or values of N satisfying

$$N\beta^{N-1} - (N-1)\beta^{N} \leq 1 - \gamma\,.$$

γ/β	.50	.70	.80	.90	.95	.98	.99
.70	5	8	12	24	49	122	244
.80	5	9	14	29	59	149	299
.90	7	12	18	38	77	194	388
.95	8	14	22	46	93	236	473
.98	9	17	27	56	114	290	581
.99	11	20	31	64	130	330	661

These values are taken from Harold Freeman (1963): *Introduction to Statistical Inference*, Addison-Wesley Publishing Co., Inc., Table A-13. This table was originally prepared by L.G. Dion (1951): *An Approximate Solution of Wilk's Tolerance Limit Equation.* M.I.T. thesis (S.B.).

Appendix IV: Order Statistics for Tolerance Limits

Values of $k = r + s$ such that one may assert with confidence at least γ that 100β percent of the population lies between the rth smallest and the sth largest in a random sample of size N drawn from that continuous population. That is, values k satisfying

$$I_\beta(N - k + 1, k) \leq 1 - \gamma$$

or

$$I_{1-\beta}(k, N - k + 1) \geq \gamma$$

where I is the incomplete beta function.

	β															
	$\gamma = 0.75$				$\gamma = 0.90$				$\gamma = 0.95$				$\gamma = 0.99$			
N	.50	.75	.90	.95	.50	.75	.90	.95	.50	.75	.90	.95	.50	.75	.90	.95
50	22	10	3	2	20	9	2	1	19	8	2	–	16	6	1	–
55	25	12	4	2	23	10	3	1	21	9	2	–	19	7	1	–
60	27	13	4	2	25	11	3	1	24	10	2	1	21	8	1	–
65	30	14	5	2	27	12	4	1	26	11	3	1	23	9	2	–
70	32	15	5	2	30	13	4	1	28	12	3	1	25	10	2	–
75	35	16	6	2	32	14	4	1	30	13	3	1	27	10	2	–
80	37	17	6	3	34	15	5	2	33	14	4	1	30	11	2	–
85	39	19	7	3	37	16	5	2	35	15	4	1	32	12	3	–
90	42	20	7	3	39	17	5	2	37	16	5	2	34	13	3	1
95	44	21	7	3	41	18	6	2	39	17	5	2	36	14	3	1
100	47	22	8	3	44	20	6	2	42	18	5	2	38	15	4	1
110	51	24	9	4	48	22	7	3	46	20	6	2	43	17	4	1
120	56	27	10	4	53	24	8	3	51	22	7	2	47	19	5	1
130	61	29	11	5	58	26	9	3	56	25	8	3	52	21	6	2
140	66	31	12	5	62	28	10	4	60	27	8	3	56	23	6	2
150	71	34	12	6	67	31	10	4	65	29	9	3	61	26	7	2

These values are taken from Paul N. Somerville (1958): Tables for obtaining non-parametric tolerance limits. *Ann. Math. Statist.*, **29**, 599–601.

Appendix V: Upper Confidence Bound for $P(Y < X)$

Values of $\delta_{\lambda,\alpha}$ such that $Q(\delta_{\lambda,\alpha}; \lambda) = 1 - \alpha$ where

$$
\begin{aligned}
Q(\delta; \lambda) &= 1 - \lambda e^{-2(1-\lambda)\delta^2} - (1-\lambda)e^{-2\lambda\delta^2} \\
&\quad -2(2\pi)^{\frac{1}{2}}\lambda(1-\lambda)\delta e^{-2\lambda(1-\lambda)\delta^2}\left\{\Phi(2\lambda\delta) - \Phi\left(2(1-\lambda)\delta\right)\right\}
\end{aligned}
$$

α	λ				
	.1	.2	.3	.4	.5
.10	4.1185	3.2027	2.8501	2.6928	2.6468
.05	4.6115	3.5667	3.1641	2.9844	2.9317
.01	4.5700	4.2745	3.7770	3.5524	3.4870
.005	5.9300	4.5405	4.0050	3.7665	3.6960
.001	6.6800	5.0980	4.4880	4.2150	4.1360

These values are taken from Table of Z.W. Birnbaum and R.C. McCarty (1958): "A distribution-free upper confidence bound for $P(Y < X)$ based on independent samples of X and Y", *Annals of Mathematical Statistics*, **28**, 558–562.

Appendix VI: Confidence Limits for Distribution

Confidence limits for the population distribution function or the acceptance limits for the Kolmogorov-Smirnov tests of goodness of fit. That is, the following table gives values of ϵ such that

$$P\left[F(x) \leq \min\left\{F_N(x) + \epsilon, 1\right\},\ \text{all } x\right] = 1 - \alpha$$

or

$$P\left[F(x) \geq \max\left\{F_N(x) - \epsilon, 1\right\},\ \text{all } x\right] = 1 - \alpha$$

or

$$P\left[F_N(x) - \epsilon \leq F(x) \leq F_N(x) + \epsilon,\ \text{all } x\right] = 1 - 2\alpha\,.$$

Sample Size	α					
N	.10*	.075*	.05	.025	.01	.005
1	.900	.925	.950	.975	.990	.995
2	.684	.726	.776	.842	.900	.929
3	.565	.597	.642	.708	.784	.829
4	.494	.525	.564	.642	.689	.734
5	.446	.474	.510	.563	.627	.669
6	.410	.436	.470	.521	.577	.618
7	.381	.405	.438	.486	.538	.577
8	.358	.381	.411	.457	.506	.543
9	.339	.360	.388	.432	.480	.514
10	.332	.342	.368	.409	.457	.486
11	.307	.326	.352	.391	.437	.468
12	.295	.313	.338	.375	.419	.450
13	.284	.302	.325	.361	.404	.433
14	.274	.292	.314	.349	.390	.418
15	.266	.283	.304	.338	.377	.404
16	.258	.274	.295	.328	.366	.391
17	.250	.266	.286	.318	.355	.380
18	.244	.259	.278	.309	.346	.370
19	.237	.252	.272	.301	.337	.361
20	.231	.246	.264	.294	.329	.352
25	.208	.22	.238	.264	.295	.316
30	.190	.20	.218	.242	.270	.290
35	.176	.19	.202	.224	.251	.269
40	.165		.189	.210	.235	.252
50	.148		.170	.188	.211	.226
60	.136		.155	.172	.193	.207
70	.126		.144	.160	.179	.192
80	.118		.135	.150	.167	.180
90	.111		.127	.141	.158	.169
100	.106		.121	.134	.150	.161
∞	$1.073/N^{\frac{1}{2}}$	$1.138/N^{\frac{1}{2}}$	$1.224/N^{\frac{1}{2}}$	$1.358/N^{\frac{1}{2}}$	$1.517/N^{\frac{1}{2}}$	$1.628/N^{\frac{1}{2}}$

*The values for $N <= 35$, $\alpha = 0.10$ and 0.075 are taken from Massey, F. J. Jr. (1951, Table 1). The rest of the values in the appendix, rounded to three decimal places are taken from Miller (1956, Table 1).

These values for α = .05, .025, .01 and .005 are taken from Leslie H. Miller (1956), "Tables of Percentage Points of Kolmogorov Statistics", *J. Amer. Statist. Assoc.*, **51**, 111–121, and the values for α = .10 and .075 and for $N \leq 35$ are taken from F.J. Massey, Jr. (1951), "The Kolmogorov-Smirnov test for goodness of fit", *J. Amer. Statist. Assoc.*, **46**, 68–78.

Bibliography

[1] Abrahmson, I.G. (1965). On the stochastic comparison of tests of hypotheses. University of Chicago, unpublished dissertation.

[2] Adichie, J.N. (1967a). Asymptotic efficiency of a class of nonparametric tests for regression parameters. *Ann. Math. Statist.*, **38**, 884–893.

[3] Adichie, J.N. (1967b). Estimates of regression parameters based on rank tests. *Ann. Math. Statist.*, **38**, 894–904.

[4] Ahmad, I.A. and P.E. Lin (1980). On the Chernoff-Savage theorem for dependent sequences *Ann. Inst. Statist. Math. Part A*, 211–222.

[5] Aitken, A.C. (1935). On least squares and linear combination of observations. *Proc. Roy. Soc. Edinb.*, **55**, 42–48.

[6] Aitken, A.C. (1945). Studies in mathematics. IV. on linear approximations by least squares. *Proc. Roy. Soc. Edinb.*, **62**, 138–146.

[7] Aiyar, R.J. (1981). Asymptotic efficiency of rank tests of randomness against auto-correlation. *Ann. Inst. Statist. Math.*, **33**, No. 2, 255–262.

[8] Albers, W., P.J. Bickel, and van W.R. Zwet (1976). Asymptotic expansion for the power of distribution-free tests in the one-sample case. *Ann. Statist.*, **4**, 108–156.

[9] Anderson, T.W. and D.A. Darling (1952). Asymptotic theory of certain 'goodness of fit' criteria based on stochastic processes. *Ann. Math. Statist.*, **23**, 193–212.

[10] Anderson, T.W. and D.A. Darling (1954). A test of goodness of fit. *J. Amer. Statist. Assoc.*, **49**, 765–769.

[11] Andrews, F.C. (1954). Asymptotic behavior of some rank tests for analysis of variance. *Ann. Math. Statist.*, **25**, 724–736.

[12] Andrews, F.C. and D.R. Truax (1964). Locally most powerful rank test for several sample problems. *Metrika*, **8**, 16–24.

[13] Ansari, A.R. and R.A. Bradley (1960). Rank-sum tests for dispersion. *Ann. Math. Statist.*, **31**, 1174–1189.

[14] Bahadur, R.R. (1960a). Simultaneous comparison of the optimum and sign tests of a normal mean. *Contributions to Probability and Statistics* (eds. I. Olkin *et al.*). Stanford University Press, Stanford, California, 79–88.

[15] Bahadur, R.R. (1960b). Stochastic comparison of tests. *Ann. Math. Statist.*, **31**, 276–295.

[16] Bahadur, R.R. (1960c). On the asymptotic efficiency of tests and estimates. *Sankhyā*, **22**, 229–252.

[17] Bahadur, R.R. (1967). Rates of convergence of estimates and test statistics. *Ann. Math. Statist.*, **38**, 303–324.

[18] Bahadur, R.R. (1971). *Some Limit Theorems in Statistics.* SIAM Publication #4, Philadelphia, Pennsylvania.

[19] Barlow, R.E., D.J. Bartholomew, J.M. Bremmer and H.D. Brunk (1972). *Statistical Inference under Order Restrictions.* John Wiley & Sons, New York.

[20] Barnett, V.D. (1966). Order statistics estimators of the location of the Cauchy distribution. *J. Amer. Statist. Assoc.*, **61**, 1205–1218.

[21] Bartholomew, D.J. (1959a). A test of homogeneity for ordered alternatives. *Biometrika*, **46**, 36–48.

[22] Bartholomew, D.J. (1959b). A test of homogeneity for ordered alternatives II. *Biometrika*, **46**, 328–335.

[23] Bartholomew, D.J. (1961a). Ordered tests in the analysis of variance. *Biometrika*, **48**, 325–332.

[24] Bartholomew, D.J. (1961b). A test of homogeneity of means under restricted alternatives. *J. Roy. Statist. Soc.*, B, **23**, 239–281.

[25] Bartlett, M.S. (1937). Properties of sufficiency and statistical tests. *Proc. Roy. Soc. Ser. A*, **160**, 268–282.

[26] Bateman, G. (1948). On the power function of the largest run as a test for randomness in a sequence of alternatives. *Biometrika*, **35**, 97–112.

[27] Bell, C.B. (1960). On the structure of distribution-free statistics. *Ann. Math. Statist.*, **31**, 703–709.

[28] Bell, C.B. and K.A. Doksum (1965). Some new distribution-free statistics. *Ann. Math Statist.*, **36**, 203–204.

[29] Bell, C.B. and P.J. Smith (1969). Some nonparametric tests for the multivariate goodness of fit, multi-sample independence and symmetry problems. *Multivariate Analysis II* (ed. P.R. Krishnaiah). Academic Press, New York.

[30] Bell, C.B. D. Blackwell, and L. Breiman (1960). A note on the completeness of order statistics. *Ann. Math. Statist.*, **31**, 794–797.

[31] Bennet, C.A. (1952). Asymptotic properties of ideal linear estimators. Unpublished dissertation. University of Michigan, Ann Arbor, Michigan.

[32] Bhapkar, V.P. (1961). A nonparametric test for the problem of several samples. *Ann. Math. Statist.*, **32**, 1108–1117.

[33] Bhuchongkul, S. (1964). A class of nonparametric tests for independence in bivariate populations. *Ann. Math. Statist.*, **35**, 138–149.

[34] Bhuchongkul, S. and M.L. Puri (1965). On the estimation of contrasts in linear models. *Ann. Math. Statist.*, **36**, 198–202.

[35] Bickel, P.J. (1967). Some contributions to the theory of order statistics. *Proc. Fifth Berkeley Symposium on Mathematical Statistics and Probability* University of California Press at Berkeley and Los Angeles **1**, 575–591.

[36] Bickel, P.J. (1972). On some analogous to linear combinations of order statistics in the linear model. *Ann. Statist.*, **1**, 597–616.

[37] Birnbaum, Z.W. (1950). On the distribution of Kolmogorov's statistic for finite sample size. *Proc. of the Seminar on Scientific Computation* (IBM Corporation, New York, Nov. 1949), 33–66.

[38] Birnbaum, Z.W. (1952). Numerical tabulation of the distribution of Kolmogorov's statistic for finite sample size. *J. Amer. Statist. Assoc.*, **47**, 425–441.

[39] Birnbaum, Z.W. (1953a). Distribution-free tests of fit for continuous distribution functions. *Ann. Math. Statist.*, **24**, 1–8.

[40] Birnbaum, Z.W. (1953b). On the power of one-sided test of fit for continuous probability functions. *Ann. Math. Statist.*, **24**, 484–489.

[41] Birnbaum, Z.W. (1956). On the use of the Mann-Whitney statistic. *Proceedings of the Third Berkeley symposium on Mathematical Statistics and Probability*, **1**, University of California Press, 13–17.

[42] Birnbaum, Z.W. and O.M. Klose (1957). Bounds for the variance of the Mann-Whitney statistic. *Ann. Math. Statist.*, **28**, 933–945.

[43] Birnbaum, Z.W. and R.C. McCarty (1958). A distribution-free upper confidence bound for $P(Y < X)$, based on independent samples of X and Y. *Ann. Math. Statist.*, **29**, 558–562.

[44] Birnbaum, Z.W. and H. Rubin (1954). On distribution-free statistics. *Ann. Math. Statist.*, **25**, 593–598.

[45] Birnbaum, Z.W. and F.H. Tingey (1951). one-sided confidence contours for probability distribution functions. *Ann. Math. Statist.*, **22**, 592–596.

[46] Birnbaum, Z.W. and H.S. Zuckerman (1949). A graphical determination of sample size for Wilks' tolerance limits. *Ann, Math. Statist.*, **20**, 313–316.

[47] Blom, G. (1958). *Statistical Estimates and Transformed Beta Variables.* John Wiley & Sons - Almqvist and Wiksell.

[48] Blomquist, N. (1950). On a measure of dependence between two random variables. *Ann. Math. Statist.*, **21**, 593–600.

[49] Blum, J.R. and L. Weiss (1957). Consistency of certain two-sample tests. *Ann. Math. Statist.*, **28**, 242–246.

[50] Blum, J.R., J. Kiefer, and M. Rosenblatt (1961). Distribution-free tests of independence based on sample distribution function. *Ann. Math. Statist.*, **32**, 485–489.

[51] Bochner, S. (1955). *Harmonic Analysis and the Theory of Probability*, Univ. of California Press.

[52] Bowker, A.H. (1946). Computation of factors for tolerance limits on a normal distribution when the sample is large. *Ann. Math. Statist.*, **17**, 238–239.

[53] Brillinger, D.R. (1966). An external property of the conditional expectation. *Biometrika*, **53**, 594–595.

[54] Brown, G.W. and A.M. Mood (1951). On median tests for linear hypotheses. *Proceedings of the Second Berkeley Symposium on Math. Statist and Probl.*. University of California Press, Berkeley (ed. J. Neyman), pp. 159–166.

[55] Brunk, H.D. (1955). Maximum likelihood estimates of monotone parameters. *Ann. Math. Statist.*, **29**, 607–616.

[56] Brunk, H.D. (1958). On the estimation of parameters restricted by inequalities. *Ann. Math. Statist.*, **29**, 437–454.

[57] Capon, J. (1961). Asymptotic efficiency of certain locally most powerful rank tests. *Ann. Math. Statist.*, **32**, 88–100.

[58] Chacko, V.J. (1963). Testing homogeneity against ordered alternatives. *Ann. Math. Statist.*, **32**, 945–956.

[59] Chanda, K.C. (1963). On the efficiency of the two-sample Mann-Whitney test for discrete populations. *Ann. Math. Statist.*, **24**, 612–617.

[60] Chapman, D.G., and H. Robbins (1951). Minimum variance estimation without regularity assumptions. *Ann. Math. Statist.*, **22**, 581–586.

[61] Chernoff, H. (1952). A measure of asymptotic efficiency for tests of a hypothesis based on the sum of observations. *Ann. Math. Statist.*, **23**, 493–507.

[62] Chernoff, H. and I.R. Savage (1958). Asymptotic normality and efficiency of certain non-parametric test statistcs. *Ann. Math. Statist.*, **32**, 972–994.

[63] Chernoff, H., J.T. Gastwirth, and M.V. Johns Jr. (1967). Asymptotic distribution of linear combinations of functions of order statistics with application to estimation. *Ann. Math. Statist.*, **38**, 52–72.

[64] Cifarelli, D.M., and E. Ragazzini (1990). Some contributions to the theory of monotone dependence. Tech. Rep. 90.17 CNR-LAMA, Milano.

[65] Cifarelli, D.M., P.L. Conti, and E. Ragazzini (1996). On the asymptotic distribution of a general measure of monotone dependence. *Ann. Statist.*, **24**, No. 3, 1386–1399.

[66] Clark, C.E. and T.G. Williams (1958). Distribution of the members of an ordered sample. *Ann. Math. Statist.*, **29**, 862–870.

[67] Clemmens, A.E. and Z. Govindarajulu (1990). A certain locally most powerful test for one-way random effects model: null distribution and power consideration. *Comm. Statist.*, **19**, No. 11 (special issue), 4139–4151.

[68] Clemmens, A.E. and Z. Govindarajulu (1994). Locally most powerful rank tests for random effects in two-way experiments. *Multivariate Analysis and Its Applications* (eds. T.W. Anderson, K.T. Fang and I. Olkin). IMS Lecture Notes Monograph Series, Vol. **24**, pp. 427–437.

[69] Cox, D.R. and A. Stuart (1955). Some quick sign tests for the trend in location and dispersion. *Biometrika*, **42**, 80–95.

[70] Craig, A.T. (1943). A note on the best linear estimate. *Ann. Math. Statist.*, **14**, 88–90.

[71] Cramér, H. (1946). *Mathematical Methods of Statistics*, Princeton University Press.

[72] Cramér, H. (1970). *Random Variables and Probability Distributions* (3rd Edition). Cambridge: Cambridge University Press.

[73] Crouse, C.F. (1964). Note on Mood's test. *Ann. Math. Statist.*, **35**, 1825–1826.

[74] Crouse, C.F. (1969). A multiple comparison rank procedure for one-way analysis of variance. *S. African Statist. J.*, **3**, 35–48.

[75] Daniels, H.E. (1954). A distribution-free test for regression parameters. *Ann. Math. Statist.*, **25**, 499–513.

[76] Dantzig, D. van (1951). On the consistency and the power of Wilcoxon's two sample test. *Koninklijke Nederlandse Akademie van Wetenschappen Proceedings*, Series A, **54**, No. 1, 1–9.

[77] Danziger, L. and S.A. Davis (1964). Tables of distribution-free tolerance limits. *Ann. Math. Statist.*, **35**, 1361–1365.

[78] Darling, D.A. (1955). The Cramér-Smirnov test in the parametric case. *Ann. Math. Statist.*, **26**, 1–20.

[79] Darling, D.A. (1957). The Kolmogorov-Smirnov, Cramér-von-Mises test. *Ann. Math. Statist.*, **28**, 823–838.

[80] Darling, D.A. and T.W. Anderson (1952). Asymptotic theory of certain goodness of fit criterion based on stochastic processes. *Ann. Math. Statist.*, **23**, 193–212.

[81] David, F.N. (1947). A power function for tests for randomness in a sequence of alternatives. *Biometrika*, **34**, 335–339.

[82] David, H.T. (1958). A three-sample Kolmogorov-Smirnov test. *Ann. Math. Statist.*, **29**, 842–851.

[83] David, H.T. (1963). The sample mean among the extreme normal order statistics. *Ann. Math. Statist.*, **34**, 33–55.

[84] David, F.N. and N.L. Johnson (1954). Statistical treatment of censored data. Part I, fundamental formulae. *Biometrika*, **41**, 228–240.

[85] Deshpande, J.V. (1965). A nonparametric test based on U-statistics for the problem of several samples. *J. Indian Statist. Assoc.*, **3**(1), 20–29.

[86] Diananda, P.H. (1953). Some probability limit theorems with statistical applications. *Proc. Cambridge Philos. Soc.*, **49**, 239–246.

[87] Diananda, P.H. (1954). The central limit theorem for m-dependent variables asymptotically stationary to second order. *Proc. Cambridge Philos. Soc.*, **50**, 287–292.

[88] Diananda, P.H. (1955). The central limit theorem for m-dependent variables. *Proc. Cambridge Philos. Soc.*, **51**, 92–95.

[89] Dion, L.G. (1951). An approximate solution of Wilks' tolerance limit equation. M.I.T. thesis.

[90] Dixon, W.J. (1940). A criterion for testing the hypothesis that two samples are from the same population. *Ann. Math. Statist.*, **11**, 199–204.

[91] Doksum, K. (1967). Robust procedure for some linear models with one observation per cell. *Ann. Math. Statist.*, **38**, 878–883.

[92] Donsker, M.D. (1952). Justification and extension of Doob's heuristic approach to the Kolmogorov-Smirnov theorems. *Ann. Math. Statist.*, **23**, 277–281.

[93] Doob, J.L. (1949). Heuristic approach to the Kolmogorov-Smirnov theorems. *Ann. Math. Statist.*, **20**, 393–403.

[94] Downton, F. (1953). A note on ordered least squares estimation. *Biometrika*, **40**, 457–458.

[95] Downton, F. (1954). Least squares estimates using ordered observations. *Ann. Math. Statist.*, **25**, 303–316.

[96] Duran, B.S. (1996). A survey of nonparametric tests for scale. *Communications in Statistics – Theory and Methods*, **A5**, 1287–1312.

[97] Durbin, J. (1951). Incomplete blocks in ranking experiments. *British Journal of Psychology*, **4**, 85–90.

[98] Durbin, J. (1961). Some methods of constructing exact tests. *Biometrika*, **48**, 41–55.

[99] Dwass, M. (1953). On the asymptotic normality of certain rank order statistics. *Ann. Math. Statist.*, **24**, 303–306.

[100] Dwass, M. (1955). On the asymptotic normality of some statistics used in nonparametric tests. *Ann. Math. Statist.*, **26**, 334–339.

[101] Dwass, M. (1956a). On the distribution of ranks and of certain rank order statistics. *Ann. Math. Statist.*, **27**, 862.

[102] Dwass, M. (1956b). The large-sample powers of rank order tests in the two-sample problem. *Ann. Math. Statist.* **27**, 352–374.

[103] Dwass, M. (1957). Modified randomization tests for nonparametric hypotheses. *Ann. Math. Statist.*, **28**, 181–187.

[104] Dwass, M. (1960). Some k-sample rank order tests. *Contributions to Probability and Statistics.* Stanford University Press, Stanford, California, 198–202.

[105] Dwyer, P.S. (1958). Generalization of a Gaussian Theorem. *Ann. Math. Statist.*, **29**, 106–117.

[106] Dwyer, P.S. and M.S. MacPhail (1948). Symbolic matrix derivatives. *Ann. Math. Statist.*, **19**, 517–534.

[107] Dyer, D.D. (1973). Estimation of the scale parameter of the chi-distribution based on sample quantiles. *Technometrics*, **15**, 489–496.

[108] van Eaden, C. (1957). Maximum likelihood estimation of partially or completely ordered parameters. *Indagationes Mathematicae*, **19**, 128–211.

[109] Ederer, F., M.J. Podgor, and the Diabetic Retinopathy Study Research Group (1984). Assessing possible late treatment effects in stopping a clinical trial early: A case study. Diabetic Retinopathy Study Report No. 9 *Controlled Clinical Trials.* **5**, 373–381.

[110] Elandt, R.C. (1962). Exact and approximate power of the nonparametric test of tendency. *Ann. Math. Statist.*, **33**, 471–481.

[111] van Elteren, P. and G.E. Noether (1959). The asymptotic efficiency of the χ_r^2-test for a balanced incomplete block design I and II. *Biometrika*, **46**, 475–477.

[112] Esscher, F. (1924). On a method of determining correlation from the rank of variates. *Skandinavisk Aktuarietidskrift*, **7**, 201–219.

[113] Farlie, D.J.G. (1961). The asymptotic efficiency of Daniels' generalized correlation coefficients. *J. Roy. Statist. Soc., Ser. B*, **23**, 128–142.

[114] Feller, W. (1948). On the Kolmogorov-Smirnov limit theorems for empirical distributions. *Ann. Math. Statist.*, **19**, 177–189.

[115] Fisher, R.A. (1935). *Statistical Methods for Research Workers.* Edinburgh: Oliver and Boyd, Section 24, Example 9.

[116] Fisher, R.A. (1936). Coefficient of racial likeness and the future of craniometry. *J.R. Anthrop. Inst.*, **66**, 57–63.

[117] Fisz, M. (1960). Some nonparametric tests for the k-sample problem. *Colloquium Math.*, **7**, 289–296.

[118] Fix, E. and J.L. Hodges Jr. (1951). Discriminatory analysis, nonparametric discrimination: consistency properties. USAF School of Aviation Medicine, Project No. 21-49-004, Report No. 4.

[119] Foster, F.G. and A. Stuart (1954). Distribution-free tests in time series based on the breaking of records. *J. Roy. Statist. Soc. (B)*, **16**, 1–22.

[120] Fraser, D.A.S. (1951). Sequentially determined statistically equivalent blocks. *Ann. Math. Statist.*, **22**, 372–381.

[121] Fraser, D.A.S. (1953). Completeness of order statistics. *Can. J. Math.*, **6**, 42–45.

[122] Fraser, D.A.S. (1953). Non-parametric tolerance regions. *Ann. Math. Statist.*, **24**, 44–55.

[123] Fraser, D.A.S. (1957a). Most powerful rank-type tests. *Ann. Math. Statist.*, **28**, 1040–1043.

[124] Fraser, D.A.S. (1957b). *Nonparametric Methods in Statistics.* John Wiley & Sons, Inc., New York.

[125] Fraser, D.A.S. and I. Guttman (1956). Tolerance regions. *Ann. Math. Statist.*, **27**, 162–179.

[126] Fraser, D.A.S. and R. Wormleighton (1951). Non-parametric estimation IV. *Ann. Math. Statist.*, **22**, 294–298.

[127] Fréchet, M. (1937). Generalités sur les probabilités. Variables aléatoires. *Traité de calcul des probabilités* (edited by E. Borel), Gauthier-Villars, Paris.

[128] Freeman, H. (1963). *Introduction to Statistical Inference.* Addison-Wesley Publishing Co.

[129] Freund, J.E. and A.R. Ansari (1957). Two-way rank-sum tests for variances. Tech. Report 34, Department of Statistics, Virginia Polytechnic Institute.

[130] Friedman, M. (1937). The use of ranks to avoid the assumption of normality implicit in the analysis of variance. *J. Amer. Statist. Assoc.*, **32**, 675–701.

[131] Friedman, M. (1940). A comparison of alternative tests of significance for the problem of m rankings. *Ann. Math. Statist.*, **11**, 86–92.

[132] Gaenssler, P. and W. Stute (1979). Empirical processes: A survey of results for independent and identically distributed random variables. *Ann. Prob.*, **7**, 193–243.

[133] Gastwirth, J.L. and S.S. Wolfe (1968). An elementary method for obtaining lower bounds on the asymptotic power of rank lists. *Ann. Math. Statist.*, **39**, 2128–2130.

[134] Gavin, G. G. (1977). Functions of order statistics and tests of fit, *South African Statist. J.*, **11**, No. 2, 99–118.

[135] Gibbons, J.D. (1971). *Nonparametric Statistical Inference* (Second Edition), New York: Marcel Dekker.

[136] Gibbons, J.D., and S. Chakrabarti (1992). *Nonparametric Statistical Inference* (Third Edition). Marcel Dekkar Inc. New York.

[137] Gibbons, J.D., and S. Chakraborti (2003). *Nonparametric Statistical Inference* (Fourth Edition). Marcel Dekker, Inc., New York.

[138] Gini, C. (1914). Di una misura della relazioni tra le graduatorie di due caratteri. Tipografia cecchini. Roma.

[139] Girschick, M., F. Mosteller, and L.J. Savage (1946). Unbiased estimates for certain binomial sampling problems with applications. *Ann. Math. Statist.*, **17**, 13–23.

[140] Gnedenko, B.V. (1962). *The Theory of Probability.* Fourth Edition. Chelsea, New York. Problem #16 on p. 134.

[141] Godwin, H.J. (1949). On the estimation of dispersion by linear systematic statistics. *Biometrika*, **36**, 92–100.

[142] Gokhale, D.V. (1968). On the asymptotic efficiencies of a class of rank tests for independence of two variables. *Ann. Inst. Statist. Math.*, **20**, 255–261.

[143] Govindarajulu, Z. (1960). Central limit theorems and asymptotic efficiency for one-sample nonparametric procedures. Tech. Rep. No. 11, University of Minnesota. Department of Statistics 66 pages.

[144] Govindarajulu, Z. (1962). Exact lower moments of order statistics in samples from the chi-distribution (1 d.f.). *Ann. Math. Statist.*, **33**, 1292–1307.

[145] Govindarajulu, Z. (1963). On moments of order statistics and quasi-ranges from normal populations. *Ann. Math. Statist.*, **34**, 633–651.

[146] Govindarajulu, Z. (1965). Asymptotic normality of linear functions of order statistics in one and multi-samples. Tech-Report No. 216 U.S. Air Force Office Grant No. AF-AFOSR-741-65 (Univ. of California, Berkeley) 1–26.

[147] Govindarajulu, Z. (1966). Characterization of normal and generalized truncated normal distributions using order statistics. *Ann. Math. Statist.*, **37**, 1011–1015.

[148] Govindarajulu, Z. (1966). Normal approximations to the classical discrete distributions. *Sankhyā* Ser. A **27**, 143–172.

[149] Govindarajulu, Z. (1968). Asymptotic normality of linear combinations of functions of order statistics II. *Proc. Nat. Acad. Sci.*, **59**, 713–719.

[150] Govindarajulu, Z. (1968). Certain general properties of unbiased estimates of location and scale parameters based on ordered observation. *SIAM J. Math.*, **16**, 533–551.

[151] Govindarajulu, Z. (1975). Characterization of the exponential distribution using lower moments of the order statistics. *Statistical Distributions in Scientific Work*, **3** Characterizations and Applications (G.P. Patil, S. Kotz and J.K. Ord, editors) Reidel, Ulricht and Boston, 117–129.

[152] Govindarajulu, Z. (1975). Locally most powerful rank order test for one-way random effects model. *Studia Scientiarum Mathematicarum Hungarica*, **10**, 47–60.

[153] Govindarajulu, Z. (1975). Robustness of Mann-Whitney-Wilcoxon test to dependence in the variables. *Studia Scientiarum Mathematicarum Hungarica*, **10**, 39–45.

[154] Govindarajulu, Z. (1976). Asymptotic normality and efficiency of a class of test statistics. *Essays in Probability and Statistics* (eds. S. Ikada *et al.*) Shinko Tsusho Col., Ltd., Tokyo, chapter 34, pp. 535–558.

[155] Govindarajulu, Z. (1977). A note on distribution-free tolerance limits. *Naval Research Logistics Quarterly*, **24**, 381–384.

[156] Govindarajulu, Z. (1980). A note on asymptotic efficiency of Chernoff-Savage class of tests. *Indian Journal of Mathematics.*, **22**, No. 1, 81–87.

[157] Govindarajulu, Z. (1980). Asymptotic normality of linear combinations of functions of order statistics in one and several samples. *Colloquia Mathematica Societatis Janos Bolyai.* No. 32. *Nonparametric Statistical Inference.* Budapest (Hungary), pp. 315–349.

[158] Govindarajulu, Z. (1985). Asymptotic normality and efficiency of c-sample nonparametric procedures under random allocation. *Journal of National Academy of Mathematics*, **3**, 151–181.

[159] Govindarajulu, Z. (1991). Robustness of Mann-Whitney-Wilcoxon test for scale to dependence in the variables. *Topics in Statistical Dependence.* (eds. H.N. Block, A.R. Sampson and T.H. Savits). IMS Lecture Notes. Monograph Series No. 16, 237–250.

[160] Govindarajulu, Z. (1995). A Class of asymptotically distribution-free test procedures for equality of marginals under multivariate dependence. *Amer. J. Math. and Mgmt. Sci.*, **15**, Nos. 3 and 4, 375–394.

[161] Govindarajulu, Z. (1997). A class of asymptotically distribution-free tests for equality of marginals in multivariate populations. *Mathematical Methods of Statistics*, **6**, 92–11.

[162] Govindarajulu, Z. and J.V. Deshpande (1972). Random effects model: nonparametric case. *Ann. Inst. Statist. Math.*, **24**, 165–170.

[163] Govindarajulu, Z. and A.P. Gore (1977). Locally most powerful and other tests for independence in multivariate populations. *Colloquia Mathematica Societitatis*, Janos Bolyai #21. *Analytic Function Methods in Probability Theory.* Debrecen (Hungary), 99–121.

[164] Govindarajulu, Z. and G.D. Gupta (1978). tests for homogeneity of scale against ordered alternatives. Transactions of 8th Prague Conference on Information Theory, Statistical Decision Functions, Random Processes (eds. J. Kozesnik *et al.*). Academia Publishing House, Prague, Vol. **A**, 235–245.

[165] Govindarajulu, Z. and H.S. Haller (1972). c-sample rank order statistics. *J. Indian Statist. Assoc.*, **10**, 17–35.

[166] Govindarajulu, Z. and H.S. Haller (1977). c-sample tests of homogeneity against ordered alternatives. *Proceedings of the Symposium to Honour Jerzy Neyman.* PWN - Polish Scientific Publishers, Warszawa, 91–102.

[167] Govindarajulu, A. and M. Joshi (1968). Best linear unbiased estimation of location and scale parameters of Weibull distribution using ordered observations. *Rep. Statist. Appl. Res.*, **15**, 57–70.

[168] Govindarajulu, Z. and S.H. Mansouri-Ghiassi (1986). On nonparametric tests for ordered alternatives in two-way lay-outs. *Adavances in Order Restricted Statistical Inference.* Springer-Verlag Lecture Notes, **37**, 153–168.

[169] Govindarajulu, Z. and I.D. Shetty (1988). Locally most powerful nonparametric tests for ordered scale alternatives. *Proceedings of the International Conference on Advances in Multivariate Analysis* (Das Gupta and Ghosh, eds.). Indian Statistical Institute, Calcutta, India, 445–464.

[170] Govindarajulu Z., L. LeCam and M. Raghavachari (1967). Generalizations of theorems of Chernoff and Savage on the asymptotic normality of test statistics. *Proc. Fifth Berkeley Symposium on Math. Statist. and Prob.*, Univ. of Calif. Press at Berkeley and Los Angeles, **1**, 609–638.

[171] Greenberg, V.L. (1964). *Robust Inference on Some Experimental Designs.* Doctoral dissertation submitted to the University of California at Berkeley.

[172] Gregory, G. (1977). Large sample theory for U-statistics and tests of fit. *Ann. Statist.*, **5**, No. 1, 110–123.

[173] Gumbel, E.J. (1958). *Statistics of Extremes.* John Wiley and Sons, New York.

[174] Gupta, A.K. (1952). Estimation of the mean and standard deviation of a normal population from a censored sample. *Biometrika*, **39**, 260–273.

[175] Gupta, G.D. and Z. Govindarajulu (1980). Nonparametric tests of randomness against auto-correlated normal alternatives. *Biometrika*, **67**, 375–379.

[176] Gupta, S.S. and B.K. Shah (1965). Exact moments and percentage points of the order statistics and the distribution of the range from the logistic distribution. *Ann. Math. Statist.*, **36**, 907–920.

[177] Guttman, I. (1957). On the power of optimum tolerance regions when sampling from normal distributions. *Ann. Math. Statist.*, **28**, 773–778.

[178] Guttman, I. (1959). Optimum tolerance regions and power when sampling from some non-normal universes. *Ann. Math. Statist.*, **30**, 926–938.

[179] Hájek, J. (1961). Some extensions of the Wald-Wolfowitz-Noether Theorem. *Ann. Math. Statist.*, **32**, 506–523.

[180] Hájek, J. (1962). Asymptotically most powerful rank order tests. *Ann. Math. Statist.*, **33**, 1124–1147.

[181] Hájek, J. (1964). Brownian Bridge. Lecture delivered at the IMS Summer Institute held at Michigan State University (unpublished).

[182] Hájek, J. (1968a). Asymptotic normality of sample linear rank statistics under alternatives. *Ann. Math. Statist.*, **30**, 325–346.

[183] Hájek, J. (1968b). Locally most powerful rank test of independence. *Studies in Math. Statist., Theory and Applications.* Akad. Kiadó, Budapest, 45–51.

[184] Hájek, J. and Z. Sidák (1967). *Theory of Rank Tests.* Academic Press, New York, p. 31.

[185] Hájek, J., Z. Sidák, and P.K. Sen (1999). *Theory of Rank Tests.* Second Edition. Academic Press, San Diego, CA.

[186] Haller, H.S. Jr. (1968). *Optimal c-sample Rank Order Procedures for Selection and Tests against Slippage and Ordered Alternatives.* Ph.D. Dissertation, Case Institute of Technology.

[187] Halmos, P.R. (1946). The theory of unbiased estimation. *Ann. Math. Statist.*, **18**, 34–43.

[188] Halperin, M. (1960). Extension of the Wilcoxon-Mann-Whitney test to samples censored at the same fixed point. *J. Amer. Statist. Assoc.* **55**, 125–138. Errata, ibid. p. 755.

[189] Hammersley, J.M. and K.W. Morton (1954). The estimation of location and scale parameters from grouped data. *Biometrika*, **41**, 296–301.

[190] Hanson, D.L. and D.B. Owen (1963). Distribution-free tolerance limits, elimination of the requirement that cumulative distribution functions be continuous. *Technometrics*, **5**, no. 4, 518–522.

[191] Harter, H.L. (1961). Expected values of normal order statistics *Biometrika*, **48**, 151–165. Correction, ibid. **48**, 476.

[192] Hartley, H. and H.A. David (1954). Universal bounds for mean range and extreme observations. *Ann. Math. Statist.*, **25**, 85–89.

[193] Hettmansperger, T.P. (1975). Non-parametric inference for ordered alternatives in a randomized block design. *Psychometrika*, **40**, 53–62.

[194] Hettmansperger, T.P. and J.W. McKean (1998). *Robust nonparametric Statistical Methods.* Edward Arnold, London; co-published by Wiley & Sons Inc., New York.

[195] Hill, B.M. (1960). A relation between Hodges bivariate sign test and a nonparametric test of Daniels. *Ann. Math. Statist.*, **31**, 1190–1196.

[196] Hoadley, B.A. (1967). On the probability of large deviations of functions of several empirical cdf's. *Ann. Math. Statist.*, **38**, 360–381.

[197] Hodges, J.L. and E.L. Lehmann (1956). The efficiency of some parametric competitors of the t-test. *Ann. Math. Statist.*, **27**, 324–335.

[198] Hodges, H.L. Jr. and E.L. Lehmann (1963). Estimates of location based on rank tests. *Ann. Math. Statist.*, **34**, 598–611.

[199] Hodges, J.L. Jr. and E.L. Lehmann (1970). Deficiency. *Ann. Math. Statist.*, **41**, 783–801.

[200] Hoeffding, W. (1948a). A class of statistics with asymptotically normal distribution. *Ann. Math. Statist.*, **19**, 293–325.

[201] Hoeffding, W. (1948b). A nonparametric test of independence. *Ann. Math. Statist.*, **19**, 546–557.

[202] Hoeffding, W. (1951). Optimum nonparametric tests. *Proc. Second Berkeley Symp. Math. Statist. and Prob.* University of California Press, Berkeley, 83–92.

[203] Hoeffding, W. (1952). The large-sample power of tests based on permutations of observations. *Ann. Math. Statist.*, **23**, 169–192.

[204] Hoeffding, W. (1953). On the distribution of the expected values of order statistics. *Ann. Math. Statist.*, **24**, 93–100.

[205] Hoeffding, W. and H. Robbins (1948). The central limit theorem for dependent random variables. *Duke Math. J.*, **15**, 773–780.

[206] Hollander, M. (1967). Rank tests for randomized blocks when the alternatives have an *a prior* ordering. *Ann. Math. Statist.*, **38**, 867–877.

[207] Hollander, M. and D.A. Wolfe (1973). *Nonparametric Statistical Methods.* New York: John Wiley & Sons.

[208] Hotelling, H. and M.R. Pabst (1936). Rank correlation and tests of significance involving no assumption of normality. *Ann. Math. Statist.*, **7**, 29–43.

[209] Huskova, M. (1970). Asymptotic distribution of simple linear rank statistics for testing symmetry. *Z. Wahrseheinticheitstheorie verw. geb.*, **14**, 308–322.

[210] Jaeckel, L.A. (1972). Estimating regression coefficients by minimizing the dispersion of the residuals. *Ann. Math. Statist.*, **43**, 1449–1458.

[211] Jirina, M. (1952). Sequential estimation of distribution-free tolerance limits. *Cekoslovackii Matematiiceskii Zurnal*, **2**(77), 221–232. [English translation in *Inst. of Math. Statist.* and *Amer. Math Soc.* (1961)].

[212] Johnson, N.L. (1949). Systems of frequency curves generated by methods of translation. *Biometrika*, **36**, 149–176.

[213] Jonckheere, A.R. (1954a). A distribution-free k-sample test against ordered alternatives. *Biometrika*, **41**, 135–145.

[214] Jonckheere, A.R. (1954b). A test of significance for the relation between m rankings and k ranked categories. *Brit. J. Statist. Psych.*, **1**, 93–100.

[215] Jones, H.L. (1953). Approximating the mode from weighted sample values. *J. Amer. Statist. Assoc.*, **48**, 113–127.

[216] Jung, J. (1955). On linear estimates defined by a continuous weight function. *Arkiv For Matematik*, **3**(15), 199–209.

[217] Jurecková, J. (1971). Nonparametric estimate of regression coefficients. *Ann. Math. Statist.*, **42**, 1928–1338.

[218] Kac, M., J. Kiefer, and J. Wolfowitz (1955). On tests of normality and other tests of goodness of fit based on distance methods. *Ann. Math. Statist.*, **26**, 189–211.

[219] Kemperman, J.H.B. (1956). Generalized tolerance limits. *Ann. Math. Statist.*, **27**, 180–186.

[220] Kendall, M.G. (1940). Note on the distribution of quantiles for large samples. *J. Roy. Statist. Soc.* (B), **7**, 83–85.

[221] Kendall, M.G. (1954). Two problems in sets of measurements. *Biometrika*, **41**, 560–562.

[222] Kendall, M.G. (1960). *The Advanced Theory of Statistics*, Vol. **2**, Third Edition, Hafher Publishing Co., New York.

[223] Kendall, M.G. (1962). *Rank Correlation Methods*, Third Edition. Charles Griffin, London.

[224] Kendall, M.G. and J.D. Gibbons (1990). *Rank Correlation Methods*, 5th Edition. Edward Arnold, London.

[225] Kiefer, J. (1959). k-sample analogues of the Kolmogorov, Smirnov and Cramér von-Mises tests. *Ann. Math. Statist.*, **30**, 420–447.

[226] Kiefer, J. and J. Wolfowitz (1958). On the deviations of the empiric distribution function of vector chance variables. *Trans. Amer. Math. Soc.*, **87**, 173–186.

[227] Klotz, J. (1962). Nonparametric tests for scale. *Ann. Math. Statist.*, **33**, 498–517.

[228] Knoke, J.D. (1975). Testing for randomness against autocorrelated alternatives: the parametric case. *Biometrika*, **62**, 571–575.

[229] Kolmogorov, A.N. (1933). Sulla determinasione empirica di una legge di distribuzione. *Giornale dell' Istituto Italiano degli Attuari*, **4**, 83–91.

[230] Kolmogorov, A.N. (1941). Confidence limits for an unknown distribution function. *Ann. Math. Statist.*, **12**, 461–463.

[231] Konijn, H.S. (1956). On the power of certain tests for independence in bivariate populations. *Ann. Math. Statist.*, **27**, 300–323. Correction: *Ann. Math. Statist.*, **29**, (1958), 935.

[232] Koul, H.L. (1969). Asymptotic behavior of Wilcoxon type confidence regions on multiple regression. *Ann. Math. Statist.*, **40**, 1950–1979.

[233] Krishna Iyer, P.V. (1958). A theorem on factorial moments and its applications. *Ann. Math. Statist.*, **29**, 254–261.

[234] Kruskal, W.H. (1952). A nonparametric test for the several sample problem. *Ann. Math. Statist.*, **23**, 525–540.

[235] Kruskal, W.H. and W.A. Wallis (1952). Use of ranks in one-criterion variance analysis. *J. Amer. Statist. Assoc.*, **47**, 583–621. Errata, ibid, **48** (1953), 910.

[236] Kudo, A. (1963). A multivariate analogue of the one-sided test, *Biometrika*, **50**, 403–418.

[237] Kulkarni, S.R. (1969). On the optimal asymptotic test for the effects of cloud seeding on rainfall (II): The case of variable effects. *The Aust. J. Statist. Math.*, **11**, No. 1, 39–51.

[238] Kulkarni, S.R. (1970). Locally asymptotically most powerful test about the effects of K treatments. *Ann. Inst. Statist. Math.*, **22**, 145–158.

[239] Kunisawa, K., H. Makabe, and H. Morimura (1951). Tables of confidence bands for the population distribution function. *Reports of Statist. App. Res.*, JUSE, **4**, 18–20.

[240] Laubsher, N.F., F.E. Steffens, and E.M. De Lange (1968). Exact critical values for Mood's distribution-free test statistic for dispersion and its normal approximation. *Technometrics*, **10**, 497–508.

[241] Lehmann, E.L. (1950). *Notes on Theory of Estimation.* Mimeographed Lectures. Univ. of California.

[242] Lehmann, E.L. (1951a). A general concept of unbiasedness. *Ann. Math. Statist.*, **22**, 587–591.

[243] Lehmann, E.L. (1951b). Consistency and unbiasedness of certain nonparametric tests. *Ann. Math. Statist.*, **22**, 165–179.

[244] Lehmann, E.L. (1953). The power of rank tests. *Ann. Math. Statist.*, **24**, 23–43.

[245] Lehmann, E.L. (1959). *Testing Statistical Hypotheses.* New York: Wiley & Sons, pp. 1, 191, 217.

[246] Lehmann, E.L. (1963a). Robust estimation in analysis of variance. *Ann. Math. Statist.*, **34**, 957–966.

[247] Lehmann, E.L. (1963b). Asymptotically nonparametric inference: An alternative approach to linear models. *Ann. Math. Statist.*, **34**, 1494–1506.

[248] Lehmann, E.L. (1964). Asymptotically nonparametric inference in some linear models with one observation per cell. *Ann. Math. Statist.*, **35**, 726–734.

[249] Lehmann, E.L. (1966). Some concepts of dependence. *Ann. Math. Statist.*, **37**, 1137–1153.

[250] Lehmann, E.L. and H. Scheffé (1950 and 1955). Completeness, similar regions, and unbiased estimation:

Part I: *Sankhyā*, **10**, 305–340.
Part II: *Sankhyā*, **15**, 219–236.

[251] Lehmann, E.L. and C. Stein (1949). On the theory of some nonparametric hypotheses. *Ann. Math. Statist.*, **20**, 28–45.

[252] Levene, H. (1952). On the power function of tests of randomness based on runs up and down. *Ann. Math. Statist.*, **23**, 34–56.

[253] Levene, H. and J. Wolfowitz (1944). The covariance matrix of runs up and down. *Ann. Math. Statist.*, **15**, 58–69.

[254] Lieblein, J. (1955). On moments of order statistics from the Weibull distribution. *Ann. Math. Statist.*, **26**, 330–333.

[255] Lillifors, H.W. (1967). On the Kolmogorov-Smirnov test for normality with mean and variance unknown. *J. Amer. Statist. Assoc.*, **62**, 399–402.

[256] Lindgren, B.W. (1976). *Statistical Theory.* New York: McMillan Co.

[257] Lloyd, E.N. (1952). Least squares estimation of location and scale parameters using order statistics. *Biometrika*, **39**, 88–95.

[258] Loève, M. (1960). *Probability Theory*, 2nd Edition. van Nostrand Co., New York.

[259] Ludwig, O. (1960). Uber erwartungswerte und varianzen von ranggrozen in kleinen stichproblem. *Metrika*, **3**, 218–233.

[260] Madow, W.G. (1948). On the limiting distributions of estimates based on samples from finite universe. *Ann. Math. Statist.*, **19**, 535–545.

[261] Malmquist, S. (1954). On certain confidence contours for distribution functions. *Ann. Math. Statist.*, **25**, 532–533.

[262] Maniya, G.M. (1949). Generalization of the criterion of A.N. Kolmogorov. *Doklady Akad. Nauk USSR(NS)*, **69**, 495–497.

[263] Mann, H.B. (1945). On a test for randomness based on signs of differences. *Ann. Math. Statist.*, **16**, 193–199.

[264] Mann, H.B. and D.R. Whitney (1947). On a test of whether one of two random variables is stochastically larger than the other. *Ann. Math. Statist.*, **18**, 50–60.

[265] Mansouri-Ghiassi, S.H. and Z. Govindarajulu (1986). An asymptotically distribution-free test for ordered alternatives in two-way lay-outs. *J. Statist. Planning and Inference*, **13**, 329–249.

[266] Marascuilo, L.A. (1966). Large-sample multiple comparisons. *Psychol. Bull.*, **64**, 280–290.

[267] Marshall, A.W. (1958). The small sample distribution of $N\omega_N^2$. *Ann. Math. Statist.*, **29**, 307–309.

[268] Mason, D.M. (1977). *Almost Sure Linearity of Rank Statistics with a Rate and The Asymptotic Normality of Linear combination of order Statistics.* PhD Dissertation, Univ. of Washington, Seattle.

[269] Mason, D.M. (1979). Asymptotic normality of linear combinations of order statistics with a smooth score function. Department of Statistics, Univ. of Kentucky, Tech. Rep. No. 149, 1–18.

[270] Massey, F.J. Jr. (1950). A note on the estimation of a distribution function by confidence limits. *Ann. Math. Statist.*, **21**, 116–119.

[271] Massey, F.J. Jr. (1951). The Kolmogorov-Smirnov test for goodness of fit. *J. Amer. Statist. Assoc.*, **46**, 68–78.

[272] McKay, A.T. (1935). The distribution of the difference between the extreme observation and the sample mean of n from a normal universe. *Biometrika*, **27**, 466–471.

[273] Miller, L.H. (1956). Tables of percentage points of Kolmogorov statistics. *J. Amer. Statist. Assoc.*, **51**, 111–121.

[274] Mitra, S.K. (1957). Tables for tolerance limits for a normal population based on sample mean and range or mean range. *J. Amer. Statist. Assoc.*, **52**, 88–94.

[275] Mood, A.M. (1940). The distribution theory of runs. *Ann. Math. Statist.*, **11**, 367–392.

[276] Mood, A.M. (1950). *Introduction to Theory of Statistics.* McGraw Hill, New York.

[277] Mood, A.M. (1954). On the asymptotic efficiency of certain nonparametric two-sample tests. *Ann. Math. Statist.*, **25**, 514–522.

[278] Moore, D.S. (1968). An elementary proof of asymptotic normality of linear functions of order statistics. *Ann. Math Statist.*, **39**, 263–265.

[279] Morigutti, S. (1954). Bounds for second moments of the sample range. *JUSE* (Japanese Union of Scientists and Engineers), Vol. **3**, No. 3, 1–8.

[280] Mosteller, F. (1941). Note on an application of runs to quality control charts. *Ann. Math. Statist.*, **12**, 228–232.

[281] Mosteller, F. (1946). On some useful inefficient statistics. *Ann. Math. Statist.*, **17**, 377–407.

[282] Murphy, R.B. (1948). Non-parametric tolerance limits. *Ann. Math. Statist.*, **19**, 581–584.

[283] Nair, K.R. (1940). Table of confidence interval for the median in samples from any continuous population. *Sankhyā*, **4**, 551–558.

[284] Neyman, J. (1949). Contributions to the theory of the χ^2 test. *Proc. Berkeley Symp. Math. Statist. and Prob.* University of California Press, Berkeley, 239–273.

[285] Neyman, J. (1959). Optimal asymptotic tests of composite statistical hypotheses. The Harold Cramér Volume. Almqvist and Wiksell, Stockholm, and Wiley & Sons, New York, pp. 213–234.

[286] Neyman, J. (1967). Experimentation with weather control. *J. Royal Statist. Soc. Series A*, **130**, 285–326.

[287] Neyman, J. and E.S. Pearson (1933). On the problem of the most efficient tests of statistical hypotheses. *Phil. Trans. Roy. Soc.*, A**231**, 289–337.

[288] Neyman, J. and E.L. Scott (1965). On the use of $c(\alpha)$ optimal tests of composite hypotheses. *Proceedings of the 35th Session of the International Statistical Institute*, Beograd, Yugoslavia.

[289] Niles, R.E. (1959). The complete amalgamation into blocks, by weighted means, of a finite set of real numbers. *Biometrika*, **46**, 317–327.

[290] Noether, G.E. (1948). On confidence limits for quantiles. *Ann. Math. Statist.*, **19**, 416–419.

[291] Noether, G. (1949). On a Theorem by Wald and Wolfowitz. *Ann. Math. Statist.*, **20**, 455–458.

[292] Noether, G.E. (1951). On a connection between confidence and tolerance intervals. *Ann. Math. Statist.*, **22**, 603–604.

[293] Noether, G. (1955). On a theorem of Pitman. *Ann. Math. Statist.*, **26**, 64–68.

[294] Nugroho, S. and Z. Govindarajulu (1999). Tests for random effects in two-way experiments with one observation per cell. *Indian Journal of Mathematics* (B.N. Prasad Birth Centenary Commemoration Volume), **41**, 55–66.

[295] Nugroho, S. and Z. Govindarajulu (2002). Nonparametric tests for random effects in the balanced incomplete block design. *Statistics & Probability Letters*, **56**, 431–437.

[296] Ogawa, J. (1951). Contributions to the theory of systematic statistics I. *Osaka Math. J.*, No. 2, 175–213.

[297] Ogawa, J. (1952). Contributions to the theory of systematic statistics II, Large sample theoretical treatments of some problems arising from dosage and time-mortality curves. *Osaka Math. J.*, **4**, 41–61.

[298] Ogawa, J. (1957). A further contribution to the theory of systematic statistics. Inst. of Statist., University of North Carolina, Mimeograph Ser. 168.

[299] Olmstead, P.S. (1946). Distribution of sample arrangements for runs up and down. *Ann. Math. Statist.*, **17**, 24–33.

[300] Olmstead, P.S. and J.W. Tukey (1947). A corner test for association. *Ann. Math. Statist.*, **18**, 495–513.

[301] Orey, S. (1958). A central limit theorem for m-dependent random variables. *Duke Math. J.*, **25**, 543–546.

[302] Page, E.B. (1963). Ordered hypotheses for multiple treatments: a significance test for linear ranks. *J. Amer. Statist. Assoc.*, **58**, 216–230.

[303] Parzen, E. (1961). Mathematical considerations in the estimation of spectra. *Technometrics*, **3**, 167–190.

[304] Parzen, E. (1962). On estimation of a probability density function and mode. *Ann. Math. Statist.*, **33**, 1065–1076.

[305] Patnaik, P.B. (1949). The Noncentral chi-square and F distributions and their applications. *Biometrika*, **36**, 202–232.

[306] Paulson, E. (1943). A note on tolerance limits. *Ann. Math. Statist.*, **14**, 90–91.

[307] Pearson, K. (1900). On the criterion that a given system of deviations from the probable in the case of a correlated system of variables is such that it can be reasonably supposed to have arisen from random sampling. *Phil. Mag.* 5 Ser. **50**, 157–175.

[308] Pearson, K. (1911). On the probability that two independent distributions of frequency are really samples from the same population. *Biometrika*, **8**, 250–254.

[309] Pearson, K. (1934). *Tables of The Incomplete Beta-function.* Cambridge University Press.

[310] Pirie, W.R. and M. Hollander (1972). A distribution-free normal scores test for ordered alternatives in the randomized block design. *J. Amer. Statist. Assoc.*, **67**, 855–857.

[311] Pitman, E.J.G. (1937a). Significance tests which may be applied to samples from any populations. *Suppl. J. Roy. Statist. Soc.*, **4**, 119–130.

[312] Pitman, E.J.G. (1937b). Significance tests which may be applied to samples from any population II. The correlation coefficient tests. *Suppl. J. Roy. Statist. Soc.*, **4**, 225–232.

[313] Pitman, E.J.G. (1938). Significance tests which may be applied to samples from any population III. The analysis of variance test. *Biometrika*, **29**, 322–335.

[314] Pitman, J. (2003). Poisson-Kingman Partitions. *A Festschrift for Terry Speed.* (Ed. D.R. Goldstein). Inst. Math. Statist. Lecture Notes–Monograph Series, **30**, pages 1–30, Hayward, CA.

[315] Plackett, R.L. (1949). A historical note on the method of least squares. *Biometrika*, **36**, 458–460.

[316] Plackett, R.L. (1958). Linear estimation from censored data. *Ann. Math. Statist.*, **29**, 131–143.

[317] Podgor, M.J. and J.L. Gastwrith (1996). Efficiency robust rank tests for stratified data. *Research Developments in Prob. and Statist.* (Eds. E. Brunnen and M. Denker) VSP International Science Publishers, The Netherlands.

[318] Puri, M.L. (1964). Asymptotic efficiency of a class of c-sample tests. *Ann. Math. Statist.*, **35**, 102–121.

[319] Puri, M.L. (1965). On some tests of homogeneity of variances. *Ann. Inst. Statist. Math.*, **17**, 323–330.

[320] Puri, M.L. (1965). Some distribution-free k-sample rank tests of homogeneity against ordered alternatives. *Comm. Pure Appl. Math.*, **18**, 51–63.

[321] Puri, M.L. and P.K. Sen (1968). On Chernoff-Savage tests for ordered alternatives in randomized blocks. *Ann. Math. Statist.*, **39**, 967–972.

[322] Puri, M.L. and P.K. Sen (1969). On the asymptotic normality of one-sample Chernoff-Savage theorem for random sample sizes. *Theory of Prob. and its Application*, **14**, 163–167.

[323] Puri, M.L. and L.T. Tran (1981). Invariance principles for rank statistics for testing independence. *Contributions to Probability* (Eds. J. Gani and V. K. Rohatgi). Academic Press, New York, 267–282.

[324] Puri, M.L., P.K. Sen, and D.V. Gokhale (1970). On a class of rank order tests for independence in multi-variate distributions. *Sankhy a*, Ser. A, **32**, 271–298.

[325] Pyke, R. (1965). Spacings. *JRSS*, Ser. B, **27**, 395–447.

[326] Pyke, R. and G. Shorack (1968a). Weak convergence of a two-sample empirical process and a new approach to Chernoff-Savage Theorem. *Ann. Math. Statist.*, **39**, 755–771.

[327] Pyke, R. and G. Shorack (1968b). Weak convergence and a Chernoff-Savage Theorem for random samples. *Ann. Math. Statist.*, **39**, 1675–1685.

[328] Raghavachari, M. (1965). The two-sample scale problem when the locations are unknown. *Ann. Math. Statist.*, **36**, 1236–1242.

[329] Rao, C.R. (1948). Large sample tests of statistical hypotheses concerning several parameters with applications to problems of estimation. *Proc. Camb. Philos. Soc.*, **44**, 50–57.

[330] Rao, C.R. (1965). *Linear Statistical Inference and Its Applications.* Wiley & Sons, New York.

[331] Rao, K.S.M. (1982). Non-parametric tests for homogeneity of scale against ordered alternatives. *Annals of the Institute of Statistical Mathematics*, **34**, 327–334.

[332] Rao, U.V.R., I.R. Savage, and M. Sobel (1960). Contributions to the theory of rank order statistics: the two-sample censored case. *Ann. Math. Statist.*, **31**, 415–426.

[333] Renyi, A. (1953). On the theory of order statistics. *Acta Mathematica, Journ. Hungarian Academy of Sciences*, **4**, 191–231.

[334] Riedwyl, H. (1967). Goodness of fit. *J. Amer. Statist. Assoc.*, **62**, 390–398.

[335] Robbins, H. (1944). On distribution-free tolerance limits in random samples. *Ann. Math. Statist.*, **15**, 214–216.

[336] Rosenblatt, M. (1956). Remarks on some non-parametric estimates of a density function. *Ann. Math. Statist.*, **27**, 832–837.

[337] Rustagi, J.S. (1957). On minimizing and maximizing a certain integral with statistical applications. *Ann. Math. Statist.*, **28**, 309–328.

[338] Rustagi, J.S. (1959). On the bounds for the variance of Mann-Whitney statistic (Abstract). *Ann. Math. Statist.*, **30**, 257.

[339] Ruymgaart, F.H. and van M.C.A. Zuijlen (1977). Asymptotic normality of linear combinations of order statistics in the non i.i.d. case. Proc. Koninklijke Nederlandse Akademie van Wettensehappen Ser. A., Mathematical Sciences **80** (5), 432–447.

[340] Ruymgaart, F.H. and van M.C.A. Zuijlen (1978). Asymptotic normality of multivariate linear rank statistics in the non i.i.d. case. *Ann. Statist.*, **6**, 588–602.

[341] Sarhan, A.E. (1954, 1955, 1955). Estimation of the mean and standard deviation by order statistics.

Part I:	*Ann. Math. Statist.*, **25**, 317–328
Part II:	*Ann. Math. Statist.*, **26**, 505–511
Part III:	*Ann. Math. Statist.*, **26**, 576–93

[342] Sarhan, A.E. and E.G. Greenberg (1956, 1958a 1958b) Estimation of location and scale parameters by order statistics from singly and doubly censored samples.

Part I:	The normal distribution up to samples of size 10. *Ann. Math. Statist.*, **27**, 427–51.
Part II:	*Ann. Math. Statist.*, **29**, 79–105.
Part III:	Tech. Rep. 4, OOR Project 1597.

[343] Sarhan, A.E. and B.G. Greenberg (1957). Tables for best linear estimates by order statistics of the parameters of single exponential distributions from singly and doubly censored samples. *J. Amer. Statist. Assn.*, **52**, 58–87.

[344] Sarhan, A.E. and B.G. Greenberg (1962). *Contributions to Order Statistics* (editors). John Wiley and Sons, New York.

[345] Sarkadi, K. (1974). On the convergence of the expectation of the sample quantile. *Colloquia Mathematica Societatis Janos Bolya.*, **11**, 341–345.

[346] Sarkadi, K. (1975). The consistency of the Shaprio-Francia test. *Biometrika*, **62**, No. 2, 445–450.

[347] Sarkadi, K. (1981). The asymptotic distribution of certain goodness of fit test statistics. *The First Pannon Symposium on Math. Statist.* (ed. W. Wertz). Lecture Notes in Statistics. Vol. 8, Springer-Verlag, New York, 245–253.

[348] Saunders, S.C. (1959). Confidence bounds for an integral function of an estimate with applications to reliability theory (Abstract). *Ann. Math. Statist.*, **30**, 1278.

[349] Saunders, S.C. (1960). Sequential tolerance regions. *Ann. Math. Statist.*, **31**, 198–216.

[350] Savage, I.R. (1956). Contributions to the theory of rank order statistics – the two-sample case. *Ann. Math. Statist.*, **27**, 590–615.

[351] Savage, I.R. (1957a). On the hypothesis of tests of randomness and other hypotheses. *J. Amer. Statist. Assoc.*, **52**, 53–57.

[352] Savage, I.R. (1957b). Contributions to the theory of rank order statistics: the "trend" case. *Ann. Math. Statist.*, **28**, 968–977.

[353] Savage, I.R. (1959). Contributions to the theory of rank order statistics – the one-sample case. *Ann. Math. Statist.*, **30**, 1018–1023.

[354] Savage, I.R. (1962). *Bibliography of Nonparametric Statistics.* Harvard University Press, Cambridge, Massachusetts.

[355] Savage, I.R. (1963). Personal Communication.

[356] Savur, S.R. (1937). The use of the median in tests of significance. *Proc. Ind. Acad. Sci.* (Section A), **5**, 564–576.

[357] Scheffé, H. (1943). On a measure problem arising in the theory of non-parametric tests. *Ann. Math. Statist.*, **14**, 227–233.

[358] Scheffé, H. (1943a). On a measure arising in the theory of nonparametric tests. *Ann. Math. Statist.*, **14**, 227–233.

[359] Scheffé, H. (1959). *The Analysis of Variance.* Wiley & Sons, New York.

[360] Scheffé, H. and J.W. Tukey (1944). A formula for sample sizes for population tolerance limits. *Ann. Math. Statist.*, **15**, p. 217.

[361] Scheffé, H. and J.W. Tukey (1945). Non-parametric estimation I. Validation of order statistics. *Ann. Math. Statist.*, **16**, 187–192.

[362] Sen, P.K. (1959). On the moments of the sample quantiles. *Calcutta Statist. Assoc. Bulletin*, Vol. **9**, Nos. 33 and 34, 1–19.

[363] Sen, P.K. (1961). A note on the large sample behavior of extreme sample values from distribution with finite end-points. *Calcutta Statist. Assoc. Bulletin*, Vol. **10**, No. 39, 106–115.

[364] Sen, P.K. (1963). On weighted rank-sum tests for dependence. *Ann. Inst. Statist. Math.*, **15**, 117–135.

[365] Sen, P.K. (1966). On nonparametric simultaneous confidence regions and tests for the one criterion analysis of variance problem. *Ann. Inst. Statist. Math.*, **18**, 319–336.

[366] Sen, P.K. (1968). Estimates of the regression coefficient based on Kendall's tau. *J. Amer. Statist. Assoc.*, **63**, 1379–1389.

[367] Sen, P.K. and Z. Govindarajulu (1966). On a class of c-sample weighted rank-sum tests for location and scale. *Ann. Inst. Stat. Math.*, **18**, 87–105.

[368] Serfling, R.J. (1957). *Approximation Theorems of Mathematical Statistics.* Wiley, New York.

[369] Shapiro, S.S. and R.S. Francia (1972). Approximate analysis of variance test for normality. *J. Amer. Statist. Assoc.*, **67**, 215–216.

[370] Shapiro, S.S. and M.B. Wilk (1965). An analysis of variance test for normality. *Biometrika*, **52**, 591–612.

[371] Shapiro, S.S. and M.B. Wilk (1972). An Analysis of variance test for the exponential distribution (complete samples). *Technometrics*, **14**, 355–370.

[372] Shapiro, S.S., M.B. Wilk and H.J. Chen (1968). A comparative study of various tests for normality. *J. Amer. Statist. Assoc.*, **63**, 1343–1372.

[373] Sherman, B. (1950). A random variable related to the spacing of sample values. *Ann. Math. Statist.*, **21**, 339–361.

[374] Sherman, B. (1957). Percentiles of the ω_n statistic. *Ann. Math. Statist.*, **28**, 259–261.

[375] Shetty, Ishwar D. and Govindarajulu, Z. (1988). An LMP rank test for random effects: its asymptotic distribution under local alternatives and asymptotic relative efficiency. *Statistical theory and data analysis, II (Tokyo, 1986)*, 171–190.

[376] Shewhat, W.A. (1931). *Economic Control of Quality of Manufactured Product.* D. van Nostrand Company, New York.

[377] Shirahata, S. (1973). Locally most powerful rank tests for independence. *Bull. Math. Statist.*, **16**, 11–21.

[378] Shorack, G.R. (1967). Testing against ordered alternatives in Model I analysis of variance: normal theory and nonparametric. *Ann. Math. Statist.*, **38**, 1740–1752.

[379] Shorack, G.R. (1972). Functions of order statistics. *Ann. Math. Statist.*, **43**, 412–427.

[380] Siegel, S. and J.W. Tukey (1960). A nonparametric sum of ranks procedure for relative spread in unpaired samples. *J. Amer. Statist. Assoc.*, **55**, 529–445. Correction, ibid, **56** (1961), 1005.

[381] Slutsky, E. (1925). Über stochastiche asymptoter und grenzewerti. *Math. Annalen*, **5**, 93.

[382] Smirnov, N.V. (1935). Uber die verleilung des allagemeinen gliedes in der variationsreihe. *Metron*, **12**, 59–81.

[383] Smirnov, N.V. (1939a). On the deviation of the empirical distribution function. *Rec. Math. (Mat. Shornik) (N.S.)*, **6**, 3–26.

[384] Smirnov, N.V. (1939b). Sur les écarts de la courbe de distribution empirique. *Recueil Math. (Mat. Sbornik)*, M.S. **6**, 3–26.

[385] Smirnov, N. (1948). Tables for estimating the goodness of fit of empirical distributions. *Ann. Math. Statist.*, **19**, 279–281.

[386] Sobel, M. (1957). On a generalized Wilcoxon statistic for life testing. *Proc. of Working Conference on the Theory of Probability.* New York University, pp. 8–13.

[387] Somerville, P.N. (1958). Tables for obtaining non-parametric tolerance limits. *Ann. Math. Statist.*, **29**, 599–601.

[388] Spjøvoll, E. (1968). A note on robust estimation in analysis of variance. *Ann. Math. Statist.*, **39**, 1486–1492.

[389] Stevens, W.L. (1939). Distribution of groups in a sequence of alternatives. *Ann. Eugenics*, **9**, 10–17.

[390] Stiegler, S.M. (1969). Linear functions of order statistics. *Ann. Math. Statist.*, **40**, 770–788.

[391] Stiegler, S.M. (1974). Linear functions of order statistics with smooth weight functions. *Ann. Statist.*, **2**, 676–693.

[392] Stone, C.J. (1975). Nearest weighted estimates of a nonlinear regression function. *Proc. Comput. Sci. Statist. 8th Ann. Symp. Interface*, 413–418. Health Sciences Computer Facility, UCLA.

[393] Stone, C.J. (1977). Consistent nonparametric regression (with discussion). *Ann. Statist.*, **5**, 595–645.

[394] Stone, C.J. (1982). Optimal global rates of convergence for nonparametric regression. *Ann. Statist.*, **10**, 1040–1053.

[395] Stone, C.J., M.H. Hansen, C. Kooperberg, and Y.K. Troung (1977). Polynomial splines and their tensor products in extended linear modelling. *Ann. Statist.*, **25**, 1371–1470.

[396] Stuart, A. (1954). Asymptotic relative efficiencies of distribution-free tests of randomness against normal alternatives. *J. Amer. Statist. Assoc.*, **49**, 147–157.

[397] Stuart, A. (1956). The efficiencies of tests of randomness against normal regression. *J. Amer. Statist. Assoc.*, **51**, 285–287.

[398] Sugiura, N. (1962). On the orthogonal inverse expansion with an application to the moments of order statistics. *Osaka Math J.*, **14**, 253–263.

[399] Sugiura, N. (1964). The bivariate orthogonal inverse expansion and the moments of order statistics. *Osaka Math J.*, **1**, 45–59.

[400] Sukhatme, B.V. (1957). On certain two-sample nonparametric tests for variances. *Ann. Math. Statist.*, **28**, 188–194.

[401] Sukhatme, B.V. (1958). Testing the hypothesis that two populations differ only in location. *Ann. Math. Statist.*, **29**, 60–67.

[402] Sukhatme, S. (1972). Fredholm determinant of a positive definite kernel of a special type and its application. *Ann. Math. Statist.*, **43**, 1914–1925.

[403] Tate, R.F. (1959). Unbiased estimation: Functions of location and scale parameters. *Ann. Math. Statist.*, **30**, 341–66.

[404] Terpestra, T.J. (1952). A nonparametric k sample test and its connection with the H-test. Report 5(92) (VP2) of the Statistical Department of the Mathematical Center, Amsterdam.

[405] Terpestra, T.J. (1954). A nonparametric test for the problem of k-samples. *Proceedings Koninklijke Nederlandse Academie van Wetenschaffen* (A), **57** (*Indagationes Mathematicae*, **16**), 505–512.

[406] Terry, M.E. (1952). Some rank order tests which are most powerful against specific parameter alternatives. *Ann. Math. Statist.*, **23**, 346–366.

[407] Theil, H. (1950a, b, c). A rank-invariant method of linear and polynomial regression analysis.

(a) *Proc. Kon. Ned. Akad. v. Wetensch*, A**53**, 386–392,
(b) *ibid*, 521–525,
(c) *ibid*, 1397–1412.

[408] Thompson, W.R. (1936). On confidence ranges for the median and other expectation distributions for population of unknown distribution form. *Ann. Math. Statist.*, **7**, 122–128.

[409] Thompson, W.R. (1938). Biological applications of normal range and associated significance tests in ignorance of original distribution forms. *Ann. Math. Statist.*, **9**, 281–287.

[410] Tiku, M.L. (1965). Chi-square approximations for the distribution of goodness of fit statistics U_n^2 and W_n^2. *Biometrika*, **32**, 630–633.

[411] Tryon, P.V. and T.P. Hettmansperger (1973). A class of nonparametric tests for homogeneity against ordered alternatives. *Ann. of Statist.*, **1**, 1061–1070.

[412] Tukey, J.W. (1947). Non-parametric estimation II. Statistically equivalent blocks and tolerance regions – the continuous case. *Ann. Math. Statist.*, **18**, 529–539.

[413] Tukey, J.W. (1948). Non-parametric estimation III. Statistically equivalent blocks and tolerance regions – the discontinuous case. *Ann. Math. Statist.*, **19**, 30–39.

[414] Tukey, J.W. (1962). The future of data analysis. *Ann. Math. Statist.*, **33**, 1–67.

[415] Uspensky, J.V. (1937). *Introduction to Mathematical Probability.* New York: McGraw-Hill Book Company.

[416] van der Vaart, H.R. (1961). A simple derivation of the limiting distribution function of a sample quantile with increasing sample size. **Statistica Neerlandica**, 15 No. 3, 239–242.

[417] Vorlickova, D. (1972). Asymptotic properties of rank tests of symmetry under discrete distributions. *Ann. Math. Statist.*, **43**, 2013–2018.

[418] Wahba, G. and S. Wold, (1975). A completely automatic French curve: Fitting spline functions by cross validation. *Comm. Statist.*, **6**, 1–17.

[419] Wald, A. (1942). Setting of tolerance limits when the sample is large. *Ann. Math. Statist.*, **13**, 389–399.

[420] Wald, A. (1943). An extension of Wilks' method of setting tolerance limits. *Ann. Math. Statist.*, **14**, 45–55.

[421] Wald, A. and J. Wolfowitz (1939). Confidence limits for continuous distribution functions. *Ann. Math. Statist.*, **10**, 105–118.

[422] Wald, A. and J. Wolfowitz (1940). On a test whether two samples are from the same population. *Ann. Math. Statist.*, **11**, 147–162.

[423] Wald, A. and J. Wolfowitz (1943). An exact test for randomness in the nonparametric case based on serial correlation. *Ann. Math. Statist.*, **14**, 378–388.

[424] Wald, A. and J. Wolfowitz (1946). Tolerance limits for a normal distribution. *Ann. Math. Statist.*, **17**, 208–215.

[425] Walsh, J.E. (1958). Nonparametric estimation of sample percentage point standard deviation. *Ann. Math. Statist.*, **29**, 601–604.

[426] Walsh, J.E. (1959). Exact nonparametric tests for randomized blocks. *Ann. Math. Statist.*, **30**, 1034–1040.

[427] Walsh, J.E. (1962). Some two-sided distribution-free tolerance intervals of a general nature. *J. Amer. Statist. Assoc.*, **57**, 775–784.

[428] Welch, B.L. (1937). On the z test in randomized blocks and Latin squares. *Biometrika*, **29**, 21–52.

[429] Wellner, J.A. (1977). A Glivenko-Cantelli theorem and strong laws of large numbers for functions of order statistics. *Ann. Statist.*, **5**, 473–480.

[430] White, J. (1962). Least squares estimation for censored samples (Abstract). *Ann. Math. Statist.*, **33**, 1502.

[431] Whittle, P. (1958). On the smoothing of probability density functions. *J. Roy. Statist. Soc.*, Ser. B, **20**, 334–343.

[432] Wilcoxon, F. (1945). Individual comparisons by ranking methods. *Biometrics*, **1**, 80–83.

[433] Wilks, S.S. (1941). Determination of sample sizes for setting tolerance limits. *Ann. Math. Statist.*, **12**, 91–96.

[434] Wilks, S.S. (1942). Statistical prediction with special reference to the problem of tolerance limits. *Ann. Math. Statist.*, **13**, 400–409.

[435] Wilks, S.S. (1948). Order Statistics. *Bull. Amer. Math Soc*, **54**, 6–50.

[436] Wilks, S.S. (1962). *Mathematical Statistics.* John Wiley & Sons, New York.

[437] Wolfowitz, J. (1942). Additive partition functions and a class of statistical hypotheses. *Ann. Math. Statist.*, **13**, 247–279.

[438] Wolfowitz, J. (1944). Asymptotic distribution of runs up and down. *Ann. Math. Statist.*, **15**, 163–172.

[439] Wolfowitz, J. (1949). Non-parametric statistical inference. *Proceedings of the Berkeley Symposium on Math. Statist. and Probability.* Univ. of California Press, 93–113.

[440] Youden, W.J. (1953). Sets of three measurements. Sci. Mon., **57**, 143.

[441] van Zwet, W.R. (1980). A strong law for linear functions of order statistics. *Ann. Prob.* **8**, 986–990.

Author Index

Subject Index

Q

R

T